AF594962

Green Technologies for Production Processes

Green Technologies for Production Processes

Editors

Wei Cai
Zhigang Jiang
Conghu Liu
Yan Wang

MDPI • Basel • Beijing • Wuhan • Barcelona • Belgrade • Manchester • Tokyo • Cluj • Tianjin

Editors
Wei Cai
College of Engineering and Technology
Southwest University
Chongqing
China

Zhigang Jiang
School of Machinery and Automation Engineering
Wuhan University of Science & Technology
Wuhan
China

Conghu Liu
Sino-US Global Logistics Institute
Shanghai Jiao Tong University
Shanghai
China

Yan Wang
School of Computing, Engineering & Maths
University of Brighton
East Sussex County
UK

Editorial Office
MDPI
St. Alban-Anlage 66
4052 Basel, Switzerland

This is a reprint of articles from the Special Issue published online in the open access journal *Processes* (ISSN 2227-9717) (available at: https://www.mdpi.com/journal/processes/special_issues/green_production_process).

For citation purposes, cite each article independently as indicated on the article page online and as indicated below:

LastName, A.A.; LastName, B.B.; LastName, C.C. Article Title. *Journal Name* **Year**, *Volume Number*, Page Range.

ISBN 978-3-0365-1871-8 (Hbk)
ISBN 978-3-0365-1872-5 (PDF)

Contents

About the Editors

Wei Cai received the Ph.D. degree in mechanical engineering from Chongqing University, China, in 2018. He is a Postdoctoral Fellow at Department of Logistics and Maritime Studies from The Hong Kong Polytechnic University. Currently, he is a lecturer at Department of Mechanical Engineering, College of Engineering and Technology, Southwest University, Chongqing, China. He has been honored by Hong Kong Scholars in 2019. His fields of interest are sustainable manufacturing, intelligent manufacturing, and energy benchmarking. He has published more than 40 academic papers, H-index is 23. He is the Topic Editor of Processes, and the Guest editor of several Journals including the International Journal of Production Research and Processes.

Zhigang Jiang received the Ph. D degree in mechanical engineering from Wuhan University of Science and Technology, Hubei, China. He is currently a deputy dean of International School and a Professor of School of Machinery and Automation of Wuhan University of Science and Technology, China. His research focuses on green manufacturing and remanufacturing, energy efficient manufacturing, Life Cycle assessment. He has Hosted three projects supported by the National Natural Science Foundation, one project by National 863 plan, one sub-project by National Key Technology R&D Program, and one sub-project by the State Key Program of National Natural Science. He has authored a book, and authored or co-authored over 80 academic papers. He has been honored by: (1) 2006 Second Prize for Natural Science, the Government of Hubei Province (2) 2013 First Prize for Wuhan Science and Technology Progress; (3) 2014 Second Prize for Science and Technology Progress of Hubei province.

Conghu Liu, associate professor of Suzhou University, received Ph.D. degree from Hefei University of technology in 2016. He works as a postdoctoral in Shanghai Jiaotong University from 2016 to 2019 and is now a visiting scholar at Tsinghua University. He has hosted and participated in several national and provincial scientific research projects including the National Social Science Foundation of China, Humanities and Social Sciences project of the Ministry of education, and China Postdoctoral Science Foundation, the Anhui Natural Science Foundation. He has published more than 50 academic papers in Energy Conversion and Management, Energy, Journal of Cleaner Production, and Science of The Total Environment. His fields of interest are remanufacturing, sustainable manufacturing and management.

Yan Wang received the Ph.D. degree in Manufacturing Engineering from University of Nottingham, UK, in 2005. Currently, she is a Senior Lecturer at School of Computing, Engineering & Maths from University of Brighton. Her research focuses on green manufacturing and remanufacturing, green production, energy saving. She has hosted three projects supported by Natural Science Foundation and Foreign Commonwealth Office. She has published more than forty academic papers.

Preface to "Green Technologies for Production Processes"

Numerous pathways and narratives have been developed to shed light on how society could transform its production systems in line with the aspirational targets of the Paris Agreement and Sustainable Development Goals. Green technologies, as an emerging technical innovation, are necessary for driving the production system transformation that is called for global sustainability. Key potential benefits of green technology are that they might substantially reduce the cost of mitigating CO_2 emissions and improve the environment performance with the development of more affordable, better-performing technologies.

The scope of the Processes journal covers research in chemistry, biology, materials, and allied engineering fields. Thus, in this book, we invite articles focused on research regarding the chemistry, biology, materials, and allied engineering firms (manufacturing processes, iron and steel production processes, mining processes, power generation processes, and so on).

This book focuses on original research works about Green Technologies for Production Processes, including discrete production processes and process production processes, from various aspects that tackle product, process, and system issues in production. The aim is to report the state-of-the-art on relevant research topics and highlight the barriers, challenges, and opportunities we are facing. This book involves following topics:

(1) Energy saving and waste reduction in production processes
(2) Production of new and renewable energy devices
(3) Design and manufacturing of green products
(4) Low carbon manufacturing and remanufacturing
(5) Materials for green production
(6) Management and policy for sustainable production
(7) Technologies of mitigating CO_2 emissions
(8) Other green technologies

Wei Cai, Zhigang Jiang, Conghu Liu, Yan Wang
Editors

Editorial

Special Issue on "Green Technologies for Production Processes"

Wei Cai [1,2,*], Zhigang Jiang [3], Conghu Liu [4] and Yan Wang [5]

1 College of Engineering and Technology, Southwest University, Chongqing 400715, China
2 Faculty of Business, The Hong Kong Polytechnic University, Hung Hum, Kowloon, Hong Kong, China
3 School of Machinery and Automation Engineering, Wuhan University of Science & Technology, Wuhan 430081, China; jiangzhigang@wust.edu.cn
4 Sino-US Global Logistics Institute, Shanghai Jiao Tong University, Shanghai 200240, China; lch339@126.com
5 School of Computing, Engineering & Maths, University of Brighton, Brighton BN2 4GJ, UK; Y.Wang5@brighton.ac.uk
* Correspondence: weicai@swu.edu.cn

Citation: Cai, W.; Jiang, Z.; Liu, C.; Wang, Y. Special Issue on "Green Technologies for Production Processes". *Processes* **2021**, *9*, 1022. https://doi.org/10.3390/pr9061022

Received: 9 June 2021
Accepted: 9 June 2021
Published: 10 June 2021

1. Introduction

Numerous pathways and narratives have been developed to shed light on how society could transform its production systems in line with the aspirational targets of the Paris Agreement and Sustainable Development Goals. Green technologies, as an emerging technical innovation, are necessary for driving the production system transformation that is called for global sustainability. Key potential benefits of green technology are that they might substantially reduce the cost of mitigating CO_2 emissions and improve the environment performance with the development of more affordable, better-performing technologies.

The scope of the *Processes* journal covers research in chemistry, biology, materials, and allied engineering fields. Thus, in this Special Issue, we invite articles focused on research regarding the chemistry, biology, materials, and allied engineering firms (manufacturing processes, iron and steel production processes, mining processes, power generation processes, and so on).

This Special Issue on "Green Technologies for Production Processes" will focus on publishing original research works about Green Technologies for Production Processes, including discrete production processes and process production processes, from various aspects that tackle product, process, and system issues in production. The aim is to report the state-of-the-art on relevant research topics and highlight the barriers, challenges, and opportunities we are facing. It also welcomes studies that stimulate research discussion of moving towards production in a particular industrial sector. Topics of interest for this Special Issue include but are not limited to:

(1) Energy saving and waste reduction in production processes;
(2) Production of new and renewable energy devices;
(3) Design and manufacturing of green products;
(4) Low carbon manufacturing and remanufacturing;
(5) Materials for green production;
(6) Management and policy for sustainable production;
(7) Technologies of mitigating CO_2 emissions;
(8) Other green technologies.

2. Industry Development Issues

Zhang et al. [1] propose a model framework that uses the fuzzy decision-making test and evaluation laboratory (fuzzy DEMATEL) method to analyze the drivers of corporate environmental responsibility (CER) from the perspective of the triple bottom line (TBL) of economy, environment, and society. The results show that some effective measures to implement CER can be provided for the government, the automobile manufacturing industry, and the public to promote sustainable development of the Chinese Auto Manufacturing Industry (CAMI). Zhang et al. [2] propose a multicriteria decision-making (MCDM) method considering the circular economy in order to select the best reverse logistics provider for remanufacturing. In this article, a circularity dimension is included in the evaluation criteria. Then, analytic hierarchy process (AHP) is used to calculate the global weights of each criterion, which are used as the parameters in selecting RL providers. Finally, technique for order of preference by similarity to ideal solution (TOPSIS) is applied to rank reverse logistics providers with three different modes. Wang et al. [3] construct a sustainability evaluation method of logistics parks based on energy; analyze the input (energy, land, investment, equipment, information technology, and human resources) and output (income and waste) of logistics parks from the perspective of energy; study the characteristics of the emergy flow of logistics parks; and construct the function, structure, ecological efficiency, and sustainable development indexes of logistics parks. The results reveal the internal relationship among economic, environmental, and social benefits of logistics parks through emergy and provides theoretical and methodological support for the sustainable development of logistics parks. Zhang et al. [4] present a literature review on reverse logistics (RL) supplier selection in terms of criteria and methods and propose a three-stage typology of decision-making frameworks to understanding RL supplier selection. The results show that reverse logistics (RL) is closely related to remanufacturing and could have a profound impact on the remanufacturing industry. Ivascu [5] proposes a hierarchical framework for sustainability assessment of manufacturing industry in Romania. This hierarchical framework can be customized in detail for the specific of each organization and can be adapted in other industries, including banking, retail, and other services. It can be observed that waste management and the interests of the stakeholders are major implications that must be measured and properly motivated.

3. New Energy Issues

Wu et al. [6] construct a forecasting model based on a correlation test, the cuckoo search optimization (CSO) algorithm and extreme learning machine (ELM) method in order to forecast and analyze the developing potential of electric power substitution. The results showed that the CSO-ELM model has great forecasting accuracy. Bu and Zhang [7] propose a data-driven prospect analysis framework to evaluate the activated potential under two kinds of nearby accommodation approaches and to explore the completion prospect of this new obligated quota from provincial levels. They analyze two accommodation approach of the pathway for the former is to activate more provincial accommodation potential either via releasing system flexibility or by substituting generation right, and the pathway for the latter is to introduce trans-regional or trans-provincial accommodation and import more renewable energy power. Li et al. [8] present the current development status and GHG (greenhouse gas) mitigation effect of the straw-based biomass power plants in Anhui Province in China. They consider that the large-scale development of biomass power plants remains a challenge for the future, especially in areas of AHP with a low biomass density.

4. Process Design Issues

Lu et al. [9] propose an all-factors analysis approach on energy consumption in the blast furnace iron making process (BFIMP). The results show that the improvement of some material flows, energy flows, and operation parameters could increase the amount of the pulverized coal injection ratio (PCIR), such as sinter size, ore grade, sinter grade, M10, blast volume, blast temperature, and especially for sinter alkalinity. Zhang et al. [10] address a coincineration scheme for mixing multi-component wastes in a rotary kiln for waste disposal from pesticide production. Their results show that the rotary kiln incineration and flue gas treatment processes were successfully applied in engineering for green production of pesticides. In order to achieve carbon efficiency improvement and save costs, Hu et al. [11] establish a carbon emission and processing cost models of the grinding process. Considering the constraints of machine tool equipment performance and processing quality requirements, the grinding wheel's linear velocity, cutting feed rate, and the rotation speed of the workpiece were selected as the optimization variables, and the improved NSGA-II algorithm was applied to solve the optimization model. Considering that the process parameters of hot stamping considerably influence the product forming quality and energy consumption, Gao et al. [12] introduce the energy-economizing indices of hot stamping with multiobjective consideration of energy consumption and product forming quality to find a pathway by which to obtain optimal hot stamping process parameters. The obtained results may be used for guiding process optimization regarding energy saving and the method of manufacturing parameters selection. Hu et al. [13] based on the basic theory of fluid–solid coupling, the correlation definition between coal porosity and permeability, and previous studies on the influence of adsorption expansion, change in pore free gas pressure, and the Klinkenberg effect on gas flow in coal derive a mathematical model of the dynamic evolution of coal permeability and porosity. Numerical simulation results show that the solution can effectively guide gas extraction and discharge during mining.

5. Scheduling and Planning Issues

Considering carbon emissions, Sun et al. [14] establish a multiobjective flexible job shop scheduling problem (MO-FJSP) mathematical model with minimum completion time, carbon emission, and machine load. To solve this problem, they study six variants of the non-dominated sorting genetic algorithm-III (NSGA-III) and from the experimental results, it can be concluded that the NSGAIII-COE has significant advantages in solving the low carbon MO-FJSP. Sun et al. [15] establish a multiprocess route scheduling optimization model with carbon emissions and cost as the multiobjective in order to solve the problems of flexible process route and workshop scheduling scheme changes frequently in the multivariety small batch production mode. The optimization results under single-target and multitarget conditions are contrasted and analyzed, to guide enterprises to choose a reasonable scheduling plan, improve the carbon efficiency of the production line, and save costs. Fu et al. [16] focus on an optimal schedule for a micro energy grid considering the maximum total carbon emission allowance (MTEA). The results show that: (1) a micro energy grid can make the most use of the complementary characters of different energy sources to meet different energy demands for electricity, heat, cold, and gas; (2) the risk aversion scheduling model can represent the influence of uncertainty variables in objective functions and constraints, and provide a basis for decision makers who have different attitudes; and (3) demand response (DR) can smooth the energy load curves. Liu et al. [17] promotes a bilevel optimization planning approach for photovoltaic (PV)-storage charging stations. The results show that the proposed bilevel optimization model can provide a more economical and reasonable planning scheme than the single-level model, and can reduce the investment cost by 8.84%, operation and maintenance cost by 13.23%, and increase net revenue by 5.11%.

6. Other Issues

These are some accepted papers in the Special Issue "Green Technologies for Production Processes", besides the industry development issues, new energy issues, process design issues, and scheduling and planning issues. Huynh et al. [18] describes a green and facile method for the preparation of multibranched gold nanoparticles using hydroquinone as a reducing agent and chitosan as a stabilizer, through ultrasound irradiation to improve the multibranched shape and stability. To provide guidance towards reducing the weight of the HFC-125 storage vessel by reducing the release pressure and to reveal the effects of release pressure on the extinguishing efficiency of HFC-125, Jin et al. [19] investigate the flow and diffusion characteristics of HFC-125 under six release pressures in the present study. Results show that the degree of superheat and the injection duration both decreased with the release pressure. Lin et al. [20] analyze the factors that affect the implementation of intelligent systems in motor production lines. Their research results can be provided as a reference for production lines that acquaint with intelligent systems. Liang et al. [21] use data envelopment analysis (DEA) to analyze the production process efficiency and the effective use of input elements of greenhouse vegetables at the provincial level in China. The results reveal that many chemical fertilizers, farmyard manure, and pesticides in China are inefficient. Xia et al. [22] based on the authorized remanufacturing, construct the game model between a manufacturer and a remanufacturer. The main results are as follows: the OEM could increase its profit and change its unfavorable market competition status by authorizing remanufacturing; a franchise contract could make the sustainability supply chain optimized; when the ratio of the environment effect is greater than a certain threshold, centralized decision-making could not only increase the supply chain revenue, but also reduce the impact on the environment.

We thank all the contributing authors, reviewers, and editorial team members from the journal. It would not be possible to have this Special Issue without them. We hope that this Special Issue will further promote deep research into, and the development and application of, the related theories, methods, and technologies in order to advance the field of green production processes, and achieve lasting and sustainable development of the world economy.

Author Contributions: W.C.; Z.J.; C.L.; Y.W.: Conceptualization, Writing—Original draft preparation, Editing and Reviewing. All authors have read and agreed to the published version of the manuscript.

Funding: This research received no external funding.

Conflicts of Interest: The authors declare no conflict of interest.

References

1. Zhang, H.; Zhang, M.; Yan, W.; Liu, Y.; Jiang, Z.; Li, S. Analysis the Drivers of Environmental Responsibility of Chinese Auto Manufacturing Industry Based on Triple Bottom Line. *Processes* **2021**, *9*, 751. [CrossRef]
2. Zhang, X.; Li, Z.; Wang, Y.; Yan, W. An Integrated Multicriteria Decision-Making Approach for Collection Modes Selection in Re-manufacturing Reverse Logistics. *Processes* **2021**, *9*, 631. [CrossRef]
3. Wang, C.; Liu, H.; Yu, L.; Wang, H. Study on the Sustainability Evaluation Method of Logistics Parks Based on Emergy. *Processes* **2020**, *8*, 1247. [CrossRef]
4. Zhang, X.; Li, Z.; Wang, Y. A Review of the Criteria and Methods of Reverse Logistics Supplier Selection. *Processes* **2020**, *8*, 705. [CrossRef]
5. Ivascu, L. Measuring the Implications of Sustainable Manufacturing in the Context of Industry 4.0. *Processes* **2020**, *8*, 585. [CrossRef]
6. Wu, J.; Tan, Z.; De, G.; Pu, L.; Wang, K.; Ju, L.; Tan, Q. Multiple Scenarios Forecast of Electric Power Substitution Potential in China: From Perspective of Green and Sustainable Development. *Processes* **2019**, *7*, 584. [CrossRef]
7. Bu, Y.; Zhang, X. On the Way to Integrate Increasing Shares of Variable Renewables in China: Activating Nearby Accommo-dation Potential under New Provincial Renewable Portfolio Standard. *Processes* **2021**, *9*, 361. [CrossRef]
8. Li, H.; Min, X.; Dai, M.; Dong, X. The Biomass Potential and GHG (Greenhouse Gas) Emissions Mitigation of Straw-Based Biomass Power Plant: A Case Study in Anhui Province, China. *Processes* **2019**, *7*, 608. [CrossRef]
9. Lu, B.; Wang, S.; Tang, K.; Chen, D. An All-Factors Analysis Approach on Energy Consumption for the Blast Furnace Iron Making Process in Iron and Steel industry. *Processes* **2019**, *7*, 607. [CrossRef]
10. Zhang, B.; He, J.; Hu, C.; Chen, W. Experimental and Numerical Simulation Study on Co-Incineration of Solid and Liquid Wastes for Green Production of Pesticides. *Processes* **2019**, *7*, 649. [CrossRef]
11. Hu, M.; Sun, Y.; Gong, Q.; Tian, S.; Wu, Y. Multi-Objective Parameter Optimization Dynamic Model of Grinding Processes for Promoting Low-Carbon and Low-Cost Production. *Processes* **2019**, *8*, 3. [CrossRef]
12. Gao, M.; Wang, Q.; Li, L.; Ma, Z. Energy-Economizing Optimization of Magnesium Alloy Hot Stamping Process. *Processes* **2020**, *8*, 186. [CrossRef]
13. Hu, S.; Liu, X.; Li, X. Fluid–Solid Coupling Model and Simulation of Gas-Bearing Coal for Energy Security and Sustainability. *Processes* **2020**, *8*, 254. [CrossRef]
14. Sun, X.; Wang, Y.; Kang, H.; Shen, Y.; Chen, Q.; Wang, D. Modified Multi-Crossover Operator NSGA-III for Solving Low Carbon Flexible Job Shop Scheduling Problem. *Processes* **2020**, *9*, 62. [CrossRef]
15. Sun, Y.; Gong, Q.; Hu, M.; Yang, N. Multi-Objective Optimization of Workshop Scheduling with Multiprocess Route Considering Logistics Intensity. *Processes* **2020**, *8*, 838. [CrossRef]
16. Fu, X.; Fan, W.; Lin, H.; Li, N.; Li, P.; Ju, L.; Zhou, F.; Li, A. Risk Aversion Dispatching Optimal Model for a Micro Energy Grid Integrating Intermittent Renewable Energy and Considering Carbon Emissions and Demand Response. *Processes* **2019**, *7*, 916. [CrossRef]

17. Liu, Y.; Dong, H.; Wang, S.; Lan, M.; Zeng, M.; Zhang, S.; Yang, M.; Yin, S. An Optimization Approach Considering User Utility for the PV-Storage Charging Station Planning Process. *Processes* **2020**, *8*, 83. [CrossRef]
18. Huynh, P.T.; Nguyen, G.D.; Thi Le Tran, K.; Minh Ho, T.; Lam, V.Q.; Ngo, T.V.K. Rapid and Green Preparation of Multi-Branched Gold Nanoparticles Using Surfactant-Free, Combined Ultrasound-Assisted Method. *Processes* **2021**, *9*, 112. [CrossRef]
19. Jin, J.; Pan, R.; Chen, R.; Xu, X.; Li, Q. Flow and Diffusion Characteristics of Typical Halon Extinguishing Agent Substitute under Dif-ferent Release Pressures. *Processes* **2020**, *8*, 1684. [CrossRef]
20. Lin, Y.C.; Yeh, C.C.; Chen, W.H.; Hsu, K.Y. Implementation Criteria for Intelligent Systems in Motor Production Line Process Management. *Processes* **2020**, *8*, 537. [CrossRef]
21. Liang, Y.; Jing, X.; Wang, Y.; Shi, Y.; Ruan, J. Evaluating Production Process Efficiency of Provincial Greenhouse Vegetables in China Us-ing Data Envelopment Analysis: A Green and Sustainable Perspective. *Processes* **2019**, *7*, 780. [CrossRef]
22. Xia, X.; Zhang, C. The Impact of Authorized Remanufacturing on Sustainable Remanufacturing. *Processes* **2019**, *7*, 663. [CrossRef]

Article

Measuring the Implications of Sustainable Manufacturing in the Context of Industry 4.0

Larisa Ivascu

Department of Management, Faculty of Management in Production and Transportation, Politehnica University of Timisoara, via Victoria Square No. 2, 300006 Timisoara, Romania; larisa.ivascu@upt.ro

Received: 28 March 2020; Accepted: 11 May 2020; Published: 14 May 2020

Abstract: Sustainability is increasingly being addressed globally. The manufacturing industry faces various constraints and opportunities related to sustainable development. Currently, there are few methodological frameworks for evaluating sustainable organizational development. Assessing and improving organizational capacity is important for producers and researchers in the field and local, national, and international authorities. This research proposes a hierarchical framework for sustainability assessment of manufacturing industry in Romania. The proposed framework integrates performance elements and measures to improve all the processes and activities from the triple perspective of sustainability. Sustainability assessment captures the entire supply chain of the organization, including stakeholder interests and end-of-life directions for products. To establish the elements to be integrated in the development of the proposed framework, market research (online questionnaire-for the characterization of Industry 4.0) and the Delphi method were used to identify the categories of performance indicators that must be measured to identify organizational capacity for sustainable development. The framework was tested by an automotive manufacturing organization. A number of improvements have been identified that relate to Industry 4.0 facilities and the application of the facilities related to recovering the value of the product at the end of its life cycle. This hierarchical framework can be customized in detail for the specific of each organization and can be adapted in other industries, including banking, retail, and other services. It can be observed that waste management and the interests of the stakeholders are major implications that must be measured and properly motivated.

Keywords: sustainable process; sustainability; efficiency; sustainability strategy; scale development; sustainable supply chain management; implementation framework

---

1. Introduction

Sustainability and sustainable development as concepts have gone through different development stages since their introduction. The historical development of the concept has been carried out at various conferences and within organizations and institutions, which are currently concerned with the implementation of the principles, targets, and objectives of sustainable development [1]. The concept of sustainable development has encounter over time different criticisms and interpretations, being accepted in different fields of activity. The concept of sustainable development, throughout its evolution, has adapted to environmental and technological requirements, but the heart, principles, directions, and objectives have been preserved and are still present. Due to the fact that the environment is dynamic and new aspects come in, some sustainable development goals have been updated. The objectives of sustainable development are contained in the 2030 Agenda. At the same time, the objectives of this agenda contribute to the survival on the planet and to the increase of the standard of living [1,2].

If sustainable development initially focused more on the environmental dimension, gradually, obligations regarding the social and economic dimension have been added. The concept of sustainable development has become one appreciated by organizations due to the identification of organizational benefits and advantages [2–5].

The social and economic aspects are addressed and appreciated by the organizations involved in sustainable development [6]. Five decades ago, society was characterized by consumerism, economic growth, polluted living space, and unorganized ways of living [2–7]. The exploitation of some natural resources contributes to the entrenchment of the right to a decent living for the next generations. The needs of the population are inversely proportional to those of the organizations.

Imbalances in the environment contribute to the generation of negative effects for future generations. Pollution is a major environmental problem. Among the sources of pollution are: (1) the development of the economic environment (economic growth, traffic intensification, traffic improvement, increasing the number of inhabitants, increased tourism, etc.), (2) natural hazards (earthquakes, severe rain, volcanic eruptions, droughts, wind, etc.), and (3) technology (different networks built, new concepts, intensification of the use of information technology, construction of different attractive products for users, waste management, etc.) [8–11]. These factors contribute to the occurrence of consequences, some severe, which concern ecological problems, ecosystem instability, global climate change, natural disasters, hunger and poverty, lower quality of life, etc. [10].

The global scenario of depletion of natural resources and environmental, economic, and social imbalance motivates organizations and individuals to adopt sustainability practices in organizational processes. Sustainability was born many years ago, but few guidelines are available for its practical implementation and evaluation [8–11]. The research [3] states a considerable impact of sustainability on the environment. This research presents the impact of greenhouse gases and waste on the environment. Reference is made to the wood industry, energy, and heat generation. The researches [4,5] states that sustainability is achieved in the field of transport and adjacent industries and emphasizes the reduction of environmental impact. Previous research [6] presents sustainability studies in the airline field, and the automotive industry must be evaluated and analyzed in future research. Other research [7] presents the impact of sustainability in the fashion industry. Further research [8] presents the impact of sustainable development in the foam and chemical industry, presenting different visions. It is specified that the automotive industry has a considerable impact in the field of sustainability, without specifying these impacts. The research by Amui et al. [9] has various implications for the food industry. From the analysis of the different definitions discussed in several studies [3–9], it was observed that the approach on sustainability in manufacturing is not yet concretely developed. Sustainable development improves the conditions of companies, thus contributing to their competitiveness. Sustainable development is a voluntary approach, but it is increasingly adopted by companies. Stakeholders are interested in this concept as long as they get improved financial results (increased profit). The first direction was the one of environmental sustainability, with major implications for national and international authorities and organizations [5]. However, sustainability is now defined by three dimensions—environmental, social, and economic [10]. Pervious research [5,6] addressed sustainability in land and air transport industries. It emphasizes the importance of addressing sustainability in this industry as a result of generating a large amount of greenhouse gases. The results published in reference [7] highlight the impact of sustainability in the fashion industry and the impact it has social impact by producing the articles used by the final customer who is in the society. The importance of occupational safety and health in this industry is also mentioned. This industry focuses on re-manufacturing, reconditioning of items, use for other purposes, reuse of other people's items, recycling materials by implementing buy-back (in various stores), reducing the impact on the environment by using natural materials (bio-cotton, wool, etc.), repair and recovery of items to meet basic needs, redesign of the manufacturing process by including automatic lines, and reconditioning items to be used until the end of their life cycle. Other research [8] presents studies in the foam and chemical industry and emphasizes the importance of renewable energy sources. In the same direction as the fashion industry, the food industry is one that addresses the end customer, and the functions of sustainability are addressed. One study [9] concerning the food industry refers to the 9Rs (remanufacturing, reconditioning, reuse, recycling, reduce, repair, recover, redesign, and reconditioning). The importance of the 9Rs in the entire manufacturing process is underlined.

Another study [10] specifies the imperatives of Industry 4.0 and its importance in the current economic development. Therefore, five important reasons can be systematized. Sustainable manufacturing is one of the most important issues to address for the following five important reasons [5–10]:

1. Manufacturing generates a significant quantity of greenhouse gases alongside the energy and transport industries [5,6];
2. Manufacturing has social impact by producing the articles used by the final customer, but also on occupational safety and health [7].
3. This high impact is due to huge energy consumption and the use of physical resources [8];
4. Manufacturing needs to adopt the following functions of sustainability: remanufacturing, recondition, reuse, recycling, reduce, repair, recover, redesign, and recondition (9R) [9];
5. Manufacturing must adopt the Industry 4.0 imperatives by integrating its requirements into sustainable development [10].

Organizations implement various strategies, in accordance with the interests of their stakeholders and good practices to make their processes environmentally efficient and sufficiently socially and economically viable. Therefore, it is suggested that manufacturing integrate production processes that pursue sustainable manufacturing practices. It is imperative that a study should include all aspects of sustainability related to stakeholder involvement, the entire logistics chain, and strategies up to the end of product life [6].

The practices, methods, and tools used for sustainability assessment in the manufacturing industry are based on a pioneering roadmap for applying the imperatives of the circular economy in the context of Industry 4.0. The results of this research refer to the presentation of the relationship between circular economy and Industry 4.0, as well as the improvement of the ReSOLVE (Regenerate, Share, Optimise, Loop, Virtualise, Exchange—a framework with six action areas for businesses) framework [11]. It is an approach based on specialized literature and qualitative evaluation. Another model [12] aims to integrate technologies from Industry 4.0 integrated with circular economy (EC) practices to provide a business model. This business model is based on the reuse and recycling of ferrous materials and waste. It is based on qualitative evaluation. A study of 600 German companies claims that the opportunities of digital networks are used to a limited extent, especially for streamlining production processes [13]. This study does not present an improvement framework, but only an evaluation of the results obtained from the market research. Other research offers a synergistic and integrative circular economy-digital technologies framework based on the empirical literature [14]. The research results state that the research directions of the circular economy have been submitted, but the research and applicability of the digital technologies that allow an EC are still in the basic form. Another study presents "X" production systems (XPS) and the importance of lean manufacturing and continuous improvement principles [15]. It presents the situation of a company that has better aligned its XPS with the sustainability objectives. Following that research, the indicator panel and the evaluation framework are completed. Other research [16] takes into account the life cycle of the product, stakeholders, employees and customers, and end-of-life strategies, but also includes environmental, social, and economic aspects in a single comprehensive review on the aforementioned directions. The results highlight an integrated approach based on various research in this direction.

In summary, other research is based on sustainable value stream mapping [17], use of multi-criteria decision making [18], assessment questionnaires [17–19], indicator-based assessment [19–21], rating system [20], scoring tools [21], software tools [20], mathematical modeling [20–23] life cycle analysis [24–26], product service systems [27], and a sustainability index [27,28]. By evaluating these approaches, it can be seen that the research covers the evaluation of the sustainability via the measurement of performance taking into account certain practices and indicators. Therefore, an integrated framework for sustainability assessment, including product life cycle engineering, stakeholder interests, supplier and supply chain management, employees and customers, and end-of-life strategies is impetuous to develop. These approaches [17–28] are applied on specific business typologies without taking into account the opportunities and requirements of Industry 4.0. The framework that this research seeks to develop takes into account the behavior of different companies in the manufacturing industry and the characteristics of the Industry 4.0. Previous studies do not take into account the characteristics of Romanian manufacturing industry. This research does not use as a research method, discussions on the results obtained with experts in the field.

This paper is structured in two main directions: researching the characteristics of Industry 4.0 and of the indicators that evaluate the organizational sustainability. Finally, the hierarchical framework for sustainability assessment of manufacturing industry is pre-tested and validated. This research presents a new evaluation framework, which integrates the goals of sustainable development and those of Industry 4.0. To develop this framework, market research was conducted to identify the current needs and implications in Industry 4.0. This research was validated following discussions with manufacturers in the manufacturing industry. These debates were based on the Delphi method. Finally, a hierarchical framework was developed based on the needs identified and validated through the Delphi method. This framework is used to evaluate and improve the involvement of companies in the manufacturing in sustainable development and reduce the negative impact on the environment.

2. Research Methodology

The research methodology comprises three research directions—market research by means of a questionnaire applied to 100 manufacturing industry experts, the Delphi method involving one facilitator and 40 experts, and the author's empirical experience. All the phases of the research are progressively completed, and finally, the proposed framework is pre-tested and validated.

2.1. The Questionnaire

Marketing research helps to identify the needs and desires of the clients [29]. There are a number of tools, but a questionnaire survey is a facile, cheap, and easy to apply method. The questionnaire provides an easy to apply way to contact a number of individuals [29–35]. Various questions may be asked depending on the type of information that is to be obtained. A questionnaire can feature open-ended questions (completely unstructured, structured, describing an image), closed questions with predetermined answers (with different scales—Stapel, semantic differential, constant amount appreciation attribute), or mixed questions. If the goal is to collect motivations and opinions regarding the creativity and innovation of the respondent, then open questions will be used. If all the answer variants can be anticipated, then closed questions will be used. In other situations, mixed questions can be used [31].

In this research, the questionnaire was used to identify the characteristics and imperatives of Industry 4.0 in Romania. This research tool has been applied to shareholders throughout Romania. The informants were 100 shareholders, directors, or managers. The confidence level is 95%, and p-value > α. These results emphasize that the null hypothesis is not rejected. The 100 companies were selected from the classification of companies based on turnover, net profit, and number of employees (top 100 companies). This classification was made on the basis of data from the Trade Register and National Institute of Statistics. The application of the questionnaire was done online and was specific to each previously identified respondent. The respondents were identified according to the activity field of the company. It was intended to cover all areas of activity in the manufacturing industry. All responses were valid. The Likert scale (1—least important and 5—most important), distribution of a set score (0—poorly implemented and 100—fully implemented), and open questions were used. The questionnaire was structured in four parts, Table 1: information about the company; Industry 4.0 interpretation, facilitators, and barriers; and Industry 4.0 maturity and national technology platform. The results obtained in this research are used to develop the hierarchical framework for sustainability assessment in the manufacturing industry.

Table 1. Sustainable development and Industry 4.0.

Direction	Investigated Elements
Company information	Activity domain; Identification of the best-selling product; Number of employees; Assessment of the level of innovation of the company (0–100)'
Industry 4.0-Interpretation, facilitators and barriers	Identification of company practices in Industry 4.0; Evaluation of Industry facilitators 4.0; Barriers to Industry 4.0.
Industry 4.0-Maturity	Identifying the degree of maturity; Evaluation of a proposed model for digital maturity, including strategy, technology, operations, organization and culture, and clients.
National Technology Platform Industry 4.0.	Evaluation of the national platform Industry 4.0; Assessment of the importance of the actors of the Industry 4.0 platform: government, universities and research institutes, users/companies of Industry 4.0, and suppliers of Industry 4.0; The level of resistance to digitalization; Transforming the company into the digital era.

2.2. Delphi Method

The Delphi method is a forecast framework that includes the results of several rounds of questionnaires sent to a group of experts [15]. The results of a round are recorded, and then they are sent to the expert group, and the anonymous responses are aggregated and shared again to the expert group [15–18]. The process of applying this method is shown in the figure below. The Delphi technique is a method used to estimate the probability and outcome of future events [15]. The expert group exchanges opinions, and each expert personally provides estimates and assumptions based on their experience to a facilitator, who examines the data and develops a summary report [19–21]. Experts review the form of the report, and a new (second) report is issued. This process continues until all experts/participants agree with the developed report. This technique is an iterative one and is successfully applied in management and in different approaches to competitiveness [18]. In this area, sustainable development also has its place [32–35]. The Delphi method clarifies and extends problems to identify all areas and features that need modification [36]. In the present research, we used the Delphi method to identify all sustainable performance measures for improving organizational policies, people, processes and products. The following steps are presented in Figure 1. For the application of this method, a facilitator was identified from the automotive industry, being the manager of the processes and research-development department. The facilitator is characterized by over 30 years of experience in manufacturing, is the manager of a company with over 3000 employees, has personal involvement in sustainable development, is a good communicator, and has the capacity to analyze (as a result of the competences registered on the basis of certificates obtained at the international level).

The automotive industry has a significant percentage of manufacturing in Romania (it generates over 15% of gross domestic product [37–40]). Industries have their peculiarities and must be evaluated in a complex way [40–44]. The construction of the sample of respondents from different fields contributes to the achievement of an integrated framework for evaluating sustainable development [45–47]. Forty experts participated in the analysis rounds. Individuals with solid expertise and sustainability skills were selected. The database used for the selection of experts was the one from the application of the questionnaire (of the 100 respondents). There were six segments in the manufacturing industry identified based on the activity performed [38] in Romania. These segments, their percentages, and the targeted directions are presented in Table 2.

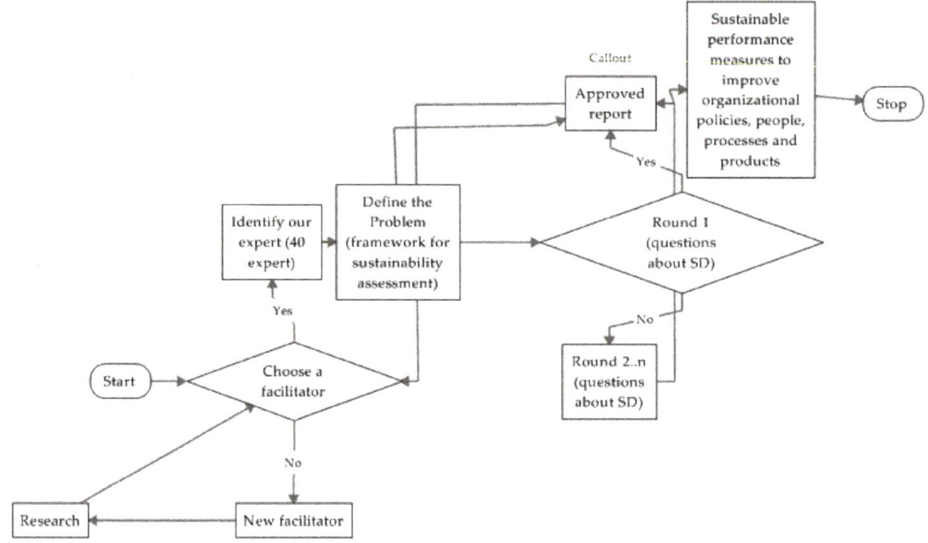

Figure 1. The stages of the Delphi method used in establishing the hierarchical framework for sustainability assessment of manufacturing industry.

Table 2. Sustainable development and Industry 4.0.

Segment of Manufacturing Industry	Number of Firms (and % of the Sample Firms)	Directions Evaluated by Experts
Automotive	35 (35%)	Domain experience Involvement in innovation Involvement in sustainable development Involvement in strategic management Strategic vision
Production of foams, chemicals, plastics, oil	9 (9%)	
Food and beverage production	15 (15%)	
Furniture production	11 (11%)	
Pharmaceutical production	7 (7%)	
Other productions (metal, electronic, non-metallic, clothing)	23 (23%)	

2.3. Empirical Experience

The interest in sustainable development over the last 10 years and the multiple studies carried out on the subject (over 150 works) have contributed to the extension of the research toward this model of manufacturing. The research carried out [48–53] has contributed to the foundation of the concept and to the identification of measures and performance indicators. They are the pillars of the hierarchical framework for sustainability assessment. The author has contributed as the main author to the development of a series of studies that have advanced the field of sustainability (Table 3).

Table 3. Research by the author in the field of sustainability and innovation.

Research	Journal	Development
[48]	Sustainability	Sustainable development model for the automotive industry. This research is based on in—depth interviews with 33 experts.
[49]	Sustainability	Integrating sustainability and lean: SLIM method and enterprise game proposed. This research is used to train students as experts in sustainability and lean.
[50]	Safety	Occupational accidents assessment by field of activity and investigation model for prevention and control. This research identifies the risks and proposes some preventive and corrective measures in the direction of sustainable development.
[51]	Sustainability	Risk indicators and road accident analysis for the period 2012–2016. This is a strategic framework for the sustainability of transport.
[52]	Sustainability	The evaluation and application of the TRIZ method for increasing eco-innovative levels in SMEs. This research tested furniture production.
[53]	Sustainability	Sustainable development and technological impact on CO_2 reducing conditions in Romania. This research contributes to the reduction of CO_2 for improving climatic conditions.

3. National and International Situation in the Manufacturing Industry

Manufacturing is the production of goods intended for use or sale with labor and machinery, instruments, processing, or chemical or biological formulation [36]. Finished products can be sold, through a distribution chain, to other producers for the production of more complex products or redistributed through the tertiary industry to final consumers [15–20]. In order to characterize manufacturing, a qualitative evaluation of the existing data series in the databases of the accredited institutes is performed. To characterize manufacturing, the following characteristics are taken into account: the number of employees, the amount of waste generated, the greenhouse gas emissions, and the level of innovation. These indicators are presented for the European Union (EU) and Romania [38–41,44,45].

The number of employed persons in the European Union decreased to 230,356,800 in the first quarter of 2019 from 231,342,700 in the fourth quarter of 2018. Of the EU employees over 15% are in the manufacturing industry.

In Romania, the number of employees in industry, construction, trade and other services in 2018 was 8,197,014, and in 2019, 8,249,779 employees. The employee is the person who exercises his activity on the basis of an employment contract in an economic or social unit—regardless of its form of ownership—or to private persons in exchange for a remuneration in the form of a salary, paid in money or nature, under commission form and others [38,39,54]. Figure 2 shows the main areas of activity and the number of employees for the period 2017–2019.

From the perspective of the quantity of waste generated, at the EU level, there were 2,116,310,000 tons in 2016 and 2,125,300,000 in 2017. From the perspective of the countries that generate these quantities, the situation is presented in Figure 3. The series is presented according to the reported data (some countries have not reported the amount of waste generated). Romania generated 176,742,421 tons of waste in 2017.

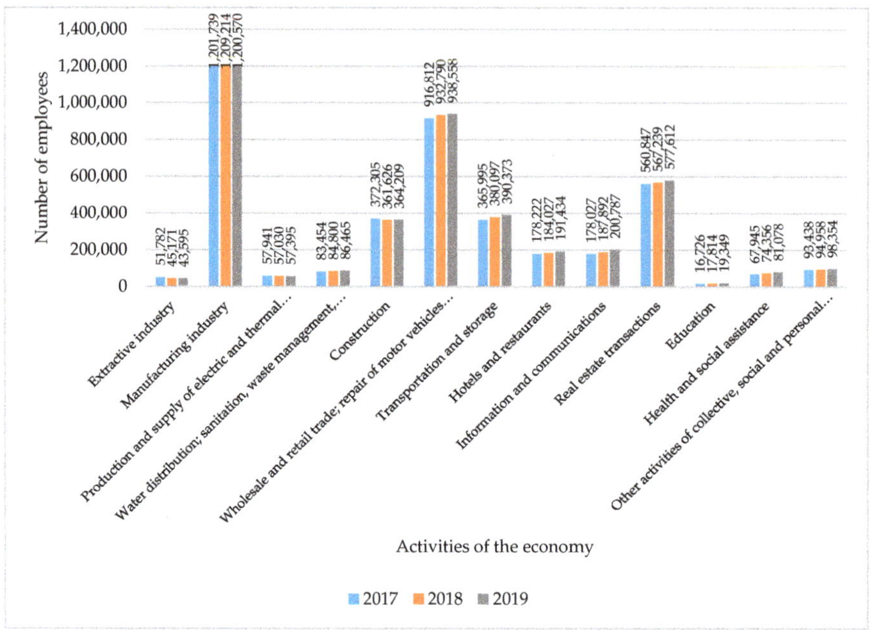

Figure 2. Situation of the number of employees in the main activities of the Romanian economy in the period 2017–2019 [44].

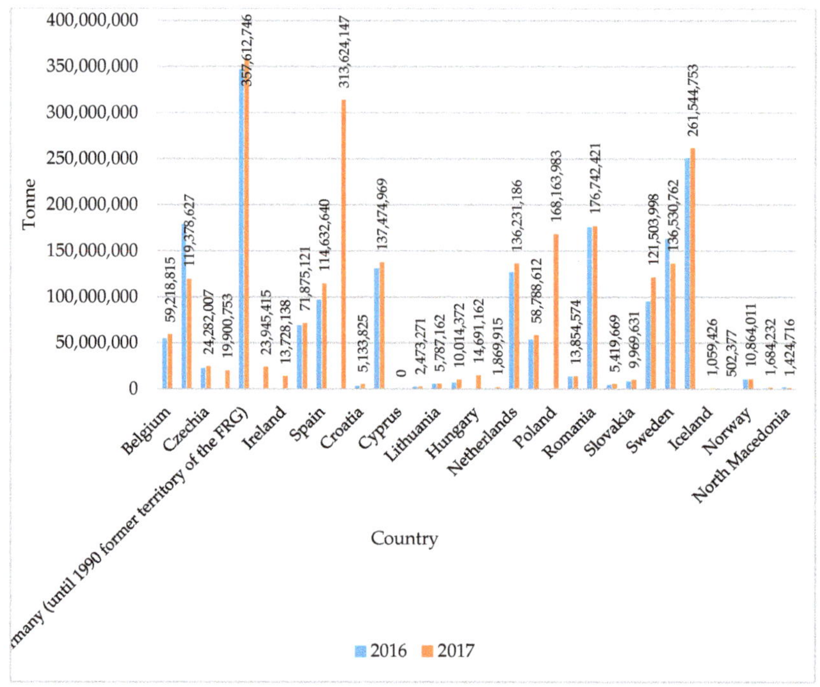

Figure 3. Amount of waste generated in the European Union (EU) for 2016–2017 (tons) [45].

Of the total amount of waste, at the EU level, 253,440,000 tons in 2016 and 258,890,000 tons in 2017 were generated by the manufacturing industry (Figure 3). The percentage of waste generated in manufacturing is over 10% of the total waste generated. For Romania, the quantity of waste generated by manufacturing was 6,727,021 tons in 2016 and 7,770,090 tons in 2017 (Figure 4.)

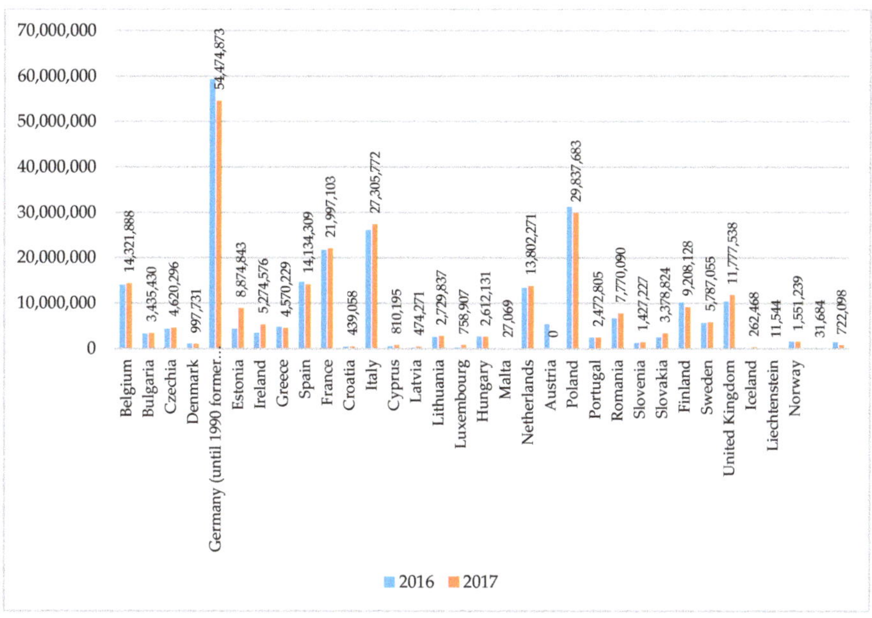

Figure 4. The quantity of waste generated by the manufacturing industry in EU countries for the period 2016–2017 (tons) [45].

From the perspective of the quantity of greenhouse gases emitted, the EU generated 4,461,685.11 tons in 2016 and 4,492,127.01 in 2017. Of this quantity, over 10% was generated by the manufacturing industry. Romania generated 115,150.66 tons in 2016 and 114,811.43 tons in 2017. The manufacturing industry in Romania generated 12,836.27 tons in 2016 and 13,105.39 tons in 2017. It can be seen that over 10% is generated by the manufacturing industry in Romania as well. The EU trend is also followed nationally.

Thus, it can be stated that the manufacturing industry is an important economic activity, with considerable contribution in EU and Romania, and this research approach is essential in this field.

4. Characteristics of Industry 4.0

Manufacturing processes are responsible for a significant portion of the consumption of natural resources and the generation of greenhouse gases. Manufacturing is defined as "the transformation of materials and information into tangible and intangible goods to satisfy the needs and desires of the buyers" [21,55–57]. The industry sector, including production, consumes almost half of the total energy delivered worldwide [15]. In the US, manufacturing absorbs more than 42% of total energy consumption [18,58]. Similarly, in China, the manufacturing sector absorbs 58% of total energy consumption [59,60]. As a result, numerous efforts have been made to reduce the environmental impact of different manufacturing processes, and several strategies have been implemented to monitor, improve, and control variables such as energy consumption [14–19], carbon emissions [15–18], the development of sustainable jobs [18–20,61], and the integration of innovative solutions [20–22,62].

Within this framework of sustainable development, Industry 4.0 appears as an industrial opportunity. The concept of Industry 4.0 began as a strategic framework for industrial production conceived and implemented by the German government in 2011 [10–24]. Industry 4.0 can be defined as a combination of technologies and value concepts applicable to organizational processes [16–18]. This is a general transformation using digital integration and intelligent engineering [26–28,61]. Industry 4.0 imperatives are in the following directions: the preparation of an intelligent, computerized, optimized manufacturing environment, which guarantees the flexibility and high efficiency of production and minimal impact on the environment [21–27,62]. Therefore, approaching Industry 4.0 in the context of sustainable development [63] is mandatory because the potential results obtained from this approach are productivity and resource efficiency [64]. For example, in previous studies [25–29], it is emphasized that Industry 4.0 encourages digitization by offering new efficient approaches to process control using the Internet of Things and integrating cyber-physical systems into manufacturing, which can improve resource and energy efficiency, and automated manufacturing concepts will increase the level of innovation and will reduce the amount of waste generated [30]. Industry 4.0 describes the progressive fusion of industrial production processes with the digital world of information technology.

Evaluating the two approaches—sustainable development and Industry 4.0 framework in the manufacturing industry—we can identify the following applications in manufacturing [22–36]. This analysis takes into account the 17 objectives and 169 goals of sustainable development (17 sustainable development goals—SDGs and 169 goals) and the definition of Industry 4.0. The entire analysis is based on studies published in the literature and are based on the needs of the manufacturing industry (see Table 4). For example, simulating different algorithms contributes to reducing poverty by proposing frameworks for improvement and identifying problems, improving living conditions, education through access to technology, identifying gaps for energy efficiency and improving conditions and outcomes for social responsibility.

Table 4. Sustainable development and Industry 4.0.

Industry 4.0 Imperatives	Research	17 SDG (Sustainable Development Goals)/169 Goals	Applicability in Manufacturing	Benefits
Internet of Things	[30]	3, 5, 7, 8, 9, 11, 12, 17 SDGs	The materials, structural elements, and components of the machine are equipped with sensors and Internet connection.	Process efficiency, data exchange between robots, increased production capacity, and increasing the level of innovation.
Radio frequency identification technology (RFID)	[31,33]	3, 5, 7, 8, 9, 11, 12, 17 SDGs	It allows the real-time visibility of the materials and goods of the manufacturing processes.	Reduces transportation errors, improves security, validates raw materials, and increases the visibility of goods in the supply chain.
Cognitive Computing	[34]	1, 3, 5, 7, 8, 9, 11, 12, 17 SDGs	Understanding tasks, workflows, and business process logic.	New cognitive technologies, scalability, productivity, and quality.
Cybersecurity	[35]	9, 11, 12, 17 SDGs	Process security.	Loss reduction.
Cloud Computing	[36]	3, 5, 7, 8, 9, 11, 12, 17 SDGs	Scalable business solutions.	Process innovation and expansion.
Mobile technologies	[37]	1, 5, 7, 8, 9, 11, 12, 15, 17 SDGs	Real-time data monitoring, collection, and processing.	Reduce time and streamline processes.
M2M (machine to machine)	[38]	3, 5, 7, 8, 9, 11, 12, 13, 17 SDGs	Communication of devices connected to the same network.	Automation of devices connected to the network to improve production efficiency.
3D Printing	[39]	3, 5, 7, 8, 9, 11, 12, 17 SDGs	Attractive and efficient presentation of new concepts.	New collaborations and the reduction of resource consumption.
Advanced Robotics	[41]	11, 12, 13, 14, 17 SDGs	Efficient automation.	Reduction of waste and greenhouse gas (GHG) quantity.
Augmented Reality	[42]	3, 5, 7, 8, 9, 16 SDGs	Testing of some products and processes in accordance with market requirements.	Efficient operations by reducing production downtime, quickly identifying problems, and maintaining all services and processes.
Simulation	[42]	1, 2, 3, 4, 5, 7, 12 SDGs	The use of software to make computer models of manufacturing systems.	Reducing gaps and improving production capacity.

5. The Manufacturing Industry in Romania

The manufacturing industry in Romania is constantly growing, being one of the most important industries therein. More than 1.2 million employees work in this industry, which is 35% of the total workforce in the economy [31]. Included in the manufacturing industry are companies in the food industry, beverage manufacturing, tobacco products, textiles and clothing, wood processing, furniture, chemical industry, auto industry, and pharmaceutical products manufacturing (companies with a NACE code between 10 and 33). The net profit balance is 542 million euros (the difference between the total net profit and the total losses), and the margin of the big factories in the local economy was 3.5% in 2019 [39–42]. The first positions are occupied by companies in the automotive industry, crude oil processing, road transport, the production of soft drinks, the manufacture of household appliances, and the manufacture of alcoholic beverages [42–44]. The characteristics of the Romanian market are a skilled labor force that is competent in the field, a low/medium level of remuneration (compared to the salary level in the European Union), a university environment open to collaborations with the industry in order to develop the required competences, adaptability, average involvement of the state authorities, and the capacity for expansion and globalization [38–44]. The process of production systematized for the manufacturing is shown in Figure 5. The definition of the process begins with the definition of the strategic elements. The next level includes the requirements, the conditions for carrying out the processes, and the current situation of the processes. The next level includes the elements related to the company's logistics and integrating the customer's requirements. Finally, we find the reverse logistics.

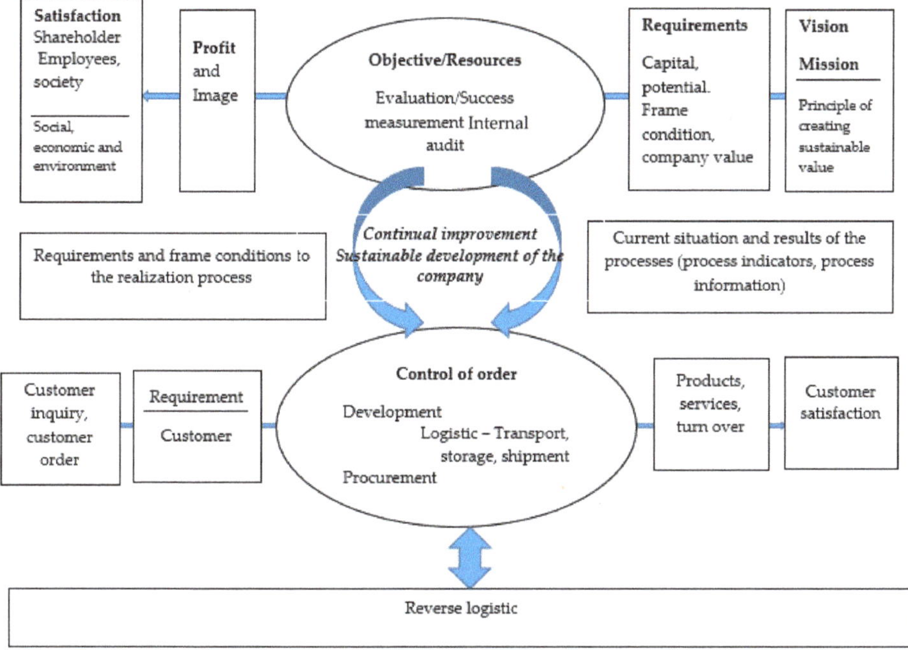

Figure 5. Process model for manufacturing industry.

6. Results

This section presents the results obtained from the research. At the end of the presentation of the results, the hierarchical framework for sustainability assessment of manufacturing industry is presented.

6.1. Industry 4.0: Characteristics, Implications, and Proposed Developments

Industry 4.0. includes the tendency of companies toward automation and data exchange in technologies and manufacturing processes that include cyber-physical systems (CPS), the Internet of Things (IoT), the Industrial Internet of Things (IIoT), cloud computing, cognitive computing and artificial intelligence, and other implications related to these fields [44–51]. New digital industrial technology presents a number of facilities for the sustainability of companies [52–58]. Below are the results obtained in the research conducted with the 100 experts from the fields related to manufacturing industry.

The information is structured in Table 5. The results obtained for each of the following four directions are presented: company information, Industry 4.0—interpretation, facilitators, and barriers; Industry 4.0—maturity; and national technology platform Industry 4.0. For each investigated element, the characteristics and indicators that will be used in the hierarchical framework formation were retained.

Table 5. Sustainable development and Industry 4.0.

Direction	Investigated Elements	Recorded Answers
Company information	Field of activity	Automotive, foam, chemicals, plastics, oil, food and beverage, furniture, pharmaceuticals, and other productions.
	Identification of the best-selling product	Metal article, automotive article, industry equipment, software, office furniture, women's clothing, rings production (jewelry).
	Number of Employees	1–200 employees (65%), >200 employees (35%)
	Assessment of the level of innovation of the company (0–100)	40% score < 50, 51 ≤ 30% ≤ 80, 20% > 80
Interpretation, facilitators and barriers	Identification of company practices in Industry 4.0	Automation of production processes (90%), Big Data (53%), Cloud Computing (23%), Internet of Things—IoT (43%), and digitization (45%).
	Evaluation of Industry facilitators 4.0	Cost reduction (75%), Competitiveness (63%), Need for higher control for top management (51%), demand from partners (83%), challenges of the era (53%), financial benefits (73%), Times of delivery (65%), increased customer satisfaction (53%), efficiency improvement (65%), flexibility (73%), reliable operation (81%), and production interruptions (78%).
	Barriers Industry 4.0	Lack of financial resources (65%), skills (75%), supply chain dimensioning (83%), organizational structure (51%), and employee resistance (87%).
Maturity	Identifying the degree of maturity	High (65%), medium (30%), low (5%).
	Evaluation of a proposed model for digital maturity that includes strategy, technology, operations, organization, and culture and clients.	Medium strategy (70%), advanced (30%) Medium technology (55%), advanced (45%) Operations average (62%), advanced (38%). Organization and culture: medium (57%), advanced (43%). Customers: average (51%), advanced (49%).
National Technology Platform Industry 4.0.	Evaluation of the national platform Industry 4.0	They use the platform (24%), do not use and did not know (76%). No company is a member.
	Assessment of the importance of the actors of the Industry 4.0 platform: government, universities and research institutes, users-companies of Industry 4.0 and suppliers of Industry 4.0	Government: medium (75%), advanced (25%). Universities and research institutes: medium (15%), advanced (85%). Users: companies in Industry 4.0-medium (5%), advanced (95%). Industry suppliers 4.0: medium (3%), advanced (97%).
	The level of resistance to digitization Transforming the company into the digitalization era	Human resources (57%), financial (33%) 100% follow the transformation of the company

For the first direction—information about the company—directions for characterizing the companies of the study respondents were targeted. Percentages were identified by major categories (as a result of responses based on a single choice from a number of variants). The second direction—interpretation, facilitators, and barriers—addressed a number of important elements for Industry 4.0. The respondents had multiple selections. In the interpretation of the data the percentage of the total respondents was used. The third direction—maturity—allowed the respondents a unique choice based on Likert scale assessment. In the interpretation of the data the percentage of the total respondents was used. The last direction, —national technology platform Industry 4.0—allowed a single selection, and the value obtained was interpreted as a percentage.

6.2. Indicators of Sustainable Development on the Dimensions of the Triple Baseline

To identify the performance measures and sustainability indicators, the Delphi method was used, with three rounds of discussions for defining their importance. The results of the first round are presented in Table 6. For the evaluation of the targeted directions, the following were used:

a. Experience—the lowest level recorded in the 40 experts is presented;
b. Involvement in innovation—the arithmetic mean is calculated;
c. Involvement in sustainable development—there are categories of indicators presented;
d. Involvement in strategic management—there are categories of indicators presented;
e. Strategic vision—there are categories of indicators presented.

For each segment of manufacturing industry, the indicators that are not found in the previous segments are filled in.

Table 6. Sustainable development and Industry 4.0.

Segment of Manufacturing Industry	Directions Evaluated by Experts	Response
Automotive	Domain experience:	>15 years
	Innovation	>70% (New technologies, big data, simulation, cloud computing for processes)
	Involvement in sustainable development	Economic performance indicators, continuous improvement, external interaction, digitalization, waste management, operations management, loss reduction, and occupational health and safety policies.
	Involvement in strategic management	Resource management, activity planning, globalization, improving the capacity for regeneration, process quality, and financial indicators.
	Strategic vision	Process planning, continuous learning, knowledge management, and corporate social responsibility.
Production of foams, chemicals, plastics, oil	Domain experience	>21 years
	Involvement in innovation	>78%
	Involvement in sustainable development	Hazardous waste management, water protection, and other elements.
	Involvement in strategic management	Collaboration with other institutions for research and development.
	Strategic vision	Sharing knowledge, copyright.
Food and beverage production	Domain experience	>30 years
	Involvement in innovation	>56%
	Involvement in sustainable development	Packaging management, customer information.
	Involvement in strategic management	Global distribution and collaboration.
	Strategic vision	Increased the capacity of Industry 4.0 implementation.
Furniture production	Domain experience	>25 years
	Involvement in innovation	>62%
	Involvement in sustainable development	Waste reuse, reverse logistics, customer created value, redesign.
	Involvement in strategic management	Defining local and national strategies.
	Strategic vision	Penetration of a new market segment.
Pharmaceutical production	Domain experience	>18 years
	Involvement in innovation	>87%
	Involvement in sustainable development	Agile manufacturing, reverse logistic, collaboration with universities, and product specifications.
	Involvement in strategic management	Customer management, globalization, waste reduction.
	Strategic vision	Merging with international companies.
Other productions	Domain experience:	>5 years
	Involvement in innovation	>64%
	Involvement in sustainable development	Supplier management, sustainable jobs, eco-design, redesign.
	Involvement in strategic management	Voice of customer, sustainable product.
	Strategic vision	Annual reporting.

After identifying all the measures and indicators in round 1, they are reviewed in round 2 to develop the final report. The whole approach is coordinated by the facilitator. These indicators were ranked on the five levels of importance. Each level includes the indicators and measures related to the three basic lines (social, economic and environmental). A structural self-interaction matrix (SSIM) was used for the selected elements from round 2. The report in round 3 was accepted by all experts (40 experts) of the target group. The definition of the hierarchical framework for sustainability assessment of manufacturing industry is presented in Table 7. All indicators agreed to by experts were divided into categories on the three dimensions of sustainability. These categories were allocated to five levels. Their levels and importance were determined by experts in the field during the rounds of the Delphi method. Shareholders believe that any implementation must be approved and accepted by them and that they will not finance techniques and technologies that are not profitable. Everyone appreciated that this is the first level of evaluation.

Table 7. Defining the hierarchical framework.

Level	Identification in the Manufacturing Industry	Social Dimension	Economical Dimension	Environmental Dimension
Level 5	Reverse logistic (recovery of raw materials)	Continuous learning	Supplier management	9R
Level 4	Customer satisfaction	Customer management	Resource optimization	Environmental health and safety
Level 3	Life cycle assessment (product, services)	Agile	Sustainable maintenance	Design for environment (lean, agile, manufacturing)
Level 2	Process of the logistics chain (development, procurement, transport, storage, shipment)	Sustainable workplaces	Knowledge and quality management	Industry 4.0
Level 1	Shareholders (satisfaction, requirement shareholders, profit and image)	Strategic element	Financial improvement	Continuous improvement

6.3. Proposed Conceptual Hierarchical Framework for Sustainability Assessment of Manufacturing Industry

For each level, indicators were defined for each dimension of sustainability. The indicators are identified based on market research (Industry 4.0) and Delphi analysis. Each level records a score calculated as the arithmetic mean of the scores recorded. For each indicator, the evaluator gives a grade from 1 to 5, depending on the degree of implementation (1 = not implemented and 5 = fully implemented). The final report after measuring the performance on sustainable development will include the score obtained on each level for each dimension, Figure 6. At the end of the evaluation, the value of the levels is calculated as the arithmetic mean of the five evaluated levels.

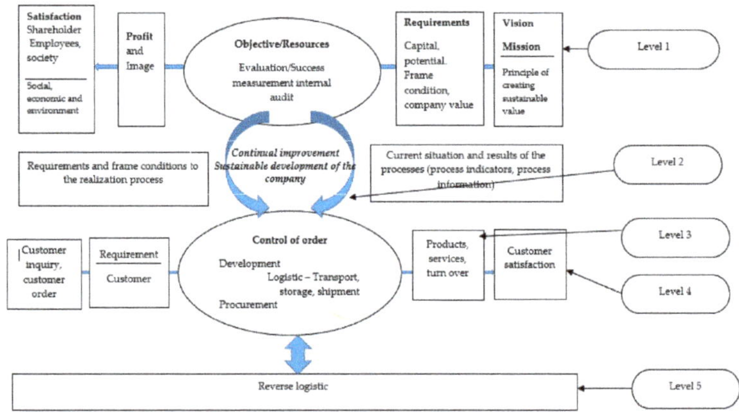

Figure 6. Hierarchical framework for sustainability assessment of manufacturing industry.

6.4. Empirical Testing

The result of the empirical testing is shown in Figure 7. It can be observed that if the indicator level is lower than 2, it returns to the initial phase for level improvement. The value of the total score registered for a company highlights the involvement in the sustainable development. If the value L is less than 10, then the company is at the limit of the level of sustainability and it is recommended to improve all indicators of the five levels. If the value is between 10 and 15, the involvement is average, and the recommendations refer to the implementation of some directions of Industry 4.0 in order to increase the level of competitiveness. If the score value is greater than 15, then the situation of the company is favorable.

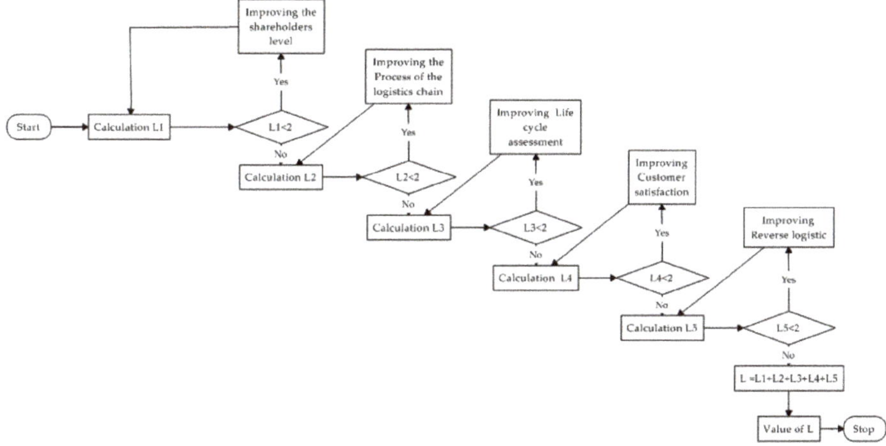

Figure 7. Empirical evaluation based on the proposed hierarchical framework.

This hierarchical model was conceived in the form of a continuous loop. The evaluation of a company does not go further if the evaluation of the level indicators does not exceed 2. If the value is not 2, it returns to the previous level to improve certain indicators that have received low scores. When each level is satisfied, the end is reached by measuring the company's implications. At each level, depending on the value, different recommendations are received.

6.5. Validation and Future Research Approach

The validation was performed on a company in the automotive industry, and the results obtained are presented in the following process, Table 8. For the evaluated company, the five levels were completed. For each level, the value being less than 2, a series of recommendations were received which are presented in Table 8. The whole evaluation was based on the algorithm presented in Figure 7.

Table 8. The results obtained during the evaluation.

Level	Improvements Implemented Following Empirical Testing
Level 5	Application of 9Rs (remanufacturing, reconditioning, reuse, recycling, reduce, repair, recover, redesign, and reconditioning) in any situation of the company
Level 4	Evaluation of requirements regarding international standards
Level 3	Improving the materials used and recovering the value at the end of the life cycle
Level 2	Imposing improvements regarding Industry 4.0
Level 1	Involving shareholders in establishing financial indicators

7. Discussion

Following the conducted research, various aspects related to the need for sustainable development and Industry 4.0 can be discussed. From the perspective of Industry 4.0, it can be seen that not all companies in Romania have accessed the national platform for Industry 4.0. Among the most important facilitators of Industry 4.0 are cost reduction, competitiveness, demand for partners, financial benefits, flexibility, reliable operation, and production interruptions. Barriers mainly refer to the lack of financial resources, competencies, the dimensioning of the supply chain, and the resistance of employees.

From the perspective of sustainable development, following the application of the Delphi method, a series of measures and indicators have been identified that contribute to the assessment of sustainable development and to the proposal of improvement measures.

The proposed model takes into account the imperatives of Industry 4.0 for manufacturing, the objectives and indicators of sustainable development. Five levels of evaluation are developed to systematically identify the measures required to improve the company's situation. Compared to existing methods, the present model reaches a final form during critical rounds of brainstorming analysis. These were performed during the Delphi method.

A five-level hierarchical model was developed starting from the interests of the shareholders to the reverse logistics. For each level, the impact of the indicators is calculated and, finally, the total value. A value of a level below score 2 requires a review of the company's behavior in those directions. Finally, the company's position toward sustainable development is obtained.

Following the test, a series of improvements of the initially proposed conceptual model were obtained. These improvements refer to improvements implemented following the empirical testing, the application of the 9Rs in any situation of the company, the evaluation of the requirements regarding the international standards, the improvement of the materials used, the recovery of the value at the end of the life cycle, the imposition of improvements regarding the Industry 4.0, and the involvement of shareholders in establishing financial indicators.

The proposed model advances the level of knowledge in the field by the fact that it has really identified the imperatives of industry 4.0 and develops a model in rounds of debates with experts in the field of manufacturing. This model is not a theoretical one but one approved by 40 experts chosen according to the top companies in Romania. The model is applicable to every industry because the evaluation is done by semi-quantitative assessments by competitiveness experts. The reference for each field is represented by the most important company from the top of the companies (made for the selection of the interviewed companies).

A set of the proposed frameworks [12–14] propose evaluations based on checklists that present certain limitations from the perspective of covering all organizational levels. These indicators were established on the basis of the specialized literature without a multi-round discussion with experts in the fields of activity.

The hierarchical model is valid and can be extended to other industries. The banking field is targeted because it is a field that has major implications in the economic and social dimensions. At the same time, this model will also identify the disruptive factors that may occur (we expect that the medical factors may affect the functioning of a system).

8. Conclusions

This research proposed a five-level framework for assessing manufacturing sustainability. The definition of these five levels offers an important stage in terms of production as well as sustainability in the manufacturing sector. The research focused on the two important directions for the competitiveness of the business environment: the sustainability and the digitization of the industry. Only experts from the manufacturing industry were involved in the research, and the experience of multiple previous researches was used. At the same time, the theoretical frameworks developed previously contributed to the foundation of the proposed and tested framework.

To characterize the industry, existing data series from accredited institutions were used. These influenced the selection of the manufacturing for the carried-out analysis. The use of market research and the involvement of experts in the research undertaken have contributed to the outline of the proposed framework. Empirical testing has led to the completion of the improved proposed framework and can be transposed into an online platform using databases and web programming facilities.

Future directions will also extend the model to other industries. Barriers encountered during the research refer to the factors that contribute to changing the conditions of the business environment. The research was carried out during 2019, and at the end of it, we started structuring and analyzing the data obtained. Since 2020, a number of factors have begun to appear that affect the economic conditions worldwide (the virus that has spread worldwide). These factors completely change the behavior of industries. This will be investigated in a future paper. The limitation of the study refers to the fact that the Romanian manufacturing characteristics are taken into account for research. In Romania, there is a need to develop a model, and that is why I focused only on these characteristics.

Funding: This work was partially supported by research grant GNaC2018-ARUT, no. 1359/01.02.2019, financed by Politehnica University of Timisoara.

Acknowledgments: The author wants to acknowledge the support of Politehnica University of Timisoara, Faculty of Management in Production and Transportation, and Management Department, Department of Industrial Engineering and Management, Faculty of Engineering for administrative and technical support, for the use of the infrastructure, and for allowing the author to create a new tool for research and didactic use. The author also acknowledges the support of industry in the region for helpful discussions and survey responses.

Conflicts of Interest: The author declares no conflict of interest.

References

1. Malthus, T.R. An essay on the principle of population as it affects the future improvement of society, with remarks on the speculations of Mr. Godwin, M. Condorcet, and other writers, 1798. In *First Essay on Population*; Macmillan: London, UK, 1926.
2. Mitcham, C. The concept of sustainable development: Its origins and ambivalence. *Technol. Soc.* **1995**, *17*, 311–326. [CrossRef]
3. United Nations. Declaration of the United Nations Conference on the Human Environment. Stockholm, Sweden, 16 June 1972. Available online: http://www.unep.org/Documents/Default.asp?DocumentID=97&ArticleID=1503 (accessed on 19 April 2020).
4. Zimmerman, L.J. The distribution of world income, 1860–1960". In *Essays on Unbalanced Growth: A Century of Disparity and Convergence*; de Vries, E., Ed.; Mouton: S-Gravenhage, The Netherlands, 1962; pp. 28–55.
5. Cioca, L.-I.; Ivascu, L.; Rada, E.C.; Torretta, V.; Ionescu, G. Sustainable Development and Technological Impact on CO_2 Reducing Conditions in Romania. *Sustainability* **2015**, *7*, 1637–1650. [CrossRef]
6. Chang, D.S.; Chen, S.H.; Hsu, C.W.; Hu, A.H. Identifying strategic factors of the implantation CSR in the airline industry: The case of Asia-Pacific airlines. *Sustainability* **2015**, *7*, 7762–7783. [CrossRef]
7. Garcia-Torres, S.; Rey-Garcia, M.; Albareda-Vivo, L. Effective Disclosure in the Fast-Fashion Industry: From Sustainability Reporting to Action. *Sustainability* **2017**, *9*, 2256. [CrossRef]
8. Siew, R.Y.J. A Review of Corporate Sustainability Reporting Tools (SRTs). *J. Environ. Manag.* **2015**, *164*, 180–195. [CrossRef]
9. Amui, L.B.L.; Jabbour, C.J.C.; de Sousa Jabbour, A.B.L.; Kannan, D. Sustainability as a Dynamic Organizational Capability: A Systematic Review and a Future Agenda toward a Sustainable Transition. *J. Clean. Prod.* **2017**, *142*, 308–322. [CrossRef]
10. Steen, B. *A Systematic Approach to Environmental Priority Strategies in Product Development (EPS): Version 2000-general System Characteristics*; Centre for Environmental Assessment of Products and Material Systems: Gothenburg, Sweden, 1999.
11. Beatriz, A.; De Sousa, L.; Charbel, J.; Chiappetta, J.; Godinho, M.; David, F. Industry 4. 0 and the circular economy: A proposed research agenda and original roadmap for sustainable operations. *Ann. Oper. Res.* **2018**. [CrossRef]

12. Luiz, D.; Nascimento, M.; Alencastro, V.; Luiz, O.; Quelhas, G.; Luiz, D.; Quelhas, G. Exploring Industry 4.0 technologies to enable circular economy practices in a manufacturing context: A business model proposal. *J. Manuf. Technol. Manag.* **2019**, *30*, 607–627.
13. Neligan, A. Digitalisation as enabler towards a sustainable circular economy in Germany. *Intereconomics* **2018**, *53*, 101–106. [CrossRef]
14. Okorie, O.; Salonitis, K.; Charnley, F.; Moreno, M.; Turner, C.; Tiwari, A. Digitisation and the Circular Economy: A Review of Current Research and Future Trends. *Energies* **2018**, *11*, 3009. [CrossRef]
15. Arena, M.; Azzone, G.; Conte, A. A Streamlined LCA Framework to Support Early Decision Making in Vehicle Development. *J. Clean. Prod.* **2013**, *41*, 105–113. [CrossRef]
16. Ahmad, S.; Wong, K. Sustainability assessment in the manufacturing industry: A review of recent studies. *Benchmarking Int. J.* **2018**, *25*, 3162–3179. [CrossRef]
17. Sangwan, K.S.; Mittal, V.K. A Bibliometric Analysis of Green Manufacturing and Similar Frameworks. *Manag. Environ. Qual. Int. J.* **2015**, *26*, 566–587. [CrossRef]
18. Baud, R. The concept of sustainable development: Aspects and their consequences from a social-philosophical perspective. *YES Youth Encount. Sustain. Summer Course Mater.* **2008**, *2*, 8–17.
19. Garetti, M.; Taisch, M. Sustainable manufacturing: Trends and research challenges. *Prod. Plan. Control* **2012**, *23*, 83–104. [CrossRef]
20. UN General Assembly. Transforming our world: The 2030 Agenda for Sustainable Development. 21 October 2015. A/RES/70/1. Available online: https://www.refworld.org/docid/57b6e3e44.html (accessed on 20 March 2020).
21. Adams, M.A.; Ghaly, A.E. An integral framework for sustainability assessment in agro-industries: Application to the Costa Rican coffee. *Int. J. Sustain. Dev. World Ecol.* **2006**, *13*, 83–102. [CrossRef]
22. Benoît, C.; Norris, G.A.; Valdivia, S.; Ciroth, A.; Moberg, A.; Bos, U.; Prakash, S.; Ugaya, C.; Beck, T. The Guidelines for Social Life Cycle Assessment of Products: Just in Time! *Int. J. Life Cycle Assess.* **2010**, *15*, 156–163. [CrossRef]
23. Mani, M.; Johansson, B.; Lyons, K.W.; Sriram, R.D.; Ameta, G. Simulation andanalysis for sustainable product development. *Int. J. Life Cycle Assess.* **2013**, *18*, 1129–1136. [CrossRef]
24. Marchese, D.; Reynolds, E.; Bates, M.E.; Morgan, H.; Clark, S.S.; Linkov, I. Resilience and sustainability: Similarities and differences in environmental management applications. *Sci. Total Environ.* **2018**, *613*, 1275–1283. [CrossRef]
25. Faulkner, W.; Badurdeen, F. Sustainable Value Stream Mapping (Sus-VSM): Methodology to visualize and assess manufacturing sustainability performance. *J. Clean. Prod.* **2014**, *85*, 8–18. [CrossRef]
26. Rajak, S.; Vinodh, S. Application of fuzzy logic for social sustainability performance evaluation: A case study of an Indian automotive component manufacturing organization. *J. Clean. Prod.* **2015**, *108*, 1–9. [CrossRef]
27. Azapagic, A. Developing a framework for sustainable development indicators for the mining and minerals industry. *J. Clean. Prod.* **2004**, *12*, 639–662. [CrossRef]
28. Veleva, V.; Ellenbecker, M. Indicators of sustainable production: Framework and methodology. *J. Clean. Prod.* **2001**, *9*, 519–549. [CrossRef]
29. Fernández, M.; Martínez, A.; Alonso, A.; Lizondo, L. A mathematical model for the sustainability of the use of cross-laminated timber in the construction industry: The case of Spain. *Clean. Technol. Environ. Policy* **2014**, *16*, 1625–1636. [CrossRef]
30. Oliva-Maza, L.; Torres-Moreno, E.; Villarroya-Gaudó, M.; Ayuso-Escuer, N. Using IoT for Sustainable Development Goals (SDG) in Education. *Proceedings* **2019**, *31*, 1. [CrossRef]
31. Herrmann, C.; Hauschild, M.; Gutowski, T.; Lifset, R. Life cycle engineering and sustainable manufacturing. *J. Ind. Ecol.* **2014**, *18*, 471–477. [CrossRef]
32. Herrmann, C.; Bergmann, L.; Thiede, S.; Halubek, P. Total Life Cycle Management: An Integrated Approach towards Sustainability. In Proceedings of the 3rd International Conference on Life Cycle Management, Zurich, Switzerland, 27–29 August 2007.
33. Chou, C.; Chen, C.; Conley, C. An approach to assessing sustainable product-service systems. *J. Clean. Prod.* **2015**, *86*, 277–284. [CrossRef]
34. Bala, G.; Bartel, H.; Hawley, J.P.; Lee, Y. Tracking "Real-time" Corporate Sustainability Signals Using Cognitive Computing. *J. Appl. Corp. Financ.* **2015**, *27*, 95–102. [CrossRef]

35. Salvado, M.; Azevedo, S.; Matias, J.; Ferreira, L. Proposal of a Sustainability Index for the Automotive Industry. *Sustainability* **2015**, *7*, 2113–2144. [CrossRef]
36. Buyya, R.; Singh Gill, S. Sustainable Cloud Computing: Foundations and Future Directions. *Bus. Technol. Digit. Transform. Strateg. Cut. Consort.* **2018**, *21*, 1–9.
37. Hsiao, M.-K. The Provision of Sustainability Information for Electronic Product Consumers Through Mobile Phone Technology. *Int. J. Eng. Res.* **2013**, *1*, 2321.
38. Hallstedt, S.I. Sustainability criteria and sustainability compliance index for decision support in product development. *J. Clean. Prod.* **2017**, *140*, 251–266. [CrossRef]
39. Wilkinson, S.; Cope, N. *Chapter 10—3D Printing and Sustainable Product Development a2–Akhgar, Mohammad Dastbazcolin Pattinsonbabak. Green Information Technology*; Morgan Kaufmann: Boston, MA, USA, 2015; pp. 161–183.
40. Gasparatos, A.; Scolobig, A. Choosing the most appropriate sustainability assessment tool. *Ecol. Econ. J.* **2012**, *80*, 1–7. [CrossRef]
41. Mayyas, A.; Qattawi, A.; Omar, M.; Shan, D. Design for sustainability in automotive industry: A comprehensive review. *Renew. Sustain. Energy Rev.* **2012**, *16*, 1845–1862. [CrossRef]
42. Petruse, R.; Grecu, V.; Chiliban, B. Augmented Reality Applications in the Transition towards the Sustainable Organization. In Proceedings of the International Conference on Computational Science and Its Applications, Beijing, China, 4–7 July 2016.
43. McAuley, J.W. Global sustainability and key needs in future automotive design. *Environ. Sci. Technol.* **2003**, *37*, 5414–5416. [CrossRef] [PubMed]
44. National Intitute of Statistics. Available online: http://statistici.insse.ro:8077/tempo-online/ (accessed on 10 March 2020).
45. European Commission (Eurostat). Available online: https://ec.europa.eu/eurostat (accessed on 21 February 2020).
46. Muhuri, P.; Shukla, A.; Abraham, A. Industry 4.0: A bibliometric analysis and detailed overview. *Eng. Appl. Artif. Intell.* **2019**, *78*, 218–235. [CrossRef]
47. Muñoz-Villamizar, A.; Santos, J.; Viles, E.; Ormazábal, M. Manufacturing and environmental practices in the Spanish context. *J. Clean. Prod.* **2018**, *178*, 268–275. [CrossRef]
48. Cioca, L.-I.; Ivascu, L.; Turi, A.; Artene, A.; Găman, G.A. Sustainable Development Model for the Automotive Industry. *Sustainability* **2019**, *11*, 6447. [CrossRef]
49. Tăucean, I.M.; Tămășilă, M.; Ivascu, L.; Miclea, Ș.; Negruț, M. Integrating Sustainability and Lean: SLIM Method and Enterprise Game Proposed. *Sustainability* **2019**, *11*, 2103. [CrossRef]
50. Ivascu, L.; Cioca, L.-I. Occupational Accidents Assessment by Field of Activity and Investigation Model for Prevention and Control. *Safety* **2019**, *5*, 12. [CrossRef]
51. Cioca, L.-I.; Ivascu, L. Risk Indicators and Road Accident Analysis for the Period 2012–2016. *Sustainability* **2017**, *9*, 1530. [CrossRef]
52. Feniser, C.; Burz, G.; Mocan, M.; Ivascu, L.; Gherhes, V.; Otel, C.C. The Evaluation and Application of the TRIZ Method for Increasing Eco-Innovative Levels in SMEs. *Sustainability* **2017**, *9*, 1125. [CrossRef]
53. Li, S.; Ruiz-Mercado, G.J.; Lima, F.V. A Visualization and Control Strategy for Dynamic Sustainability of Chemical Processes. *Processes* **2020**, *8*, 310. [CrossRef]
54. Han, C.; Dong, Y.; Dresner, M. Emerging market penetration, inventory supply, and financial performance. *Prod. Oper. Manag.* **2013**, *22*, 335–347. [CrossRef]
55. Đurđević, D.; Hulenić, I. Anaerobic Digestate Treatment Selection Model for Biogas Plant Costs and Emissions Reduction. *Processes* **2020**, *8*, 142. [CrossRef]
56. Siddiqi, M.M.; Naseer, M.N.; Abdul Wahab, Y.; Hamizi, N.A.; Badruddin, I.A.; Chowdhury, Z.Z.; Akbarzadeh, O.; Johan, M.R.; Khan, T.M.Y.; Kamangar, S. Evaluation of Municipal Solid Wastes Based Energy Potential in Urban Pakistan. *Processes* **2019**, *7*, 848. [CrossRef]
57. Henao, R.; Sarache, W.; Gómez, I. Lean Manufacturing and sustainable performance: Trends and future challenges. *J. Clean. Prod.* **2019**, *208*, 99–116. [CrossRef]
58. Liang, Y.; Jing, X.; Wang, Y.; Shi, Y.; Ruan, J. Evaluating Production Process Efficiency of Provincial Greenhouse Vegetables in China Using Data Envelopment Analysis: A Green and Sustainable Perspective. *Processes* **2019**, *7*, 780. [CrossRef]

59. Zhou, G.; Chung, W.; Zhang, Y. Measuring energy efficiency performance of China's transport sector: A data envelopment analysis approach. *Expert Syst. Appl.* **2014**, *41*, 709–722. [CrossRef]
60. Okudan, G.E.; Akman, G. Manufacturing performance modeling and measurement to assess varying business strategies. *Glob. J. Flex. Syst. Manag.* **2004**, *5*, 51–59.
61. Kwak, J.K. Analysis of Inventory Turnover as a Performance Measure in Manufacturing Industry. *Processes* **2019**, *7*, 760. [CrossRef]
62. Grados, D.; Schrevens, E. Multidimensional analysis of environmental impacts from potato agricultural production in the Peruvian Central Andes. *Sci. Total Environ.* **2019**, *663*, 927–934. [CrossRef] [PubMed]
63. García-Arca, J.; Prado-Prado, J.C.; Gonzalez-Portela Garrido, A.T. "Packaging logistics": Promoting sustainable efficiency in supply chains. *Int. J. Phys. Distrib. Logist. Manag.* **2014**, *44*, 325–346. [CrossRef]
64. De Angelis, R.; Howard, M.; Miemczyk, J. Supply chain management and the circular economy: Towards the circular supply chain. *Prod. Plan. Contr.* **2018**, *29*, 425–437. [CrossRef]

Article

Analysis the Drivers of Environmental Responsibility of Chinese Auto Manufacturing Industry Based on Triple Bottom Line

Hua Zhang [1,2], Meihang Zhang [2,3], Wei Yan [4,*], Ying Liu [5], Zhigang Jiang [3] and Shengqiang Li [2,3]

1 Academy of Green Manufacturing Engineering, Wuhan University of Science and Technology, Wuhan 430081, China; zhanghua403@163.com
2 Key Laboratory of Metallurgical Equipment and Control Technology, Ministry of Education, Wuhan University of Science and Technology, Wuhan 430081, China; zhangmeihang@wust.edu.cn (M.Z.); 18271807321@163.com (S.L.)
3 Hubei Key Laboratory of Mechanical Transmission and Manufacturing Engineering, Wuhan University of Science and Technology, Wuhan 430081, China; jiangzhigang@wust.edu.cn
4 School of Automobile and Traffic Engineering, Wuhan University of Science & Technology, Wuhan 430081, China
5 Institute of Mechanical and Manufacturing Engineering, School of Engineering, Cardiff University, Cardiff CF10 3XQ, UK; LiuY81@Cardiff.ac.uk
* Correspondence: yanwei81@wust.edu.cn; Tel.: +86-27-68862816

Abstract: The rapid increasing number of automobile products has brought great convenience to people's living, but it has also caused serious environmental issues, waste of resources and energy shortage during its whole lifecycle. Corporate Environmental Responsibility (CER) refers to the company's responsibility to avoid damage to the natural environment derived from its corporate social responsibility (CSR), and it plays an important role in solving resource and environmental problems. However, due to various internal and external reasons, it is difficult for the automobile manufacturing industry to find the key drivers for the implementation of CER. This research proposes a model framework that uses the fuzzy decision-making test and evaluation laboratory (fuzzy DEMATEL) method to analyze the drivers of CER from the perspective of the triple bottom line (TBL) of economy, environment and society. Firstly, the common drivers of CER are collected using literature review and questionnaire survey methods. Secondly, the key drivers are analyzed by using the fuzzy DEMATEL. Finally, the proposed approach was verified through a case study. The research results show that some effective measures to implement CER can be provided for the government, the automobile manufacturing industry and the public to promote sustainable development of Chinese Auto Manufacturing Industry (CAMI).

Keywords: Chinese automobile manufacturing industry; corporate environmental responsibility; fuzzy DEMATEL; triple bottom line; sustainable development; green manufacturing

Citation: Zhang, H.; Zhang, M.; Yan, W.; Liu, Y.; Jiang, Z.; Li, S. Analysis the Drivers of Environmental Responsibility of Chinese Auto Manufacturing Industry Based on Triple Bottom Line. *Processes* **2021**, *9*, 751. https://doi.org/10.3390/pr9050751

Academic Editor: Luis Puigjaner

Received: 2 April 2021
Accepted: 22 April 2021
Published: 24 April 2021

1. Introduction

According to relevant survey reports, the number of Chinese new automobiles almost exceeded 25.76 million, and the amount of Chinese vehicle ownership reached more than 260 million up to 2019. The rapid growth of the number of automobiles has brought a series of environmental issues. In the process of automobile production, use, recycling and disposal, air, soil, water, etc., will be polluted to a certain extent, resulting in waste of resources and energy shortage [1]. These environmental issues have aroused widespread concern from the government, enterprises and scholars. While the Chinese government is strengthening environmental supervision of the automotive industry, the extension of the production responsibility and Corporate Environmental Responsibility (CER) have become hot topics of concern to many automotive industry units. Therefore, many Chinese auto companies are aware of the importance of sustainable development, and have implemented some sustainable development strategies, such as CER, remanufacturing, green manufacturing and cleaner production [2–6].

In fact, due to various reasons, it is difficult for them to effectively implement sustainable development strategies, especially CER. The main reason is that some automobile companies are worried that the implementation of CER may affect their financial interests, so they are unwilling to implement CER. In addition, some companies voluntarily implement CER, but they do not know what measures should be taken. Therefore, by analyzing the drivers of CER, the fundamental reason why enterprises cannot effectively implement CER can be found, thereby promoting the change of development strategy of Chinese Auto Manufacturing Industry (CAMI) and achieving coordinated development of economic and environmental benefits.

At present, some studies focus on the combination of CER and sustainable practices, such as integrating CER into the supply chain, and the combination of sustainable development and CER. As many automobile companies pay more attention to economic interests, compared with well-known foreign automobile companies, the efficiency of CER implementation by Chinese automobile manufacturers is relatively low. So far, only a few studies have focused on the drivers of CER for Chinese automobile manufacturers and revealed the internal connection between the corporate environmental responsibility and the corporate economic benefits. The "Triple Bottom Line" was proposed by John Elkington [7], a well-known British management consultant and sustainability expert, and used it to measure his company's performance in the United States. Triple Bottom Line (TBL) theory believes that there should be three bottom lines: profit, people and the earth. TBL aims to assess the level of corporate commitment to social responsibility and its impact on the environment over time. Additionally, CER is concerned about the impact of the development of the enterprise on the environment. There is a close relationship between TBL and CER. The influencing factors of enterprises implementing CER can be analyzed from the perspective of TBL. The purpose of both is to achieve a balanced development of economic and environmental benefits. Therefore, this study attempts to identify and analyze the CER drivers of Chinese automakers from the perspective of TBL to bridge the gap. In this study, first, based on a comprehensive analysis of the existing literature, expert opinions and the opinions of Chinese auto industry managers, common drivers were identified. Second, we sent questionnaires to some automobile companies. Third, based on the results of the questionnaire survey, a quantitative and the fuzzy decision-making test and evaluation laboratory (fuzzy DEMATEL) analysis was used to determine the key drivers and classifications.

The rest of this paper is arranged as follows: Section 2 reports a literature review related to the research topic and proposes the main innovations of this paper based on comparative analysis. Section 3 describes the problem. Section 4 introduces the method used in this research, namely fuzzy DEMATEL, and proposes a model framework for analyzing the CER drivers of the Chinese automobile industry. Section 5 verifies the proposed model through case studies. Section 6 discusses in detail. Finally, Section 7 gives conclusions and future work.

2. Literature Review

2.1. *Corporate Environmental Responsibility and Triple Bottom Line*

CER, called corporate environmental responsibility, refers to a company's duties to abstain from damaging natural environments, which derives from corporate social responsibility (CSR) [8]. In recent years, CER has become an important concept and has received extensive attention from relevant researchers. This is because the implementation of CER can enhance the sustainable development ability of enterprises, improve the natural environment, and solve various social and ecological problems such as climate change and biodiversity loss [9,10]. Gunningham [11] described the development of the concept of CER and studied the debate about the relationship between CER and competitive advantage. On the basis of Carroll's CSR pyramid model, Wang Hong [12] explored the system characteristics of CER and sorted out its elements, structure and functions. Studies have shown that the implementation of CER can promote the green and sustainable development of manufacturing to a certain extent [13], and the improvement of the company's market competitiveness and profitability can be tracked through the implementation of environmental management activities [14]. The results of the study were verified by listed company A [15,16], which showed that the performance of corporate environmental responsibility (CER) by company A has a significant positive impact on the company's financial performance, but it has a lag effect, and higher environmental investment can bring higher profitability. Furthermore, many firms are discovering that there is an advantage to advocating for environmental regulations and preparing for them to be implemented before they become law. In a recent study, the researcher found that firms support climate change legislation as a means of gaining power over their competitors. Essentially, even if a new regulation hurts a firm in the short term, the firm may embrace it because they know that it will hurt their competitors even more. This allows them to come out on top in the long run [17].

The TBL proposed by Elkington [7] is an accounting framework that includes three aspects: economic (profit), social (people), and environmental (planet). It is used by many researchers to solve various problems [18,19]. The TBL considers profit using traditional measures for evaluating company profits, evaluating the company's environmental responsibility, evaluating the company's sustainability pillars [20,21], and citizens' concerns about corporate social responsibility as indicated by the company's operations [22,23]. Ahi and Searcy [24] claim that sustainability is the ability to maintain long-term welfare responsibly, manage resources so that the company can meet current needs without compromising the ability of future generations to meet their own needs. The TBL has been widely applied in many domains to promote sustainable development that also provides a co-benefit. Wu et al. [25] suggested that firms should consider stakeholders, resilience, long-term goals and current operations when evaluating sustainability strategies. Previous research also emphasized that TBL is not sufficient to achieve complete sustainability, and we must take greater steps to discuss socioeconomics, social environment and ecological efficiency [26,27]. Moreover, several studies noted that relations, resource consumption and policies must be integrated with sustainable practices to ensure the cobenefit [28].

Many scholars have conducted case studies from the perspective of TBL. Bergenwall [29] studied the differences in process design between American automakers and Toyota on the three aspects of sustainability. Gimenez [30] studied the impact of TBL on sustainable management. They proposed that the internal environmental plan has a positive impact on the three components of TBL, while internal social activities only have a positive impact on the two components of social and environmental performance. Neri et al. [31] designed a triple bottom line balanced key performance indicator set to measure the sustainability performance of industrial supply chains. Agrawal et al. [32] discussed the deployment decisions of sustainable reverse logistics in the Indian electronics industry, and studied the impact of disposal decisions on TBL, that is, the economic, environmental and social performance of reverse logistics. Hussain et al. [33] studied the relationship between corporate governance and triple bottom line sustainability performance through the perspectives of agency theory and stakeholder theory.

2.2. Drivers of CER in the Automotive Industry

The rapid growth of automobile ownership has caused a series of problems, such as climate change, emissions, pollution, etc. [34]. For CAMI, the factors that promote the effective implementation of CER through some sustainable development practices (such as CER, green manufacturing, sustainable supply chain management, manufacturer extension responsibility, life cycle analysis, and environmental certification, etc.) are called drivers. Therefore, many researchers have paid attention to the CER problem and conducted some extended studies.

Some studies discuss the relationship and importance of TBL principles and strategic decisions from the perspective of sustainable supply chains in the automotive industry. The successful implementation of sustainable supply chains can promote the implementation of CER [35–37]. Other studies illustrate the important indicators of CER implementation from the perspective of the green evaluation system of the automobile manufacturing industry [38]. There are also some studies that mainly elaborated the relationship between the implementation of CER and legislation from the aspect of government legislation. Studies have shown that government legislation is the most critical driver for the implementation of CER [39,40]. There are also other explorations of the relationship between energy certification, changes in corporate management strategies and the implementation of CER, including Cai et al. [41], which explored energy performance certification in the machinery manufacturing industry, and provided information for the implementation of energy performance certification strategies. The theoretical foundation is thus promoted to promote the active implementation of CER by automakers. Nunes [42] focuses on investigating and benchmarking the green operating plans of the automotive industry as documented in the environmental reports of selected companies. Research by Yu Cheng et al. [43] shows that Chinese automakers still have much room for improvement in terms of consumer satisfaction, resource conservation, community services and low-carbon activities. Kehbila et al. [40] systematically analyzed the motivations, obstacles and benefits of South African automobile companies participating in environmental change and provided some suggestions that may promote the effective implementation of strategic corporate environmental management. The research results show that achieving consistent compliance, reducing the daily impact on the environment, improving the working and living conditions of employees, and improving image and reputation are the most important driving forces. Babiak and Trendafilova [44] studied the motivations and pressures reported by senior managers to adopt sustainable practices in the industry. The research results show that strategic motivation and institutional pressure are the main reasons for adopting environmental management measures. Lee et al. [45] studied the driving forces for the implementation of CER and green practices in the Korean logistics industry, and pointed out that social expectations, organizational support and stakeholder pressure are important driving forces for the implementation of CER and green practices. Goli et al. [46] explained that corporate environmental responsibility (CER) involves key solutions for the success of corporate innovation.

By reviewing the existing literature, we know that domestic and foreign scholars have conducted CER research from different aspects. Based on the TBL method, some studies have been conducted on the sustainable development of the automotive industry, but mainly focus on the TBL analysis of the automotive industry supply chain order optimization, supply chain management or supplier sustainability. However, there are few studies on the implementation of CER in CAMI. From the perspective of TBL, there are fewer drivers for CER in automobile companies. This has led to companies not paying attention to the implementation of CER, and the effect of CER implementation is poor and difficult.

Therefore, this paper aims to analyze the key drivers of CAMI's implementation of CER from the perspective of TBL, and then analyze the promotion effect of key drivers on economy, environment and society, so as to improve the effect of CER implementation, and realize the coordination and sustainability development of economic and environmental benefits.

3. Identify Common Drivers of CER

As environmental problems have become more prominent, the public's awareness of environmental protection has continued to increase. When consumers buy automobiles, green features have become one of their important choices [47,48]. Some automakers realize that they should implement CER throughout the product life cycle to minimize the negative impact on the environment to meet consumer's demand for environmentally friendly products [49]. However, they do not know how to effectively implement CER to promote the sustainable development and green development of enterprises, and the implementation of CER in CAMI has always been controversial, so there is an urgent need to improve the effectiveness of CER implementation.

Therefore, this paper analyzes the drivers of CER in CAMI from the perspective of TBL (economic, social and environmental). Based on relevant literature, expert opinions and the opinions of Chinese auto industry managers, we jointly determine the common drivers for CER in China's auto industry. First, we collect CER drivers from relevant literature, and use "corporate environmental responsibility", "corporate environmental responsibility drivers" and "Chinese automobile manufacturer environmental responsibility" as search keywords. Secondly, we inquired about the main motivations for Chinese automakers to implement CER in 120 Chinese auto industry units through e-mail and telephone. We finally received replies from 82 Chinese auto companies. Third, on the basis of the above process of determining common drivers, we held an online seminar, inviting CAMI experts and managers to participate in order to solve these classifications. After discussion, we got the final result on the classification of common drivers. Through the above process, the common CER drivers in CAMI were identified and classified, as shown in Table 1. The main drivers identified include policy drivers, technology drivers, corporate internal motivations and corporate external pressure. Finally, a case verification was carried out through a Chinese automobile manufacturer.

Table 1. Common drivers of corporate environment responsibility.

No.	Main Driers	Explanation	Common Drivers
1	Policy drivers	Policy drivers can be categorized into two areas, namely compulsory aspect (regulations, standards, etc.) and incentive aspect.	Incentives (A1) Government regulations (A2) Standards (A3)
2	Technological drivers	Technological drivers can help corporate achieve the sustainable development.	Green technology import (A4) Green technology innovation (A5)
3	Corporate internal motivations	These motivations mainly focus on company level, such as its own development and the demand of company's internal staff, etc.	Top management commitment (A6) Employee demand (A7) Financial benefits (A8) Shareholders motivation (A9) Company image (A10) Competitive advantage (A11)
4	Corporate external pressure	Companies need to do things that they are unwilling to do but must do to meet the needs of the public.	Consumers demand (A12) Green supply chain pressure (A13) Societal expectation (A14) Media pressure (A15) Market trend (A16)

4. The Fuzzy DEMATEL Method of Key Drivers

The proposed model framework for analyzing the drivers of CER in China's automobile industry is shown in Figure 1. First of all, with the help of existing literature, expert opinions and industry managers' opinions, most of the drivers were collected. Second, a questionnaire containing the drivers of the five-point Likert scale was distributed to Chinese automobile companies. For the collected valid questionnaires, we averaged the survey results, discussed with experts, and finally identified and classified 16 common driving factors. Then, based on the TBL, a fuzzy direct relationship matrix was established, and the key drivers of CER in CAMI were analyzed using the DEMATEL program. Finally, the results were verified in a medium-sized automobile company through feedback from the automotive industry managers and comparison with existing literature.

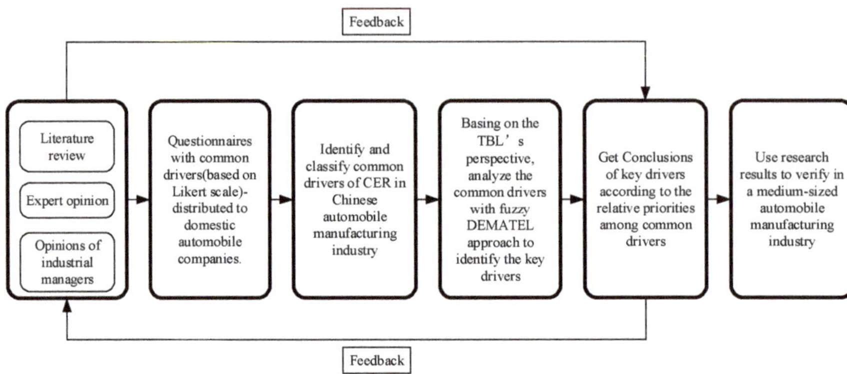

Figure 1. The model framework for analyzing the drivers of CER in Chinese auto industry.

Since the drivers for the implementation of environmental responsibility in CAMI is a complex decision-making problem, it is a common method to use multicriteria decision-making (MCDM) [50,51] method or fuzzy analytic hierarchy process [52] to make decisions. DEMATEL, as one of the MCDM approaches, firstly used by The Battelle Memorial Institute at its Geneva Research Centre in 1973, is utilized as a solution method in this paper [53]. The DEMATEL method can visualize complex causal structures by establishing and analyzing structural models between complex factors. Furthermore, it can analyze the influence relationship between complex criteria and separate the factors into cause group and effect group in which the cause group affects the effect group thus reckoning the relative weights of criteria. In this paper, since the interaction between all the drivers of CER in CAMI is relatively complex, it is necessary to use DEMATEL to help us better understand the interaction between drivers.

Although DEMATEL is a good way to deal with complex decision-making problems, the degree of mutual influence between systems is usually ambiguous, which will make language information unsuitable for expression. In order to reduce uncertainty and increase accuracy, DEMATEL is combined with fuzzy logic proposed by Zadeh [54]. It is rather effective to measure the ambiguous concepts related to human's subjective judgments with fuzzy logic [55]. Therefore, this paper uses fuzzy DEMATEL with triangular fuzzy numbers to evaluate the driving factors of CER in CAMI.

A triangular fuzzy number can be defined as a triplet $\widetilde{A} = (l, m, u)$, where l, m and u denote lower, medium, and upper numbers, respectively, to describe a fuzzy event. Additionally, the membership function $\mu_{\widetilde{A}}$ of a triangular fuzzy number can be expressed as follows:

$$\mu_{\widetilde{A}}(x) = \begin{cases} 0 & x < l \\ \frac{(x-l)}{(m-l)} & l \leq x \leq m \\ \frac{(u-x)}{(u-m)} & m \leq x \leq u \\ 0 & x > u \end{cases} \tag{1}$$

where l, m and u are real numbers and $l \leq m \leq u$.

In view of above, the model of triangular fuzzy numbers is shown in Figure 2. The correspondence between the linguistic terms and triangular fuzzy numbers can be determined by Table 2. For any of two triangular fuzzy numbers $\widetilde{A} = (l_1, m_1, u_1)$ and $\widetilde{B} = (l_2, m_2, u_2)$, the operational laws of the two triangular numbers are as shown below:

$$\begin{cases} \widetilde{A}_1 + \widetilde{A}_2 = (l_1 + l_2, m_1 + m_2, u_1 + u_2) \\ \widetilde{A}_1 - \widetilde{A}_2 = (l_1 - l_2, m_1 - m_2, u_1 - u_2) \\ \widetilde{A}_1 \times \widetilde{A}_2 = (l_1 \times l_2, m_1 \times m_2, u_1 \times u_2) \\ \widetilde{A}_1 \div \widetilde{A}_2 = (l_1 \div l_2, m_1 \div m_2, u_1 \div u_2) \\ \lambda\widetilde{A}_1 = (\lambda l_1, \lambda m_1, \lambda u_1), (k > 0) \\ \frac{\widetilde{A}}{\lambda} = (\frac{l_1}{\lambda}, \frac{m_1}{\lambda}, \frac{u_1}{\lambda}), (k > 0) \end{cases} \tag{2}$$

where l_1, m_1 and u_1 are real numbers and $l_1 \leq m_1 \leq u_1$.

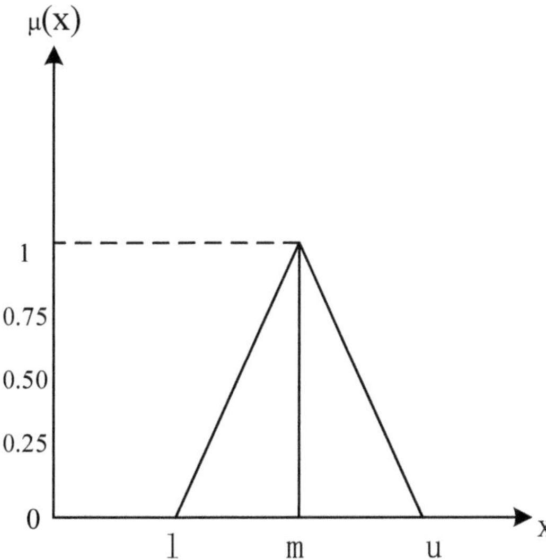

Figure 2. Triangular fuzzy number.

Table 2. Correspondence between the linguistic terms and triangular fuzzy numbers.

Linguistic Terms	Triangular Fuzzy Numbers
No influence (N)	(0,0,0.25)
Very low influence (VL)	(0,0.25,0.5)
Low influence (L)	(0.25,0.5,0.75)
High influence (H)	(0.5,0.75,1)
Very high influence (VH)	(0.75,1,1)

This section is not mandatory but may be added if there are patents resulting from the work reported in this manuscript.

The main steps of the fuzzy DEMATEL method are briefly described as follows:

Step 1: Establish the fuzzy direct relation matrix T with fuzzy linguistic terms.

Step 2: Defuzzified-Initial relation matrix F. In this step, the fuzzy direct relation matrix T is defuzzified, namely the triangular fuzzy numbers are converted to crisp numbers by centroid method, a kind of defuzzification approach. Correspondingly, the Defuzzified-Initial relation matrix F can be established by Equation (3).

$$F_g(x) = \frac{\sum_{i=1}^{n} x_i \mu_{\widetilde{A}}(x_i)}{\sum_{i=1}^{n} \mu_{\widetilde{A}}(x_i)} \tag{3}$$

Step 3: Establish the normalized direct-relation matrix X. In this step, the initial direct-relation matrix F is normalized by utilizing Equations (4) and (5). Consequently, the normalized direct-relation matrix X can be obtained.

$$K = \frac{1}{\max_{1 \le i \le n} \sum_{j=1}^{n} a_{ij}} \tag{4}$$

$$X = K \times F \tag{5}$$

Step 4: Establish the total relation matrix M. In this step, the total relation matrix M is calculated through Equation (6) where I denotes identity matrix. The element m_{ij} denotes the indirect effects that criterion i have on criterion j, and the matrix M gives the total relationship among the each pair of factors.

$$M = X(I - X)^{-1} \tag{6}$$

Step 5: Get the sum of rows and columns. In this step, the sum of rows and columns of matrix M are calculated through Equations (7) and (8). In the two equations, r_i denotes all direct and indirect influence given by criterion i to all other factors, c_j denotes the degree of influenced impact.

$$r_i = \sum_{1 \le j \le n} t_{ij} \tag{7}$$

$$c_j = \sum_{1 \le i \le n} t_{ij} \tag{8}$$

When $i = j$, $(r_i + c_j)$ denotes all effects that are given and received by criterion i. $(r_i + c_j)$ can show the degree of importance that criterion i, in the total system, namely the centrality of the element i in the problem group. Meanwhile, $(r_i - c_j)$ represents the net effect that criterion i has on the system. If $(r_i - c_j) > 0$, the element i will be classified into cause group. By contrast, if $(r_i - c_j) < 0$, it will be classified into effect group.

Step 6: Establish the cause–effect relation diagram. In the final step, the cause and effect relationship diagram is depicted according to the dataset of $(r_i - c_j)$. The horizontal axis (R + C) is obtained by adding R to C, and the vertical axis (R − C) is obtained by subtracting C from R.

Step 7: According to the results of the step 6, the cause group of the key drivers are ranked again to determine the most critical drivers. The calculation formula is:

$$R_g = \frac{(r_i - c_j)}{(r_i - c_j)_{\max}} \times 100\% \tag{9}$$

5. Case Study: An Explanation

5.1. Case Background

Due to the expanding trend of economic globalization, as well as the continuous development of smart and green technologies, CAMI must be able to respond to the market in a timely manner while taking into account economic growth, environmental protection, and the realization of social expectations. K Company is a medium-scale automobile manufacturing company in China, which has been committed to manufacturing energy-saving and environmentally friendly vehicles. However, the automobile company found that the effect of implementing CER on energy conservation and emission reduction is far behind that of well-known foreign automobile manufacturing companies. Therefore, the company realized the need to evaluate the drivers of CER to ensure the effective implementation of its CER. The main research goals of the automobile company are as follows: (1) The automobile company hopes to understand the common drivers and key drivers of CER. (2) The company hopes to understand the comprehensive impact of implementing CER on economic, environmental and social benefits. (3) Last but not least, automobile companies hope to achieve their own green development and sustainable development by effectively implementing CER throughout the entire life cycle of the car.

5.2. Results and Analysis

According to the results of the questionnaire, a fuzzy direct relationship matrix T in the form of fuzzy linguistic terms can be established. The fuzzy direct relationship matrix T is shown in Table 3. Using the centroid method, the fuzzy direct relationship matrix T is defuzzified into clear numbers, and transformed into the initial matrix F (Table 4), and the normalized direct relationship matrix X (Table 5) is established according to F, and then the total relationship matrix M is established (Table 6), and the row sum and column sum of the total relationship matrix M are calculated. Finally, a causality diagram is established, the results of (R + C) and (R − C) (Table 7) are calculated and used as the horizontal and vertical axes of the causality diagram, as shown in Figure 3.

Table 3. The fuzzy direct matrix T—the TBL's perspective.

Criteria	A1	A2	A3	A4	A5	A6	A7	A8	A9	A10	A11	A12	A13	A14	A15	A16
A1	N	VL	VL	H	H	H	N	H	H	N	N	N	VL	N	VL	H
A2	L	N	VH	L	L	H	N	L	L	N	N	VL	H	N	H	L
A3	VL	H	N	VL	VL	VL	H	L	L	N	L	L	H	L	H	L
A4	L	L	VL	N	H	H	H	L	VL	H	H	L	H	VL	VL	H
A5	H	L	L	VH	N	H	VH	VH	H	H	VH	L	L	H	L	H
A6	L	VL	VL	H	H	N	H	VL	H	H	VL	VL	L	L	VL	VL
A7	N	N	VL	L	N	H	N	VL	L	VL	N	N	VL	N	VL	N
A8	VL	VL	L	H	VH	H	H	N	VH	H	L	VL	H	N	N	L
A9	VL	VL	N	H	H	H	L	VL	N	N	L	VL	VL	H	N	VL
A10	N	N	VL	VL	VL	H	L	H	N	N	H	L	N	H	H	N
A11	N	N	VL	L	H	H	N	VH	H	H	N	H	VL	N	N	H
A12	N	VL	H	H	VH	H	N	H	H	VL	L	N	VH	L	VL	VH
A13	VL	VL	H	H	H	VL	L	L	L	N	L	H	N	H	H	H
A14	N	N	VL	VL	H	L	N	N	VL	VL	N	H	L	N	H	H
A15	VL	N	L	L	L	L	L	N	N	VL	N	H	VL	H	N	VL
A16	L	VL	VL	H	H	L	N	H	H	N	VL	H	H	H	L	N

Table 4. Initial relation matrix F—the TBL's perspective.

Criteria	A1	A2	A3	A4	A5	A6	A7	A8	A9	A10	A11	A12	A13	A14	A15	A16
A1	0.082	0.25	0.25	0.75	0.75	0.5	0.082	0.75	0.75	0.082	0.082	0.082	0.25	0.082	0.25	0.75
A2	0.5	0.082	0.928	0.5	0.5	0.75	0.082	0.5	0.5	0.082	0.082	0.25	0.75	0.082	0.75	0.5
A3	0.25	0.75	0.082	0.25	0.25	0.25	0.75	0.5	0.5	0.082	0.5	0.5	0.75	0.5	0.75	0.5
A4	0.5	0.5	0.25	0.082	0.75	0.75	0.25	0.5	0.25	0.75	0.75	0.5	0.75	0.25	0.25	0.75
A5	0.75	0.5	0.5	0.918	0.082	0.75	0.928	0.928	0.75	0.75	0.928	0.5	0.5	0.75	0.5	0.75
A6	0.5	0.25	0.25	0.75	0.75	0.082	0.75	0.25	0.75	0.75	0.25	0.25	0.5	0.5	0.25	0.25
A7	0.082	0.082	0.25	0.5	0.082	0.75	0.082	0.25	0.5	0.25	0.082	0.082	0.25	0.082	0.25	0.082
A8	0.25	0.25	0.5	0.75	0.928	0.75	0.75	0.082	0.928	0.75	0.5	0.25	0.75	0.082	0.082	0.5
A9	0.25	0.25	0.082	0.75	0.75	0.75	0.5	0.25	0.082	0.082	0.5	0.25	0.25	0.75	0.082	0.25
A10	0.082	0.082	0.25	0.25	0.25	0.75	0.5	0.75	0.082	0.082	0.75	0.5	0.082	0.75	0.75	0.082
A11	0.082	0.082	0.25	0.5	0.75	0.75	0.082	0.928	0.75	0.75	0.082	0.75	0.25	0.082	0.082	0.75
A12	0.082	0.25	0.75	0.75	0.928	0.75	0.082	0.75	0.75	0.25	0.5	0.082	0.928	0.5	0.25	0.928
A13	0.25	0.25	0.75	0.75	0.75	0.25	0.5	0.5	0.5	0.082	0.5	0.75	0.082	0.75	0.75	0.75
A14	0.082	0.082	0.25	0.25	0.75	0.5	0.082	0.082	0.25	0.25	0.082	0.75	0.5	0.082	0.75	0.75
A15	0.25	0.082	0.5	0.5	0.5	0.5	0.5	0.082	0.082	0.25	0.082	0.75	0.25	0.75	0.082	0.25
A16	0.5	0.25	0.25	0.75	0.75	0.5	0.082	0.75	0.75	0.082	0.25	0.75	0.75	0.75	0.5	0.082

Table 5. The normalized direct-relation matrix *X*—the TBL's perspective ($\times 10^{-2}$).

Criteria	A1	A2	A3	A4	A5	A6	A7	A8	A9	A10	A11	A12	A13	A14	A15	A16
A1	0.8	2.3	2.3	0.7	0.7	4.7	0.8	7	7	0.8	0.8	0.8	2.3	0.8	2.3	7
A2	4.7	0.8	8.5	4.7	4.7	7	0.8	4.7	4.7	0.8	0.8	2.3	7	8	7	4.7
A3	2.3	7	0.8	2.3	2.3	2.3	7	4.7	4.7	0.8	4.7	4.7	7	4.7	7	4.7
A4	4.7	4.7	2.3	8	7	7	4.7	4.7	2.3	7	7	4.7	7	0.2	2.3	7
A5	7	4.7	4.7	8.5	0.8	7	8.5	8.5	7	7	8.5	4.7	4.7	7	4.7	7
A6	4.7	2.3	2.3	7	7	0.8	7	2.3	7	7	2.3	2.3	4.7	4.7	2.3	2.3
A7	0.8	0.8	2.3	4.7	4.7	7	0.8	2.3	7	2.3	0.8	0.8	2.3	0.8	2.3	0.8
A8	2.3	2.3	2.3	7	2.3	7	7	0.8	4.7	7	4.7	2.3	7	0.8	0.8	4.7
A9	2.3	2.3	4.7	7	7	7	4.7	2.3	8.5	0.8	4.7	2.3	2.3	7	0.8	2.3
A10	0.8	0.8	0.8	4.7	2.3	7	4.7	7	0.8	0.8	7	4.7	0.8	7	7	0.8
A11	0.8	0.8	2.3	4.7	7	7	0.8	8.5	7	7	0.8	7	4.7	0.8	0.8	7
A12	0.8	2.3	7	7	8.5	7	0.8	7	7	2.3	4.7	0.8	8.5	4.7	2.3	8.5
A13	2.3	2.3	7	7	7	2.3	4.7	4.7	4.7	0.8	4.7	8.5	0.8	7	7	7
A14	0.8	0.8	2.3	2.3	7	4.7	0.8	0.8	2.3	2.3	0.8	8.5	4.7	0.8	7	7
A15	2.3	0.8	4.7	4.7	4.7	4.7	47	0.8	0.8	2.3	0.8	8.5	2.3	7	0.8	2.3
A16	7	2.3	2.3	7	7	4.7	0.8	7	7	0.3	2.3	8.5	7	7	4.7	0.8

Table 6. The total relation matrix M—the TBL's perspective ($\times 10^{-2}$).

Criteria	A1	A2	A3	A4	A5	A6	A7	A8	A9	A10	A11	A12	A13	A14	A15	A16
A1	6.2	6.7	8.2	16.3	15.9	13.8	7.7	14.5	15.1	6.7	7.3	8.3	10.3	7.5	8.1	14.6
A2	10.5	5.9	15.3	15.3	15.1	16.9	8.8	13.2	14.0	7.2	8.0	11.2	15.9	8.8	13.9	13.6
A3	8.0	11.4	8.1	13.1	13.2	13.0	14.1	13.2	14.0	7.1	14.4	13.6	16.0	12.3	13.9	13.7
A4	11.3	9.9	10.2	13.4	19.0	19.0	13.3	15.3	13.5	14.5	15.2	14.6	17.1	9.3	10.4	17.2
A5	15.2	11.5	14.5	24.2	16.8	22.8	19.4	21.5	20.9	16.8	18.8	17.6	18.0	17.9	14.8	20.2
A6	10.3	7.1	9.1	17.5	17.4	11.6	14.6	11.3	16.0	13.4	9.8	11.0	13.3	12.3	9.4	11.3
A7	4.4	3.7	6.3	11.0	10.9	13.1	15.8	7.5	10.1	6.6	5.3	5.9	7.6	5.5	6.3	6.0
A8	8.1	7.1	9.1	17.7	13.2	17.6	14.6	9.8	17.7	13.5	12.1	11.0	15.7	8.7	7.7	13.4
A9	7.7	6.8	10.7	16.4	16.6	16.4	11.6	10.4	9.5	7.2	11.2	10.3	10.7	13.6	7.2	10.8
A10	5.4	4.5	6.4	13.5	11.4	15.9	11.0	14.0	8.8	7.0	12.7	12.2	8.6	13.0	12.4	8.6
A11	7.1	6.0	9.6	16.5	18.2	18.3	9.4	18.1	17.2	14.2	9.1	16.1	14.4	9.5	8.1	16.5
A12	8.5	8.7	15.6	20.5	21.8	20.0	10.9	18.2	19.0	10.9	14.1	12.4	2.1	14.7	11.3	20.0
A13	9.3	8.3	15.3	19.7	19.8	15.2	13.6	15.3	16.0	8.7	13.3	19.3	12.2	16.2	15.2	18.1
A14	6.0	4.9	8.5	11.8	16.1	13.5	7.4	8.6	10.5	7.9	7.1	16.3	12.6	8.1	12.9	14.8
A15	6.9	4.8	10.2	13.1	13.4	13.1	10.6	8.0	8.5	7.6	6.7	15.5	9.9	13.1	6.6	10.0
A16	13.7	8.1	10.8	19.9	19.9	17.2	9.9	17.4	18.3	8.8	11.1	18.9	17.9	16.1	12.8	12.3

Table 7. The values of R, C, (R + C), (R − C)—the TBL's perspective.

Criteria	R	C	R + C	R − C
A1	1.672	1.386	3.058	0.286
A2	1.936	1.154	3.090	0.782
A3	1.991	1.679	3.670	0.312
A4	2.232	2.599	4.831	−0.367
A5	2.909	2.587	5.496	0.322
A6	1.954	2.574	4.528	−0.620
A7	1.260	1.927	3.187	−0.667
A8	1.970	2.163	4.133	−0.193
A9	1.771	2.291	4.062	−0.520
A10	1.654	1.581	3.235	0.073
A11	2.083	1.762	3.845	0.321
A12	2.287	2.142	4.429	0.145
A13	2.355	2.023	4.378	0.332
A14	1.670	1.866	3.536	−0.196
A15	1.580	1.710	3.290	−0.130
A16	2.331	2.211	4.542	0.120

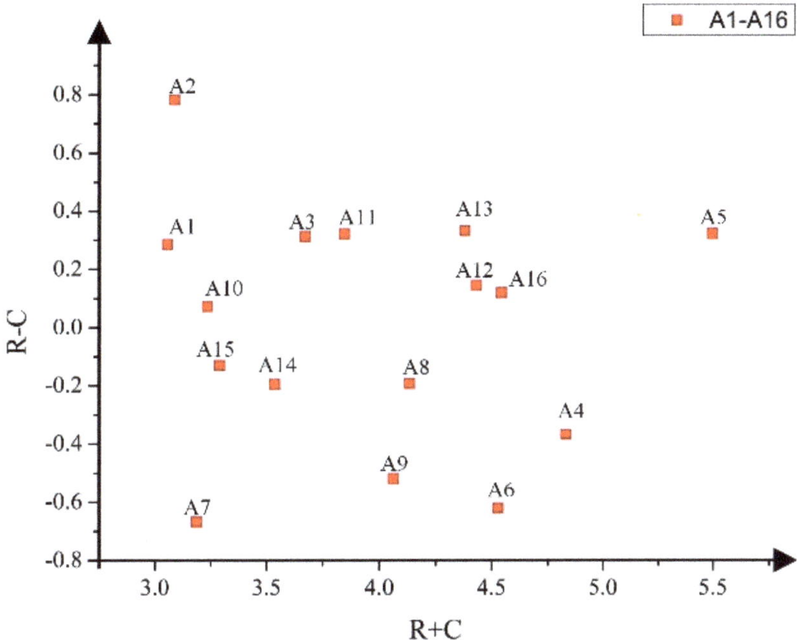

Figure 3. Cause–effect relation diagram.

The final results of our research are shown in Table 7 and Figure 3. According to Equations (7) and (8), we can know the elements in the cause group and the result group. The cause group includes drivers A1, A2, A3, A5, A10, A11, A12, A13, A16, and the effect group includes drivers A4, A6, A7, A8, A9, A14, A15. The results show that government regulations (A2), competitive advantage (A11), green supply chain pressure (A13) and green technology innovation (A5), incentive measures (A1) and standards (A3) are the six key drivers that promote the effective implementation of CER in the automotive industry. The most important driver is government regulations (A2). Half of the six key drivers are policy drivers. The fact is also true. The effective implementation of CER requires the government to formulate and supervise the implementation of environmental protection regulations and standards. Green supply chain pressure (A13), green technology innovation (A5) and competitive advantage (A11) rank second, third, and fourth among all important drivers. From the perspective of the effect group, it can be seen that media pressure (A15) ranks first in the effect group, and employee demand (A7) ranks last in the effect group.

As far as we know, there is no relevant analysis on the research directions involved in this paper, so it is impossible to give appropriate horizontal comparison results, but we have found similar experimental results in papermaking enterprises, industrial enterprises and the fashion industry. Among them, papermaking enterprises provide internal and external drivers for the company's green and sustainable development [56]. External drivers include government pressure, social pressure and economic pressure. Internal drivers include management, employees, corporate culture, the size of the company, and the financial situation. Experimental results show that economic pressure is the first driving force, and internal management and employee environmental awareness are the second driving force. The external factors for the green development of industrial enterprises [57] include policy and institutional environment, market environment and public supervision, and internal factors include the tangible and intangible resources of the enterprise. In the fashion industry [58], there are similar results. The driving factors of the sustainable fashion industry are attributed to internal driving factors (entrepreneurial direction and founder culture, integration between different companies, innovation) and external driving factors (regulation, consumer awareness, competitiveness). Obviously, these have confirmed that government supervision, policies and regulations are the most important external driving factors for the manufacturing industry to fulfill its environmental responsibilities and implement environmental behaviors, which is basically consistent with the research results of this paper.

The verification is based on feedback from experts in the CAMI and references to relevant existing literature. After verification, our research results will be submitted to K Company.

6. Discussion

According to the research results, we drew a histogram of the key drivers and sorted the key drivers according to formula (9), as shown in Figure 4 and Table 8. It can be seen from Figure 4 and Table 8 that government regulations (A2) is the most important driver, ranking first. Only when the government and regulatory agencies jointly promote the automobile manufacturing industry to perform environmental responsibilities in strict accordance with regulations and standards can the company's environmental and economic interests achieve balanced development. In addition, green supply chain pressure (A13), green technology innovation (A5), and competitive advantage (A11) rank second, third, and fourth. These are the key drivers for the company to implement CER. These identified key drivers are interrelated: the implementation of government regulations (A2) is conducive to the improvement of enterprises' green technology innovation (A5), and promotes the improvement of enterprises' competitive advantage (A11), and the green supply chain pressure (A13) can also promote enterprises to comply with the laws and regulations of higher-level government departments, and improve their compliance and compliance. The other five drivers, including standards (A3), incentives (A1), consumers demand (A12), market trend (A16), and company image (A10), are also important for implementation of CER, and they all have different degrees of each other. The following research will conduct a correlation analysis of the importance of all these key drivers.

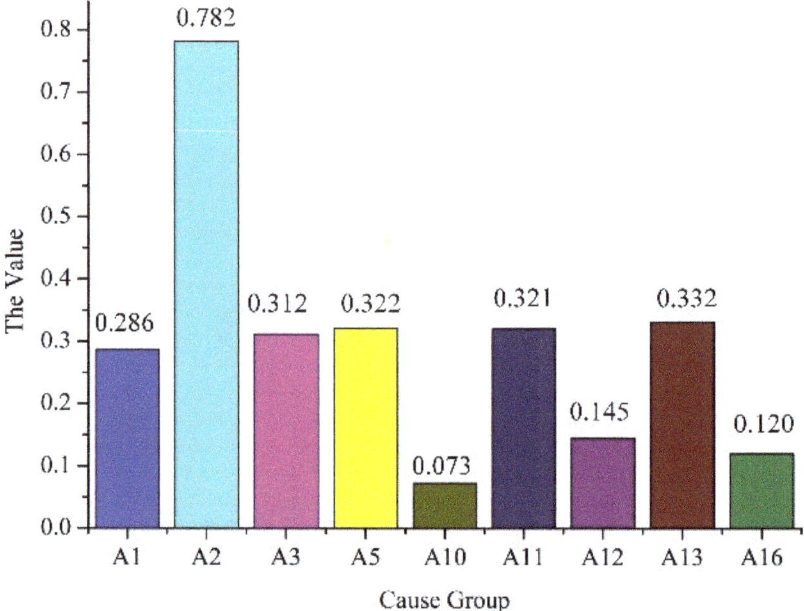

Figure 4. Causes group histogram.

Table 8. The sequencing of the key drivers.

Key Drivers	R − C	Rg	Ranking
Incentives (A1)	0.286	36.57	6
Government regulations (A2)	0.782	100.00	1
Standards (A3)	0.312	39.89	5
Green technology innovation (A5)	0.322	41.18	3
Company image (A10)	0.073	9.34	9
Competitive advantage (A11)	0.321	41.04	4
Consumers demand (A12)	0.145	18.54	7
Green supply chain pressure (A13)	0.332	42.46	2
Market trend (A16)	0.120	15.35	8

In this study, we discussed the conclusions of the study and the importance of implementing CER, and drew some management implications including:

First of all, in order to accelerate the effective implementation of CER in CAMI, the government needs to improve the implementation and supervision of environmental protection policies and regulations, and increase the penalties for violations of environmental protection regulations. At the same time, it also encourages and praises companies that effectively implement CER and promote environmental improvements to promote its excellent practices. Second, the pressure of the upstream and downstream supply chains of automobile manufacturers also provides a strong impetus to promote the implementation of CER, because downstream consumers are more inclined to choose environmentally friendly products and are willing to pay for environmental protection. Upstream suppliers need to promote the design and development of more environmentally friendly and green products in the automobile manufacturing industry, thereby promoting the green innovation of upstream suppliers. Third, the green technological innovation of CAMI is also a very key driver, which is fundamentally the way for the enterprise to improve environmental performance and economic efficiency. Innovation is the source of all development; enterprises need to improve the construction of green technology innovation talent teams and increase the investment of green innovation costs. Fourth, the establishment of high-level environmental protection standards by the government and local governments is also a driving factor for enterprises to effectively implement CER. Finally, corporate incentives, public awareness of environmental protection, market trends, and corporate image can all have a positive impact on the implementation of CER by companies.

7. Conclusions and Future Work

Based on the perspective of TBL, and through the fuzzy DEMATEL method, we identified six key drivers from 16 common drivers, including government regulations (A2), green supply chain pressure (A13), green technological innovation (A5), competitive advantage (A11), standards (A3), and incentive measures (A1). These six key drivers are critical to the implementation of CER in CAMI. The main measures include the following: automobile manufacturers should improve the level of green technology innovation, pay attention to the needs of the upstream and downstream supply chain, improve the level of standard implementation, pay attention to competitive advantages, and formulate internal incentive mechanisms on the premise of meeting the requirements of government regulations. Only by continuing to meet the needs of all relevant parties can the company's environmental and economic benefits be comprehensively improved, and the company's green and sustainable development can be promoted. Through this research, K company can better understand the importance of CER practice to its own green development and sustainable development, and it is feasible to realize the coordinated development of economy, environment and society.

The work of this paper provides a valuable reference for the research and practice of CER in CAMI, finds the key drivers for the implementation of CER, and gives some management enlightenment. It is worth noting that this study still has certain limitations.

First of all, in the process of identifying and evaluating drivers, the number of questionnaires is relatively limited. Secondly, due to the limited amount of data, the proposed method cannot be further and extensively verified. The verified situation of a small and medium-sized automobile manufacturing industry cannot fully represent the situation of China's entire automobile manufacturing industry. The research conclusions cannot be widely applied to all automobile manufacturing industries, automobile sales companies, etc. These may constitute the basic elements of future research.

Therefore, the future research direction is from the perspective of the impact of the development of artificial intelligence and smart manufacturing technology on CAMI. The research perspective of CAMI's implementation of CER can also be shifted from the perspective of TBL to other perspectives, such as technology and multiple stakeholders. Regarding model construction and selection of multicriteria decision-making methods, the existing fuzzy decision-making can be extended to more advanced intelligent decision-making models and decision-making algorithms. In addition, it may be very interesting to study the relationship between CAMI's implementation of CER on corporate sustainable development and green development, and how to improve the company's image and increase profitability.

Author Contributions: M.Z. and S.L. received the research, collected data, performed the analyses and wrote the paper. H.Z., W.Y., Z.J. and Y.L. provided support and helpful suggestions in setting up and revising the manuscript. All authors have read and agreed to the published version of the manuscript.

Funding: The authors are grateful for the research support from the National Natural Science Foundation of China (No. 51975432, 52075396).

Institutional Review Board Statement: Not applicable.

Informed Consent Statement: Informed consent was obtained from all subjects involved in the study.

Data Availability Statement: The datasets used and/or analyzed during the current study are available from the corresponding author on reasonable request.

Conflicts of Interest: The authors declare no conflict of interest.

References

1. Zhao, F.W.; Zhao, B.J. Research on the Development Strategies of New Energy Automotive Industry Based on Car Charging Stations. *Appl. Mech. Mater.* **2015**, *740*, 985–988. [CrossRef]
2. He, M.; Chen, J. Sustainable Development and Corporate Environmental Responsibility: Evidence from Chinese Corporations. *J. Agric. Environ. Ethic* **2009**, *22*, 323–339. [CrossRef]
3. Li, P.C.; Zhang, H.M. Application Situation and Development Strategies of Green Manufacturing in China's Automobile Industry. *Adv. Mater. Res.* **2014**, *1049–1050*, 945–948. [CrossRef]
4. Severo, E.A.; de Guimarães, J.C.F.; Dorion, E.C.H. Cleaner production, social responsibility and eco-innovation: Generations' perception for a sustainable future. *J. Clean. Prod.* **2018**, *186*, 91–103. [CrossRef]
5. Jiang, Z.; Jiang, Y.; Wang, Y.; Zhang, H.; Cao, H.; Tian, G. A hybrid approach of rough set and case-based reasoning to remanufacturing process planning. *J. Intell. Manuf.* **2019**, *30*, 19–32. [CrossRef]
6. Ding, Z.; Jiang, Z.; Zhang, H.; Cai, W.; Liu, Y. An integrated decision-making method for selecting machine tool guideways considering remanufacturability. *Int. J. Comput. Integr. Manuf.* **2018**, *33*, 686–700. [CrossRef]
7. Elkington, J. Accounting for The Triple Bottom Line. *Meas. Bus. Excel.* **1998**, *2*, 18–22. [CrossRef]
8. Mazurkiewicz, P. *Corporate Environmental Responsibility: Is a Common CSR Framework Possible*; World Bank: Washington, DC, USA, 2004; pp. 1–18.
9. González-Rodríguez, M.R.; Díaz-Fernández, M.C.; Biagio, S. The perception of socially and environmentally responsible practices based on values and cultural environment from a customer perspective. *J. Clean. Prod.* **2019**, *216*, 88–98. [CrossRef]
10. Nawaz, W.; Linke, P.; Koç, M. Safety and sustainability nexus: A review and appraisal. *J. Clean. Prod.* **2019**, *216*, 74–87. [CrossRef]
11. Gunningham, N. Shaping corporate environmental performance: A review. *Environ. Policy Gov.* **2009**, *19*, 215–231. [CrossRef]
12. Wang, H. Systematic analysis of corporate environmental responsibility: Elements, structure, function, and principles. *Chin. J. Popul. Resour. Environ.* **2016**, *14*, 96–104. [CrossRef]
13. Cai, W.; Liu, C.; Lai, K.-H.; Li, L.; Cunha, J.; Hu, L. Energy performance certification in mechanical manufacturing industry: A review and analysis. *Energy Convers. Manag.* **2019**, *186*, 415–432. [CrossRef]

14. Fujii, H.; Managi, S. Trends in corporate environmental management studies and databases. *Environ. Econ. Policy Stud.* **2016**, *18*, 265–272. [CrossRef]
15. Liu, Y.; Xi, B.; Wang, G. The impact of corporate environmental responsibility on financial performance—Based on Chinese listed companies. *Environ. Sci. Pollut. Res.* **2021**, *28*, 7840–7853. [CrossRef] [PubMed]
16. Shabbir, M.S.; Wisdom, O. The relationship between corporate social responsibility, environmental investments and financial performance: Evidence from manufacturing companies. *Environ. Sci. Pollut. Res.* **2020**, *27*, 39946–39957. [CrossRef] [PubMed]
17. Kennard, A. The Enemy of My Enemy: When Firms Support Climate Change Regulation. *Int. Organ.* **2020**, *74*, 187–221. [CrossRef]
18. Gong, J.Z.Y.; Wang, G. *An Investigation into Triple Bottom Line Assessment of Eco-Friendly Office Supplies: File Folders*; University of British Columbia: Vancouver, BC, Canada, 2014.
19. Casey, E.; Beaini, S.; Pabi, S.; Zammit, K.; Amarnath, A. The Triple Bottom Line for Efficiency: Integrating Systems Within Water and Energy Networks. *IEEE Power Energy Mag.* **2017**, *15*, 34–42. [CrossRef]
20. Fairley, S.; Tyler, B.D.; Kellett, P.; D'Elia, K. The Formula One Australian Grand Prix: Exploring the triple bottom line. *Sport Manag. Rev.* **2011**, *14*, 141–152. [CrossRef]
21. Wise, N. Outlining triple bottom line contexts in urban tourism regeneration. *Cities* **2016**, *53*, 30–34. [CrossRef]
22. Depken, D.; Zeman, C. Small business challenges and the triple bottom line, TBL: Needs assessment in a Midwest State, U.S.A. *Technol. Forecast. Soc. Chang.* **2018**, *135*, 44–50. [CrossRef]
23. Tseng, M.-L.; Wu, K.-J.; Ma, L.; Kuo, T.C.; Sai, F. A hierarchical framework for assessing corporate sustainability performance using a hybrid fuzzy synthetic method-DEMATEL. *Technol. Forecast. Soc. Chang.* **2019**, *144*, 524–533. [CrossRef]
24. Ahi, P.; Searcy, C. A stochastic approach for sustainability analysis under the green economics paradigm. *Stoch. Environ. Res. Risk Assess.* **2013**, *28*, 1743–1753. [CrossRef]
25. Wu, K.-J.; Liao, C.-J.; Tseng, M.; Chiu, K.K.-S. Multi-Attribute Approach to Sustainable Supply Chain Management Under Un-Certainty. *Ind. Manag. Data Syst.* **2016**, *116*, 777–800. [CrossRef]
26. Esquer-Peralta, J.; Velazquez, L.; Munguia, N. Perceptions of core elements for sustainability management systems (SMS). *Manag. Decis.* **2008**, *46*, 1027–1038. [CrossRef]
27. Palvia, P.; Baqir, N.; Nemati, H. ICT for socio-economic development: A citizens' perspective. *Inf. Manag.* **2018**, *55*, 160–176. [CrossRef]
28. Tseng, M.-L.; Chiu, A.S.; Liang, D. Sustainable consumption and production in business decision-making models. *Resour. Conserv. Recycl.* **2018**, *128*, 118–121. [CrossRef]
29. Bergenwall, A.L.; Chen, C.; White, R.E. TPS's process design in American automotive plants and its effects on the triple bottom line and sustainability. *Int. J. Prod. Econ.* **2012**, *140*, 374–384. [CrossRef]
30. Gimenez, C.; Sierra, V.; Rodon, J. Sustainable operations: Their impact on the triple bottom line. *Int. J. Prod. Econ.* **2012**, *140*, 149–159. [CrossRef]
31. Neri, A.; Cagno, E.; Lepri, M.; Trianni, A. A triple bottom line balanced set of key performance indicators to measure the sustainability performance of industrial supply chains. *Sustain. Prod. Consum.* **2021**, *26*, 648–691. [CrossRef]
32. Agrawal, S.; Singh, R.K. Analyzing disposition decisions for sustainable reverse logistics: Triple Bottom Line approach. *Resour. Conserv. Recycl.* **2019**, *150*, 104448. [CrossRef]
33. Hussain, N.; Rigoni, U.; Orij, R.P. Corporate Governance and Sustainability Performance: Analysis of Triple Bottom Line Performance. *J. Bus. Ethics* **2018**, *149*, 411–432. [CrossRef]
34. Chen, Y.; Lawell, C.-Y.C.L.; Muehlegger, E.J.; Wilen, J.E. Modeling Supply and Demand in the Chinese Automobile Industry. In Proceedings of the 2017 Annual Meeting, Chicago, IL, USA, 30 July–1 August 2017.
35. Azevedo, S.; Barros, M. The application of the triple bottom line approach to sustainability assessment: The case study of the UK automotive supply chain. *J. Ind. Eng. Manag.* **2017**, *10*, 286–322. [CrossRef]
36. Kumar, D.; Rahman, Z.; Chan, F.T.S. A fuzzy AHP and fuzzy multi-objective linear programming model for order allocation in a sustainable supply chain: A case study. *Int. J. Comput. Integr. Manuf.* **2017**, *30*, 535–551. [CrossRef]
37. Mathivathanan, D.; Kannan, D.; Haq, A.N. Sustainable supply chain management practices in Indian automotive industry: A multi-stakeholder view. *Resour. Conserv. Recycl.* **2018**, *128*, 284–305. [CrossRef]
38. Wang, H. Factor analysis of corporate environmental responsibility. *Environ. Dev. Sustain.* **2009**, *12*, 481–490. [CrossRef]
39. Dummett, K. Drivers for Corporate Environmental Responsibility (CER). *Environ. Dev. Sustain.* **2006**, *8*, 375–389. [CrossRef]
40. Kehbila, A.G.; Ertel, J.; Brent, A.C. Strategic corporate environmental management within the South African automotive industry: Motivations, benefits, hurdles. *Corp. Soc. Responsib. Environ. Manag.* **2009**, *16*, 310–323. [CrossRef]
41. Cai, W.; Lai, K.-H.; Liu, C.; Wei, F.; Ma, M.; Jia, S.; Jiang, Z.; Lv, L. Promoting sustainability of manufacturing industry through the lean energy-saving and emission-reduction strategy. *Sci. Total Environ.* **2019**, *665*, 23–32. [CrossRef]
42. Nunes, B.; Bennett, D. Green operations initiatives in the automotive industry: An environmental reports analysis and benchmarking study. *Benchmarking: Int. J.* **2010**, *17*, 396–420. [CrossRef]
43. Yucheng, C.; You, J.; Wang, R.; Shi, Y. Designing a mixed evaluating system for green manufacturing of automotive industry. *Probl. Ekorozwoju* **2016**, *11*, 73–86.
44. Babiak, K.; Trendafilova, S. CSR and environmental responsibility: Motives and pressures to adopt green management practices. *Corp. Soc. Responsib. Environ. Manag.* **2011**, *18*, 11–24. [CrossRef]

45. Lee, J.W.; Kim, Y.M. Antecedents of Adopting Corporate Environmental Responsibility and Green Practices. *J. Bus. Ethics* **2016**, *148*, 397–409. [CrossRef]
46. Goli, Y.S.; Ye, J.; Ye, Y.; Kalgora, B. Impact of Ecological Related Innovations Enhancing the Efficiency of Corporate Environmental Responsibility. *Am. J. Ind. Bus. Manag.* **2020**, *10*, 191–217. [CrossRef]
47. Chen, Y.-S. The Drivers of Green Brand Equity: Green Brand Image, Green Satisfaction, and Green Trust. *J. Bus. Ethics* **2009**, *93*, 307–319. [CrossRef]
48. Ho, C.-W. Does Practicing CSR Makes Consumers Like Your Shop More? Consumer-Retailer Love Mediates CSR and Behavioral Intentions. *Int. J. Environ. Res. Public Heal.* **2017**, *14*, 1558. [CrossRef] [PubMed]
49. Jahanshahi, A.A.; Brem, A. Antecedents of Corporate Environmental Commitments: The Role of Customers. *Int. J. Environ. Res. Public Heal.* **2018**, *15*, 1191. [CrossRef]
50. Sevinç, A.; Eren, T. Determination of KOSGEB Support Models for Small- and Medium-Scale Enterprises by Means of Data Envelopment Analysis and Multi-Criteria Decision Making Methods. *Process* **2019**, *7*, 130. [CrossRef]
51. Wang, H.; Jiang, Z.; Zhang, H.; Wang, Y.; Yang, Y.; Li, Y. An integrated MCDM approach considering demands-matching for reverse logistics. *J. Clean. Prod.* **2019**, *208*, 199–210. [CrossRef]
52. Hamdia, K.M.; Arafa, M.; Alqedra, M. Structural damage assessment criteria for reinforced concrete buildings by using a Fuzzy Analytic Hierarchy process. *Undergr. Space* **2018**, *3*, 243–249. [CrossRef]
53. Wu, W.; Lan, L.W.; Lee, Y. Factors hindering acceptance of using cloud services in university: A case study. *Electron. Libr.* **2013**, *31*, 84–98. [CrossRef]
54. Zadeh, L.A. *Fuzzy Sets, Fuzzy Logic, and Fuzzy Systems: Selected Papers by Lotfi A Zadeh*; State University of New York at Binghamton: New York, NY, USA, 1996; pp. 394–432. [CrossRef]
55. Lin, R.-J. Using fuzzy DEMATEL to evaluate the green supply chain management practices. *J. Clean. Prod.* **2013**, *40*, 32–39. [CrossRef]
56. He, Z.-X.; Shen, W.-X.; Li, Q.-B.; Xu, S.-C.; Zhao, B.; Long, R.-Y.; Chen, H. Investigating external and internal pressures on corporate environmental behavior in papermaking enterprises of China. *J. Clean. Prod.* **2018**, *172*, 1193–1211. [CrossRef]
57. Li, X.; Du, J.; Long, H. Green Development Behavior and Performance of Industrial Enterprises Based on Grounded Theory Study: Evidence from China. *Sustainability* **2019**, *11*, 4133. [CrossRef]
58. Dicuonzo, G.; Galeone, G.; Ranaldo, S.; Turco, M. The Key Drivers of Born-Sustainable Businesses: Evidence from the Italian Fashion Industry. *Sustainability* **2020**, *12*, 10237. [CrossRef]

Article

Implementation Criteria for Intelligent Systems in Motor Production Line Process Management

Yao-Chin Lin, Ching-Chuan Yeh, Wei-Hung Chen * and Kai-Yen Hsu

Department of Information Management, Yuan Ze University, Taoyuan City 32003, Taiwan; imyclin@saturn.yzu.edu.tw (Y.-C.L.); s1049202@mail.yzu.edu.tw (C.-C.Y.); s1086216@mail.yzu.edu.tw (K.-Y.H.)

* Correspondence: wehchen123@gmail.com or s999202@mail.yzu.edu.tw

Received: 7 April 2020; Accepted: 30 April 2020; Published: 3 May 2020

Abstract: In this study, the factors that affect the implementation of intelligent systems in motor production lines are analyzed. A motor production line located in Vietnam is used as the research object. The research methods include secondary data collection, field study, and interviews. This study demonstrates the following: firstly, the implementation of intelligent systems in motor production lines is heading toward Industry 4.0. Secondly, it is proposed that three functional systems—robot arm, image recognition, and big data analysis—can be introduced in the motor production line. This study analyzes the process involved in coil and motor production lines and attempts to combine intelligent system functions. It is expected that in the future, manpower will be reduced, production line productivity will increase, and intelligent production lines will be proposed. The factors that affect the introduction of intelligent systems in motor production lines are improved, and the importance of intelligent systems, which has been rarely considered in previous studies, is highlighted. In the implementation criteria of the intelligent system in the process management of the motor production line, this study provides some suggestions (to coil and motor assembly line) for the production process management. These suggestions can be provided as a reference for production lines that acquaint with intelligent systems.

Keywords: intelligent systems; motor production lines; process management

1. Introduction

This section elaborates on the research background related to this research, how to introduce the application of intelligent systems in the motor production line and analyze from the perspective of process management.

1.1. Research Background

In the past, as a result of underdeveloped technology, production was costly and consumed considerable resources including time; Also, the output was low, which caused failure in meeting customers' product demand on schedule. Moreover, there was no system capable of monitoring the production line to implement quality control in real time. The low quality of goods indirectly caused credit problems with customers. This led to the rise of the so-called industrial revolution, which gradually solved the problems related to resources, time, and output. In the last decade, Germany first proposed the concept of Industry 4.0, which integrates the current Internet of Things (IoT), IT, virtual reality, and other technologies for developing smart factories and smart grids to, in turn, achieve smart cities. For the English abbreviation of this study, please see Appendix A.

Similar to those in Germany and Japan, companies in Taiwan that are confronted with problems in production have gradually implemented intelligent production lines. In this study, the research object is Solen company, which is committed to the manufacture of specialized and automated motor products as well as complex key electromagnetic components. The factories in operation are located in Taoyuan, Kaohsiung, Taiwan and Ho Chi Minh, Vietnam. The motor manufacturing plant in Ho Chi Minh City is an original design model manufacturer commissioned by the original factory of Japan's M company to manufacture and deliver motors. In this study, the motor assembly and coil assembly production lines are surveyed. It is found that the production line employs considerable manpower to inspect the motor appearance and test motor performance. A semi-automatic production mode, which uses both humans and machines, is employed in processes operated by personnel; hence, labor cost is substantial.

It has been reported in literature that artificial intelligence [1] can be utilized to collect and analyze the fault mechanism of motor insulation, and image recognition can assist in the inspection of the motor's rotor appearance [2]. Artificial intelligence can also collect vibration signals through machine learning-based techniques [3] to evaluate sound data and implement automatic fault detection functions. If cyber-physical systems (CPSs) used in the aviation, automotive, chemical process, and infrastructure manufacturing industries can be introduced into the production line, then costs may be reduced and production capacity may increase, thereby achieving a fully automated model. Moreover, the growth of the world's population has gradually changed in recent years; hence, if intelligent systems can be employed to reduce manpower in the production line, then the problem of insufficient manpower may also be resolved in the future.

1.2. Research Objectives

The purpose of this study is to explore the application of intelligent systems in motor and coil assembly production lines as well as to determine the available intelligent systems. The research questions are as follows.

(1) What is the current status of "SM" (Solen Vietnam Motor Production Line) motor and coil assembly lines in Vietnam, and are there intelligent systems in place?

At present, there are seven employees in the motor assembly line. Each employee is assigned to a workbench to assemble the motor and test its operation on a machine. After completion, the motor is transferred to the next workbench for final visual inspection before shipment. All of these are manually implemented.

(2) After understanding the current status of the "SM" motor and coil assembly lines in Vietnam, what intelligent system may be implemented in the future to support the motor and coil assembly lines?

In this study, secondary data collection, field survey, and interview are conducted to determine the intelligent system that may be utilized to support the production line.

1.3. Literature Review

In this study, we collected relevant literature on intelligent systems, Industry 4.0 and motor production lines; In particular, this paper analyzes the motor production process in nine steps.

1.3.1. Intelligent Systems and Industry 4.0

In an information-developed society, the use of intelligent systems is necessary to aid companies in resolving complex problems and facilitating the allocation of manpower. The interpretation of smart systems varies from industry to industry; the perspectives of experts also vary. Many people quote Prof. John McCarthy (recognized as the father of artificial intelligence) who said, "artificial intelligence is to make machines behave like intelligent behaviors shown by humans." It is akin to putting human wisdom on a system so that it can autonomously make reasonable evaluations and implement external changes. Simply put, an intelligent system must be able to aid people to perceive changes, analyze judgments, and perform tasks [4].

Currently, many intelligent systems can directly send and receive information, analyze received information, and perform tasks automatically. Moreover, with the development of 5G technology, these functions can be performed anytime and anywhere to rapidly process tasks and produce outputs for the enterprise or users.

1. According to some scholars [1], CPS is a network and engineering system that can integrate both cybernetic and physical worlds. That is, it can be combined with physical sensors and other integrated systems in the field of virtual computer control called "virtual-integrated systems." Some embedded devices, such as the IoT and sensor networks, have systems similar to CPS. Embedded devices, however, emphasize device performance, whereas the CPS is a combined system of physical devices and network with emphasis on the relationship between their interaction [5].
2. The CPS is typically used in automation and sensor systems (e.g., robots, autonomous driving systems, monitoring systems, and process control systems) and in manufacturing [6]. This means that the physical devices in the manufacturing system can generate the same virtual model through the CPS. The data generated by the physical system are analyzed through the network and thereafter applied to the virtual model in real time in order to accurately present the current status of the physical system. The advantage of the virtual model is that it can be optimized through big data and artificial intelligence, which in turn can be used in manufacturing systems [7].

With the advent of globalization, the trade competition among countries has become increasingly intense. In this era, the Internet as well as science and technology have rapidly developed. Industry 4.0 was first proposed in the German Industry Fair in 2011 as a high-tech industrial plan. Thereafter, plans for the U.S. manufacturing partners, i.e., Japanese Industry 4.1J, Korean Manufacturing Innovation 3.0, and Made in China 2025, were made public.

In literature, it has been reported that [8] Industry 4.0 is the result of the fourth industrial revolution that has not been fully completed. Similar to other historical periods, the introduction of structural changes to production processes can lead to significant innovations and changes, which will also considerably impact production as well as influence and ultimately generate new business models. Industry 4.0 is comprised of technologies, equipment, and processes that allow a self-sufficient production model. This model can run the supply chain in an integrated manner at multiple stages and production process levels as well as operate with minimal human effort.

Industry 4.0 also represents the process of automating high-tech integration with manufacturing. Its contents include the integration of sales and products, and production and sales processes are customized. To obtain various data for the production process, an IoT framework that assists production is established. The big data information analysis technology is utilized for enabling operators to understand engineering and product-related data from the machine computer and identify the project and its stage of implementation. The objective is to improve the existing quality of the product and make appropriate decisions. The concept of Industry 4.0 includes the use of paradigms and technologies to support big data, augmented reality, robotics, CPSs, cloud computing, and the industrial Internet to promote the shift from traditional factories to intelligent production [9].

1.3.2. Intelligent Application for Motor Production Line

The motor production process involves approximately nine steps [10], and the research on this aspect can also utilize the concept of Industry 4.0, as follows.

1. Housing and Rotor Cage Production

Production starts from casing fabrication; however, no direct solution for the casting of motor casings has been found. During the casting process, the voltage or temperature can be collected and stored by sensors. Thereafter, machine learning technology in combination with artificial intelligence can be used to optimize the parameters in the casting process, predict the casting quality, and determine the appropriate casting material variables.

In addition, simulation methods can be employed to model the solidification of the liquid metal in the mold. The core can then be assembled by subordinate robotic arms, thereby reducing manpower and the probability of error. The asynchronous motor's rotor is manufactured by die-casting, which is also a method applicable to casing fabrication. Note that if the quantity of rotors to be manufactured is small, then it is necessary to prepare copper rods to be welded to the short-circuit ring.

2. Laminated Core Production

The data collected from the laminated core manufacturing process using machine learning algorithms can be utilized to predict the quality of raw materials required for production. In the case of electrical board production, machine learning algorithms can classify the materials for electrical steel plates based on the electromagnetic properties of microstructures. During the cutting of electrical boards, by monitoring the cutting force, audio signals, or vibration signals through sensors, machine learning algorithms can be employed to detect and predict errors and build models. Virtual reality smart devices can also be used to assist in the cutting and welding operations. Using different techniques to cut out electrical boards, sensors can be used to monitor the punching forces, audio signals, or vibration signals. For the laser-cutting process, data mining through the use of machine learning algorithms can afford the potential of detecting errors and predicting key events. The quality inspection of cuts and the entire laminated core can be performed through computer vision.

3. Insulation and Impregnation

In the motor manufacturing process, many of the steps involve providing insulation to motor parts. There are different alternatives for insulating the raw stator materials used in the slot, including the application of insulating paper powder coating or injection-molded polymers. Machine learning may be employed to aid in modeling this process and determining the best process parameters. Big data can also be combined to determine quality-related parameters and generate predictive models. After winding, the entire coil is insulated. In this aspect, machine learning can also identify the appropriate impregnation process simulation support.

4. Winding

The technology of Industry 4.0 may be employed to set the sensor in the winding machine and guide the machine in creating a precisely positioned coil. Rodriguez et al. [11] proposed that a machine learning system can be used to optimize the wire contour generated by an automatic winding machine so that the coil contour is wound as uniformly as possible. Apart from accurate and non-destructive winding, it is also important to maintain the required coil resistance during enameled winding. This can also be optimized by machine learning by simulating the winding parameters (e.g., wire tension and winding speed). Combined with the image recognition technology, the use of a robot to wind and install coils can facilitate production and detect the cause of failure at any time.

5. Contact technology

The existing method relative to contact technology directly solves the problem of crimping in motor production. It was reported in [12] that a standard OPC Unified Architecture condition monitoring system may be used to track the thermal crimping process (OPC Please See Table A1). In addition, Mayr et al. [13] studied the application of machine learning algorithms in the ultrasonic field for crimping. First, quality indicators (e.g., resistance and extraction force) can be estimated based on process parameters. Second, visual or auditory features can be used to classify joint quality.

In quality management, machine learning-based models are used as quality estimators, eliminating the necessity of quality management measures, such as random checks. For example, convolutional neural networks use visual features to predict joint quality. A comparison between deterministic models and machine learning methods shows that machine learning technology is more powerful, easier to automate, and more accurate. It can also detail the machine learning-based process control methods for the real-time measurement of parameters in the near future [14].

In addition to crimping, the potential of using machine learning in the laser welding of hairpin windings is studied. The application of laser welding contacts is particularly suitable for hairpin windings because of the large number of contact points. The use of machine learning can predict the quality of weld processing based on machine parameters. For the subsequent quality assessment, the combined image data can be used to detect and classify weld defects based on their severity. In the future, these applications will be incorporated into a quality monitoring system, which may also contain data from the upstream process. In particular, the burrs formed during the cutting process and residues resulting from peeling considerably influence the welding results [15].

6. Shaft Production

Al-Zub Ardi [16] indicated that machine learning can be employed for quality management, process control, and predictive maintenance in the cutting process. The proposed model focuses on the prediction of surface roughness and cutting forces as well as the estimation of tool life and wear. In addition, the application of machine learning technology combined with edge and cloud computing allows the analysis of large data streams. Computer vision can be used for process monitoring or quality inspection of machined parts through machine learning algorithms. Different methods have also demonstrated that the OPC Unified Architecture condition monitoring system can aid in optimizing machining processes [17,18]. Moreover, compared with traditional machining centers, robots are convenient for machining operations because of their flexibility and relatively low investment costs. Finally, weight reduction can be achieved through the AM (Additive Manufacturing) of the shaft [19].

7. Permanent Magnet Rotor Production

Various methods are employed in the production of motors, including permanent magnet synchronous motor rotors and those used in related fields. For example, image recognition can be employed for the visual inspection of magnetic surfaces after fabrication. In addition, sensors can be used for magnetic field measurements, which can be utilized to test a single magnet or the entire rotor. Magnetic field measurements also provide the basis for selective magnet assembly. Magnetic deviation has a critical influence on the operating characteristics of the motor, and it can be applied to optimize the deviation compensation magnet arrangement. Apart from traditional algorithms, machine learning techniques can be employed to develop optimal magnet assembly strategies. For each batch of magnets, a magnet or magnet stack is selected and installed according to an algorithm to minimize the deviation from the simulated magnetic field of an ideal rotor [20–22].

8. Final Assembly and Testing

In the final motor assembly, various connection processes are involved, including press-in operations, gluing, shrinking, tightening, and welding. Data mining methods can be utilized to study the relationship between force–displacement and current curves in a press operation. The testing process naturally generates a considerable amount of data; hence, data-driven methods, such as machine learning, can be used to check multiple areas. For stator testing, machine learning methods can be used to evaluate and classify the fault mechanisms in electrical insulation. The quality characteristics measured herein can also be used as labels for predictive models based on machine learning that utilize parameters from previous processes, such as winding, insulation, and contact. In addition, several methods directly solve the end-of-line test of the motor, e.g., image recognition technology can support stator inspection. Machine learning technology capabilities can also be utilized for evaluating vibration signals [3] as well as acoustic data to analyze motors [23] and produce automatic fault detection functions.

9. Overall Process

Apart from the application scenarios of Industry 4.0 technology in each sub-process, there are other methods related to the overall process that can include various steps in the value chain. For instance, with regard to the development of motors and related production systems, semantic technology can considerably improve the cross-domain information exchange. In addition to pure knowledge management, from the perspective of a configurator, knowledge-based systems can be used to automate simple engineering tasks. Thus far, it can only be found in motor design; however, it can also be employed in the engineering design of related production systems. For example, simulations can aid in analyzing the cost–benefit ratio of alternative production technologies.

With regard to production, knowledge-based systems can aid in optimizing individual jobs, and machine learning algorithms will be useful in processing big data and making fault predictions for the entire production line. Data mining can therefore facilitate the detection and rapid response to deviations within the assembly line. In this case, machine-to-machine communication technology performs an important function. For example, wireless communication technologies, such as Bluetooth 4.0 can be used to identify and locate tools and objects [24]. In the human–machine interface field, virtual reality-based auxiliary systems can support complex assembly processes [25]. In motor production, simple tasks can be handled with sensitive lightweight robots and automated guided vehicles. Another approach involves developing machine learning-based controllers for robots that can reduce the programming effort for assembly tasks [26]. In this case, machine or robot control can be transferred to the cloud and be provided as a service [27].

With the foregoing, the relationship between the motor production process and Industry 4.0 is established. Moreover, the application of the motor production process of the research object is summarized in Table 1.

In summary, the literature summarizes the intelligent systems that can be used in motor manufacturing, and also supplements the system that is lacking in existing literature. For the industry, it also provides a practical intelligent system model.

Table 1. Comparison table of motor production process and Industry 4.0.

STEP	INDUSTRY 4.0
1. Housing and Rotor Cage Production	Use sensors to monitor production data (specifications, quality, etc.) and store them on a computer. Then, use artificial intelligence to analyze and predict the required materials to avoid excessive waste. For example, if the specifications and quality can be predicted in advance, it is set at the time of production, and when the rotor of the production line is cut, the excess parts can be cut out to save raw material consumption.
2. Laminated Core Production	When cutting electrical boards, the sensors installed on the machine receive cutting force and sound and vibration frequency data. Then, use artificial intelligence and CPS virtual operation models to simulate and obtain more accurate data to aid in production. For example, when the rotor of the production line is cut, the above methods can be used to make the machine operate more accurately, and the accuracy of the finished product specifications can be improved.
3. Insulation and Impregnation	In the selection of washer (insulation parts) materials, artificial intelligence is used for simulation and analysis to find the most suitable materials and specifications. Then, production and assembly are performed to find the most suitable materials and specifications. To provide a better insulation effect, accurate installation position is necessary.
4. Winding	In the winding step, install the sensor to return data, and use artificial intelligence analysis to find the best winding data (number of turns and winding speed, etc.). The image recognition system can make production faster and more accurate and inspect the appearance at any time.
5. Contact	Part of the rotor spot welding is equipped with an intelligent system. In addition to returning production data (resistance, temperature, etc.) to monitor the entire welding process, it can also optimize the welding quality. If image recognition is added, weld defects can be found. The severity is classified for subsequent analysis and optimization.
6. Shaft Production	Use artificial intelligence to analyze and optimize shaft production data (quality, specifications, etc.) to find the best production method. It can also be used in the assembly of the bearing, installation of sensors, and monitoring of parameters (size, assembly force, shaft radius, etc.) for computer analysis and optimization to reduce the number of defective products.
7. Permanent Magnet Rotor Production	The permanent magnet assembly step uses image recognition to visually inspect the magnet surface after production. The machine is equipped with a sensor during the magnetization step of the permanent magnet. After magnetization, the position and magnetic field of positive and negative poles are automatically checked to avoid affecting the motor operation.
8. Final Assembly and Testing	In the motor assembly line, image recognition (e.g., brush appearance inspection and motor press-fit appearance inspection) is added to check the appearances. The data generated during the motor performance test can also be stored and analyzed using artificial intelligence in order to identify problems and aid the motor designer.
9. Overall Process	Using the machine-to-machine technology, the intelligent system installed on each machine is connected to exchange and upload data in real time. Data management is performed through knowledge-based systems to speed up the entire operation and management processes.

2. Materials and Methods

2.1. Research Architecture and Methods

In this study, secondary data collection, field study, and interviews are performed to explore the selection of intelligent systems to be applied to Solen company, which is Vietnam's motor production line. In terms of secondary data collection, this study has searched for related research papers, journals, and reports on existing smart manufacturing cases; production line standard operating procedure manuals are also obtained. From these, feasible methods for "SM" motor production line are initially evaluated.

The main author and coauthor of this study also visited the Solen company Ho Chi Minh City factory in Vietnam on 19–22 September 2019 to conduct a field survey. This allowed the authors to observe the actual operation of the motor production line staff and understand each step of the operations. Based on this field visit, the processes and systems that smart systems may import from the technology of Industry 4.0 are initially assessed. (See Table 2)

Table 2. Research methods and data types.

RESEARCH METHOD	DATE	DESCRIPTION	COLLECTED DATA
Secondary data collection	July 2019–February 2020	Information on research issues from existing papers, journals, cases, and company materials is obtained. After analysis, the feasible methods for "SM" motor production line are summarized, integrated, and initially evaluated.	Papers, journals, books: 28 Company production line operation process description SOP(Standard Operating Procedure): 1 copy Production line working time statistics
Field study	19–22 September 2019	Actual operation status of production line is monitored by observing motor production line and collecting data.	Production line photographs
Interview	27 November 2019, 13:00–14:00	Two practical experts were invited to laboratory for interview.	Record interviews, verbatim files, photographs
	31 January 2020,14:00–15:15 On February	Visited Taiwan headquarters of Solen company, interviewed three practical experts, collected questions related to the implementation and design of production line. Questions raised by interviewees are asked to obtain helpful information for research. Two experts are responsible for the production line process and machine design. One expert is responsible for the actual production line execution.	Record interview files, verbatim files, photographs
	14 February 2020, from 14:00 to 14:30	Visited Solen company Taiwan Headquarters, and a practical expert was interviewed.	Record interview files, verbatim files, photographs

A three-stage interview is conducted to collect expert opinions. 1. On 27 November 2019, two practical experts were invited to the laboratory for interview. 2. In order to understand the current situation of the company's production line automation, three persons in charge of the motor production line in the Taiwan headquarters of Solen company were interviewed on 31 January 2020 and from 14 February 2020. They were interviewed regarding feasible solutions for preliminary evaluation as well as the company's views on the introduction of intelligent systems and future planning. To ensure the correctness of the information, the consent of the respondents to record and photograph the interviews was obtained. It is also treated anonymously after the interview to protect respondents.

In this research, secondary data collection, field study, and expert interviews are mainly conducted to study the "SM" motor production line in Vietnam. Based on the observations of researchers of the actual production line and personnel interviews to understand existing production process, the smart machine tools (smart systems) necessary for smart manufacturing are introduced. Finally, the intelligence of the motor and coil assembly processes of this production line is analyzed. The research architecture is shown in Figure 1. In this study, there are three stages of research steps, Step 1: Secondary data collection, to widely collect data on the "SM" production line. Step 2: Field study, go to the production line for research. Step 3: Interview, this study conducted a total of 3 interviews to obtain a prosperity of research data.

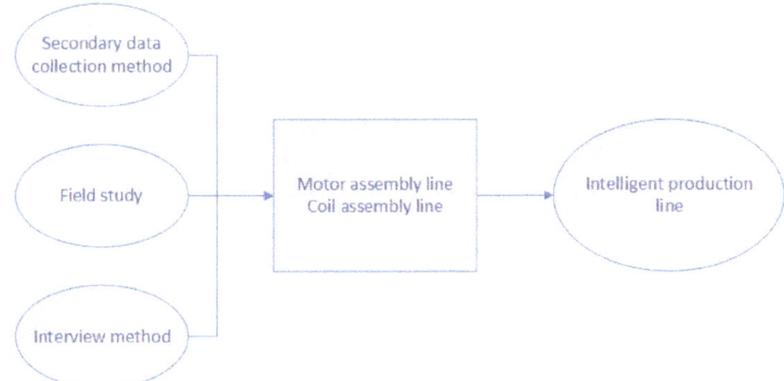

Figure 1. Research architecture diagram.

2.2. Research Objects

The "SM" motor production lines in Vietnam are mainly divided into engineering, coil, and motor assembly production lines. The engineering production line is mainly focused on the assembly of rotors, sleeves, bearings, and other components, as well as magnet production. The coil assembly line is primarily responsible for winding, spot welding, cutting of commutators, washer assembly, bearing assembly, and distance inspection as well as voltage, insulation, and impedance testing. The motor assembly line deals with coils, brushes, washer, and motor assembly, as well as visual inspection. The layout of the "SM" motor production line in Vietnam is shown in Figure 2. In total, there are three production lines and two assembly lines. Each rectangle represents one workbench, each person is assigned to a workbench and is responsible for handling 1–2 operating procedures.

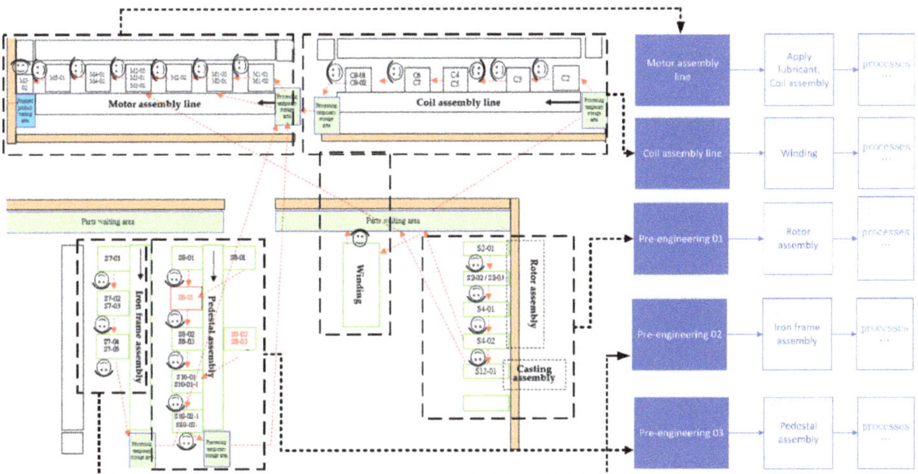

Figure 2. Production line layout.

The operation processes of the coil and motor assembly lines are shown in Figures 3 and 4, respectively.

Figure 3. Coil assembly line.

The coil assembly line in Figure 3 is composed of eight main operating procedures as follows.

1. Winding

    (1) Check if the rotor is rusty and there is a mounting sleeve; there must be five grooves. After confirmation, ensure that the machine is inserted according to the direction.
    (2) Press the button for the machine to start winding.
    (3) Use a wooden shaft to number the parts.
    (4) Ensure that the copper wire is not cracked. The diameter of the rotor should not exceed those of the casing and spot welding of the rotor.

Figure 4. Motor assembly line.

2. Commutator spot welding

    (1) Place the coil into the machine according to the direction.
    (2) Press the button to allow the machine to start the spot welding.
    (3) Check whether the appearance of parts is complete and if there are five grooves. The NG product is placed in the NG box, and good products proceed to the next step.
    (4) Photographs are taken and saved for every 1000 products.

3. Commutator cutting
   (1) Place the parts of the machine according to the direction.
   (2) Start the button to allow machine cutting.
   (3) Check whether the appearance of parts is intact; put the NG product in the NG box.
   (4) Install the bolt gauge tool to the commutator after cutting.
4. Pneumatic rectification
   (1) Place the parts in the machine.
   (2) Clean the parts with a brush.
   (3) Use a magnifying glass to check the appearance of parts and for foreign objects. Put the NG product into the NG box; the good products proceed to the next step.
5. Air pressure removal
   (1) The parts are placed into the machine and aligned with the air-jet holes for air-jet cleaning.
   (2) Check the parts for foreign matter and whether the parts appear complete and bright. Place the NG product in the NG box; the good products proceed to the next step.
6. Washer group voltage withstand, absolute resistance, and impedance tests
   (1) Place the washer into the machine.
   (2) Press the switch to initiate the washer punch.
   (3) Put the parts into the machine for voltage withstand, absolute resistance, and impedance tests. Put the NG products into the NG box; good products proceed to next step.
7. Bearing assembly
   (1) Place the bearing and coil in the machine in the same direction.
   (2) Press the switch to initiate punching.
   (3) Check whether the parts are intact and correctly aligned. Place the NG product into the NG box; good products proceed to the next step.
8. Bearing distance check
   (1) Place the parts of the inspection machine as directed.
   (2) Press the button to initiate machine check.
   (3) Check if the parameters are within range. Put the NG product into the NG box; good products proceed the next step.

The motor assembly line in Figure 4 is composed of nine main operating procedures as follows.

1. Lubricant application
   (1) Align the iron frame with the coil cut groove and place it in the dispenser.
   (2) Press the switch to dispense the machine.
   (3) Check whether the parts have a sufficient amount of glue. The NG product is placed into the NG box, and the good products proceed to the next step.
2. Coil assembly
   (1) Place the parts of the machine according to the direction.
   (2) Clip the iron frame into the coil with the white assistive device and take it out.
   (3) Check whether the parts are installed. Place the NG product into the NG box; good products proceed to the next step.

3. Brush assembly

    (1) Align the red part of the brush with the machine.
    (2) Press the switch to open the terminal.
    (3) Assemble the parts and coils into the finished product according to the working procedure.
    (4) To provide a gap in the iron frame, check the positioning point. The base assembly must be accurately positioned. The NG product is placed in the NG box; good products proceed to the next step.

4. Brush appearance inspection

    (1) Assemble the machine parts.
    (2) Inspect the parts for damage through closed-circuit television. The NG products are placed in the NG box, and the good products proceed to the next step.

5. Washer enrollment and inspection

    (1) Put the small and large washers on the machine and install the parts.
    (2) Press the button to connect the washer.
    (3) Put the parts into the machine according to the indicated direction, and capture image.
    (4) Ensure that no washer is missing based on the image. Put the NG product into the NG box, and good products proceed to the next step.

6. Motor press assembly

    (1) Check whether the size of the washer is correct, and the iron frame cover must fully fit the iron frame.
    (2) Align the machine parts.
    (3) Press the button to initiate pressing.
    (4) After installation, ensure that there are eight cut-outs in the parts. Place the NG product into the NG box; good products proceed to the next step.

7. Clearance confirmation

    (1) Place the parts of the inspection machine.
    (2) Press the button for the machine to detect the gap. Place the NG box out of the specification range; the good products proceed to the next step.

8. Motor performance check and lighting

    (1) Align the parts of the inspection machine.
    (2) Press the button for the machine to transport the parts until they touch the terminals.
    (3) Press the switch to start measurement and check the 88 items of motor performance data.
    (4) Check if the data are within range. Put the NG product into the NG box; good products will be automatically engraved with a number. Proceed to the next step.

9. Final visual inspection

    (1) Use a tool to check if the screw pattern is correct.
    (2) Use a tool to check whether the laser marking (seal) is sufficient.
    (3) Check if the motor rotation direction is correct.
    (4) Put the motor into the machine to check whether the terminal is defective. Put the NG product into the NG box and put the good product into the box according to its number. Finally, pack and ship.

2.3. Interview Outline

Interviews on the design and execution dimensions of the production line have been conducted.

1. Interview questions related to production line automation.
   (1) Is each part numbered and recorded in the computer to manage all parts? If so, how does it work?
   (2) Is there a sensor installed on the production unit to detect, for example, the current production quantity or whether the parts of the unit itself are faulty? If so, what is the current function of the sensor? If none, have you considered installing it?
   (3) Will the photographs captured by the CCTV (Closed Circuit Television, is a photographic equipment and equipment for capturing images to improve quality, etc.) and machine-measured data be stored? If yes, are they classified and used for other applications?
   (4) In the motor assembly line, how many photographs are taken by the CCTV brush appearance inspection, and which of these can provide basic information for image recognition in the future?
   (5) Does the company currently have any expectations? To what extent has the production line been automated, such as the introduction of robotic arms? If so, can the reduced labor and machine costs after the introduction be maintained or increased to production capacity?
   (6) In the previous interview, the two managers mentioned that the identification of imported images requires huge data to train the AI and generate a model. How much data are currently stored in the factory? Do the images identify the manufacturer of the device?
   (7) In terms of data collection, it is evident that good data must be retained, but in order to introduce image recognition or other machine learning models, considerable NG data are also necessary. Are there any plans for retaining NG data?
2. Production line interview questions
   (1) Will the photographs captured by the CCTV and machine-measured data be stored? If yes, are they classified and used for other applications?
   (2) To what extent is the production line currently automated, and which parts are produced using both machinery and labor?
   (3) The images captured by the CCTV are used for appearance inspection. How is it judged whether the product is OK or NG? Some products have missing corners but can still be used, what is the criterion for making this decision?
   (4) Production line standard operating procedure.

3. Results

The research results based on field study, interviews, literature, and secondary data analysis are compiled.

3.1. Motor Production Line Analysis

Approximately three years ago, the cooperation with company M was implemented. The company went to Vietnam several times to review the motor production line and check problems related to failure mode, design, and equipment. Approximately 5–8 persons come from Japan every visit, which is approximately 1–2 weeks every two to three months. Starting September 2019, the products have been manufactured through small-scale production. Thus far, the total output is approximately 200,000 motors, and considerable performance test data are available. The failure mode verification along the production line involves checking the 200,000 motors. In the production, the selection of materials is gradually adjusted, and the testing process is improved. One example is the defect reported by a customer that was not observed in the production line testing but was observed during use. The problem was traced in the assembly press-fitting.

Status of production lines at the "SM" Motor Plant in Vietnam: In the pre-production line, there are three production lines, each of which has 3–6 workbenches. Each workbench has 1–2 operating procedures; only one person is responsible for each workbench. Table 3 summarizes the status of the first production line.

Table 3. Status of coil assembly line production.

WORKBENCH	OPERATION STATUS
1	Responsible for winding. Check whether the rotor parts are fully installed, put the rotor into the machine (four rotors at a time), automatically wind and mark the number, and check the appearance of parts.
2	Responsible for spot welding of commutators. Place coils in the machine for spot welding, check the appearance of parts, take photographs for every 1000 pieces produced.
3	Responsible for commutator cutting. Put parts into the machine for cutting and check the cutting status of parts.
4	Responsible for air pressure rectification and air pressure removal. Clean the parts with a brush, check the foreign matter residue with a magnifying glass, put the parts into the air pressure cleaner for cleaning, and check the appearance of parts.
5	Responsible for washer group voltage resistance, absolute resistance, and impedance tests. Put the washer into the machine for punching, and then into the test machine for voltage resistance, absolute resistance, impedance tests.
6	Responsible for bearing assembly and bearing distance inspection. Put the bearings and coils into the machine for stamping, and then place the parts into the inspection machine to check the bearing distance.

There is one coil assembly production line in the "SM" Motor Plant in Vietnam. The production line has six workbenches. Each workbench has 1–2 working procedures, and only one person is responsible for each workbench.

The motor assembly line in the "SM" motor plant in Vietnam only has one production line. The production line has seven workbenches, each of which has 1–3 operating procedures and handled by one person. The status of the motor assembly line is summarized in Table 4.

3.2. Process Analysis of Smart Manufacturing Line

1. Smart manufacturing status of production lines at the "SM" Motor Plant in Vietnam.

Sensors have been installed in the machines of the production line of the "SM" Motor Plant in Vietnam. Apart from detecting machine failure, the sensors are mainly used to detect the operator's negligence. When an operation error occurs, a warning is given, and the machine is automatically stopped to avoid operator injury and product damage. Sensors are installed to record the number of operations in the coil, commutator cutting, and spot welding. They can be used to push back the total production volume (e.g., finished products and defective products) so that the person in charge of the production line can record and check the output. In the motor performance test, the data after passing the test are supposed to be retained; however, these are not retained because of limited storage space, and those that have already been stored are not used.

Table 4. Status of motor assembly line production.

WORKBENCH	OPERATION STATUS
1	Responsible for dispensing and coil assembly. Align the iron frame with the coil and place it in the dispenser. Then, place the parts in the machine, use the auxiliary tools on the machine to snap the iron frame into the coil, and check whether the installation is complete.
2	Responsible for the assembly and appearance inspection of the brush assembly. Put the parts into the machine and assemble with the finished product of the previous workbench, then move the parts in front of the CCTV, and the personnel will check its appearance through CCTV.
3	Responsible for the assembly and inspection of the washer. Put the washer on the machine to join, move the parts to another machine to take pictures, and check whether the washer is missing according to the inspection.
4	Responsible for motor assembly press-fitting and clearance inspection. Put the parts into the machine and press the motor. Finally, check if there are eight gaps in the parts. Then, put the parts into the testing machine to check the gap.
5	Responsible for motor performance inspection and lightning engraving. Put the parts into the testing machine to check the motor performance. The machine automatically engraves the parts that passed the inspection and stores the test data.
6	Responsible for the final visual inspection. Use tools to check whether the screw pattern is sufficient. Check whether the laser marking (seal) is sufficient. Check the direction of motor rotation and appearance of the terminal.
7	Responsible for packaging and shipping.

The production line frontliners are responsible for implementing withstand voltage test and insulation size inspection procedures. Some production lines have introduced a special model, which has an arm that can automatically put parts into the machine for testing. The workflow is a semi-automated state as personnel operate the machine. The Solen company have already introduced automatic arms in other production lines. The main plan is to automatically take the rotor to the intermediate process after rotor assembly. This can save time and allow the allocation of manpower to where it is more necessary. The company will keep the production line in a semi-automated state in the short term. At most, a small number of intelligent systems will be introduced because of cost considerations. If future orders increase or there are other influencing external factors, then there will be opportunities to plan for more work processes, such as adding more functions (e.g., robotic arms) and intelligent systems.

2. What intelligent systems can be imported into the production line?

This study recommends three possible intelligent systems.

The first is an image recognition system. There are many processes in the production line that check the appearance of parts. If the image recognition system can be used to automatically check the parts, the manpower responsible for visual inspection can be reduced. At this time, however, it is not feasible to acquire an image recognition system because of the lack of background data. According to certain existing conditions and negotiations with equipment manufacturers, the company's practical experts can first install the CCTV for each part that should be identified. The images can then be stored as background information for future image recognition.

The second is the use of a robot arm. In the same production line, the arm is still used to pick up parts and operate the machine. Only the conveyor belt is employed to transfer the parts to the next workbench, and the finished product is transported through different production lines; all these processes are performed by humans. If robot arms are introduced, then a more efficient production can be achieved.

The third recommendation is the use of a big data analysis system. In the production line, there are numerous operational processes that should be tested; however, not all test data are retained. If most of the available data are retained for analysis, these can be employed to facilely identify problems and fix errors as soon as they occur.

4. Discussion

Based on secondary data (SOP and working time chart), it is found that the manpower currently used in the production line has been reduced from 50 to 25. According to the analysis of secondary data and expert interviews, the statistical table of the working time of the production line. Expert experience indicates the total working time of the original design, which was 600 seconds, and the total working time after the introduction of the intelligent system was 306.1 seconds. This also proves the efficiency of the current production line. In the future, related intelligent systems, such as big data and image recognition systems, will be introduced. It is expected that the production line manpower will be reduced to 4–5 persons; it will then become a demonstration production line of Industry 4.0 and resolve the manpower shortage in various countries.

1. Implementation of intelligent systems

The introduction of systems, such as big data and image recognition, can make the production line intelligent as well as link data that are originally independent in each machine. Such a connection can achieve the machine-to-machine effect and reduce manpower. This study achieved the effectiveness of the intelligent system through the improvement of the process of the coil and motor assembly line. Based on the list in Table 3, there are currently six workbenches in the coil assembly line, and each workbench is assigned to a person to operate the machine. With the introduction of a robotic arm and an image recognition system, the number of personnel will be reduced from six to two. One personnel will be responsible for setting the winding, commutator spot welding, and cutting machines, and the other will be responsible for the second half of the pressure cleaning of parts and testing machine operation. Both workers are responsible for checking the machine at any time and immediately report any problematic situation.

Based on the list in Table 4, there are currently seven workbenches in the motor assembly line. Each workbench is assigned to a worker to operate the machine. A total of seven workers are in this production line. If robot arms, image recognition systems, and big data systems are introduced, the number of workers may be reduced to three. One person is responsible for the assembly and inspection of coils, brushes, and washers; the other is responsible for the assembly, press-fitting, and performance testing of the motor, and final visual inspection. The first two workers should check the machine at any time. If a problem is detected in the machine, this should be reported immediately. The third personnel will be responsible for packaging and shipping the finished product. In the future, it should be evaluated whether the introduction of automatic packaging machines will be useful to the entire production line.

2. Changes in process management

After the introduction of the intelligent system, the process in each production line will inevitably change and follow a systematic process. The following describes the expected changes in the processes of the coil and motor assembly lines after the introduction of the intelligent system. It is mainly divided into three parts.

(1) Robot arm

The main purposes of introducing a robot arm is to change the original process where humans get the parts and put them into the machine. Moreover, with the robot arm, the number of machine operators will be reduced. The following will explain the function of the two production lines after the robot arm is introduced and replaces manpower. The process sequence is as follows. Put the rotor in the winding machine, the coil in the spot welder, the commutator in the cutting machine, the bolt gauge tool on the cut commutator, and the washer in the test machine, bearing, and coil. Put the stamping machines and bearings into the gap inspection machines. In Figure 5, the processing sequence of the part is as follows. Put the coil into the machine, the brush into the machine for assembly, the brush into the machine to assemble the washer, the motor into the machine for stamping, the motor into the gap machine, and the motor into the performance inspection machine.

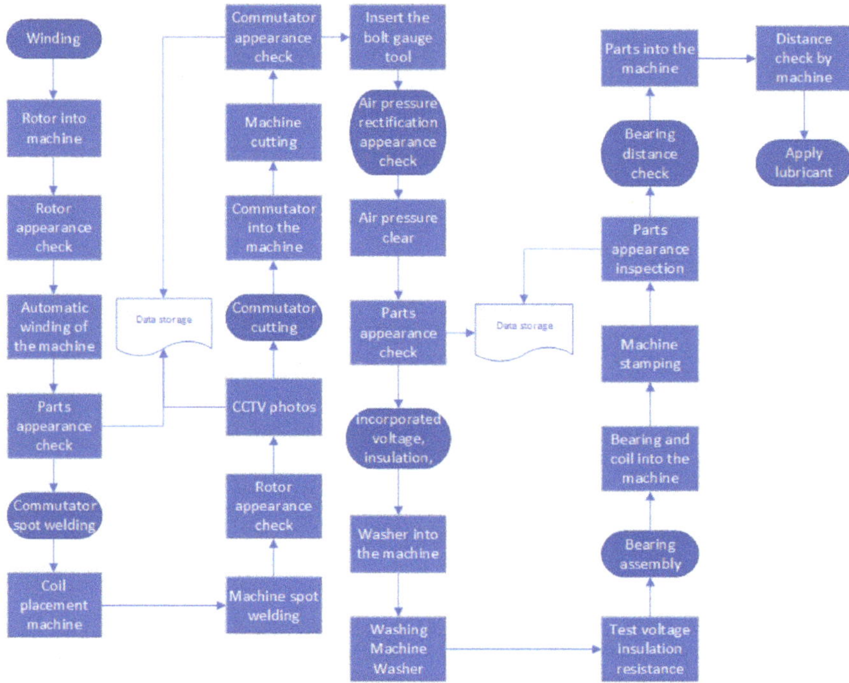

Figure 5. Coil assembly intelligent line.

At present, to replace manpower, robotic arms may be introduced only to the robotic assembly line. Taking parts from the coil assembly line to the motor assembly line is labor-intensive. Future research may start on investigating the delivery of parts among the production lines.

(2) Image recognition

The main purpose of introducing image recognition is to replace the original process where humans visually inspect parts, thereby increasing efficiency while maintaining product quality. After image recognition is introduced to the two production lines, the intelligent system will automatically perform visual inspection. The process sequence of visual inspection is as follows: the appearance of the rotor before and after winding, the appearance of the commutator spot welding, the appearance of the commutator cutting, the appearance of the cleanliness of the commutator pressure, and the appearance of the rear assembly of the bearing. In Figure 6, according to the sequence, the fitting conditions of the washer and iron frame are checked, the appearance of brushes is inspected, the appearance of rear parts assembled by the washer is checked, the eight gaps of the parts are visually inspected, and the final appearance is checked. Image recognition allows the machine to accurately identify people and things in an image faster and more efficiently than humans. It also further improves the product quality and manufacturing efficiency of the industry.

Figure 6. Motor assembly intelligent line.

(3) Big data analysis

The image recognition system includes big data analysis. In addition to the identification of parts based on the original images provided, it can also perform image data analysis. It can automatically capture, analyze, classify, and understand useful information from a single image or a series of images. The images provide information to achieve better productivity and quality.

The motor assembly line is shown in Figure 6. In the motor performance inspection process, the inspection machine is provided by a partner manufacturer. A total of 88 motor performance data points are tested, and the information on a product that passed the inspection is retained. According to the opinions of the three practical experts interviewed, these data are only stored and not used for any purpose. In the future, these 88 data points can be studied and analyzed. Models can also be created to improve the production line (e.g., increasing production capacity or identifying process problems and optimizations).

5. Conclusions

In this study, it is demonstrated that the implementation of intelligent systems (i.e., robotic arms, image recognition, and big data analysis) can aid the coil and motor assembly lines; it also has a practical reference value. Firstly, the introduction of a big data analysis system is reported. In the motor performance testing part of the motor assembly line, big data analysis is introduced to aid the motor production line by collecting 88 motor data points generated by the machine. The motor coil assembly production line and image recognition system included in the motor assembly line introduce big data analysis, and the system then makes judgments based on the pre-set images. Subsequently, the system classifies and analyzes the acquired images to allow the image recognition system to make more accurate and quick judgments or even identify the defects that may cause motor problems. According to the research argument of Mayr et al. [10], the Industry 4.0 technology can significantly optimize motor production, particularly with the use of big data analysis. It has considerable potential and can be used in a wide range of motor production. The conclusion (i.e., the introduction of big data analysis into the motor production line) is found to be consistent. Secondly, Mayr et al. [10] only reported that the introduction of an intelligent system impacts the motor production, but the changes in the production process were not discussed. This study therefore proposes a change in the intelligent system for the production process. Moreover, a robotic arm and an image recognition system are introduced. As a result, manpower is reduced, and the operation process is optimized.

Finally, Mayr et al. [10] theoretically confirmed that the Industry 4.0 technology can optimize motor production but did not discuss practicality. This means that the introduction of the technology may cause practical impossibility. Through interviews conducted with practical experts, this research evaluates the theoretical results obtained by literature analysis as well as the possibility of implementation based on the inputs of practical experts. Among the many Industry 4.0 technologies, the robotic arm, image recognition, and big data analysis are obtained. These three techniques can be practically used in the company's production line.

As a result of this research, the "SM" production line can be introduced with a robotic arm and image recognition system to make the current production line more intelligent. Research question 1 has been answered. The research question 2 is also answered through the research method. The coil assembly line can be turned into an automated production line when using an intelligent system.

Limitations of Study

This study only investigates and proves the feasibility of the introduction of intelligent systems to the coil and motor assembly lines of the Solen company. The three intelligent systems that can be used in the steps of the motor production line are proposed; some parts, however, have not yet been explored. Firstly, 88 big data points are not analyzed. In the future, the correlation between the values and 88 motor test data points will be determined. This should correspond to the operating process of the motor production line so that it can be used to build a model to facilitate the company's continuous optimization, monitoring, and management of the production line. Secondly, the study results do not indicate whether the intelligent systems can be used in other motor production lines or whether the production line of the manufacturing industry can achieve other effects (e.g., manpower reduction and production process optimization). These can be included in a future research to supplement the contents of this study.

Author Contributions: Conceptualization, Y.-C.L., C.-C.Y. and W.-H.C.; methodology, Y.-C.L. and W.-H.C.; validation, Y.-C.L., and W.-H.C.; formal analysis, W.-H.C. and K.-Y.H.; investigation, W.-H.C. and K.-Y.H.; resources, Y.-C.L. draft preparation, W.-H.C. and K.-Y.H.; writing—review and editing, Y.-C.L. and W.-H.C.; visualization, W.-H.C. and K.-Y.H.; supervision, Y.-C.L. and C.-C.Y.; project administration, W.-H.C.; funding acquisition, Y.-C.L. and C.-C.Y. All authors have read and agreed to the published version of the manuscript.

Funding: This research received no external funding.

Acknowledgments: This research was funded by the Taiwan Ministry of Science and Technology (MOST 108-2745-8-155-001).

Conflicts of Interest: The authors declare no conflict of interest.

Appendix A

Table A1. English Abbreviations.

Acronym	Definition
ANN	Artificial Neural Network
CCTV	Closed Circuit Television Photographic equipment on the production line and equipment for capturing images to improve quality, etc
CNC	Computer Numerical Control
CPS	Cyber Physical System
IoT	Internet of Things
OPC	OPC is the interoperability standard for the secure and reliable exchange of data in the industrial automation space and in other industries. It is platform independent and ensures the seamless flow of information among devices from multiple vendors. The OPC Foundation is responsible for the development and maintenance of this standard. Source: https://opcfoundation.org/
OPC UA	OPC Unified Architecture
SM	Solen Company Vietnam Motor Production Line

References

1. Guedes, A.S.; Silva, S.M.; Filho, B.C.; Conceição, C.A. Evaluation of electrical insulation in three-phase induction motors and classification of failures using neural networks. *Electr. Power Syst. Res.* **2016**, *140*, 263–273. [CrossRef]
2. De Oliveira, J.F.; Cassimiro, B.; Pacheco, A.; Demay, M.B. Detection of defects in the manufacturing of electric motor stators using vision systems. In Proceedings of the Electric Connectors in 12th IEEE International Conference on Industry Applications (INDUSCON), Curitiba, PR, Brazil, 20–23 November 2016; pp. 1–6.
3. Sun, J.; Wyss, R.; Steinecker, A.; Glocker, P. Automated fault detection using deep belief networks for the quality inspection of electromotors. *Tm-Technisches Messen* **2014**, *81*, 255–263. [CrossRef]
4. Manyika, J.; Bughin, J. The promise and challenge of the age of artificial intelligence. *Etz El-Ektrotechnik Automation* **2018**, *5*, 4–6.
5. Edward, L. Cyber Physical Systems: Design challenges, University of California, Berkeley Technical Report. In Proceedings of the IEEE International Symposium on Object and Component-Oriented Real-Time Distributed Computing (ISORC), Orlando, FL, USA, 5–7 May 2008.
6. Lee, J.; Bagheri, B.; Kao, H. A cyber-physical systems architecture for Industry 4.0-based manufacturing systems. *Manuf. Lett.* **2015**, *3*, 18–23. [CrossRef]
7. Lee, J.; Lapira, E.; Bagheri, B.; Kao, H. Recent advances and trends in predictive manufacturing systems in big data environment. *Manuf. Lett.* **2013**, *1*, 38–41. [CrossRef]
8. Castelo-Branco, I.; Cruz-Jesus, F.; Oliveira, T. Assessing Industry 4.0 readiness in manufacturing: Evidence for the European Union. *Comput. Ind.* **2019**, *107*, 22–32. [CrossRef]
9. Andez-Carames, T.M.; Blanco-Novoa, O.; Froiz-Míguez, I.; Fraga-Lamas, P. Towards an autonomous Industry 4.0 warehouse: A UAV and blockchain-based system for inventory and traceability applications in big data-driven supply chain management. *Sensors* **2019**, *19*, 10.
10. Mayr, A.; Weigelt, M.; von Lindenfels, J.; Seefried, J.; Ziegler, M.; Mahr, A.; Urban, N.; Kühl, A.; Hüttel, F.; Franke, J. Electric Motor Production 4.0–Application Potentials of Industry 4.0 Technologies in the Manufacturing of Electric Motors. In Proceedings of the 2018 8th International Electric Drives Production Conference (EDPC), Schweinfurt, Germany, 4–5 December 2018; pp. 1–13.
11. Rodriguez, A.; Vrancx, P.; Nowe, A.; Hostens, E. Model-free learning of wire winding control. In Proceedings of the 2013 9th Asian Control Conference (ASCC), Istanbul, Turkey, 23–26 June 2013; pp. 1–6.
12. Fleischmann, H.; Spreng, S.; Kohl, J.; Kißkalt, D.; Franke, J. Distributed condition monitoring systems in electric drives manufacturing. In Proceedings of the 6th International Electric Drives Production Conference (EDPC), Nuremberg, Germany, 30 November–1 December 2016; pp. 52–59.

13. Mayr, A.; Meyer, A.; Seefried, J.; Weigelt, M.; Lutz, B.; Sultani, D.; Hampl, M.; Franke, J. Potentials of machine learning in electric drives production using the example of contacting processes and selective magnet assembly. In Proceedings of the 7th International Electric Drives Production Conference, Wuerzburg, Germany, 5–6 December 2017; pp. 196–203.
14. Weigelt, M.; Mayr, A.; Seefried, J.; Heisler, P.; Franke, J. Conceptual design of an intelligent ultrasonic crimping process using machine learning algorithms. *Procedia Manuf.* **2018**, *17*, 78–85. [CrossRef]
15. Mayr, A.; Lutz, B.; Weigelt, M.; Gläßel, T.; Kißkalt, D.; Masuch, M.; Riedel, A.; Franke, J. Evaluation of machine learning for quality monitoring of laser welding using the example of the contacting of hairpin windings. In Proceedings of the 2018 8th International Electric Drives Production Conference (EDPC), Schweinfurt, Germany, 4–5 December 2018; pp. 1–7.
16. Al-Zubaidi, S.S.; Ghani, J.A.; Haron, C. Application of ANN in milling process: A review. *Model. Simul. Eng.* **2011**, *4*, 1–7. [CrossRef]
17. Ayatollahi, I.; Kittl, B.; Pauker, F.; Hackhofer, M. Prototype OPC UA server for remote control of machine tools. In Proceedings of the IN-TECH 2013 International Conference on Innovative Technologies, Budapest, Hungary, 10–13 September 2013; pp. 73–76.
18. Wang, W.; Zhang, X.; Li, Y. Open CNC machine tools state data acquisition and application based on OPC specification. *Procedia CIRP* **2016**, *56*, 384–388. [CrossRef]
19. Lammers, S.; Adam, G.; Schmid, H.J.; Mrozek, R.; Oberacker, R.; Hoffmann, M.J.; Quattrone, F.; Ponick, B. Additive manufacturing of a lightweight rotor for a permanent magnet synchronous machine. In Proceedings of the 2016 6th International Electric Drives Production Conference (EDPC), Nuremberg, Germany, 30 November–1 December 2016; pp. 41–47.
20. Murakami, Y. Machine Learning Apparatus and Method for Learning Arrangement Position of Magnet in Rotor and Rotor Design Apparatus Including Machine Learning Apparatus. 2016. Available online: https://patents.google.com/patent/US20170093256 (accessed on 10 January 2020).
21. Coupek, D.; Lechler, A.; Verl, A.W. Cloud-based control strategy: Downstream defect reduction in the production of electric motors. *IEEE Trans. Ind. Appl.* **2017**, *53*, 5348–5353. [CrossRef]
22. Colledani, M.; Coupek, D.; Verl, A.W.; Aichele, J.; Yemane, A. A cyberphysical system for quality-oriented assembly of automotive electric motors. *CIRP J. Manuf. Sci. Technol.* **2018**, *20*, 12–22. [CrossRef]
23. Lettau, U. Condition monitoring für die Akustikprüfung. *Etz-Elektrotechnik Automation* **2013**, *5*, 4–6.
24. Brecher, C.; Pallasch, C.; Hoffmann, N.; Obdenbusch, M. Identifikation und lokalisierung von werkzeugen und objekten. *ZWF Zeitschrift für Wirtschaftlichen Fabrikbetrieb* **2016**, *111*, 749–753. [CrossRef]
25. Berndt, S.; Sauer, S. Digitale Fabrik, Montage, instandhaltung: Visuelle ssistenz Unterstützung bei der Durchführung komplexer Montageaufgaben. *wt Werkstatttechnik Online* **2012**, *102*, 162–163.
26. Marvel, J.; Newman, W.S.; Gravel, D.; Zhang, F.E.; Wang, J.; Fuhlbrigge, T. Automated learning for parameter optimization of robotic assembly tasks utilizing genetic algorithms. In Proceedings of the 2008 IEEE International Conference on Robotics and Biomimetics, Bangkok, Thailand, 22–25 February 2008; pp. 179–184.
27. Vick, A.; Guhl, J.; Krüger, J. Model predictive control as a service-concept and architecture for use in cloud-based robot control. In Proceedings of the 21st International Conference on Methods and Models in Automation and Robotics (MMar), Miedzyzdroje, Poland, 29 August–1 September 2016; pp. 607–612.

Article

An All-Factors Analysis Approach on Energy Consumption for the Blast Furnace Iron Making Process in Iron and Steel industry

Biao Lu [1], Suojin Wang [1], Kai Tang [1] and Demin Chen [1,2,*]

1 School of Civil Engineering and Architecture, Anhui University of Technology, Ma'anshan 243032, China
2 The State Key Laboratory of Refractories and Metallurgy, Wuhan University of Science and Technology, Wuhan 430080, China
* Correspondence: chendemin@ahut.edu.cn

Received: 18 August 2019; Accepted: 5 September 2019; Published: 8 September 2019

Abstract: The blast furnace iron making process (BFIMP) is the key of the integrated steel enterprise for energy saving due to its largest energy consumption proportion. In this paper, an all-factors analysis approach on energy consumption was proposed in BFIMP. Firstly, the BFIMP composition and production data should be collected. Secondly, the material flows and energy flows analysis models could be established based on material balance and the thermal equilibrium. Then, the all influence factors (mainly including material flows, energy flows and operation parameters) on energy consumption were obtained. Thirdly, the main influence factors, which influenced the coke ratio (CR) and the pulverized coal injection ratio (PCIR), were obtained by using the partial correlation analysis (PCA) method, because CR and PCIR were the key energy consumption performance in BFIMP. Furthermore, anall-factors analysis result could be achieved by a multivariate linear model (MLR), which was established through these main influence factors. The case study showed that the PCIR was the most effective parameter on CR; when it was increased by 1% (0.84 kg/t), the CR would reduce by 0.507 kg/t. Therefore, the increase in PCIR consumption is the key measure to realize energy saving for BFIMP. The results showed that the improvement of some material flows, energy flows and operation parameters could increase the amount of PCIR, such as sinter size, ore grade, sinter grade, M10, blast volume, blast temperature and especially for sinter alkalinity. Moreover, theall-factors analysis approach on energy consumption can widely be used in various BFIMPs, too.

Keywords: BFIMP; all-factors analysis approach; material flows; energy flows; operation parameters

1. Introduction

Theblast furnace iron making process (BFIMP) represents the most relevant process on the main route for ore-based production of iron in the steelmaking industry [1]. Meanwhile, the iron and steel industry is known for having high energy consumption and high pollution [2]. Around the world approximately 5% of global energy is consumed by the iron and steel industry [3–5], its CO_2 emission accounts for approximately 7% of the total anthropogenic CO_2 emissions [6,7]. Therefore, many novelty methods and technologies of energy saving have emerged in manufacturing fields [8], including the iron and steel industry.

Currently, blast furnace-basic oxygen furnace (BF-BOF) is one of the major production patterns [9]. Moreover, the whole iron making system (including coking, sintering, iron making and other processes) accounts for 70–75% of the total energy consumption in the integrated steel enterprise, whereas BFIMP is more than 50% [10]. Therefore, the BFIMP is one of the most energy-intensive processes in the iron and steel industry [11–14]. RY Yin [15] pointed out that material flows and energy flows are the most basic component. Consequently, material flows and energy flows analysis can also be applied to BFIMP. Related research mainly includes as follows:

Firstly, some optimization models were established based on material flows and energy flows analysis. The improvement of energy efficiency and energy saving is the focus of these optimization models [16]. An optimization model [17] was established based on material balance and energy balance in BFIMP. In this model, the exergy loss minimization was taken as the optimization target. Then, the measures of energy saving wereput forward. Bo Zhou [18] developed the principal component analysis, which could analyze material flows, energy flows and operation parameters in the process of blast furnace (BF) smelting. Furthermore, this model was applied to detect the early abnormality in the iron-making process. Moreover, on the foundation of material flows and energy flows of the oxygen BF with top gas recycling, and a model, which comprised the oxygen BF, the top gas removal process and the preheating units, was established [19]. Then, energy consumption and carbon emission of the integrated steel mill was analyzed based on this model. While, S.B. Kuang [20] proposed a complex function, which was integrated with HM (hot metal) yield and useful energy of the BF. Then the optimal cost distribution of raw materials (namely "generalized optimal construct") was obtained, the influence of some parameters, such as oxygen enrichment ratio, blast temperature and pulverized coal dosage, on the optimization results were further analyzed.

Secondly, the mechanism of the smelting process, which was based on the material or energy evolutionary process, was studied. Due to the complexity of the BFIMP, the numerical simulation method has been applied more widely [21–24]. Y.S. Shen [25] and Yansong Shen [26] simulated the flow and combustion of a ternary coal blend under simplified BF conditions by a three-dimensional computational fluid dynamics (CFD) model. Meanwhile, the effect of the coke reaction index on the reduction and permeability of the ore layer in the BF lumpy zone under the non-isothermal condition was analyzed through a CFD model. Then, the reasonable control of the coke reaction index, which was one of the key factors for BF low-carbon, was pointed out [27]. Moreover, José Adilson de Castro [28] focused on modeling the simultaneous injection of pulverized coal and charcoal into the BF through the tuyeres with oxygen enrichment. The results indicated that the productivity of the BF could be increased up to 25% with simultaneous injection combined with oxygen enrichment. Additionally, the means of simulation, the test procedure was also used to detect reactions in the furnace. Mineral matter of tuyere level cokes was quantified using a personal computer quantitative X-ray diffraction analysis software and examined using a scanning electron microscope in a working BF. At the same time, the apparent CO_2 reaction rates were measured using a fixed bed reactor [29].

Generally, the energy saving or energy efficiency of BFIMP had been explored through material flows and energy flows in the above studies. The energy efficiency of BFIMP has been improved in a large extent. Unfortunately, there were still some deficiencies in the following two aspects. (i) The influence of operation parameters on energy consumption was not involved. (ii) The influence intensity of these parameters on energy consumption was not clear in BFIMP. Additionally, data-driven methodologies have been wildly applied to various thermal equipment in the iron and steel industry due to rapid developments of industrial automation and information systems [30]. Therefore, an all-factors analysis approach, which can analyze the influence of all parameters (material flows, energy flows and operation parameters) on the energy consumption of BFIMP, is proposed based on material balance, thermal equilibrium and data-driven methodologies in this paper. Furthermore, the key influence factors can be achieved by the application of the proposed approach. Then, the corresponding energy saving measures can be put forward effectively. Therefore, the proposed model can provide support for the formulation of the reasonable production plan and the operation management in BFIMP. In addition, the proposed model can also widely be used in various BFIMPs, too.

2. Methods

The all-factors analysis approach mainly includes (as shown in Figure 1):

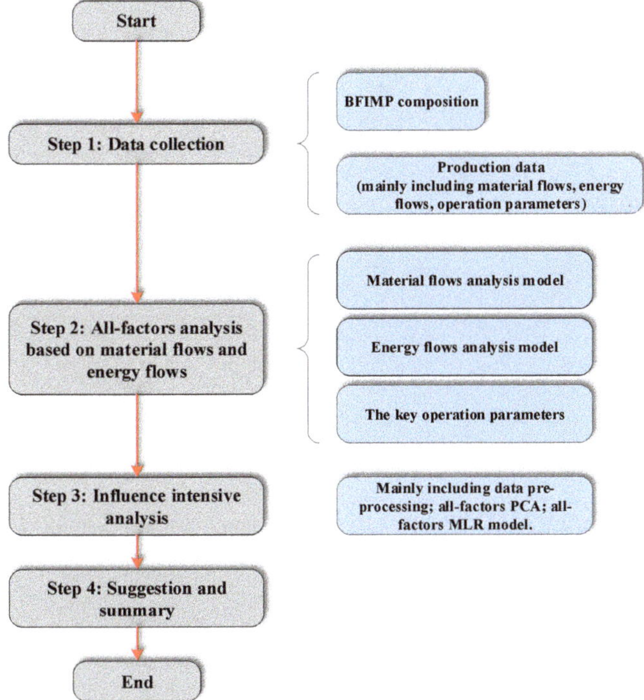

Figure 1. The research route of the all-factors analysis approach.

(1) Data collection:

The BFIMP composition and production data can be achieved through data collection.

(2) All-factors analysis based on material flows and energy flows:

In general, energy consumption is affected by many factors in BFIMP, mainly including material flows, energy flows and operation parameters. Consequently, material flows and energy flows analysis model should be established based on material balance and thermal equilibrium. Moreover, operation parameters [31], which represent the coupling quality between material flows and energy flows, should also be listed.

(3) Influence intensity analysis on energy consumption in BFIMP:

All-factors analysis approach, which mainly includes data pre-processing, all-factors partial correlation analysis (PCA) and all-factors multivariate linear model (MLR) model, is an effective influence intensity method on energy consumption in BFIMP.

(4) Suggestion and summary:

Some suggestion and summary, which can achieve improvement of energy efficiency, should be put forward based on the influence intensity analysis.

2.1. Data Collection

Data collection mainly includes the following aspects:

(1) The BFIMP composition:

The BFIMP composition should be clarified firstly. Therefore, production process investigation should be carried out.

Generally, the BFIMP is composed of the BF body and six auxiliary equipment systems, which includes the charging system (CS), blast system (BS), gas purification system (GPS), fuel injection system (FIS), top power generation system (TPGS) and slag treatment system (STS). The TPGS and STS are subsequent processes of the by-product. The material flows and energy flows proportion of BF body, CS, BS, GPS and FIS accounts for more than 90% of the total amount for BFIMP. Consequently, the TPGS and STS will not be considered in this paper due to their seldom proportion.

(2) Production data:

As discussed in the previous section, there are three kinds of parameters (material flows, energy flows and operation parameters), which affect energy consumption in BFIMP. These data can be collected through various computer detection systems or working records in BFIMP. Especially, a computer detection system can acquire and store these kinds of parameters regularly, such as the production management system and energy management system.

2.2. All-Factors Analysis Based on Material Flows and Energy Flows

2.2.1. Material Flows Analysis Model

Material flows analysis model is established on the basis of the material balance in BFIMP (as shown in Figure 2).

$$\sum_{j=1}^{m} P_{x,i} + P_a + P_{fg} = P_g + P_{hot} + P_{slag} + P_{Lo1}. \quad (1)$$

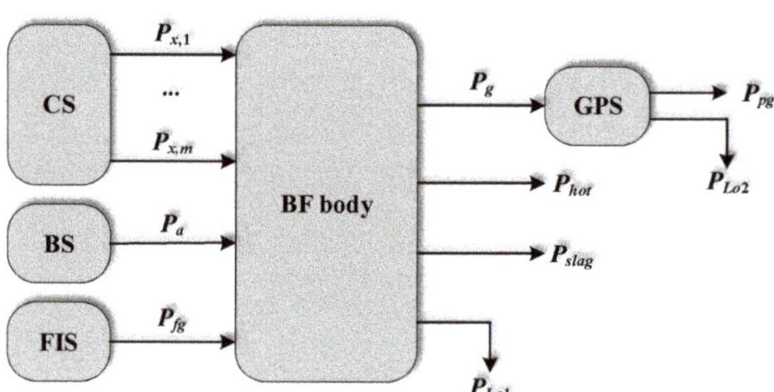

Figure 2. Material flows analysis model

In which,

$P_{x,i}$: The amount of various materials, t/t;
P_a and P_{fg}: The amount of blast and fuel injection into the furnace, respectively, t/t;
P_g, P_{hot} and P_{slag}: The mount of gas, hot metal and slag, respectively, t/t, and $P_g = P_{pg} + P_{Lo2}$;
P_{pg}: The mount of gas after purification, t/t;
P_{Lo1} and P_{Lo2}: The loss amount of various systems, t/t.

2.2.2. Energy Flows Analysis Model

The energy flows analysis model is established based on the thermal equilibrium in BFIMP. In this paper, the BF body and BS, which are the major energy consumption regions, were only in consideration (as shown in Figure 3).

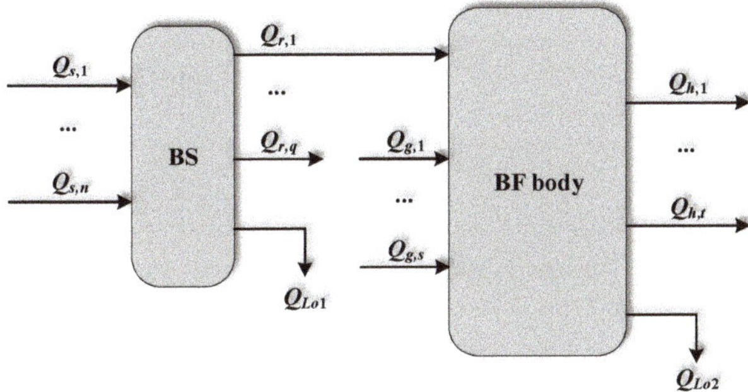

Figure 3. Energy flows analysis model.

Energy flows analysis model of BS is:

$$\sum_{j=1}^{n} Q_{s,j} = \sum_{k=1}^{q} Q_{r,k} + Q_{Lo1}. \tag{2}$$

Energy flows analysis model of BF Body is:

$$\sum_{l=1}^{s} Q_{g,l} + Q_{r,1} = \sum_{r=1}^{t} Q_{h,r} + Q_{Lo2}. \tag{3}$$

In which,

$Q_{s,j}$ and $Q_{r,k}$: The input heat items and the output heat items of the BS, kgce/t (kgce: Kilogram coal equivalent);

$Q_{g,l}$ and $Q_{h,r}$: The input heat items and the output heat items of the BF Body, kgce/t;

Q_{Lo1} and Q_{Lo2}: The loss heat items of the BS and the BF Body, kgce/t.

2.2.3. The Key Operation Parameters

Several factors such as sintering grade and the quality of coke could be measured by using the proposed model. Additionally, operation parameters, which directly reflect the coupling quality between material flows and energy flows, have an important impact on the energy consumption in BFIMP, too. Therefore, these operation parameters should also be sought out, such as the blast temperature and blast pressure.

Generally, these influence factors on energy consumption should be divided into three categories: Material flows factors (the name of the material variable starts with '*P*', as shown in Figure 2); energy flows factors (the name of the energy variable starts with '*Q*', as shown in Figure 3) and operation parameters (the name of the operation variable starts with '*C*', such as blast volume and blast temperature).

2.3. All-FactorsAnalysis on Energy Consumption in BFIMP

2.3.1. Data Pre-Processing

The data pre-treatment mainly included as follows:

(1) Some data were recorded manually. Inevitably, there would be some mistakes in the recording process, such as unrecorded, omitted and incorrectly annotated. Therefore, these data should be eliminated or modified.

(2) The study was discussed based on the normal production [5] in this paper, so the data of equipment overhauls and failures should be stripped out.
(3) Other abnormal data, which was caused by the damage of the detection or data transmission equipment, should also be eliminated.

2.3.2. PCA

A simple correlation analysis (SCA) is a common method of statistical analysis between two random variables. However, all variables may affect each other in general when the number of variables is more than two items. Unfortunately, this mutual influence is not taken into account in SCA. Consequently, SCA was not applicable to the all-factors analysis on energy consumption in this paper, whereas there is an effective way to avoid this problem: PCA [32].This method can achieve the actual relevance of any two variables while eliminating the influence of other variables. Therefore, partial correlation coefficient (PCC) between energy consumption and other parameters can be obtained through the definition of partial correlation algorithm.

2.3.3. MLR Model

First of all, relevant variables should be redefined. P_x represents the xth variable of material flows, the total number of material flows variables is M after all-factors PCA processing. Q_y represents the yth variable of energy flows and the total number of energy flows variables is N after all-factors PCA processing. C_z represents the zth variable of operation parameters and the total number of operation parameters variables is R after all-factors PCA processing. In addition, the number of samples is S, and $i = 1, 2, \cdots, S$. Therefore, there are the following two regression models. A simple example of the PCC between e and P_1 is given to describe calculation process.

$$e_i = c_0 + c_{P,2} \cdot P_{i,2} + \cdots + c_{P,M} \cdot P_{i,M} + c_{Q,1} \cdot Q_{i,1} + \cdots + c_{Q,N} \cdot Q_{i,N} + c_{C,1} \cdot C_{i,1} + \cdots + c_{C,R} \cdot C_{i,R} + \varepsilon'_i. \quad (4)$$

$$P_{i,1} = d_0 + d_{P,2} \cdot P_{i,2} + \cdots + d_{P,M} \cdot P_{i,M} + d_{Q,1} \cdot Q_{i,1} + \cdots + d_{Q,N} \cdot Q_{i,N} + d_{C,1} \cdot C_{i,1} + \cdots + d_{C,R} \cdot C_{i,R} + \varepsilon''_i. \quad (5)$$

In which,

e_i: Energy consumption of the ith group sample, kgce/t;
$P_{i,1}$: The 1st material flows variable of the ith group sample;
c_0 and d_0:Constant term;
$c_{P,2}, \cdots, c_{P,M}$, $c_{Q,1}, \cdots, c_{Q,N}$, $c_{C,1}, \cdots, c_{C,R}$, $d_{P,2}, \cdots, d_{P,M}$, $d_{Q,1}, \cdots, d_{Q,N}$, $d_{C,1}, \cdots$ and $d_{C,R}$: Regression coefficient;
$\varepsilon'_i$and ε''_i: Error term.

Then, the two fitting models can be achieved by the least square method. Meanwhile, the residuals are as follows between them:

$$u_i = e_i - \left(\hat{c_0} + \hat{c_{P,2}} \cdot P_{i,2} + \cdots + \hat{c_{P,M}} \cdot P_{i,M} + \hat{c_{Q,1}} \cdot Q_{i,1} + \cdots + \hat{c_{Q,N}} \cdot Q_{i,N} + \hat{c_{C,1}} \cdot C_{i,1} + \cdots + \hat{c_{C,R}} \cdot C_{i,R}\right). \quad (6)$$

$$v_i = P_{i,1} - \left(\hat{d_0} + \hat{d_{P,2}} \cdot P_{i,2} + \cdots + \hat{d_{P,M}} \cdot P_{i,M} + \hat{d_{Q,1}} \cdot Q_{i,1} + \cdots + \hat{d_{Q,N}} \cdot Q_{i,N} + \hat{d_{C,1}} \cdot C_{i,1} + \cdots + \hat{d_{C,R}} \cdot C_{i,R}\right). \quad (7)$$

In which,

u_i: The residual of the ith group sample between e_i and its fitting model;
v_i: The residual of the ith group sample between $P_{i,1}$ and its fitting model;

Then, simple correlation coefficient between u vector ($u = (u_1, u_2, \cdots, u_S)'$) and v vector ($v = (v_1, v_2, \cdots, v_S)'$) can be obtained by calculation. This coefficient, which is denoted $r_{e,P1}$ (as shown in Equation (8)), is called the PCC between e and P_1. The calculation process of the other PCC is so as well.

$$r_{e,P1} = \frac{Cov(u,v)}{\sqrt{Var[u] \cdot Var[v]}}. \tag{8}$$

In which,

$Cov(u,v)$: The covariance between the u vector and v vector;
$Var[u]$: The variance of the u vector;
$Var[v]$: The variance of the v vector.

In this paper, the PCC between e and influence factors can be obtained by the SPSS software package due to its powerful statistical calculations function. Meanwhile, the significance level (p value) of them can also be achieved by SPSS software. There is a higher significance level between e and an influence factor if their p value is less than 0.05, and vice versa. Consequently, all influence factors with a high significance level can be achieved through related data processing. MLR model between e and these main influence factors can be established.

3. Results and Discussion

3.1. Data Sources and Related Instructions

In this paper, the data source is the production data of a steel enterprise's BFIMP from 2013 to 2014.In order to ensure the validity of the discussion, the pretreatments of these data should be carried out before application. After data pretreatment processes, the effective samples 104 groups of invalid samples were eliminated from the data of the 730 groups, and 626 groups were obtained.

3.2. Material Flows Analysis and Energy Flows Analysis Results

Material flows analysis model and energy flows analysis model could be achieved by the modeling method and sample data, which was mentioned in Sections 2.2.1 and 2.2.2. The analysis results are shown as follow.

3.2.1. Material Flows Analysis Results

Material flows analysis is based on the material balance between raw materials and output products. Since the amount of the gas mud, which is produced by the circulation cooling, is seldom, this part could be ignored. Then, the analytical results are shown in Figure 4.

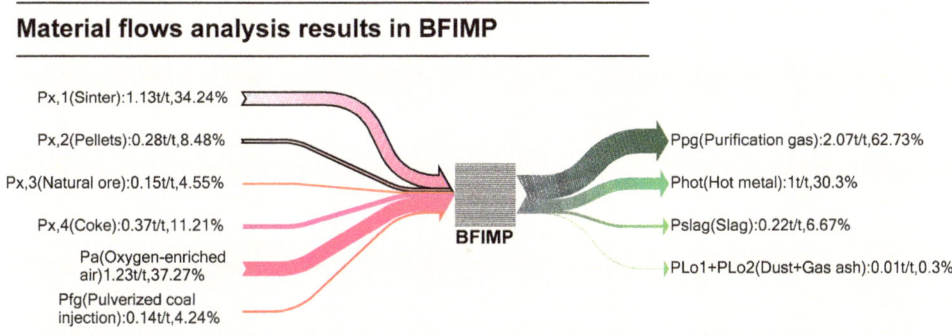

Figure 4. Material flows analysis results in blast furnace iron making process (BFIMP).

(i) The material flows (including input items and output items) are all listed out through the material balance in BFIMP.

(ii) The proportions of material flow input items and output items are all clearly indicated. For example, the amount of sinter ($P_{x,1}$), coke ($P_{x,4}$) and oxygen-enriched air (P_a) accounted for about 83% in all material flow input items. Meanwhile, the purification gas (P_{pg}) and hot metal (P_{hot}) accounted for about 93% in all material flow output items. Therefore, these items should be given more attention.

3.2.2. Energy Flows Analysis Results

(1) Energy flows analysis of the BS:

The energy flows analysis results of the BS are shown in Figure 5.

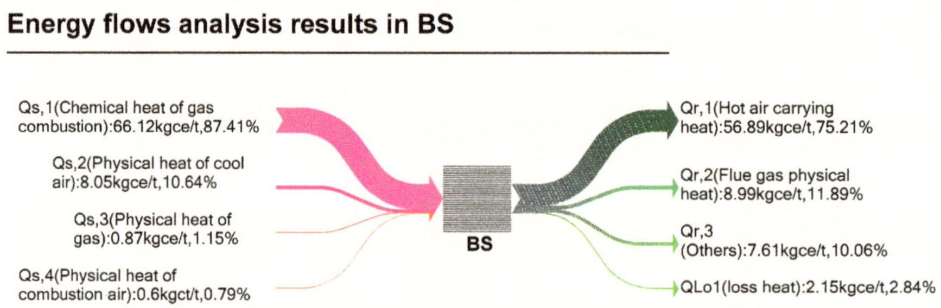

Figure 5. Energy flows analysis results in the blast system (BS).

(i) The energy flow input items and output items are all listed out through thermal equilibrium in BS.

(ii) The proportions of energy flow input items and output items are all clearly indicated in BS. For example, chemical heat of gas combustion ($Q_{s,1}$) is the main input item (accounted for about 87%) among all energy flow input items in BS. The 75% amount of heat quantity has carried out by hot air carrying heat ($Q_{r,1}$) in all energy flow output items.

(2) Energy flows analysis of the BF body:

The energy flows analysis results of the BF body are shown in Figure 6.

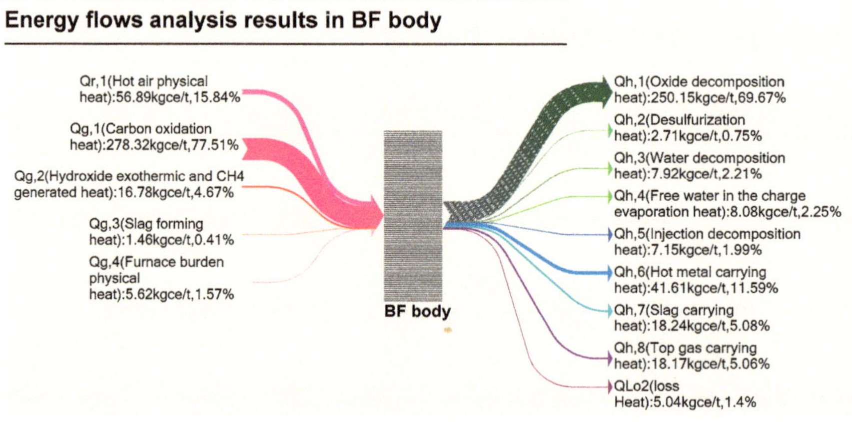

Figure 6. Energy flows analysis results in the blast furnace (BF) body.

(i) The energy flow input items and output items are all listed out through thermal equilibrium in the BF body.

(ii) The proportions of energy flow input items and output items are all clearly indicated in the BF body. For example, the amount of hot air physical heat ($Q_{r,1}$) and carbon oxidation heat ($Q_{g,1}$) accounted for 93% in all energy flow input items. The oxide decomposition heat ($Q_{h,1}$) and hot metal carrying heat ($Q_{h,6}$) accounted for 81% in all energy flow output items.

3.3. All-Factors Analysis on Energy Consumption in BFIMP

3.3.1. PCA

As shown in Figure 6, the carbon oxidation heat ($Q_{g,1}$), which accounted for 77.51% of the total heat consumption, is the main energy source in the BF body. Coke and pulverized coal injection are the main carriers of carbon oxidation heat [33–35]. Therefore, these two parameters can reflect the energy consumption for BFIMP.

Usually, the percentage of coke in total material consumption is called CR, the percentage of pulverized coal injection in total material consumption is called PCIR. It has been proved that that the PCIR improvement and CR reduction are the most effective energy saving measures in BFIMP [36]. Therefore, the influence factors analysis on PCIR and CR will be carried out in this paper. According to the material flows analysis results (as shown in Figure 4) and energy flows analysis results (as shown in Figure 5; Figure 6), the influence factors on CR and PCIR can be achieved. In addition, operation parameters have an important impact on CR and PCIR, too. Then, three kinds of parameters are shown in Table 1.

Table 1. Influence factors analysis on coke ratio (CR) and pulverized coal injection ratio (PCIR) in BFIMP by partial correlation analysis (PCA).

Classification of Influence Factors	Influence Factors	CR		PCIR	
		p Value	Correlation Degree	*p* Value	Correlation Degree
Constant		0.002		0.000	
MaterialFlows Parameters	Sinter grade	0.44	−0.062	0.000	0.299
	Sinter size	0.023	−0.185	0.203	−0.104
	Pellet grade	0.506	−0.054	0.800	0.021
	Ore grade	0.047	−0.161	0.002	0.253
	Sinter alkalinity	0.001	0.277	0.000	0.334
	Sinter tumbler index	0.000	−0.419	0.758	−0.025
	Sinter screening index	0.135	−0.122	0.571	−0.046
	Clinker ratio	0.426	0.065	0.695	0.032
	Slag ratio	0.021	0.187	0.334	−0.079
	PCIR	0.000	−0.598		
Energy Flows Parameters	Coke size	0.343	0.077	0.131	0.123
	Coke ash	0.471	−0.059	0.175	−0.110
	Coke volatile	0.227	−0.098	0.093	−0.136
	Coke sulfur	0.811	0.020	0.165	0.113
	M40	0.641	0.038	0.904	0.010
	M10	0.579	−0.045	0.013	−0.201
	Coal ash	0.534	−0.051	0.075	−0.144
	Coal volatile	0.146	−0.118	0.468	0.059
Operation Parameters	Blast volume	0.411	0.067	0.000	0.398
	Blast temperature	0.004	0.233	0.001	0.273
	Blast pressure	0.719	−0.029	0.071	0.146
	Oxygen enrichment ratio	0.449	0.062	0.000	0.319
	Permeability	0.353	0.076	0.947	−0.005
	Top gas pressure	0.103	−0.133	0.341	0.078
	Top temperature	0.199	−0.105	0.000	0.457

Noted: M40, resistance to crushing of coke; M10, abrasion strength of coke.

As shown in Table 1, the significant influence factors on the CR mainly included: Sinter size, ore grade, sinter alkalinity, sinter tumbler index, slag ratio, PCIR and blast temperature, due to their lower p values (≤0.05), whereas the PCIR is the best influence factors among them. The other influence factors were weakly correlated with CR due to their higher p values (>0.05). Moreover, the significant influence factors on the PCIR mainly included: Sinter grade, ore grade, sinter alkalinity, M10, blast volume, blast temperature, oxygen enrichment ratio and top temperature, due to their lower p values (≤0.05).

3.3.2. MLR Models on CR and PCIR

The CR and the PCIR prediction models could be established through MLR models, based on the high correlation influence factors (as shown in Table 1). On the one hand, these prediction models have higher precision. On the other hand, this method could reduce the prediction models complexity due to a reduction in the number of variables.

Then, the fitting degrees of the MLR models were 95% and 94% respectively. In addition, these models were validated by actual production data, too. The fitting coefficients are shown in Table 2. Standardized coefficients (as shown in Table 2), which were calculated through the normalization method, eliminated the influence of dimensional differences among various parameters. Therefore, standardized coefficients could qualitatively reflect the influence intensity of each parameter on CR and PCIR. As shown in Table 2, the PCIR had the highest correlation with the CR among the main factors. Meanwhile, sinter grade had the highest correlation with the PCIR.

Table 2. Multivariate linear model (MLR) modelson CR and PCIR.

Classification of Influence Factors	Influence Factors	CR			PCIR		
		Unstandardized Coefficients	Standardized Coefficients	Correlation Degree	Unstandardized Coefficients	Standardized Coefficients	Correlation Degree
Constant		996.644			−2148.358		
Material Flows Parameters	Sinter grade				21.425	0.539	0.425
	Sinter size	−1.019	−0.053	−0.117			
	Ore grade	−0.403	−0.093	−0.193	0.960	0.241	0.365
	Sinter alkalinity	55.195	0.281	0.516	97.098	0.538	0.429
	Sinter tumbler index	−11.523	−0.277	−0.477			
	Slag ratio	0.267	0.139	0.264			
	PCIR	−0.603	−0.554	−0.725			
Energy Flows Parameters	M10				−7.424	−0.135	−0.223
Operation Parameters	Blast volume				0.068	0.485	0.577
	Blast temperature	0.125	0.147	0.289	0.285	0.365	0.414
	Oxygen enrichment ratio				7.766	0.305	0.321
	Top temperature				0.254	0.330	0.486

4. Discussion

Furthermore, a quantitative analysis was adopted to evaluate the influence intensity of the factors (independent variables) on the dependent variables (CR or PCIR). Generally, every independent variable was divided into 100 parts between minimum and maximum, which was achieved using historical production data (as shown in Table 3). Then, the variation of the dependent variable, which was caused by the change of independent variable 1%, could be calculated through MLR models. Quantitative influence intensity of significant factors on CR and PCIR are shown in Figures 7 and 8.

Table 3. The range of the influence factors (independent variable).

Classification of Influence Factors	Influence Factors	Min Value ($x_{min,i}$)	Max Value ($x_{max,i}$)	1% Increment ($x_{max,i}$-$x_{min,i}$)/100
Material Flows Parameters	Sinter grade (%)	55.7	57.8	0.021
	Sinter size (mm)	25	30.3	0.053
	Ore grade (%)	41.6	89.6	0.48
	Sinter alkalinity (%)	1.8	2.3	0.005
	Sinter tumbler index (%)	76.9	79.9	0.03
	Slag ratio (kg/t)	286	339	0.53
	PCIR (kg/t)	87.9	171.9	0.84
Energy Flows Parameters	M10 (%)	4.6	6.7	0.021
Operation Parameters	Blast volume (m3/min)	5955	6664	7.09
	Blast temperature (°C)	1086	1250	1.64
	Oxygen enrichment ratio (%)	0.8	4.6	0.038
	Top temperature (°C)	119	288	1.69

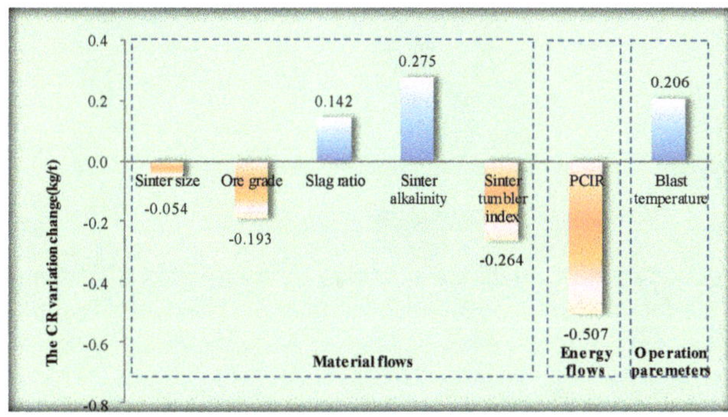

Figure 7. The influence intensity of significant factors on CR.

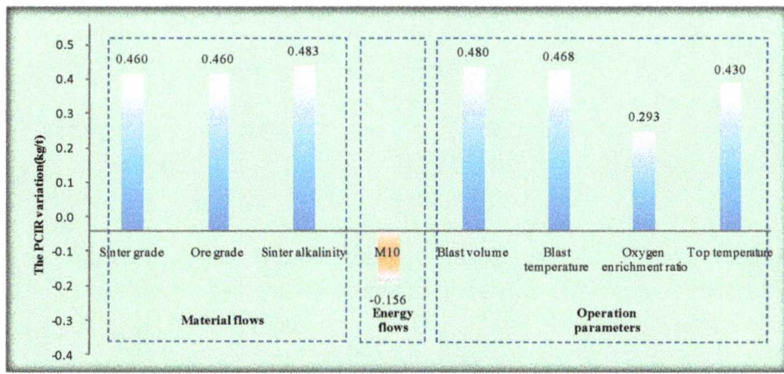

Figure 8. The influence intensity of significant factors on PCIR.

Slag ratio, sinter alkalinity and blast temperature had a positive influence on CR (as shown in Figure 7), the other factors had a negative influence. Meanwhile, the influence intensity of PCIR was the greatest among them on CR. CR would reduce by 0.507 kg/t when PCIR increased by 1% (0.84 kg/t), whereassinter size was the weakest.

As shown in Figure 8, M10 had a negative influence on PCIR among the main factors. The other factors had a positive influence. The influence intensity of sinter alkalinity was the greatest among them on PCIR. PCIR would increase by 0.483 kg/t when sinter alkalinity was promoted to 1% (0.005%), whereas M10 was the weakest.

As shown previously, some factors not only affected the CR, but also affected the PCIR, such as ore grade, sinter alkalinity and blast temperature. For example, CR would decrease and PCIR would increase with ore grade increasing. Furthermore, CR would continue to fall due to the improvement of PCIR (as shown in Figures 7 and 8). Therefore, ore grade had a comprehensive effect on CR. Meanwhile, it was the same for sinter alkalinity and blast temperature. In order to further analyze this problem, the influence factors, which affected PCIR, were converted to CR. Consequently, the comprehensive influence intensity on CR could be achieved (as shown in Figure 9).

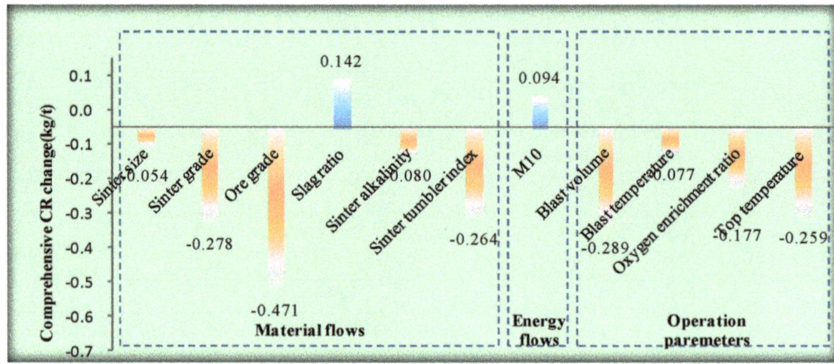

Figure 9. The comprehensive influence intensity of significant factors on CR.

In general, the slag ratio and M10 had a positive influence on CR among influence factors, whereas the rest of the factors had a negative influence (as shown in Figure 9). Meanwhile, the influence intensity of the ore grade was the strongest; the sinter size was the weakest among the material parameters. CR would reduce by 0.471 kg/t, when the ore grade increased by 1% (0.48 kg/t). CR would only reduce by 0.054 kg/t, when the sinter size increased by 1% (0.053 mm). It is worth mentioning that ore grade and sinter alkalinity had a positive influence not only on PCIR, but also on CR. Although CR would increase with the improvement of ore grade and sinter alkalinity, CR would reduce with the improvement of PCIR, which was also caused by increasement of ore grade and sinter alkalinity. Therefore, ore grade and sinter alkalinity had a negative influence on CR finally after conversion calculation.

M10, which belongs to the unique energy flows parameter, had an impact on CR. CR would increase by 0.094kg/t when the M10 increased by 1% (0.021%).

Among the operation parameters, the influence intensity of blast volume was the strongest. CR would reduce by 0.289 kg/t when blast volume increased by 1% (7.09 m^3/min), whereas blast temperature was the weakest. As before, blast temperature had also a positive influence on the CR and PCIR (as show in Figures 7 and 8). Eventually, blast temperature had a negative influence on CR after conversion calculation (as shown in Figure 9). CR would reduce by 0.077 kg/t when blast temperature increased by 1% (1.64°C).

In summary, there were many influence factors, which determined the amount of PCIR and CR. Therefore, it is a very important energy saving direction to determine how to improve these influence factors in BFIMP. Then, three kinds of parameters (material flows, energy flows and operation parameters) will be further discussed.

(1) Material flows improvement:

The improvement of sinter size, ore grade, sinter grade, sinter tumbler index and sinter alkalinity could improve the permeability in the BF body. Moreover, the production status could be more stable. Therefore, the improvement of these factors provides favorable conditions for increasing the PCIR. Meanwhile, the improvement of ore grade and sinter grade is conducive to a reduction of the slag ratio. Moreover, the amount of PCIR and CR will also drop.

(2) Energy flows improvement:

M10 is a physical performance index, which can reflect the coke abrasion strength. The decrease of M10 is beneficial to charge column permeability for gas in the BF body, too. Therefore, if the amount of pulverized coal injection is further increased, the quantity of coke could be dropped.

(3) Operation parameters improvement:

The increase of blast volume, blast temperature and oxygen enrichment ratio can effectively maintain a high temperature in the combustion zone. Moreover, they are also conducive to accelerating the pulverized coal decomposition and combustion. In addition, the improvement of blast volume and blast temperature is also favorable to blow the hearth center of the BF body (especially for pyknic-type BF). Furthermore, the increase of blast temperature, which can improve the heat input of the BF body, can reduce the demand for coke or pulverized coal. Meanwhile, the combustion efficiency of coke and pulverized coal are improved due to the increase of the oxygen enrichment ratio. Then, the amount of flue gas is further reduced. Therefore, the improvements of blast volume, blast temperature and oxygen enrichment ratio are very effective measures of energy saving in BFIMP. Additionally, top temperature, which directly can reflect gas distribution and blast furnace status, is determined by heat exchange between blast furnace gas and furnace charge. CR will drop with top temperature increasing. Therefore, top temperature, which is controlled in higher areas within allowable range, is also an effective energy saving measure.

The analysis indicates the following findings.

(1) The effective energy saving measures could be achieved through the all-factors analysis approach.
(2) The case study shows that there were 26 influence factors on energy consumption in BFIMP. Nevertheless, the seven influence factors were highly correlated with CR, and the eight influence factors had highly correlation on PCIR through the PCA analysis.
(3) The PCIR improvement was the most effective measure for reducing CR, and the increase of sinter alkalinity was the most favorable for PCIR in the case BFIMP.

5. Conclusions

An all-factors analysis approach on energy consumption for BFIMP in the iron and steel industry was proposed in this paper. Then, the all-factors analysis approach was successfully applied to a case BFIMP. The key influence factors on energy consumption were identified in this case BFIMP. Lastly, some suggestions on energy conservation were put forward. Generally, the major contributions of this paper are described as follows:

(1) An all-factors analysis approach on energy consumption was proposed in BFIMP in this paper. This method mainly included four steps: Data collection, all-factors analysis based on material flows and energy flows, influence intensity analysis on energy consumption in BFIMP and suggestion and summary. The availability of this method was validated by a case BFIMP.
(2) The case study shows that the improvement of PCIR was conducive to a reduction of CR very much. The increase of sinter alkalinity was the most effective measure for PCIR.
(3) The proposed all-factors analysis approach on energy consumption, which provided an effective measure of energy saving, could widely be used in various BFIMPs, too.

Author Contributions: Conceptualization: D.C. and B.L.; methodology, B.L.; validation, D.C. and S.W.; formal analysis, D.C. and S.W.; investigation, D.C., B.L. and S.W.; resources, D.C. and B.L.; data curation, B.L. and Kai Tang; writing—original draft preparation, D.C. and B.L.; writing—review and editing, Biao Lu; visualization, B.L., K.T. and S.W.; supervision, B.L. and S.W.; funding acquisition, D.C. and B.L.

Funding: This research was funded by the National Nature Science Foundation of China (Grant NO. 51804002) and the Scientific Researching Fund Projects for Young Teacher of Anhui University of Technology (Grant NO. QZ201614).

Acknowledgments: The authors would like to thank XuShi for help with administrative and technical support.

Conflicts of Interest: The authors declare no conflict of interest.

References

1. Mauricio, R.; Mikko, H.; Henrik, S. Principal Component Analysis of Blast Furnace Drainage Patterns. *Processes* **2019**, *7*, 519. [CrossRef]
2. Lu, B.; Tang, K.; Chen, D.; Han, Y.; Wang, S.; He, X.; Chen, G. A Novel Approach for Lean Energy Operation Based on Energy Apportionment Model in Reheating Furnace. *Energy* **2019**, *182*, 1239–1249. [CrossRef]
3. Matsuda, K.; Tanaka, S.; Endou, M.; Iiyoshi, T. Energy saving study on a large steel plant by total site based pinch technology. *Appl. Therm. Eng.* **2012**, *43*, 14–19. [CrossRef]
4. Lu, B.; Chen, G.; Chen, D.; Yu, W. An energy intensity optimization model for production system in iron and steel industry. *Appl. Therm. Eng.* **2016**, *100*, 285–295. [CrossRef]
5. Chen, D.; Lu, B.; Chen, G.; Yu, W. Influence of the production fluctuation on the process energy intensity in iron and steel industry. *Adv. Prod. Eng. Manag.* **2017**, *12*, 75–87. [CrossRef]
6. Chen, W.; Yin, X.; Ma, D. A bottom-up analysis of China's iron and steel industrial energy consumption and CO2 emissions. *Appl. Energy* **2014**, *136*, 1174–1183. [CrossRef]
7. Li, Y.; Zhu, L. Cost of energy saving and CO2 emissions reduction in China's iron and steel sector. *Appl. Energy* **2014**, *130*, 603–616. [CrossRef]
8. Cai, W.; Liu, C.; Lai, K.; Li, L.; Cunha, J.; Hu, L. Energy performance certification in mechanical manufacturing industry: A review and analysis. *Energy Convers. Manag.* **2019**, *186*, 415–432. [CrossRef]
9. Emi, T. Optimizing Steelmaking System for Quality Steel Mass Production for Sustainable Future of Steel Industry. *Steel Res. Int.* **2014**, *85*, 1274–1282. [CrossRef]
10. Liu, X.; Chen, L.; Feng, H.; Qin, X.; Sun, F. Constructal design of a blast furnace iron-making process based on multi-objective optimization. *Energy* **2016**, *109*, 137–151. [CrossRef]
11. Yilmaz, C.; Wendelstorf, J.; Turek, T. Modeling and simulation of hydrogen injection into a blast furnace to reduce carbon dioxide emissions. *J. Clean. Prod.* **2017**, *154*, 488–501. [CrossRef]
12. Hou, Q.; Dianyu, E.; Kuang, S.; Li, Z.; Yu, A.B. DEM-based virtual experimental blast furnace: A quasi-steady state model. *Powder Technol.* **2017**, *314*, 557–566. [CrossRef]
13. Jin, P.; Jiang, Z.; Bao, C.; Lu, Y.; Zhang, J.; Zhang, X. Mathematical Modeling of the Energy Consumption and Carbon Emission for the Oxygen Blast Furnace with Top Gas Recycling. *Steel Res. Int.* **2016**, *87*, 320–329. [CrossRef]
14. Zhou, P.; Yuan, M.; Wang, H.; Chai, T. Data-Driven Dynamic Modeling for Prediction of Molten Iron Silicon Content Using ELM with Self-Feedback. *Math. Probl. Eng.* **2015**, *2015*, 326160. [CrossRef]
15. Yin, R. The essence function and future development model of steel manufacturing process. *Sci. China Technol. Sci.* **2008**, *38*, 1365–1377.
16. Zetterholm, J.; Ji, X.; Sundelin, B.; Martin, P.M.; Wang, C. Model development of a blast furnace stove. *Energy Procedia* **2015**, *75*, 1758–1765. [CrossRef]
17. Liu, X.; Chen, L.; Qin, X.; Sun, F. Exergy loss minimization for a blast furnace with comparative analyses for energy flows and exergy flows. *Energy* **2015**, *93*, 10–19. [CrossRef]
18. Zhou, B.; Ye, H.; Zhang, H.; Li, M. Process monitoring of iron-making process in a blast furnace with PCA-based methods. *Control Eng. Pract.* **2016**, *47*, 1–14. [CrossRef]
19. Jin, P.; Jiang, Z.; Bao, C.; Hao, S.; Zhang, X. The energy consumption and carbon emission of the integrated steel mill with oxygen blast furnace. *Resour. Conserv. Recy.* **2017**, *117*, 58–65. [CrossRef]
20. Kuang, S.B.; Li, Z.Y.; Yan, D.L.; Qi, Y.H.; Yu, A.B. Numerical study of hot charge operation in iron making blast furnace. *Miner. Eng.* **2014**, *63*, 45–56. [CrossRef]
21. Liao, J.; Yu, A.B.; Shen, Y. Modelling the injection of upgraded brown coals in an ironmaking blast furnace. *Powder Technol.* **2017**, *314*, 550–556. [CrossRef]
22. Dong, Z.; Wang, J.; Zuo, H.; She, X.; Xue, Q. Analysis of gas–solid flow and shaft-injected gas distribution in an oxygen blast furnace using a discrete element method and computational fluid dynamics coupled model. *Particuology* **2017**, *32*, 63–72. [CrossRef]
23. Miao, Z.; Zhou, Z.; Yu, A.B.; Shen, Y. CFD-DEM simulation of raceway formation in an ironmaking blast furnace. *Powder Technol.* **2017**, *314*, 542–549. [CrossRef]
24. Yeh, C.-P.; Du, S.-W.; Tsai, C.-H.; Yang, R.-J. Numerical analysis of flow and combustion behavior in tuyere and raceway of blast furnace fueled with pulverized coal and recycled top gas. *Energy* **2012**, *42*, 233–240. [CrossRef]

25. Shen, Y.S.; Yu, A.B. Modelling of injecting a ternary coal blend into a model ironmaking blast furnace. *Miner. Eng.* **2016**, *90*, 89–95. [CrossRef]
26. Shen, Y.; Yu, A.; Austin, P.; Zulli, P. Modelling in-furnace phenomena of pulverized coal injection in ironmaking blast furnace: Effect of coke bed porosities. *Miner. Eng.* **2012**, *33*, 54–65. [CrossRef]
27. Zhao, H.; Bai, Y.; Cheng, S. Effect of Coke Reaction Index on Reduction and Permeability of Ore Layer in Blast Furnace Lumpy Zone Under Non-Isothermal Condition. *J. Iron Steel Res. Int.* **2013**, *20*, 6–10. [CrossRef]
28. De Castro, J.A.; de Mattos Araújo, G.; de Oliveira da Mota, I.; Sasaki, Y.; Yagi, J. Analysis of the combined injection of pulverized coal and charcoal into large blast furnaces. *J. Mater. Res. Technol.* **2013**, *2*, 308–314. [CrossRef]
29. Gupta, S.; Ye, Z.; Kim, B.; Kerkkonen, O.; Kanniala, R.; Sahajwalla, V. Mineralogy and reactivity of cokes in a working blast furnace. *Fuel Process. Technol.* **2014**, *117*, 30–37. [CrossRef]
30. Jiang, S.; Shen, X.; Zheng, Z. Gaussian Process-Based Hybrid Model for Predicting Oxygen Consumption in the Converter Steelmaking Process. *Processes* **2019**, *7*, 352. [CrossRef]
31. Bahgat, M.; Halim, K.S.A.; Heba, A.E.-K.; Mahmoud, I.N. Blast Furnace Operating Conditions Manipulation for Reducing Coke Consumption and CO2 Emission. *Steel Res. Int.* **2012**, *83*, 686–694. [CrossRef]
32. SJung, S.; Chang, W. Clustering stocks using partial correlation coefficients. *Phys. A Stat. Mech. Appl.* **2016**, *462*, 410–420.
33. Shen, Y.S.; Yu, A.B.; Austin, P.R.; Zulli, P. CFD study of in-furnace phenomena of pulverised coal injection in blast furnace: Effects of operating conditions. *Powder Technol.* **2012**, *223*, 27–38. [CrossRef]
34. Lomas, H.; Roest, R.; Gupta, S.; Pearson, R.A.; Fetscher, R.; Jenkins, D.R.; Pearce, R.; Kanniala, R.; Merrick, M.R. Petrographic analysis and characterisation of a blast furnace coke and its wear mechanisms. *Fuel* **2017**, *200*, 89–99. [CrossRef]
35. Gasparinia, V.M.; de Castro, L.F.A.; Quintas, A.C.B.; de Souza Moreira, V.E.; Viana, A.O.; Andrade, D.H.B. Thermo-chemical model for blast furnace process control with the prediction of carbon consumption. *J. Mater. Res. Technol.* **2017**, *6*, 220–225. [CrossRef]
36. Zou, C.; She, Y.; Shi, R. Particle size-dependent properties of a char produced using a moving-bed pyrolyzer for fueling pulverized coal injection and sintering operations. *Fuel Process. Technol.* **2019**, *190*, 1–12. [CrossRef]

Article

The Impact of Authorized Remanufacturing on Sustainable Remanufacturing

Xiqiang Xia [1] and Cuixia Zhang [2,*]

1 Business School, Zhengzhou University, Zhengzhou 450001, China; xqxia@zzu.edu.cn
2 School of Mechanical and Electronic Engineering, Suzhou University, Suzhou 234000, China
* Correspondence: cuixiazhang@126.com

Received: 20 August 2019; Accepted: 23 September 2019; Published: 27 September 2019

Abstract: Remanufacturing could effectively solve resource shortage and environment crisis and achieve sustainable development of the economy. The original equipment manufacturer (OEM) could not only focus on its core business (i.e., producing new products), but also get profit from remanufacturing through the intellectual property rights. Based on the authorized remanufacturing, the game model between a manufacturer and a remanufacturer was constructed. Based on the game model, the impact of authorized remanufacturing on sustainable remanufacturing is analysed, and the coordination mechanism between manufacturer and remanufacturer is given. The main results are as follows: the OEM could increase its profit and change its unfavourable market competition status by authorizing remanufacturing; a franchise contract could make the sustainability supply chain optimized; when the ratio of the environment effect is greater than a certain threshold, centralized decision-making could not only increase the supply chain revenue, but also reduce the impact on the environment.

Keywords: remanufacturing; sustainability supply chain; authorized remanufacturing; game

1. Introduction

In the 21st century, with rapid economic development, the world is facing an increasingly serious shortage of resources as well as environmental problems, prompting all countries to alter the existing mode of economic development and look for a sustainable mode of economic development to achieve the recycling of resources and to minimize the adverse effect of products on the environment [1]. For the manufacturing industry, an effective way to achieve recycling economic development is remanufacturing; for instance, compared with new engines, remanufactured engines can not only reduce their environmental impact by 80% but also save on costs, energy, and raw materials by 50%, 60%, and 70%, respectively [2].

However, because of the lack of remanufacturing infrastructure and expertise or the low marginal profitability of remanufactured products, original manufacturers are reluctant to engage in remanufacturing. As third-party remanufacturers engage in remanufacturing, the emergence of remanufactured products will result in reduced sales of new products in the market and reduced profits for the original manufacturer as well, which will prompt the original manufacturer to use intellectual property protection to limit the development of the remanufacturing industry. Meanwhile, none of the countries in the world will allow remanufacturers to infringe on intellectual property simply because remanufacturing is able to generate good economic and environmental benefits [3]. Authorized remanufacturing can ease the conflict between the original manufacturer and the remanufacturer so that the two can share the remanufacturing profit, manifesting plenty of successful cases in the remanufacturing industry. For example, the NCR (National Cash Register Company) Corporation of the United States paid $960,000 to buy a remanufacture license from the Lemelson Medical and

Educational Research Fund to legally engage in remanufacturing operations; IT companies such as Cisco allow independent third-party companies to remanufacture their used IT(Internet Technology) equipment and charge a license fee to the consumers who switch to buying the remanufactured products, which is essentially an authorized remanufacturing model [4]. Therefore, studying the impact of authorized remanufacturing on the manufacturing/remanufacturing supply chain is of great practical significance. Meanwhile, the comparison of the effects of decentralized decision-making and centralized decision-making under authorized remanufacturing on the remanufacturing supply chain can provide a scientific basis for the original manufacturers and remanufacturers making remanufacturing decision.

Currently, authorized remanufacturing has been seldom investigated. Zou et al. [5] compared the effects of authorized remanufacturing and outsourced remanufacturing on the manufacturing/remanufacturing supply chain and revealed under what conditions the original manufacturer should choose authorized remanufacturing. Zhu et al. [6] constructed a manufacturing/remanufacturing game model based on authorized remanufacturing to investigate the government subsidies strategy, and the results indicated that through remanufacturing, the original manufacturer could transfer not only remanufacturing proceeds but also government subsidies. Based on authorized remanufacturing, Xiong et al. [7,8] studied the original manufacturer's production decision-making and found that only when the remanufacturing saves more costs is the original manufacturer willing to engage in remanufacturing by authorizing a third party to remanufacture; further, by comparing the impacts of two methods of authorized remanufacturing (authorizing a retailer to remanufacture vs. authorizing a third-party to remanufacture) on the profits of manufacturing/remanufacturing supply chain members, they found that authorizing a third-party remanufacturer is more beneficial to the original manufacturer. Based on a network model, Li [9] analyzed the pattern of the effect of authorized remanufacturing on the closed-loop supply chain of remanufacturing. Wang [10] studied the optimal decision of the closed-loop supply chain under authorized remanufacturing based on the optimization model.

Contract analysis has received considerable attention in supply chain literature. Lind et al. [11] through several case studies of European firms, discuss various inter-organizational relationships that remanufacturers maintain, among their key finding is a new relationship called "raman-contracts" where the OEM (Original Equipment Manufacturer) collaborates with the remanufacturing activities. Matsumoto et al. [12] studies the remanufacturing practices of firms in Japan, through case studies of eleven remanufacturers, the authors attempt to determine the motives behind remanufacturing and identify several factors such as economic benefits, environmental benefits, and quality control among others. Kaya [13] discusses a linear contract with transfer payments to coordinate a remanufacturing based channel. Govindan et al. [14] discuss the case of coordination of a reverse supply chain using revenue sharing contracts. Yalabik et al. [15] build on the above to develop an economic model of remanufacturing with the assumption of lease contracting.

Although the existing studies have made great achievements, there are still some issues to be addressed. For examples, Zhu et al. [6] studied only the government subsidies under authorized remanufacturing but neglected the coordination strategy of the supply chain under authorized remanufacturing; Xiong et al. [7,8] addressed the choice of authorized remanufacturing strategies and supply chain coordination but lacked the analyses on the effects of decentralized decision-making and centralized decision-making on consumer surplus, social surplus, and the environment; although the effect of authorized remanufacturing on the remanufacturing supply chain has been examined, the coordination mechanism has not been addressed [5,9,10]. The remanufacturing contract of supply chain is studied, such as [11–15]. Therefore, in this study, a game model of manufacturing/remanufacturing was constructed based on authorized remanufacturing to compare the effects of decentralized decision-making and centralized decision-making on the competition and coordination mechanisms of the manufacturing/remanufacturing supply chain, to provide a scientific basis for the decision-making of original manufacturers and remanufacturers.

This paper contains five sections. In Section 1, the research background and literature review are given; in Section 2, the model is described; in Section 3, the model solution is presented; in Section 4, the model analysis is performed; and in Section 5, conclusions are drawn.

2. Model

2.1. Description of the Problem

In reference to the literature [5,6], a single cycle game model comprising an original manufacturer and a remanufacturer is constructed to study the mechanism of the impact of authorized remanufacturing on manufacturing/remanufacturing supply chain competition, based on which the effects of centralized decision-making and decentralized decision-making on manufacturing/remanufacturing supply chain are compared, and the supply chain coordination mechanism is further investigated. In the case of decentralized decision-making, the original manufacturer is responsible only for producing new products, and at the same time, the original manufacturer charges the remanufacturer the patent fees for the remanufactured products through the protection of intellectual property (called the authorized remanufacturing expense in this study), the decision-making variables of which are the retail price per new product unit and the licensing fee of the remanufactured product. The remanufacturer is responsible for the production of the remanufactured product; however, it needs to pay the original manufacturer a certain licensing fee per unit of remanufactured product for the remanufacturing rights, whose decision-making variable is the recycling rate of the used product.

In the case of decentralized decision-making, the sequence of the decision-making model is as follows. First, the original manufacturer makes a decision on the licensing fee on the basis of the remanufactured product and the retail price per new product unit, and then the remanufacturer makes a decision on the recycling rate of the used product. According to Stackelberg's inverse solution, the remanufacturer determines the recycling rate of the used product, and then the original manufacturer determines the retail price per new product unit and the licensing fee on the basis of the remanufactured product.

2.2. Model Function

2.2.1. Model Demand Function

In reference to the literature [16,17], let the consumer's willingness to pay (WTP) for the product be θ, and θ obeys the uniform distribution of [0, 1], i.e., $f(\theta) \sim [0,1]$. Let the retail price per remanufactured product unit relative to the consumer's lowest accepted price (referred to as discount price in this study) of the retail price per new product unit be δ; then, the consumer surplus of the consumer purchasing the new product is $U_n = \theta - p_n$, and that of the consumer purchasing the remanufactured product is $U_r = \delta\theta - p_r$. Only when the consumer surplus of the consumer purchasing a new product unit is greater than that of the consumer purchasing a remanufactured product unit, i.e., $U_n > U_r$, is the consumer willing to purchase the new product, and the WTP range of the new product is $\Theta_n = \{\theta : U_n > \max\{U_r, 0\}\}$. Similarly, the WTP range of the remanufactured product is $\Theta_r = \{\theta : U_r > \max\{U_n, 0\}\}$. Further, the market demand for the new products and remanufactured products are, respectively, as follows:

$$q_n^i = \int\limits_{\theta\in\Theta_n} f(\theta)\mathrm{d}\theta = \frac{1-\delta-p_n^i+p_r^i}{1-\delta},\ q_r^i = \int\limits_{\theta\in\Theta_r} f(\theta)\mathrm{d}\theta = \frac{\delta p_n^i - p_r^i}{\delta(1-\delta)}$$

Ultimately, we obtain:

$$p_n^i = 1 - q_n^i - \delta q_r^i,\ p_r^i = \delta(1 - q_n^i - q_r^i),\ i \in \{D,\ C,\ F\}$$

2.2.2. Model Recycling Function

According to the literature [18], the costs of recycling the used product increase with the amount of the used product in the manner of a convex function. Without loss of generality, the cost function of previous related studies, i.e., $k\left(\tau^i q_n^i\right)^2$, is used here. Wherein, τ is the recycling rate, k is the ratio parameter, and $i \in \{D,\ C,\ F\}$.

3. Model Solution

In the case of decentralized decision-making, the profit functions of the original manufacturer and the remanufacturer are, respectively, as follows:

$$\pi_n^D = (p_n^D - c)q_n^D + zq_r^D = (1 - q_n^D - \delta q_r^D - c)q_n^D + zq_r^D \tag{1}$$

$$\begin{aligned} \pi_r^D &= (p_r^D - c + s - z)q_r^D - k(\tau^D q_n^D)^2 \\ &= [\delta(1 - q_n^D - q_r^D) - c + s - z]q_r^D - k(\tau^D q_n^D)^2 \end{aligned} \tag{2}$$

in which, $(p_n^D - c)q_n^D$ and zq_r^D respectively, represent the profit from the sale of the new product and the profit through authorizing remanufacturing, and $(p_r^D - c + s - z)q_r^D$ represents the remanufacturer's profit from the sale of the remanufactured product under the authorization of remanufacturing, while $k(\tau^D q_n^D)^2$ represents the costs of recycling the used product.

In the case of centralized decision-making, the profit function of the manufacturing/remanufacturing supply chain is as follows:

$$\pi^C = (p_n^C - c)q_n^C + (p_r^C - c + s)q_r^C - k(\tau^C q_n^C)^2 \tag{3}$$

Conclusion 1. *(a) with respect to the recycling rate of the used product (τ^D), Formula (2) is a concave function, and by integrating the recycling rate obtained from Formula (2) into Formula (1), Formula (1) becomes a concave function with respect to q_n^D, z; (b) with respect to q_n^C, q_r^C, Formula (3) is a concave function.*

Proof. (a) by integrating $q_r^D = \tau q_n^D$ into Formula (2), then we obtain:

$$\pi_r^D = [\delta(1 - q_n^D - \tau^D q_n^D) - c + s - z]\tau^D q_n^D - k(\tau^D q_n^D)^2$$

By taking the first and second order partial derivatives on π_r^D with respect to τ^D, we obtain:

$$\frac{\partial \pi_r^D}{\partial \tau^D} = [\delta(1 - q_n^D - 2\tau^D q_n^D) - c + s - z]q_n^D - 2k\tau^D q_n^{D2} \tag{4}$$

$$\frac{\partial^2 \pi_r^D}{\partial \tau^{D2}} = -2(\delta + k)q_n^{D2} \tag{5}$$

From $\frac{\partial^2 \pi_r^D}{\partial \tau^{D2}} = -2(\delta + k)q_n^{D2} < 0$, it indicates that Formula (2) is a concave function with respect to the recycling rate of the used product (τ^D).

By integrating $q_r^D = \tau^D q_n^D$ and $\tau^{D*} = \frac{\delta - c + s - z - \delta q_n^D}{2(\delta + k)q_n^D}$ into Formula (2), we obtain:

$$\pi_n^D = (1 - q_n^D - \delta\frac{\delta - c + s - z - \delta q_n^D}{2(\delta + k)} - c)q_n^D + z\frac{\delta - c + s - z - \delta q_n^D}{2(\delta + k)} \tag{6}$$

By taking the first and second order partial derivatives on Formula (6) with respect to q_n^D, w^D, we obtain:

$$\frac{\partial \pi_n^D}{\partial q_n^D} = 1 - \frac{(2 - \delta)\delta + 2k}{\delta + k}q_n^D - \delta\frac{\delta - c + s}{2(\delta + k)}$$

$$\frac{\partial \pi_n^D}{\partial z} = \frac{\delta - c + s}{2(\delta + k)} - \frac{z}{\delta + k}$$

$$\frac{\partial^2 \pi_n^D}{\partial q_n^{D2}} = -\frac{(2-\delta)\delta + 2k}{\delta + k}, \ \frac{\partial^2 \pi_n^D}{\partial q_n^D \partial w^D} = 0, \ \frac{\partial \pi_n^D}{\partial w^D \partial q_n^D} = 0, \ \frac{\partial^2 \pi_n^D}{\partial w^{D2}} = -\frac{1}{\delta + k},$$

The second-order Hessian matrix of Formula (6) with respect to q_n^D, w^D is:

$$H = \begin{bmatrix} -\frac{(2-\delta)\delta+2k}{\delta+k} & 0 \\ 0 & -\frac{1}{\delta+k} \end{bmatrix}$$

The diagonal elements of the Hessian matrix are negative, and $|H| = \begin{vmatrix} -\frac{(2-\delta)\delta+2k}{\delta+k} & 0 \\ 0 & -\frac{1}{\delta+k} \end{vmatrix} = \frac{(2-\delta)\delta+2k}{(\delta+k)^2} > 0$, indicating that Formula (6) is a concave function with respect to q_n^D, z, i.e., Formula (1) is a concave function with respect to q_n^D, z.

Similarly, the proof of (b) can be established. From Conclusion 1, Conclusion 2 can be drawn. □

Conclusion 2. *Under both decision-making models, the Nash equilibrium solutions of the unit retail prices, the sales volumes, the sales profits of the new product and the remanufactured product, the unit outsourcing cost of the remanufactured product, and the recycling rate of the used product are shown as Table 1.*

Table 1. Optimal equilibrium solutions under different decision models.

	D	C
p_n^{i*}	$\frac{1+c}{2}$	$\frac{1+c}{2}$
p_r^{i*}	$\delta[\frac{1}{2} + \frac{(1+2k-\delta)c-(1-\delta)s}{2(\delta+2k-\delta^2)}]$	$\delta[\frac{1}{2} + \frac{(1+k-\delta)c-(1-\delta)s}{2(\delta+k-\delta^2)}]$
q_n^{i*}	$\frac{1}{2} - \frac{2kc+\delta(c+s)}{2(2\delta+2k-\delta^2)}$	$\frac{1}{2} - \frac{kc+\delta s}{2(\delta+k-\delta^2)}$
q_r^{i*}	$\frac{\delta c-c+s}{2(2\delta+2k-\delta^2)}$	$\frac{\delta c-c+s}{2(\delta+k-\delta^2)}$
π_n^{i*}	$\frac{(1-c)^2}{4} + \frac{(\delta c-c+s)^2}{4(2\delta+2k-\delta^2)}$	–
π_r^{i*}	$\frac{(\delta+k)(\delta c-c+s)^2}{4(2\delta+2k-\delta^2)^2}$	–
π_C^*	–	$\frac{(1-c)^2}{4} + \frac{(\delta c-c+s)^2}{4(\delta+k-\delta^2)}$
z^*	$\frac{\delta-c+s}{2}$	–
τ^{i*}	$\frac{\delta c+s-c}{2\delta+2k-\delta^2-2kc-\delta(c+s)}$	$\frac{\delta c+s-c}{\delta+k-\delta^2-ck-\delta s}$

4. Model Analysis

Based on Conclusion 2, Conclusions 3 and 4 are drawn.

Conclusion 3. *(I) $q_n^* > q_n^{D*}$, $q_n^* > q_n^{C*}$; (II) $\pi_n^{D*} > \pi_n^*$.*

Proof. When there is no remanufacturing in the market, the original manufacturer's decision function is

$$\pi_n = (p_n - c)q_n = (1 - q_n - c)q_n \tag{7}$$

Similar to those of Conclusion 2, we can obtain the optimal Nash equilibrium solutions for Formula (7) as follows:

$$p_n^* = \frac{1+c}{2}, \ q_n^* = \frac{1-c}{2}, \ \pi_n^* = \frac{(1-c)^2}{4}$$

From Table 1, we have

$$q_n^* - q_n^{D*} = \frac{\delta(\delta c - c + s)}{2(2\delta + 2k - \delta^2)} > 0,$$

$$q_n^* - q_n^{C*} = \frac{\delta(\delta c + s - c)}{2(\delta + k - \delta^2)} > 0$$

Therefore, $q_n^* > q_n^{D*}$, $q_n^* > q_n^{C*}$. □

Under authorized remanufacturing, the original manufacturer's profit is $\pi_n^{D*} = \frac{(1-c)^2}{4} + \frac{(\delta c - c + s)^2}{4(2\delta + 2k - \delta^2)}$, thus $\pi_n^{D*} - \pi_n^* = \frac{(\delta c - c + s)^2}{4(2\delta + 2k - \delta^2)} > 0$, i.e., $\pi_n^{D*} > \pi_n^*$.

Conclusion 3 shows that when the original manufacturer does not authorize remanufacturing, the remanufactured product would reduce not only the sales of the new product but also the profitability of the original manufacturer, i.e., the emergence of remanufacturing would be detrimental to the original manufacturer. However, when the original manufacturer adopts the authorized remanufacturing, although the sales volume of the new product is reduced and the sales revenue of the new product is also reduced, through authorizing remanufacturing, the original manufacturer can benefit by transferring remanufacturing and, due to the transferred remanufacturing benefit, is greater than the revenue reduction caused by the reduced sales of the new product; ultimately, the original manufacturer's revenue increases, i.e., by adopting authorized remanufacturing, it is possible to change the original manufacturer's unfavourable competitive status.

Conclusion 4. *The cost savings made on the remanufactured product unit affects the Nash Equilibrium solutions:*

$$\text{(I)}\ \frac{\partial p_n^{D*}}{\partial s} = \frac{\partial p_n^{C*}}{\partial s} = 0,\ \frac{\partial p_r^{D*}}{\partial s} < 0,\ \frac{\partial p_r^{C*}}{\partial s} < 0;$$

$$\text{(II)}\ \frac{\partial q_n^{D*}}{\partial s} = \frac{\partial q_n^{C*}}{\partial s} < 0,\ \frac{\partial q_r^{D*}}{\partial s} > 0,\ \frac{\partial q_r^{C*}}{\partial s} > 0;$$

$$\text{(III)}\ \frac{\partial \tau^{D*}}{\partial s} > 0,\ \frac{\partial \tau^{C*}}{\partial s} > 0;$$

$$\text{(IV)}\ \frac{\partial z^*}{\partial s} < 0;$$

$$\text{(V)}\ \frac{\partial \pi_n^{D*}}{\partial s} > 0,\ \frac{\partial \pi_r^{D*}}{\partial s} > 0,\ \frac{\partial \pi^{C*}}{\partial s} > 0.$$

Proof. Table 1 shows that

$$\frac{\partial p_n^{D*}}{\partial s} = \frac{\partial p_n^{C*}}{\partial s} = 0,\ \frac{\partial p_r^{D*}}{\partial s} = -\frac{\delta(1-\delta)s}{2(\delta + 2k - \delta^2)} < 0$$

i.e.,

$$\frac{\partial p_r^{C*}}{\partial s} = -\frac{\delta(1-\delta)s}{2(\delta + k - \delta^2)} < 0$$

(a) is valid; similarly, the validities of other conclusions can be established. Therefore, Conclusion 4 is proven.

According to the literature [19,20], when the original manufacturer does not adopt authorized remanufacturing, the original manufacturer's profit is negatively correlated with the cost saving made on the remanufactured product, i.e., the greater the cost savings made on the unit remanufactured product, the smaller is the profit of the original manufacturer. However, when the original manufacturer adopts authorized remanufacturing, although the sales volume of the new product is negatively correlated with the cost savings made on the remanufactured product unit, its revenue is positively correlated with the cost savings made on the remanufactured product unit, i.e., the greater the cost savings made on the remanufactured product unit, the greater is the revenue the original manufacturer generates. Therefore, the original manufacturer should strive to reduce the cost savings made on the remanufactured product; for example, in the initial stage of product design, dismantlability and remanufacture ability should be taken into account. □

Conclusion 5. $\pi_C^* > \pi_n^{D*} + \pi_r^{D*}$, *i.e., in the case of decentralized decision-making, the marginal loss will affect the supply chain, making its return lower than that in the case of centralized decision-making.*

The proof of Conclusion 5 is similar to that of Conclusion 3 and hereby omitted.

Conclusion 5 shows that in the case of decentralized decision-making, a "marginal effect" will arise, lowering the revenue of the remanufacturing supply chain. Therefore, it is necessary to investigate the coordination contract of the remanufacturing supply chain in the case of decentralized decision-making so that the profit of the remanufacturing supply chain in the case of decentralized decision-making is not less than that in the case of centralized decision-making. Next, based on the franchise contract, the coordination contract is presented.

4.1. Franchise Contract

A franchise contract mainly refers to authorization by the original manufacturer to remanufacturers to remanufacture by charging a licensing fee (z) on the basis of the remanufactured product unit and to coordinate the remanufacturing supply chain, and z is generally smaller than z^D, so the production cost of the remanufactured product is reduced, and thus, the sales of the remanufactured product are promoted. However, in this case, the remanufacturer must first pay a fee (F) to the original manufacturer. Assuming the terms that the remanufacturer provides for the original manufacturer is (z, F), i.e., remanufacturing can generate a lower license fee per remanufactured product unit, but the remanufacturer needs to pay a fixed fee (F) as a profit share, then the original manufacturer considers whether to accept the contract; if it accepts the contract, then the manufacturer makes the optimal decision on the contract. Based on the above ideas, we can have the following decision model.

The profit functions of the original manufacturer and the remanufacturer are, respectively, as follows:

$$\pi_n^f = (1 - q_n^f - \delta q_r^f - c)q_n^f + z^f q_r^f + F \tag{8}$$

$$\begin{aligned} \pi_r^f = \; & (\delta - \delta q_n^f - \delta q_r^f - c - z^f + s)q_r^f \\ & -k(\tau^f q_n^f)^2 - F \end{aligned} \tag{9}$$

From Formulas (8) and (9), we obtain the decision models of the original manufacturer and the remanufacturer, respectively, as follows:

$$\begin{aligned} \max \pi_r^f = \; & (\delta - \delta q_n^f - \delta q_r^f - c - z^f + s)q_r^f \\ & -k(\tau^f q_n^f)^2 - F \end{aligned} \tag{10}$$

s.t.

$$\begin{aligned} q_n^f, z^f \in \operatorname{argmax} \pi_n^f = (1 - q_n^f - \delta q_r^f - c)q_n^f + z^f q_r^f + F \\ \pi_n^{f*} \geq \pi_n^{D*} \end{aligned} \tag{11}$$

By solving the above models, we have:

$$\tau^{f*} = \frac{\delta c + s - c}{\delta + k - \delta^2 - ck - \delta s},\ p_n^{f*} = \frac{1+c}{2},\ p_r^{f*} = \delta[\frac{1}{2} + \frac{(1+k-\delta)c - (1-\delta)s}{2(\delta + k - \delta^2)}]$$

$$q_n^{f*} = \frac{1}{2} - \frac{kc + \delta s}{2(\delta + k - \delta^2)},\ q_r^{f*} = \frac{\delta c - c + s}{2(\delta + k - \delta^2)},\ z^{f*} = \frac{\delta^2(1 - \delta - s) + \delta k(1 - c)}{2(\delta + k - \delta^2)}$$

$$F^* = \frac{\delta(1-\delta)(1-c)^2}{4(\delta + k - \delta^2)} + \frac{(\delta c - c + s)^2}{4(2\delta + 2k - \delta^2)} - \delta\frac{(1-c)(1-\delta-s)}{4(\delta + k - \delta^2)} - \delta\frac{[1 - \delta - s + k(1-c)](\delta c - c + s)}{4(\delta + k - \delta^2)^2}$$

$$\pi_n^{f*} = \pi_n^{D*} = \frac{(1-c)^2}{4} + \frac{(\delta c - c + s)^2}{4(2\delta + 2k - \delta^2)},$$

$$\pi_r^{f*} = \pi^{C*} - \pi_n^{f*} = \pi^{C*} - \pi_n^{D*}$$

The coordination contract of the supply chain is thus obtained.

Conclusion 6. *The remanufacturer's adoption of the franchise contract can cause the supply chain to achieve coordination.*

Proof. Based on the results obtained by the above solutions, we have:

$$\pi_r^{f*} + \pi_n^{f*} = \pi_r^{f*} + \pi_n^{D*} = \pi^{C*} \tag{12}$$

According to Formula (12), it shows the supply chain has achieved coordination. □

4.2. The Effects of Two Decision Models on Consumer Surplus and Social Surplus

The consumer surpluses in the cases of decentralized decision-making and centralized decision-making are:

$$S^D = \frac{(q_n^{D*} + \delta q_r^{D*})^2 + \delta(1-\delta)q_r^{D*}}{2} \tag{13}$$

$$S^C = \frac{(q_n^{C*} + \delta q_r^{C*})^2 + \delta(1-\delta)q_r^{C*}}{2} \tag{14}$$

Social surpluses in the cases of decentralized decision-making and centralized decision-making are:

$$S^{CD} = S^D + \pi_n^{D*} + \pi_r^{D*} \tag{15}$$

$$S^{CC} = S^C + \pi_C^* \tag{16}$$

According to Table 1 and Formulas (13)–(16), Conclusion 5 can be obtained.

Conclusion 7. *The effects of the two decision-making models on consumer surplus and social surplus are* $S^C > S^D$, $S^{CC} > S^{DC}$.

Proof. From Table 1 and Formulas (13) and (14), we have:

$$S^D = \frac{(1-c)^2}{8} + \frac{\delta(1-\delta)(\delta c - c + s)}{4(2\delta + 2k - \delta^2)}$$

$$S^C = \frac{(1-c)^2}{8} + \frac{\delta(1-\delta)(\delta c - c + s)}{4(\delta + k - \delta^2)}$$

Thus,

$$S^D - S^C = -\frac{\delta(1-\delta)(\delta + k)(\delta c - c + s)}{4(\delta + k - \delta^2)(2\delta + 2k - \delta^2)} < 0$$

i.e., $S^C > S^D$.

Then, based on Conclusion 4, we have $\pi_C^* > \pi_n^{D*} + \pi_r^{D*}$, thus

$$S^{CC} = S^C + \pi_C^* > S^{CD} = S^D + \pi_n^{D*} + \pi_r^{D*}$$

i.e., $S^{CC} > S^{DC}$. □

Conclusion 7 shows that based on Table 1, in the case of centralized decision-making, although the unit retail prices of the two products are reduced and due to the competition in the market, the sales of the new product decrease while those of the remanufactured products increase, while the consumer surplus and social surplus resulting from the increase of the remanufactured product are greater than those resulting from the decrease of the new product, ultimately leading to increased consumer surplus and social surplus in the case of centralized decision-making, i.e., in the case of centralized decision-making, the remanufactured product significantly increases the consumer surplus and social surplus. Therefore, the emergence of the remanufactured product is conducive not only to environmental protection but also to increasing the consumer surplus and social surplus.

4.3. Effects of Decentralized Decision-Making and Centralized Decision-Making on the Environment

For the sake of discussion, let

$$e^D = e_n q_n^{D*} + e_r q_r^{D*},\ e^C = e_n q_n^{C*} + e_r q_r^{C*}$$

Based on Table 1, we have:

$$\begin{aligned} e^D &= e_n q_n^{D*} + e_r q_r^{D*} \\ &= \tfrac{e_n}{2} - \tfrac{2kc+\delta(c+s)}{2(2\delta+2k-\delta^2)} e_n \\ &+ \tfrac{\delta c-c+s}{2(2\delta+2k-\delta^2)} e_r \end{aligned} \tag{17}$$

$$\begin{aligned} e^C &= e_n q_n^{C*} + e_r q_r^{C*} \\ &= \tfrac{e_n}{2} - \tfrac{kc+\delta s}{2(\delta+k-\delta^2)} e_n \\ &+ \tfrac{\delta c-c+s}{2(\delta+k-\delta^2)} e_r \end{aligned} \tag{18}$$

Based on Formulas (17) and (18), we can draw Conclusion 8.

Conclusion 8. *The effects of two decision-making models on the environment are as follows:*

(1) When $\frac{e_r}{e_n} > \delta$, in the case of centralized decision-making, the supply chain has a greater impact on the environment, i.e., centralized decision-making reduces the "marginal effect" of the supply chain, but it is not conducive to environmental protection;

(2) When $\frac{e_r}{e_n} < \delta$, in the case of decentralized decision-making, the supply chain has a greater impact on the environment, i.e., centralized decision-making not only can effectively increase the profit of the supply chain but also can reduce the impact on the environment;

(3) When $\frac{e_r}{e_n} = \delta$, the environmental impacts in the cases of centralized decision-making and decentralized decision-making are identical; however, centralized decision-making can effectively increase the profit of the supply chain.

Proof. From Formulas (17) and (18), we have:

$$e^C - e^D = \frac{(\delta + k)(\delta c + s - c)(e_r - \delta e_n)}{2(\delta + k - \delta^2)(2\delta + 2k - \delta^2)} \tag{19}$$

According to Formula (19), when $\frac{e_r}{e_n} > \delta, e^C > e^D$;
If $\frac{e_r}{e_n} < \delta$, then $e^C < e^D$;
If $\frac{e_r}{e_n} = \delta$, then $e^C = e^D$. □

With the literature [21,22], Conclusion 8 shows that when the ratio of the environmental impact of the unit remanufactured product and that of the unit new product is equal to or less than a certain threshold, centralized decision-making is not only conducive to the revenue increase of the supply chain but also effectively reduces the environmental impact of the two products; in this case, the supply chain must be coordinated, and if the original manufacturer is unwilling to cooperate, the government needs to introduce relevant laws and regulations to encourage the development of the remanufacturing industry, while asking the original manufacturer to fulfil its corporate social responsibilities.

5. Conclusions

In this study, we constructed a manufacturing/remanufacturing game model to investigate the effect of authorized remanufacturing on the competition mechanism of the manufacturing/remanufacturing supply chain and its coordination mechanism. Based on this model, the influences of authorized remanufacturing on the sales profit, consumer surplus, social surplus and environment were studied under decentralized decision-making and centralized decision-making, and at the same time, based on a franchise contract, the coordination mechanism of the manufacturing/remanufacturing supply chain was investigated to realize the supply chain coordination.

This paper's main contributions are summarized as follows: Firstly, the impact of authorizing remanufacturing on remanufacturing is analysed. The OEM could increase its profit and change its unfavourable market competition position through authorizing remanufacturing. And, the OEM's profit is positive correlation to cost saving of remanufacturing. Thus, The OEM should strive to increase the cost saving of remanufacturing. Such as, The OEM could design for remanufacturing in the product design stage in order to improve the dismantling instructions and reduce the dismantling costs of the used product. Furthermore, basing on the authorizing remanufacturing, the impact of remanufacturing on the environment, the consumer surplus and the social surplus is comparatively analysed, and the boundaries of remanufacturing being beneficial to environment is established. Also, the coordination contract is given to make supply chain of remanufacturing get the overall optimal.

Our research has some limitations and future research can be done at least in the following several avenues. Firstly, the consumer's willingness to pay (WTP) for the product is assumed to obey uniform distribution in [0, 1]. Thus, obeying normal distribution is more practical and suitable to explain realistic economic activities. In addition, the authorizing remanufacturing is one of remanufacturing models, there are also two more, i.e., independent remanufacturing and outsourcing remanufacturing. So, we can comparatively analyse the advantages and disadvantages among the three model. Lastly, the coordination mechanism is only focused on the revenue of remanufacturer, the revenue of manufacturer should be considered.

Author Contributions: X.X. and C.Z. designed the study. C.Z. conducted the analysis. X.X. wrote the paper.

Funding: This research was funded by the National Natural Science Foundation of China (71702174).

Conflicts of Interest: The authors declare no conflict of interest.

Abbreviations

n	Original manufacturer
r	Remanufacturers
c	Manufacturing costs per new product unit
k	The recycling scale parameters of waste product
s	The cost saving of manufacturing costs of the remanufactured product relative to the manufacturing costs per new product unit
D	Represents the decentralized decision-making model
C	Represents the centralized decision-making model
f	Represents the coordinated decision-making model

τ^i The recycling rate of used products when i is in $\{D, C, f\}$
e_n The impact of the production per unit of new product on the environment
e_r The impact of the production per unit of remanufactured product on the environment ($e_n > e_r$)
z The costs of licensing fees per unit of remanufactured product in the case of decentralized decision-making
p_n Selling price of new product made by manufacturer when the remanufactured product is not exist on the market
q_n Quantity of new products made by manufacturer when the remanufactured product is not exist on the market
p_n^i Selling price of new product made by manufacturer when i is in $\{D, C, f\}$
q_n^i Quantity of new products made by manufacturer when i is in $\{D, C, f\}$
p_r^i Selling price of remanufactured product made by remanufacturer when i is in $\{D, C, F\}$
q_r^i Quantity of remanufactured product made by manufacturer when i is in $\{D, C, f\}$
S^i The consumer surplus of supply chain when i is in $\{D, C\}$
S^{iC} The social surplus of supply chain when i is in $\{D, C\}$
δ Compared with the retail price per new product unit, the ratio of the retail price per remanufactured product unit and the retail price per new product unit perceived by consumers, also known as the relative discount, and according to the actual situation, $0 \leq \delta \leq 1$
π_n The profit of the manufacturer when the remanufactured product is not exist on the market
π_n^i The profit of the manufacturer when i is in $\{D, C, f\}$
π_r^i The profit of remanufacturer when i is in $\{D, C, f\}$
π_C The profit of supply chain when it is the centralized decision-making model

References

1. Liu, C.; Cai, W.; Dinolov, O.; Zhang, C.; Rao, W.; Jia, S.; Chan, F.T. Emergy based sustainability evaluation of remanufacturing machining systems. *Energy* **2018**, *150*, 670–680. [CrossRef]
2. Xu, B.-S.; Liu, S.-C.; Shi, P.-J. Contribution of remanufacturing engineering and surface engineering to cycle economy. *China Surf. Eng.* **2006**, *19*, 1–6.
3. Hashiguchi, M.S. Recycling efforts and patent rights protection in the United States and Japan. *Colum. J. Envtl. L.* **2008**, *33*, 169–195.
4. Zou, Z. *Study on a Non-Independent Third Party Remanufacturing Game Model*; Dalian University of Technology: Dalian, China, 2016.
5. Zou, Z.-B.; Wang, J.-J.; Deng, G.-S.; Chen, H. Third-party remanufacturing mode selection: Outsourcing or authorization? *Transp. Res. Part E Logist. Transp. Rev.* **2016**, *87*, 1–19. [CrossRef]
6. Zhu, Q.H.; Xia, X.Q.; Li, H.Y. A game model between a manufacturer and a remanufacturer based on government subsidies and patent fees. *J. Syst. Eng.* **2017**, *32*, 8–18.
7. Xiong, Z.K.; Shen, C.R.; Peng, Z.Q. Closed-loop supply chain coordination research with remanufacturing under patent protection. *J. Manag. Sci. China* **2011**, *14*, 76–85.
8. Xiong, Z.-K.; Shen, C.-R.; Peng, Z.-Q. A remanufacturing strategy for the closed-loop supply chain under patent protection. *J. Ind. Eng. Eng. Manag.* **2012**, *26*, 159–165.
9. Li, D. *Study on Closed-Loop Remanufacturing Supply Chain Network Equilibrium*; Southwest University of Finance and Economics: Chengdu, China, 2013.
10. Wang, J. Differential price coordination of closed-loop supply chain with remanufacturing under patent protection. *Oper. Res. Manag. Sci.* **2013**, *22*, 89–96.
11. Lind, S.; Olsson, D.; Sundin, E. Exploring inter-organizational relationships in automotive component remanufacturing. *J. Remanufactur.* **2014**, *4*, 5. [CrossRef]
12. Matsumoto, M.; Umeda, Y. An analysis of remanufacturing practices in Japan. *J. Remanufactur.* **2011**, *1*, 2. [CrossRef]
13. Govindan, K.; Popiuc, M.N. Reverse supply chain coordination by revenue sharing contract: A case for the personal computers industry. *Eur. J. Op. Res.* **2014**, *233*, 326–336. [CrossRef]
14. Kaya, O. Incentive and production decisions for remanufacturing operations. *Eur. J. Op. Res.* **2010**, *201*, 442–453. [CrossRef]

15. Yalabik, B.; Chhajed, D.; Petruzzi, N.C. Product and sales contract design in remanufacturing. *Int. J. Prod. Econ.* **2014**, *154*, 299–312. [CrossRef]
16. Wu, C.-H. Strategic and operational decisions under sales competition and collection competition for end-of-use products in remanufacturing. *Int. J. Prod. Econ.* **2015**, *169*, 11–20. [CrossRef]
17. Örsdemir, A.; Kemahlıoğlu-Ziya, E.; Parlaktürk, A.K. Competitive quality choice and remanufacturing. *Prod. Oper. Manag.* **2014**, *23*, 48–64. [CrossRef]
18. Jacobs, B.W.; Subramanian, R. Sharing responsibility for product recovery across the supply chain. *Prod. Oper. Manag.* **2012**, *21*, 85–100. [CrossRef]
19. Sun, H.; Liu, C.; Chen, J.; Gao, M.; Shen, X. A novel method of sustainability evaluation in machining processes. *Processes* **2019**, *7*, 275. [CrossRef]
20. Liu, C.; Cai, W.; Jia, S.; Zhang, M.; Guo, H.; Hu, L.; Jiang, Z. Emergy-based evaluation and improvement for sustainable manufacturing systems considering resource efficiency and environment performance. *Energy Convers. Manag.* **2018**, *177*, 176–189. [CrossRef]
21. Cai, W.; Liu, C.; Lai, K.H.; Li, L.; Cunha, J.; Hu, L. Energy performance certification in mechanical manufacturing industry: A review and analysis. *Energy Convers. Manag.* **2019**, *186*, 415–432. [CrossRef]
22. Cai, W.; Lai, K.H.; Liu, C.; Wei, F.; Ma, M.; Jia, S.; Jiang, Z.; Lv, L. Promoting sustainability of manufacturing industry through the lean energy-saving and emission-reduction strategy. *Sci. Total Environ.* **2019**, *665*, 23–32. [CrossRef] [PubMed]

Article

Energy-Economizing Optimization of Magnesium Alloy Hot Stamping Process

Mengdi Gao [1], Qingyang Wang [1,*], Lei Li [2] and Zhilin Ma [1]

[1] School of Mechanical and Electronic Engineering, Suzhou University, Suzhou 234000, China; gaomengdi@ahszu.edu.cn (M.G.); mazhilin018@163.com (Z.M.)

[2] School of Mechanical Engineering, Hefei University of Technology, Hefei 230000, China; lei_li@hfut.edu.cn

* Correspondence: wangqingyang@ahszu.edu.cn; Tel.: +86-1780-557-5586

Received: 24 December 2019; Accepted: 23 January 2020; Published: 5 February 2020

Abstract: Reducing the mass of vehicles is an effective way to improve energy efficiency and mileage. Therefore, hot stamping is developed to manufacture lightweight materials used for vehicle production, such as magnesium and aluminum alloys. However, in comparison with traditional cold stamping, hot stamping is a high-energy-consumption process, because it requires heating sheet materials to a certain temperature before forming. Moreover, the process parameters of hot stamping considerably influence the product forming quality and energy consumption. In this work, the energy-economizing indices of hot stamping are established with multiobjective consideration of energy consumption and product forming quality to find a pathway by which to obtain optimal hot stamping process parameters. An energy consumption index is quantified by the developed models, and forming quality indices are calculated using a finite element model. Response surface models between the process parameters and energy-economizing indices are established by combining the Latin hypercube design and response surface methodology. The multiobjective problem is solved using a multiobjective genetic algorithm (NSGA-II) to obtain the Pareto frontier. ZK60 magnesium alloy hot stamping is applied as a case study to obtain an optimal combination of parameters, and compromise solutions are compared through stamping trials and numerical simulations. The obtained results may be used for guiding process optimization regarding energy saving and the method of manufacturing parameters selection.

Keywords: energy-economizing; hot stamping; lightweight material; magnesium alloy; process parameters

---

1. Introduction

Rapid economic development has recently accelerated increases in the consumption of energy, especially in industrial sectors, which is causing a series of environmental problems, such as greenhouse gas (GHG) emissions [1]. The GHG emissions in China reach 1.03×10^9 T, of which 10–15% come from automobiles [2]. An efficient way to improve the energy efficiency and driving range of vehicles is mass reduction. Some lightweight materials, such as magnesium alloys, aluminum alloys, and ultra-high-strength steels, have been rapidly increasing in quantity and have been applied to the automotive industry [3]. Vehicles can conserve significant amounts of energy by using these lightweight materials [4].

Hot stamping was developed to manufacture the structural components of automobiles by using lightweight materials to achieve decreased weight, improved safety, and enhanced crashworthiness [5]. Hot stamping can improve the formability of these lightweight materials, which overcomes the limits of traditional cold stamping [6]; it is particularly suitable for manufacturing complex parts [7]. Moreover, the high forming precision and reduced springback of hot stamping have led to its wide application in the automotive and aircraft industries.

However, the high-temperature forming conditions of blanks in the hot stamping process consume additional energy. Heating the sheet material to a certain temperature improves the energy consumption of hot stamping compared with that of cold stamping [8]. The energy consumption and environmental effects of the hot stamping industry warrant attention. Process energy consumption models of hot stamping have been developed to qualify its energy consumption [9]. The associated carbon footprint is identified to analyze its environmental effects [10]. All the results show that oven curing accounts for a large amount of the products' energy consumption and environmental impact. Therefore, one efficient pathway towards energy-economizing hot stamping is promoting the energy efficiency in the blank heating process. The thermal transfer and loss principle is basic for energy efficiency improvement in oven curing. To study the thermal aspects of the hot stamping process, based on the identification of the heat transfer coefficient in hot stamping [11], Abdulhay et al. proposed the heat-transfer modeling of all heat-transfer modes occurring during the hot stamping phases [12].

A suitable stamping temperature is very important in optimizing the energy efficiency of hot stamping. The product forming quality and energy consumption are determined by process parameters. Many researchers have focused on the optimization of these parameters to solve the product forming quality problem of hot stamping. Xiao et al. found that decreasing forming temperature and increasing forming speed can improve the formability of AA7075 through a hot uniaxial tensile test [13]. More than one object should be considered in multiobjective optimization methods, such as the weighted sum method [14], the global criterion-based method [15], and genetic algorithms [16], which are widely applied in the hot stamping to solve the conflict between different evaluation indices of forming quality. Kitayama et al. used sequential approximate optimization with radial basis function network to optimize the parameter of blank holder force trajectory, with the aim of reducing the products' springback [17]. Zhou et al. focused on numerical simulations, together with the combination of response surface methodology (RSM) and nondominated sorting genetic algorithm II (NSGA-II) to optimize aluminum alloy hot stamping [18]. In order to reduce the influence of the stochastic property of process parameters on forming quality, Xiao et al. integrated multiobjective stochastic approaches, such as RSM, NSGA-II, and Monte Carlo simulations (MCSs), to obtain the optimal process parameters of aluminum hot stamping [19].

The above analysis reveals that the energy consumption and environmental soundness of the hot stamping process are of serious concern. But research on how to reduce the energy consumption of the hot stamping process is still insufficient. Process optimization is an excellent way to solve this problem. But the process parameters have been optimized using numerous methods and multiobjective optimization methods to enhance only the product forming quality. Energy consumption should not be neglected in the parameters optimization of the hot stamping process. In particular, hot stamping optimization with consideration for energy saving is worth studying due to its relevance to energy-efficient manufacturing. However, there is still a lack of effective methods regarding both the optimization of energy consumption and in terms of improving the forming quality at present. In this study, a novel parameter optimization method for the hot stamping process is proposed with the multiobjective improvement of forming quality and process energy consumption.

2. Framework and Method

Different process parameters determine the amount of energy consumption and product forming quality of hot stamping. Product forming defects, such as wrinkles, cracks, or high energy consumption, will appear when sheet materials are processed under inappropriate process parameters. This can be avoided by adjusting and selecting appropriate process parameters in hot stamping. Process optimization is an efficient way to reduce the energy consumption and avoid forming defects in the hot stamping process.

A descriptive flowchart depicting the major steps of the proposed methodology is shown in Figure 1, which provides an outline for this study. A hot stamping process contains heating and forming processes; energy and material are supplied for the process with different processing parameters.

In order to achieve an environmentally-friendly manufacturing process, energy reduction is essential. Process energy consumption and forming quality are considered in the process optimization method of hot stamping, which are taken as the energy-economizing indices. These indices of hot stamping are regarded as the optimization objectives. The forming quality and energy consumption under different process parameters are qualified by the developed models or simulation and experiments. The range and constraints of the optimized process parameters should be initially determined in accordance with the forming performance and requirements. Then, sample points should be selected in the design space for experiments, and the corresponding simulations or experiments be conducted to obtain the corresponding evaluation index values of each sample point. Our model was developed to study the relationship between the process parameters and each index, which was solved using a multiobjective genetic algorithm (NSGA-II) and offers feasible optimized solutions. The comparison between the numerically-predicted technical parameters of the stamps and the real experimental results demonstrate the applicability of the method. The multiobjective optimization method can improve the energy efficiency of hot stamping while maintaining the required quality of the stamps. The proposed energy-economizing can provide a reference for industrial manufacturing and production, especially for vehicle production.

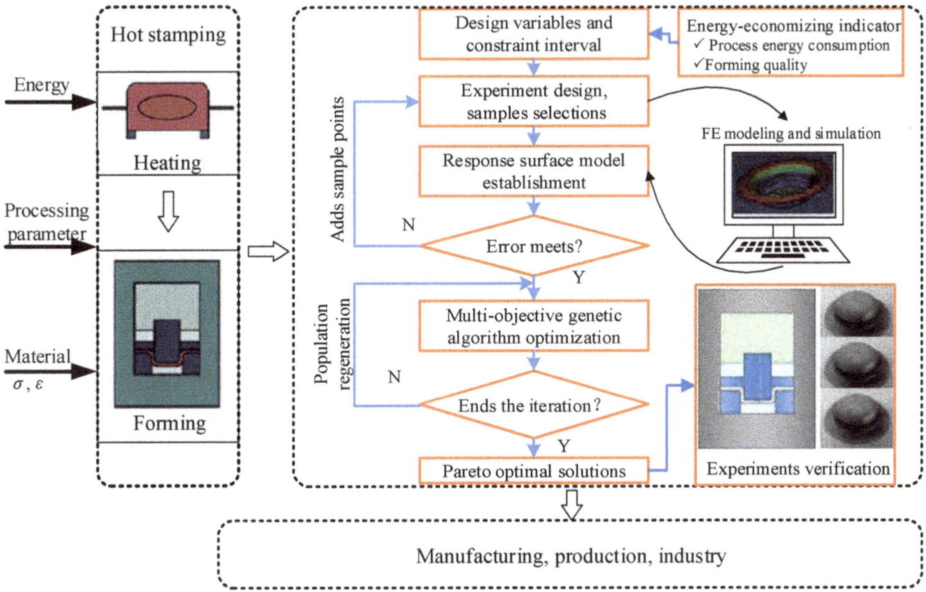

Figure 1. Methodological framework.

3. Energy-Economizing Indices of Hot Stamping

3.1. Process Energy Consumption Indices

The hot stamping process begins by heating a sheet material to a given range of temperatures to improve formability. Therefore, the energy consumption of hot stamping is associated with the energy used for heating and that used for forming. The total energy consumption of hot stamping, E, can be calculated as

$$E = E_{\text{heating}} + E_{\text{forming}}, \tag{1}$$

where E_{heating} is the heating energy consumption and E_{forming} is the forming energy consumption.

After being heated in a furnace, the sheet material is delivered to a piece of forming equipment and immediately formed in a closed tool. Thus, the heat is transferred to the environment and tools in those processes; this transfer is described as heat loss, ΔQ. The heat loss includes convection and radiation losses with ambient air during the transfer to the forming equipment, and the losses by heterogeneous approaches of convection, radiation, and conduction in the forming process [12]. Therefore, the heat loss (ΔQ) can be calculated as

$$\Delta Q = Q_{\text{convection}} + Q_{\text{radiation}} + Q_{\text{conduction}}, \quad (2)$$

where $Q_{\text{convection}}$, $Q_{\text{radiation}}$, and $Q_{\text{conduction}}$ are the heat losses by convection, radiation, and conduction, respectively, and can be calculated by as follows:

$$Q_{\text{convection}} = Q(R_h, \Delta T, \Delta t) = Ah\Delta T\Delta t, \quad (3)$$

$$Q_{\text{radiation}} = kF_r A\left((T + \Delta T)^4 - T^4\right)\Delta t, \quad (4)$$

$$Q_{\text{conduction}} = Q(R_\lambda, \Delta T, \Delta t) = A\lambda\frac{\Delta T}{t_0}\Delta t, \quad (5)$$

where A is the heat transfer area, h is the convection coefficient, $R_h = 1/Ah$ is the convection resistance, T is the temperature, ΔT is the temperature difference between hot and cold fluid object, k is the Boltzmann constant, F_r is the radiation shape factor, λ is the thermal conductivity, and $R_\lambda = \delta/A\lambda$ is the thermal resistance, δ is the thickness, t is the time.

The total heat, Q, can be calculated by Equation (6), which combines the heat in the blank and the heat loss.

$$Q = Q_{\text{blank}} + \Delta Q, \quad (6)$$

where Q_{blank} is the heat absorbed by sheet metal.

The energy consumption used for heating can be calculated as

$$E_{\text{heating}} = \frac{Q}{1 - \eta_{\text{loss}}(T, i)}, \quad (7)$$

where $\eta_{\text{loss}}(T, i)$ is thermal efficiency loss caused by the different heating temperatures and methods and i denotes the three heating means, namely, radiation, induction, and conduction.

The forming stage in hot stamping is similar to that in cold stamping, as shown in Figure 2. The heated sheet material is placed on a die with a blank holder to avoid wrinkle defects. The forming press controls the punch to draw the blank into the die and to form it into the desired shape with the elastic–plastic deformation of the sheet metal and the contact friction between the sheet material and dies [20]. Therefore, the energy consumption of the forming stage in the hot stamping process can be quantified from two perspectives. The direct ways are measuring and estimating the energy consumption of the forming equipment [21], which can be described as

$$E_{\text{forming}} = \int_{t_{\text{start}}}^{t_{\text{end}}} F(t)v(t)dt/\eta(t), \quad (8)$$

where F is the output force of the actuator of the forming equipment, which varies with the working conditions; v is the punch speed, and η is the energy efficiency in the stamping process.

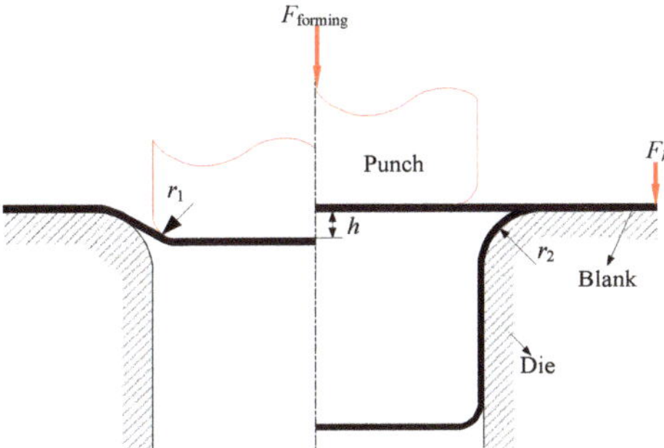

Figure 2. Forming process of hot stamping.

The second method involves consideration of the deformation process of the sheet material, and then analysis and modeling of the process energy consumption with the related parameters. The energy consumption of the sheet metal forming process may be divided into that required for plastic deformation, bending, and frictional energies [8]. These analytical quantification methods of calculating the process energy consumption are detailed in Reference [22].

3.2. Forming Quality Indices

The thinning and thickening of critical elements are used to indicate whether fractures have occurred in the blank forming process. Springback cannot be regarded as a forming quality index because the springback of hot stamping is considerably less than that of cold stamping. A sheet metal will crack if its thickness is less than a certain critical value. The rupture distance refers to the vertical distance between the strain point of dangerous elements and the forming limit curve (FLC), as shown in Figure 3. When the main strain of a region element of the formed part is above the FLC ($\varphi(\varepsilon_2)$) or the safety marginal curve ($\Phi(\varepsilon_2)$), this region of the shaped part will likely fracture. A long distance from the safety marginal curve $\Phi(\varepsilon_2)$ indicates a high rupture tendency. Therefore, the average distance between the main strain of all elements and the $\Phi(\varepsilon_2)$ curve can be used to quantify the fracture. Similarly, when the main strain of an area element is below the wrinkle limit curve ($\psi(\varepsilon_2)$), this area of the formed part will show a wrinkling trend; the farther the point from the $\psi(\varepsilon_2)$ curve, the higher the trend of wrinkling. The fracture distance and wrinkling trend are mainly used to predict the product forming quality in the finite element (FE) simulation, but the FLC limit diagram cannot be directly used to quantify the product forming quality in the actual stamping production process.

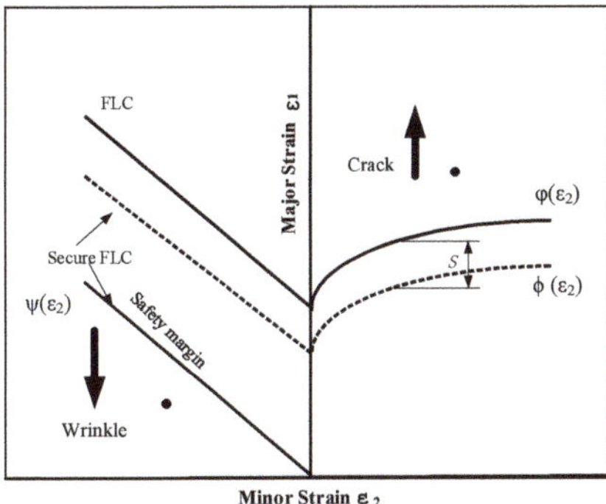

Figure 3. Definition diagram of cracking and wrinkling quantification criteria. $\Phi(\varepsilon_2)$ is the safety marginal curve, $\psi(\varepsilon_2)$ is the wrinkle limit curve, and $\varphi(\varepsilon_2)$ is the FLC.

Therefore, in the actual production and the experiment, thinning and thickening can be used to represent the possibility of fracture and wrinkling, respectively. These indices are calculated using Equations (9) and (10). An excessively large thinning rate of a sheet metal in a certain area means that the fracture trend is too large, whereas a disproportionately high thickening rate indicates a possible wrinkling phenomenon.

$$\Delta_{thinning} = (t_0 - t_{min})/t_0 \times 100\%, \tag{9}$$

$$\Delta_{thickening} = (t_{max} - t_0)/t_0 \times 100\%, \tag{10}$$

where t_0 is the thickness of the blank, t_{min} is the minimum thickness of the sheet material, t_{max} is the maximum thickness of the sheet material, $\Delta_{thinning}$ is the thinning rate, and $\Delta_{thickening}$ is the thickening rate.

4. Multiobjective Optimization for Hot Stamping Process

4.1. Optimization Variables

The considered process parameters are regarded as optimization variables in the hot stamping optimization process. Many process parameters affect the energy consumption and forming quality of stamping parts, such as stamping speed, blank holder force, friction conditions, tool gap, and draw-bead geometry parameters. The influence of the draw-bead is usually not considered because the draw-bead is typically used only in stamping extremely complex-shaped parts. In addition, the shape parameters of the draw-bead are complex, and have independent design criteria for some simple stamping processes. The clearance between the punch and die substantially affects the springback of the forming parts, which usually accounts for 110–120% of the sheet metal thickness. The die gap is usually ignored as a design variable because of the small springback in the hot stamping process. No quantitative analysis exists for the effect of friction on product forming quality because the friction conditions in an actual stamping workshop are determined by the lubricating oil, die material, and coating. Therefore, in hot stamping optimization, three process parameters, namely, stamping speed, blank holder force, and forming temperature, should be considered. A certain range of process parameters should be experimentally analyzed to obtain the optimal process parameters.

After the optimization variables are determined, a certain sample point is selected in the design space for experiment. The surrogate model between each process parameter and its corresponding index value is established on the basis of RSM, and the process parameters are optimized on the basis of the multiobjective genetic algorithm.

4.2. Sample Selection

The Latin hypercube design (LHD) is an efficient sampling method; it is advantageous in sampling efficiency and running time due to its reduced number of iterations. This method is especially suitable for computer simulation experiments. A good feature of this method is that sample points are selected from the entire design space. Therefore, the optimal process parameters can be determined systematically and accurately with less experiments by using this method. LHD evenly divides the design space of each variable into several layers, and these layers obtain some special points through random combination to determine the design matrix. Each level of each parameter is sampled only once.

4.3. Optimization Model and Solution Approach

The energy-economizing optimization aims to obtain a set of process parameters that will produce stamping parts with good forming quality (no cracks, wrinkling, or noticeable thickening and thinning) and low energy consumption. The objective function and constraints of the optimization process can be given as

$$F = \min(y_1, y_2, y_3)$$
$$\text{s.t.}\begin{cases} h_i(\mathrm{x}) = 0 \ (i = 1, 2 \cdots n) \\ g_j(\mathrm{x}) \geq 0 \ (j = 1, 2 \cdots n) \\ \mathrm{x}_m^{\min} < \mathrm{x}_m < \mathrm{x}_m^{\max} \ (m = 1, 2 \cdots n) \end{cases}, \tag{11}$$

where y_1, y_2, and y_3 are the response functions of the energy-economizing indices, namely, energy consumption, thinning, and thickening, respectively, and x_m are the design variables, namely, blank holder force, stamping speed, and forming temperature.

Determining the mapping function through theoretical derivation is difficult due to the complex relationship between process parameters and target quantity. RSM is an effective method of building substitute models. The RSM polynomial regression model adopts a quadratic regression equation. On the basis of experimental data, the coefficient of the regression equation is obtained through the least squares method to construct the functional relationship between the optimization variable and target quantity. The commonly used second-order polynomial response surface model can be represented by

$$y = a_0 + \sum_{i=1}^{n} a_i x_i + \sum_{i=1}^{n} a_{ii} x_i^2 + \sum_{i=2}^{n} \sum_{j=1}^{i-1} a_{ij} x_i x_j, \tag{12}$$

where x_i, x_j represents the design variable; n is the minimum number of samples; a_0 is the minor error; a_i, a_{ii}, and a_{ij} are the polynomial coefficients; and y represents the response of the energy-economizing indices of the stamping process.

NSGA-II is a multiobjective genetic algorithm that ensures good convergence and robustness by introducing a fast nondominated sorting algorithm, elite strategy, and congestion degree comparison operator, and other strategies [23]. The multiobjective optimization problem of hot stamping can be solved by programming NSGA-II.

5. Hot Stamping Process Optimization of ZK60 Magnesium Alloy for Energy Saving

On the basis of the proposed energy-economizing optimization method of hot stamping, the tube-shaped part, whose common profile is shown in Figure 4, is selected as a case to investigate the proposed method. Table 1 presents the tool size and relevant parameters of the sheet metal. The stamping material is ZK60 magnesium alloy, which is a lightweight material with strong deformation capability and good heat treatment strengthening effect. Three variable process parameters, namely, stamping speed (v), blank holder force (F_{h}), and forming temperature (T), are considered in the simulation process. The range of each process parameter is set as follows: stamping speed 2–11 mm/s, blank holder force 3–9 kN, and forming temperature 175–250 °C (the range of forming temperature is determined on the basis of existing research results in Reference [24]).

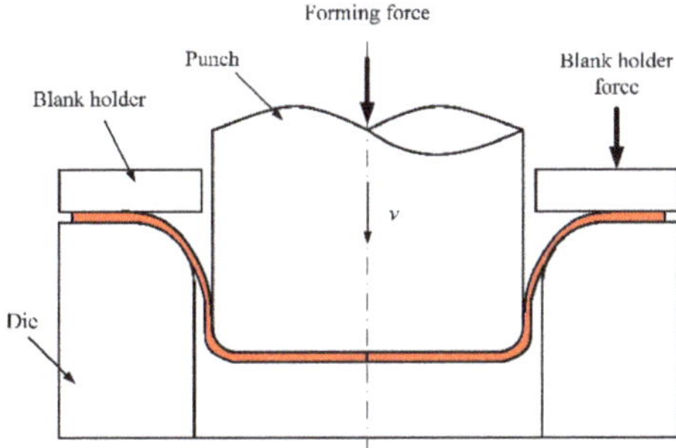

Figure 4. Stamping of tube-shaped part.

Table 1. Parameters for drawing processes of tube-shaped parts.

Blank Diameter d_p (mm)	Sheet Metal Thickness t_0 (mm)	Drawing Height h (mm)	Punch Radius r_1 (mm)	Die Radius r_2 (mm)	Clearance δ (mm)	Friction Coefficient μ
100.0	1.0	20.0	7.0	8.0	1.2	0.12

On the basis of the aforementioned experimental design method, the LHD sampling method is used to sample the point in the range of each process parameter, and the least number of sample points required by the response surface model is determined through the least squares method. The number of samples n can be determined in terms of the equation $n = (m + 1)(m + 2)/2$ of the number of considered process parameters m. Given that the number of process parameters m is 3, the number of samples n is at least 10. Eighteen design sample points are selected to improve the accuracy of the response surface model, and Figure 5 illustrates their value distribution.

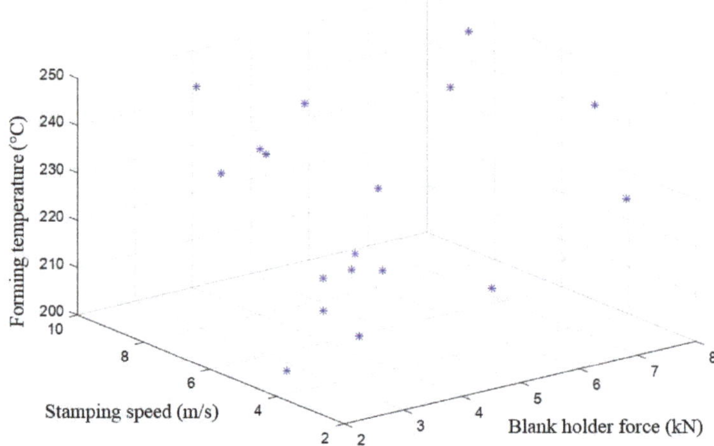

Figure 5. Sample distribution of LHD in the range of each process parameter.

5.1. Material Properties Testing

Accurate mechanical property parameters and an equivalent stress model are vital for analyzing the hot deformation of magnesium alloy. The goal is to identify the mechanical properties of ZK60 magnesium alloy under different temperatures, which are used for the theoretical calculation of energy consumption and simulation. Table 2 shows the chemical composition of ZK60 magnesium alloy. A series of hot unidirectional tensile experiments is performed on an MTS810 system material testing machine (Figure 6) at different temperatures. The testing samples are designed on the basis of the GB/T4338-2006 high-temperature tensile testing method of metallic materials, and the thickness of the samples is 2 mm, as shown in Figure 7.

Table 2. ZK60 magnesium alloy chemical composition (%).

Element	Si	Fe	Cu	Mn	Al	Zn	Ni	Zr
Quality score w	0.0014	0.003	0.0011	0.008	0.0014	5.5	0.00048	0.53

Figure 6. MTS810 material experimental system.

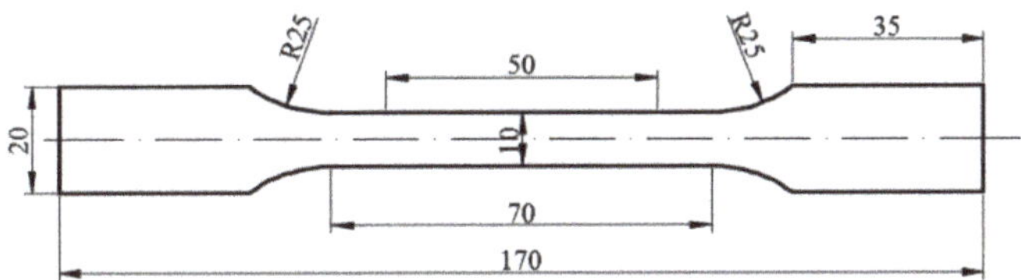

Figure 7. ZK60 magnesium alloy tensile sample.

The hot unidirectional tensile experiments are performed under invariable temperatures of 175 °C, 200 °C, 225 °C, and 250 °C. The samples are kept in an environmental cabinet for 15 min after being clamped at different temperatures. The samples are then stretched with different speeds to obtain the stress-strain curve of ZK60. Three groups of experiments are conducted under the same parameters, and the average values are regarded as the experimental results under the corresponding conditions. Figure 8 exhibits the obtained true stress–strain curve of ZK60 magnesium alloy.

A flow stress mathematical model of ZK60 magnesium alloy is established on the basis of the Fields–Backofen equation in consideration of the softening factor proposed by Zhang et al. [25]. In accordance with the experimental stress–strain curves of ZK60 magnesium alloy under various temperatures, the corresponding flow stress mathematical model is developed as

$$\sigma = 746\varepsilon^{0.1101}\dot{\varepsilon}^{0.0262}\exp(-0.00665T - 0.94781\varepsilon), \tag{13}$$

where σ is the stress of the investigated ZK60 magnesium alloy, ε is the strain of the ZK60 magnesium alloy, and $\dot{\varepsilon}$ is the strain rate of ZK60 magnesium alloy.

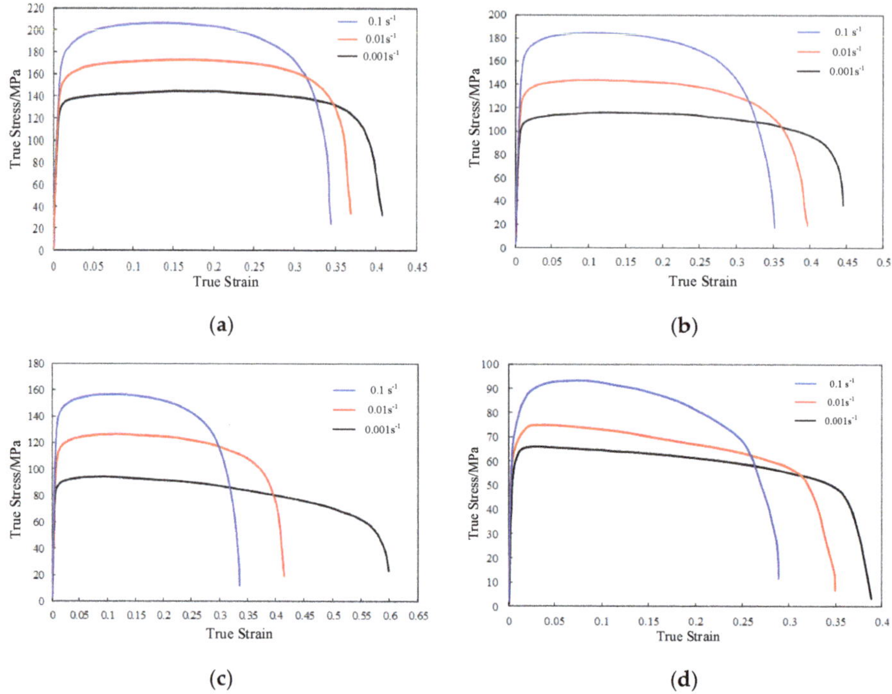

Figure 8. True stress-strain curves for ZK60 magnesium alloy deformed at different temperatures and strain rates: (**a**) 175 °C; (**b**) 200 °C; (**c**) 225 °C; (**d**) 250 °C.

5.2. FE Modeling and Simulation for Hot Stamping

The FE simulation model is established in accordance with the actual hot stamping process, as shown in Figure 9. The punch is placed above the sheet metal, the die is at the bottom, and the sheet metal lies above the die and below the blank holder. In the simulation process, the mold is regarded as a rigid body without elastic deformation; thus, it cannot be meshed. A Belytschko–Tsay shell element with five Gaussian thickness dimension integration points is used to divide the sheet metal mesh, and the thickness of the sheet metal is set as 1 mm. The initial unit number of the sheet metal is 3925, and the grid refinement level is set as 2. The kinematic relations of the tools and blank are as follows:

(1) Figure 9 depicts the initial positions of the die and blank, and the sheet metal is above the die.
(2) The blank holder is close to the sheet metal at a certain speed (it can be set to different speeds, but it maintains a constant speed throughout the stamping process), which is called "holding."
(3) When the blank holder is finished, the sheet metal is fixed, and the punch starts to act. The "stamping" process begins when the punch contacts the sheet metal.
(4) In the last stage of the simulation, the concave convex die is in a closed state, followed by the quenching stage.

In accordance with the friction condition between the blank and tools in the actual stamping process, the friction coefficient is set as 0.12 in the simulation. The blank is heated via in-mold heating. This heating method is advantageous in that the variation of temperature is relatively small; therefore, the temperature distribution of the sheet is uniform and has good formability. However, the energy consumption used for heating will be considerably higher than that for furnace heating. The simulations are performed on the basis of the design points in Figure 5, and the corresponding energy consumption is calculated through a previous energy consumption theoretical model. Table 3 shows the results.

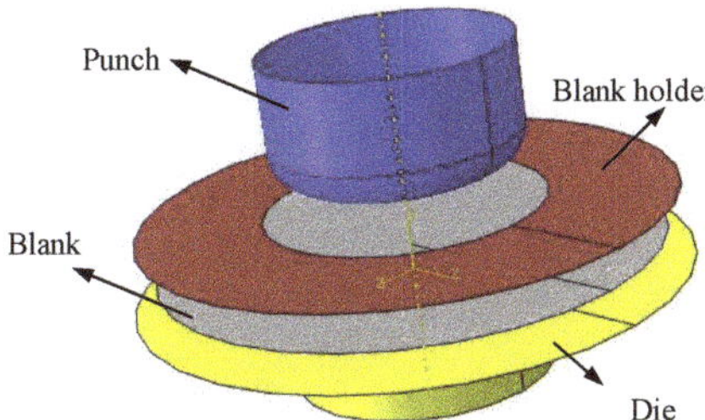

Figure 9. Numerical simulation model of the tube-shaped part.

Table 3. Experimental design points and corresponding results.

Run	F_h (kN)	v (mm/s)	T (°C)	Energy Consumption (J)	Thinning (%)	Thickening (%)
1	2.1	4.7	250	46,212.27	5.09	8.1
2	2.5	3.5	225	32,165.84	4.3	8
3	2.8	5.1	200	28,919.08	4.32	7.3
4	3.1	8.3	250	48,726.87	5.96	7.6
5	3.5	4.3	225	33,066.23	9.62	6.8
6	3.8	6.3	250	47,463.92	5.78	7
7	4.2	9.5	225	37,347.17	9.64	7.8
8	4.5	5.9	200	29,853.32	7.71	6.8
9	4.8	7.5	200	31,439.71	5.08	7.4
10	5.2	9.9	225	37,667.46	7.75	8
11	5.5	7.1	225	35,622.90	7.01	7.1
12	5.8	5.5	250	46,978.99	16.12	6
13	6.2	9.1	200	32,932.06	6.12	7.7
14	6.5	8.7	200	32,596.56	6.04	7.5
15	6.9	3.1	250	44,981.23	6.74	6.2
16	7.2	6.7	200	30,775.07	5.84	7
17	7.5	7.9	250	48,664.86	8.59	6.4
18	7.9	3.9	225	32,835.24	11.86	5.7

5.3. Process Parameters Optimization

On the basis of the above results, a second-order regression model is used to obtain the response surface for each objective function. Let y_1 be the energy consumption of hot stamping, y_2 the thinning rate of the stamping parts, and y_3 the thickening rate of the stamping parts. The developed model can then be described as

$$\begin{aligned} y_1 &= 261054 + 76.8495x_1 + 2643.08480x_2 - 2455.30317x_3 + 0.26751x_1x_2 \\ &-0.17018x_1x_2 - 6.13451x_2x_3 + 0.14412x_1{}^2 - 31.95144x_2{}^2 + 6.31017x_3{}^2 \end{aligned}, \tag{14}$$

$$\begin{aligned} y_2 &= -185.68990 - 0.037136x_1 + 0.58991x_2 + 1.68914x_3 - 0.16617x_1x_2 + 3.71659\times10^{-3}x_1x_3 \\ &+4.68506\times10^{-3}x_2x_3 + 0.069271x_1{}^2 - 0.055531x_2{}^2 - 3.83894\times10^{-3}x_3{}^2 \end{aligned}, \tag{15}$$

$$\begin{aligned} y_3 &= 0.78922 - 0.39606x_1 + 0.29588x_2 + 0.059743x_3 + 0.036733x_1x_2 - 2.74212\times10^{-3}x_1x_3 \\ &-3.66213\times10^{-3}x_2x_3 + 0.050898x_1{}^2 + 0.041834x_2{}^2 - 6.34584\times10^{-5}x_3{}^2 \end{aligned}, \tag{16}$$

where x_1 is the blank holder force, x_2 is the stamping speed, and x_3 is the forming temperature.

Hot stamping process optimization aims to obtain a set of process parameters that will produce stamping parts with reduced thickness variations and low energy consumption. Therefore, the objective function and constraint conditions in the optimization process can be expressed as

$$F = \min(y_1, y_2, y_3)$$
$$s.t.\begin{cases} 2 \le x_1 \le 8 \\ 3 \le x_2 \le 10 \\ 200 \le x_3 \le 250 \end{cases}, \tag{17}$$

where y_1, y_2, and y_3 are the objective functions. The goal is to minimize the stamping energy consumption and thickness variations in hot stamping.

NSGA-II is used to solve the multiobjective optimization problem in Equation (17). A series of considered efficient solutions constituting the Pareto frontier is obtained, as shown in Figure 10. The results show that the formability and energy consumption of sheet metals are contradictory. For ZK60 magnesium alloy hot stamping, formability improves with the increase in forming temperature, but the energy consumption of hot stamping increases significantly with the rise of heating temperature. The thinning and thickening rates are also contradictory. With increasing blank holder force, the thinning rate of stamping parts increases gradually, whereas the wrinkling trend of the sheet metal decreases. On the basis of the forming requirements (the thinning and thickening rates of stamping must be less than 10%, and the energy consumption of stamping should be relatively small), the following two groups of compromise solutions are selected, and Table 4 presents the corresponding index values.

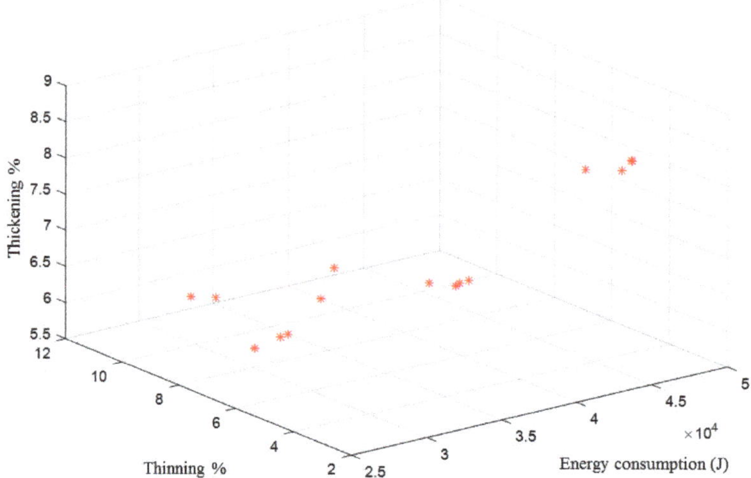

Figure 10. Pareto optimal solutions of ZK60 magnesium alloy hot stamping.

Table 4. Compromise solutions and their corresponding indices.

Compromise Solutions	Process Parameters			Indices		
	F_h (kN)	v (mm/s)	T (°C)	Energy Consumption (J)	Thinning (%)	Thickening (%)
Solution 1	8.0	3.0	225	31,785.57	6.2	5.7
Solution 2	4.7	3.3	200	26,190.36	4.8	5.9

Figure 11 shows the simulation results of the thickness variation distribution under different compromise solutions. The forming quality of stamping parts under the two groups of process parameters is good and can meet the usage requirements. The thicknesses of the stamping parts vary greatly in terms of punch and die radii. From the straight wall section to the flange area, from the bottom to the top, the thickness variation distribution of the parts presents a gradual increase trend at the parameters of the two compromise solutions, and the thickening phenomenon is shown in the flange area. A comparison of the color distribution of the thickness variation of the stamping parts in the simulation results indicates that the thickness variation distribution of the stamping parts obtained at the parameters of compromise solution 1 is slightly more uniform than that at the parameters of compromise solution 2. The energy consumption at the parameters of compromise solution 2 can be reduced by 17.6% in comparison with those in compromise solution 1. Therefore, considering all energy-economizing indices of hot stamping, the indices of energy consumption and thinning in compromise solution 2 are better than those in compromise solution 1.

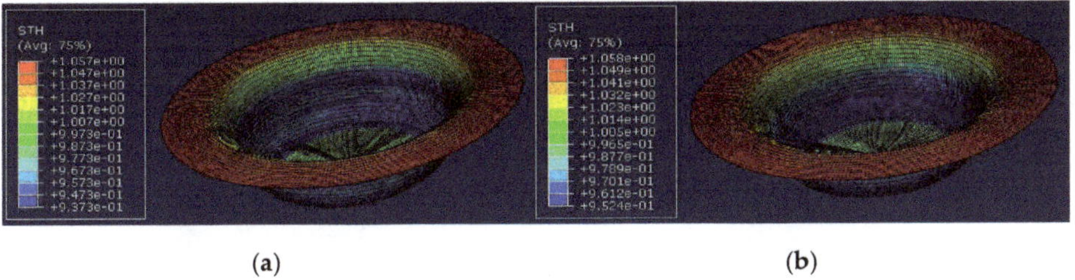

Figure 11. Thickness variation distribution demonstrated by the numerical simulation under different compromise solutions: (**a**) compromise solution 1, (**b**) compromise solution 2.

5.4. Stamping Experiments Verification

The corresponding experiment is conducted with experimental equipment to verify the feasibility of the optimization results for the hot stamping process, as shown in Figure 12. In accordance with the obtained compromise solutions, the experiments are performed as follows:

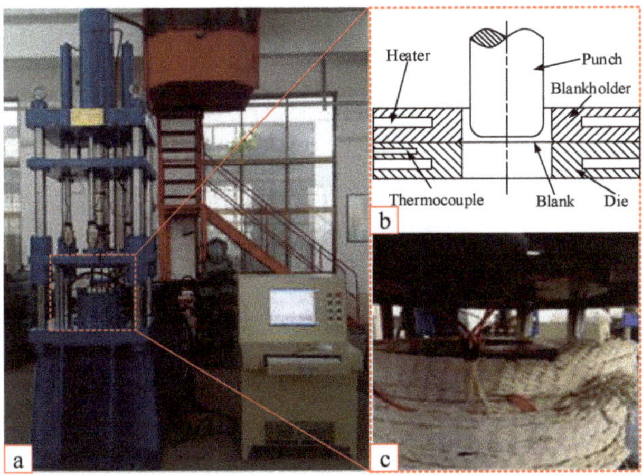

Figure 12. Equipment and die structures for hot stamping process of tube-shaped parts: (**a**) 100 T partitioned VBHF hydraulic machine, (**b**) die structure, (**c**) physical die.

The temperature in the die was raised to 225 °C by internal heating, as shown in Figure 12b,c. Then, the prepared ZK60 magnesium alloy sheet material is placed into the die, and the mold is closed. The heat preservation time is set as 10 min to heat the blank fully. The process parameters of the forming press are set as follows: 8 kN blank holder force and 3 mm/s stamping speed (Table 4). The stamping experiment is then conducted. Three groups of experiments are conducted in accordance with the above experimental steps. Then, to reduce the temperature of the die to 200 °C and repeat above experimental steps, the process parameters are set as follows: 4.7 kN blank holder force and 3.3 mm/s stamping speed (Table 4).

Figure 13 exhibits the obtained stamping parts at the parameters of the two compromise solutions. The forming quality of the obtained stamping parts is good, and no evident defects are observed. However, slight wrinkling is identified in the flange area of the stamping parts under the process parameters of compromise solution 2 because of its large thickening rate. Nevertheless, the slight wrinkling will not affect the final use of the product because the flange area of the obtained parts is cut off in actual production. Therefore, compromise solution 2 is the optimal combination of process parameters from a comprehensive perspective.

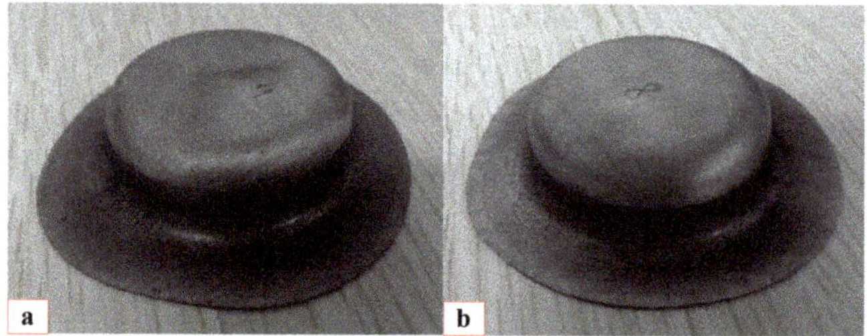

Figure 13. Hot stamping parts of experimentally stamped ZK60 magnesium alloy: (**a**) Obtained part at the parameters of compromise solutions 1; (**b**) Obtained part at the parameters of compromise solutions 2.

The thickness variation of a stamping part is an important index of the forming quality. Therefore, the thickness variation of the stamping parts obtained by the hot drawing experiment under the process parameters of comprise solution 2 is compared with that of the simulation results, as shown in Figure 14. The thickness variation distribution of each shell element in the FE model can be read in the postprocessing of the simulation. The thickness variation distribution of the stamping parts is measured along the symmetrical section of the parts by a micrometer. The results show that the thickness variation distribution obtained by the FE method is consistent with that obtained by the experiment, which further verifies the validity and rationality of the FE process simulation of hot stamping.

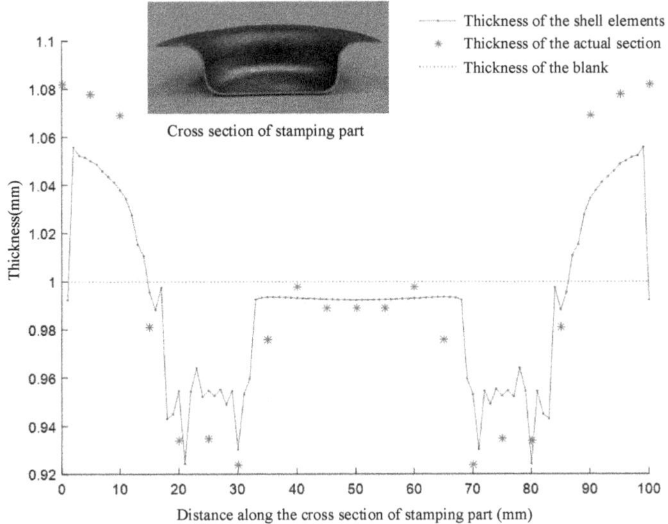

Figure 14. Thickness variation along the symmetrical section of the experimentally obtained parts and simulation result.

6. Conclusions

Hot stamping is developed and widely applied in vehicle production according to the lightweight demands of automobiles. But hot stamping is energy intensive due to the high-temperature forming conditions of blanks. To reduce the energy consumption and improve the energy efficiency of hot stamping, an energy-economizing optimization method for hot stamping is proposed. In this method, the process parameters are optimized to reduce the energy consumption of hot stamping while maintaining the required forming quality.

In this study, the mathematical modelling, simulation, and optimization of ZK60 magnesium alloy sheets hot stamping process were addressed in the aim of improving the energy efficiency and forming quality of the stamping parts. For this purpose, a new multiobjective optimization method is proposed and tested in a real industrial size case study experiment. The model is developed and solved using a multiobjective genetic algorithm (NSGA-II), and offers feasible optimized solutions. The comparison between the numerically-predicted technical parameters of the stamps and the real experimental results demonstrates the applicability of the method. Under the optimized conditions in the case study, a substantial energy consumption reduction, i.e., 17.6%, is shown. This method may serve as a reference for finding proper process parameters to solve the energy inefficiency and high energy consumption problems associated with the metal forming fields.

Author Contributions: Conceptualization, M.G. and L.L.; methodology, M.G.; software, L.L.; validation, M.G., Q.W. and L.L.; formal analysis, Z.M.; investigation, Z.M.; resources, Q.W.; data curation, Z.M.; writing—original draft preparation, M.G.; writing—review and editing, Q.W.; visualization, L.L.; supervision, L.L.; project administration, Q.W.; funding acquisition, M.G. and Q.W. All authors have read and agreed to the published version of the manuscript.

Funding: This research was funded by [key projects of natural science research in colleges and universities of Anhui province China] grant number [KJ2018A0451], [Anhui Major Science and Technology Project] grant number [18030901023], [Suzhou College Scientific Research Foundation Project] grant number [2017JB03], [Suzhou Engineering Research Center for Collaborative Innovation of Mechanical Equipment] grant number [SZ2017ZX07], [Suzhou College Teacher Application Ability Development Workstation] grant number [2018XJYY01], [Opening Project of Suzhou University Research Platform] grant number [2019kyf21, 2019ykf26, 2019ykf27], and [Suzhou College Teacher Application Ability Development Workstation] grant number [2018XJYY01].

Conflicts of Interest: We declare that we have no financial or personal relationships with other people or organizations that can inappropriately influence our work. We confirm that none of the material in the paper, in whole or in part, has been published or is under consideration for publication elsewhere. All the authors listed have approved the manuscript.

References

1. Liu, C.; Zhu, Q.; Wei, F.; Rao, W.; Liu, J.; Hu, J.; Cai, W. An integrated optimization control method for remanufacturing assembly system. *J. Clean. Prod.* **2019**, 119261. [CrossRef]
2. Yang, X. More Efforts are Needed to Promote New Energy. Available online: https://www.energy.people.com.cn/n1/2019/0307/c71661-30961766.html (accessed on 21 January 2020).
3. Whalen, S.; Overman, N.; Joshi, V.; Varga, T.; Graff, D.; Lavender, C. Magnesium alloy ZK60 tubing made by Shear Assisted Processing and Extrusion (ShAPE). *Mater. Sci. Eng. A* **2019**, *755*, 278–288. [CrossRef]
4. Singh, V.P.; Patel, S.K.; Kumar, N.; Kuriachen, B. Parametric effect on dissimilar friction stir welded steel-magnesium alloys joints: A review. *Sci. Technol. Weld. Join.* **2019**, *24*, 653–684. [CrossRef]
5. Karbasian, H.; Tekkaya, A.E. A review on hot stamping. *J. Mater. Process. Technol.* **2010**, *210*, 2103–2118. [CrossRef]
6. Gao, M.; He, K.; Li, L.; Wang, Q.; Liu, C. A review on energy consumption, energy efficiency and energy saving of metal forming processes from different hierarchies. *Processes* **2019**, *7*, 357. [CrossRef]
7. Bariani, P.F.; Bruschi, S.; Ghiotti, A.; Michieletto, F. Hot stamping of AA5083 aluminium alloy sheets. *CIRP Ann.-Manuf. Technol.* **2013**, *62*, 251–254. [CrossRef]
8. Gao, M.; Huang, H.; Wang, Q.; Liu, Z.; Li, X. Energy consumption analysis on sheet metal forming: Focusing on the deep drawing processes. *Int. J. Adv. Manuf. Technol.* **2018**, *96*, 3893–3907. [CrossRef]
9. Gao, M.; Liu, Z.; Li, L. Energy consumption analysis focusing on hot stamping of sheet metal. *J. Plast. Eng.* **2017**, *24*, 74–81.
10. Shi, C.W.P.; Rugrungruang, F.; Yeo, Z.; Gwee, K.H.K.; Ng, R.; Song, B. Identifying carbon footprint reduction opportunities through energy measurements in sheet metal part manufacturing. In *Glocalized Solutions for Sustainability in Manufacturing*; Springer: Berlin/Heidelberg, Germany, 2011; pp. 389–394.
11. Bosetti, P.; Bruschi, S.; Stoehr, T.; Lechler, J.; Merklein, M. Interlaboratory comparison for heat transfer coefficient identification in hot stamping of high strength steels. *Int. J. Mater. Form.* **2010**, *3*, 817–820. [CrossRef]
12. Abdulhay, B.; Bourouga, B.; Dessain, C. Experimental and theoretical study of thermal aspects of the hot stamping process. *Appl. Therm. Eng.* **2011**, *31*, 674–685. [CrossRef]
13. Xiao, W.; Wang, B.; Zheng, K. An experimental and numerical investigation on the formability of AA7075 sheet in hot stamping condition. *Int. J. Adv. Manuf. Technol.* **2017**, *92*, 3299–3309. [CrossRef]
14. Marler, R.T.; Arora, J.S. The weighted sum method for multi-objective optimization: New insights. *Struct. Multidiscip. Optim.* **2010**, *41*, 853–862. [CrossRef]
15. Costa, N.R.; Pereira, Z.L. Multiple response optimization: A global criterion-based method. *J. Chemom.* **2010**, *24*, 333–342. [CrossRef]
16. Zhou, J.; Wang, B.; Lin, J.; Fu, L. Optimization of an aluminum alloy anti-collision side beam hot stamping process using a multi-objective genetic algorithm. *Arch. Civ. Mech. Eng.* **2013**, *13*, 401–411. [CrossRef]
17. Kitayama, S.; Huang, S.; Yamazaki, K. Optimization of variable blank holder force trajectory for springback reduction via sequential approximate optimization with radial basis function network. *Struct. Multidiscip. Optim.* **2013**, *47*, 289–300. [CrossRef]
18. Zhou, J. A Method to Optimize Aluminum Alloy Door Impact Beam Stamping Process Using NSGA-II. *Mater. Sci. Forum* **2013**, *773–774*, 89–94. [CrossRef]
19. Xiao, W.; Wang, B.; Zhou, J.; Ma, W.; Yang, L. Optimization of aluminium sheet hot stamping process using a multi-objective stochastic approach. *Eng. Optim.* **2016**, *48*, 2173–2189. [CrossRef]
20. Liang, Y. Research and Application on Key Process Experiment of High Strength Steel for Hot Forming. Ph.D. Thesis, Da Lian University of Technology, Dalian, China, 15 December 2013.
21. Gao, M.; Huang, H.; Li, X.; Liu, Z. Carbon emission analysis and reduction for stamping process chain. *Int. J. Adv. Manuf. Technol.* **2017**, *91*, 667–678. [CrossRef]

22. Li, L.; Huang, H.; Zhao, F.; Zou, X.; Lu, Q.; Wang, Y.; Liu, Z.; Sutherland, J.W. Variations of Energy Demand With Process Parameters in Cylindrical Drawing of Stainless Steel. *J. Manuf. Sci. Eng.-Trans. ASME* **2019**, *141*, 091002. [CrossRef]
23. Deb, K.; Agrawal, S.; Pratap, A.; Meyarivan, T. *A Fast Elitist Non-dominated Sorting Genetic Algorithm for Multi-objective Optimization: NSGA-II*; Springer: Berlin/Heidelberg, Germany, 2000; pp. 849–858.
24. Zhang, S.; Song, G.; Song, H.; Cheng, M. Deformation Mechanism and Warm Forming Technology for Magnesium Alloys Sheets. *J. Mech. Eng.* **2012**, *48*, 28–34. [CrossRef]
25. Zhang, X.; Cui, Z.; Ruan, X. Warm Forging of Magnesium Alloys: The Formability and Flow Stress of AZ 31B. *J. Shanghai Jiaotong Univ.* **2003**, *37*, 1874–1877.

 processes 

Article

Modified Multi-Crossover Operator NSGA-III for Solving Low Carbon Flexible Job Shop Scheduling Problem

Xingping Sun, Ye Wang, Hongwei Kang *, Yong Shen *, Qingyi Chen and Da Wang

School of Software, Yunnan University, Kunming 650000, China; sunxp@ynu.edu.cn (X.S.); wangye@mail.ynu.edu.cn (Y.W.); devas9@ynu.edu.cn (Q.C.); wangda@mail.ynu.edu.cn (D.W.)
* Correspondence: hwkang@ynu.edu.cn (H.K.); sheny@ynu.edu.cn (Y.S.)

Abstract: Low carbon manufacturing has received increasingly more attention in the context of global warming. The flexible job shop scheduling problem (FJSP) widely exists in various manufacturing processes. Researchers have always emphasized manufacturing efficiency and economic benefits while ignoring environmental impacts. In this paper, considering carbon emissions, a multi-objective flexible job shop scheduling problem (MO-FJSP) mathematical model with minimum completion time, carbon emission, and machine load is established. To solve this problem, we study six variants of the non-dominated sorting genetic algorithm-III (NSGA-III). We find that some variants have better search capability in the MO-FJSP decision space. When the solution set is close to the Pareto frontier, the development ability of the NSGA-III variant in the decision space shows a difference. According to the research, we combine Pareto dominance with indicator-based thought. By utilizing three existing crossover operators, a modified NSGA-III (co-evolutionary NSGA-III (NSGA-III-COE) incorporated with the multi-group co-evolution and the natural selection is proposed. By comparing with three NSGA-III variants and five multi-objective evolutionary algorithms (MOEAs) on 27 well-known FJSP benchmark instances, it is found that the NSGA-III-COE greatly improves the speed of convergence and the ability to jump out of local optimum while maintaining the diversity of the population. From the experimental results, it can be concluded that the NSGA-III-COE has significant advantages in solving the low carbon MO-FJSP.

Keywords: multi-objective optimization; flexible job shop scheduling problem; low carbon; genetic algorithm; multi-crossover operator; co-evolution

Citation: Sun, X.; Wang, Y.; Kang, H.; Shen, Y.; Chen, Q.; Wang, D. Modified Multi-Crossover Operator NSGA-III for Solving Low Carbon Flexible Job Shop Scheduling Problem. *Processes* **2021**, *9*, 62. https://doi.org/10.3390/pr9010062

Received: 4 December 2020
Accepted: 26 December 2020
Published: 29 December 2020

1. Introduction

The flexible job scheduling problem (FJSP) is an extension of the classic job scheduling problem (JSP) and is closer to the actual production environment. In the scheduling process of the FJSP, processing operations can be processed on all optional machines. The assignable machine expands the search range of feasible solutions and also increases the complexity and the difficulty of solving the problem. The FJSP is a complex NP-hard problem, and its solution time increases exponentially as the problem size increases.

With the increasingly prominent energy crisis and environmental pollution, manufacturing has gradually become one of the hot spots in modern manufacturing. Manufacturing has adopted a new sustainable manufacturing model that has attracted widespread attention from industry and academia. In the manufacturing process of an enterprise, workshop scheduling is an important factor in the manufacturing process. It not only affects the production efficiency and the economic benefits of the enterprise but also is closely related to the social responsibility of the enterprise. Therefore, it has important theoretical and practical significance to conducting research on flexible job scheduling with the goal of protecting the environment and saving energy.

In the past few decades, the single-objective FJSP (SO-FJSP), which has been extensively studied in the literature, has usually sought to minimize the total completion time [1–6]. However, many realistic scheduling problems often need to optimize multiple objectives at the same time, and these objectives usually conflict with each other. The main method used to solve the MO-FJSP is the MOEA, which can be roughly divided into two categories: the weighting method and the Pareto method. The weighting method solves the MO-FJSP by assigning different weights to each objective and transforming the multi-objective problem into a single-objective problem. The Pareto method solves the MO-FJSP based on the Pareto dominance relationship and generates a set of Pareto optimal solutions. As a Pareto method, the non-dominated sorting genetic algorithm-II (NSGA-II) [7] is an effective method to solve various multi-objective optimization problems in recent years. Z.-Q. Jiang et al. [8] used the NSGA-II algorithm, which optimizes mutation strategies to solve the multi-objective FJSP of strategy. Yuan Y. and Xu H. [9] proposed a new memetic algorithm that combines the memetic algorithm with the NSGA-II to solve the FJSP with the goal of minimizing completion time, total workload, and critical workload. Bandyopadhyay and Bhattacharyaput [10] proposed a modified NSGA-II with a new mutation algorithm for a parallel machine scheduling problem and proved the effectiveness of the algorithm. The research goal of all the improved algorithms is to solve the MO-FJSP more effectively.

The FJSP is widely present in various manufacturing processes. It has received extensive attention from researchers. A large number of research results have appeared [8–19]. However, in these studies, the objective function of the problem is rarely to minimize carbon emissions or total energy consumption. The FJSP with these objectives has not attracted attention. The existing FJSP research focuses on the relationship between carbon emissions or energy consumption and time [16–19]. The machine load is rarely optimized as a target problem. The completion time and the machine load are also two conflicting issues. The price of minimizing the total completion time is the long-term overload of high-performance machines. Therefore, it is necessary to optimize machine load as an objective.

This paper establishes an MO-FJSP model targeting carbon emissions, the completion time, and the machine load and modifies NSGA-III [20]. By studying the differences in exploration and developmental capabilities of different NSGA-III variants in the MO-FJSP decision space, indicator-based thought is introduced into NSGA-III, a genetic model of multiple populations and multiple crossover operators is established, and a new evolutionary mechanism is proposed. We apply this evolutionary mechanism to NSGA-III and propose the co-evolutionary NSGA-III (NSGA-III-COE). Then, calculation experiments are carried out on 27 well-known FJSP benchmark instances [21,22]. Quantities of experiments in this paper prove that the NSGA-III-COE achieves good results in solving the low carbon MO-FJSP and verifies the advantages and the competitiveness of the NSGA-III-COE in solving the low carbon MO-FJSP.

2. Mathematical Modeling of the MO-FJSP

Before building the mathematical model and the assumptions are listed below.

2.1. Assumptions

The machining process satisfies the following assumptions and constraints. These assumptions and constraints are common in the FJSP literature [5–12].

1. Each job can be processed on multiple machines.
2. All machines are available at the initial moment.
3. Each job can be processed at the initial moment.
4. Each machine can only process one job at a time.
5. In a given time, a machine can only process one job.
6. The process of each job can only be processed in a given order.
7. Each process has a processing time, and the processing times of these processes are different.
8. The processing time of a job's process varies with the machine.
9. The processing time of the process on the processing machine is known.

2.2. Mathematical Model

The mathematical formulas and constraints are as follows:

$$T = \min T_{\max} = \min\left\{\sum_{p=1}^{N_n}\{T_p\}\right\}, \tag{1}$$

$$W = \min W_{\max} = \min\left\{\sum_{h=1}^{N_m}\{W_h\}\right\}, \tag{2}$$

$$C = \min C_{\max} = \min\left\{\sum_{p=1}^{N_n}\sum_{h=1}^{N_m}\sum_{q=1}^{N_p}\left(T_{sh}C_{sh} + T_{pqh}C_{pqh}\right)\right\}. \tag{3}$$

$$S_{pq}, T_{pqh} \geq 0, J_p \in J; q = 1,2,\cdots,N_p; h = 1,2,\cdots,N_m, \tag{4}$$

$$\sum_{h=1}^{M_{pq}} \sigma_{pqh} = 1, J_p \in J; q = 1,2,\cdots N_p, \tag{5}$$

$$S_{p(q+1)} \geq S_{pq} + T_{pqh}, J_p \in J; M_h \in M_{pq}; q = 1,2,\cdots,N_p - 1, \tag{6}$$

$$\sum_{p=1}^{N_n}\sum_{q=1}^{N_p} T_{pqh}\sigma_{pqh} \leq W_h, J_p \in J; M_h \in M_{pq}, \tag{7}$$

Objective (1) represents the objective function that minimizes the maximum completion time, objective (2) represents the objective function that minimizes the total machine load, and objective (3) represents the objective function that minimizes total carbon emissions during processing. Constraint (4) indicates that the start time and the processing time of the process are greater than or equal to 0, constraint (5) indicates that each process can only select one machine from the set of candidate processing machines, constraint (6) indicates that each job must be processed in the given order, and constraint (7) represents the total machine load.

2.3. Chromosome Encoding

The FJSP needs to select processing machines for each process and sort the processes allocated on each machine. According to the characteristics of the FJSP, this paper adopts the two-dimensional chromosome encoding method based on the combination of process coding and machine coding [9]. The following uses an FJSP instance to illustrate chromosome encoding. The processing time of an FJSP instance with three jobs, three machines, and seven processes is shown in Table 1.

Table 1. Machine processing schedule of the flexible job shop scheduling (FJSP) instance. '-' means that the process cannot be processed by this machine.

Job	Operation	M_1	M_2	M_3
J_1	O_{11}	2	3	-
	O_{12}	1	1	-
	O_{13}	3	2	1
J_2	O_{21}	-	1	3
	O_{22}	-	1	1
J_3	O_{31}	2	-	1
	O_{32}	2	-	1

Table 2 is a set of randomly generated chromosome codes corresponding to the instances in Table 1. Pro is a process-based code used to determine the processing order, and Mac is a machine-based code used to allocate the processing machines for each process. Each column of genes can be interpreted as (O_{pq}, M_h), that is, Pro is the processing sequence of the process $O_{21} \rightarrow O_{11} \rightarrow O_{12} \rightarrow O_{22} \rightarrow O_{31} \rightarrow O_{13} \rightarrow O_{32}$, and the corresponding Mac is the machine on which the process is processed $(M_2, M_1, M_2, M_2, M_3, M_3, M_3)$. The gene $(2,2)$ in the first column of Table 2 can be interpreted as (O_{21}, M_2), and the gene $(2,2)$ in the fourth column can be interpreted as (O_{22}, M_2). That is, the first process of the second job is processed on machine 2, and the required processing time is $T_{212} = 1$; the second process of the first job is processed on machine 2, and the required processing time is $T_{222} = 1$ (the processing time is found in Table 2).

Table 2. A set of chromosome code representation instances.

Pro	2	1	1	2	3	1	3
Mac	2	1	2	2	3	3	3

2.4. Chromosome Decoding

Chromosome decoding allocates a period of time for each operation on the designated machine according to the sequence of the process in the chromosome. Take the FJSP instance shown in Table 1 as an instance to decode the chromosomes in Table 2. There are two different decoding schemes; the first is to assign the machine processing according to the sequence of the process chromosome Pro to the Mac chromosome. The scheduling Gantt chart is shown in Figure 1a. The second is when processing a process in the Pro chromosome, first obtain the machine selected in the Mac chromosome for the process, and then scan the machine from left to right to determine the idle time interval between the processing processes and insert the current process until the time period that can be processed is found. The second scheduling Gantt chart is shown in Figure 1b. The second decoding scheme allows process scheduling to search for the earliest available idle time interval on a specified machine, which can effectively reduce the production cycle. Therefore, this paper uses the second decoding scheme.

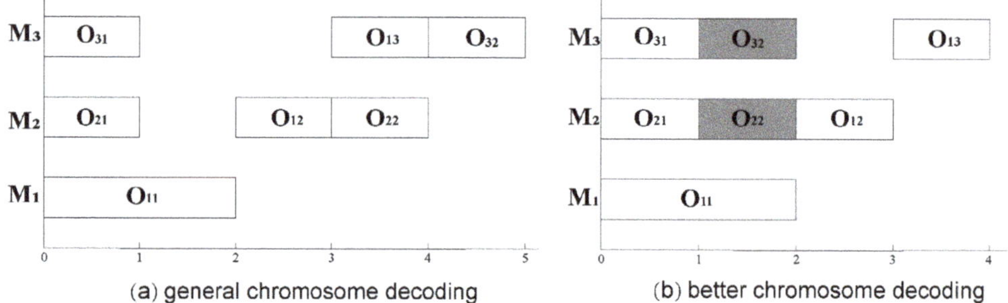

Figure 1. Chromosome decoding corresponds to the Gantt chart.

3. Modification of the NSGA-III

3.1. Introduction to the NSGA-III

The NSGA-III replaces the crowding distance selection operation in the NSGA-II with a reference point-based selection operation and uses well-distributed reference points to maintain the diversity of the population. This is the reason why this paper selects it. In addition, the NSGA-III is the most widely used MOEA in the existing literature. Next, we briefly describe the main procedures of the NSGA-III.

The NSGA-III first defines a set of reference points, then randomly produces an initial population containing N individuals, and then iterates until the termination condition is met. P_i is the population in the ith generation, and Q_i is the population generated by P_i after the reproduction phase. In order to select N individuals from population $R_i (R_i = P_i \cup Q_i)$ into the next generation, non-dominated sorting is used to divide the individuals in R_i into several different non-dominated layers $(F_1, F_2, \cdots)$ and to add the non-dominated layers into S_i in order. S_i is determined to be selected. It is the population of the $(i + 1)$th generation P_{i+1}. Assuming that F_l is the last non-dominated layer where the population size of S_i is larger than N for the first time, use the reference point to find the optimal number of remaining P_{i+1} individuals in F_l and join the next generation population P_{i+1}.

3.2. Study of NSGA-III Variants

The original NSGA-III used simulated binary crossover (SBX) [23] to generate individual offspring. This section calls this method the NSGA-III-SBX to distinguish it from other NSGA-III variants studied in this section. In this paper, we introduce cycle crossover (CX) [24], order-based crossover (OBX) [25], order-crossover (OX) [26], partially mapped crossover (PMX) [27], and position-based crossover (PBX) [25] into the NSGA-III, replacing the original SBX operator and forming 5 NSGA-III variants, namely, NSGA-III-CX, NSGA-III-OBX, NSGA-III-OX, NSGA-III-PBX, and NSGA-III-PMX.

In order to study the search performance of all NSGA-III variant algorithms in the decision space, we use part of three sets of well-known FJSP benchmark instances, including ka3, ka4, and ka5 in the Kacem instance [22] and mk4, mk5, and mk7 in the BRdata instance [21], to conduct exploration and testing. These six instances are representative from simple to complex. In the experiments in this section, we use these six benchmark instances to explore the NSGA-III variants mentioned in this section and propose a modified NSGA-III based on the research results.

Table 3 lists the parameters used in this section to study the different variants of NSGA-III, and we use uniform parameter values for all variants. In preliminary research, we found that the widely used MOEAs are prone to fall into local optimum on FJSP. Some studies in the literature have found that a larger mutation probability can effectively help the population jump out of the local optimum [10]. Thus, we use a high mutation probability. In order to explore whether the performance of different variants is related to population size, we use two population sizes in our research, 200 and 300. When the number of iterations reaches the set maximum number of iterations, the algorithm is terminated. To ensure a fair comparison, for each benchmark instance, all variants are run independently with the same initial population 30 times, and the average of 30 experiments is taken for comparison.

Table 3. Parameter setting of the non-dominated sorting genetic algorithm-III (NSGA-III) variant algorithm.

Parameter	Value
Crossover probability (P_c)	0.95
Mutation probability (P_m)	0.05

In our experiments, we use the generational distance (GD) [28] and the inverted generational distance (IGD) [29] as evaluation indicators to evaluate the convergence of the non-dominated solution set and the comprehensive performance of the algorithm. They can be expressed as follows.

GD: Assuming that P is the solution set obtained by the algorithm and P^* is a set of uniformly distributed reference points sampled from the Pareto front (PF), the *GD* of solution set P is defined as follows:

$$GD(P, P^*) = \frac{1}{|P|}\sqrt{\sum_{y \in P} \min_{x \in P^*}\left(dis(x,y)^2\right)}, \tag{8}$$

$dis(x, y)$ represents the Euclidean distance between point y in solution set P and point x in reference set P^*. *GD* only evaluates the convergence of the solution set. The smaller the *GD* value is, the better the convergence of the algorithm is.

IGD: Assuming that P is the solution set obtained by the algorithm and P^* is a set of uniformly distributed reference points sampled from the Pareto front (PF), then the IGD value of solution set P is defined as follows:

$$IGD(P, P^*) = \frac{1}{|P^*|} \sum_{x \in P^*} \min_{y \in P} dis(x, y), \tag{9}$$

$dis(x, y)$ represents the Euclidean distance between point x in reference set P^* and point y in solution set P. If $|P^*|$ is large enough to fully represent the Pareto front, then the *IGD* can comprehensively measure the convergence and the diversity of the solution set. If we want to obtain a smaller *IGD*, the solution set must be close enough to the Pareto front in the target space.

When calculating the *GD* and the *IGD*, a reference set is needed. Since the actual Pareto front of the benchmark instance is unknown, the reference set used in the calculation of the *GD* and the *IGD* in this paper is formed by collecting all the non-dominated solutions found during the runtime of all implemented algorithms.

Next, we compare the search behavior of different NSGA-III variants on the MO-FJSP. Simply put, we want to understand the algorithm search process for solution sets in the decision space as well as which algorithm is better at exploration and which algorithm is better at development. Knowing these can help us design more effective evolutionary mechanisms. First, we randomly initialize a population for each instance and then perform 200 and 300 iterations.

Figures 2 and 3 depicts the trajectories of the GDs obtained from the six NSGA-III variants as the number of iterations increases when the population sizes are 200 and 300, respectively. Figures 2 and 3 show that, although the population sizes are different, the same NSGA-III variant shows very similar convergence trends on these benchmark instances, and the convergence of different NSGA-III variants is significantly different due to the different complexities of the benchmark instances. As the complexity of the benchmark instance increases, NSGA-III-CX, NSGA-III-OBX, and NSGA-III-PBX show better convergence, and the GDs of these three algorithms decrease in a similar way during the evolutionary process. This phenomenon indicates that the initial population is a randomly distributed solution in the decision space, and then the optimal solution is searched continuously. NSGA-III-CX, NSGA-III-OBX, and NSGA-III-PBX can search for better solutions faster than other algorithms. The above experimental results show that the three NSGA-III variants, NSGA-III-CX, NSGA-III-OBX, and NSGA-III-PBX, can explore more optimal solutions more effectively in the decision space.

In order to study the development capabilities of different NSGA-III variants, we conduct a similar experiment. The difference from the previous experiment is that we replace the initial population with a population that is already closer to the Pareto front, which can be obtained through iteration by any multi-objective evolutionary algorithm. In this experiment, only the three NSGA-III variants with better exploration capabilities are used. The purpose is to study the abilities of NSGA-III-CX, NSGA-III-OBX, and NSGA-III-PBX to develop better solutions. The three NSGA-III variant algorithms use the population close to the Pareto front as the initial population to perform 100 iterations. As before, use two population sizes, 200 and 300.

Figures 4 and 5 depicts the trajectories of IGDs obtained from the three NSGA-III variants as the number of iterations increases when the initial population is close to the Pareto front when the population size is 200 and 300 respectively. We compare Figures 4 and 5 first. Similar to the previous experiment, the same NSGA-III variant showed very similar ability to develop better solutions when the population size was different. However, the situation in Figures 4 and 5 is very different from that in Figures 2 and 3. In Figures 4 and 5, from the beginning, as the number of iterations increases, a certain algorithm will reduce IGD faster, while the IGD of other algorithms will decrease more slowly. Since the initial population is a population closer to the Pareto frontier, an algorithm with a faster IGD decline has a better ability to develop better solutions in the decision space. In Figures 4e and 5e, on the instance mk5, the NSGA-III-CX has the best ability to develop better solutions. NSGA-III-OBX has the best development capability on other instances. It is speculated from this that when faced with different decision spaces, the crossover operator with better development capabilities may change. The research in this section can help us design more effective evolutionary mechanisms to solve low carbon FJSP.

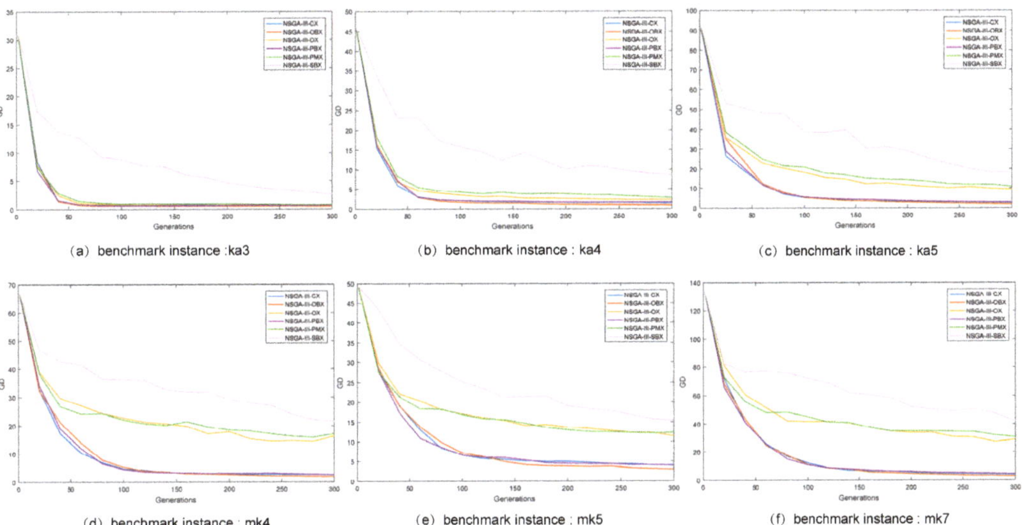

Figure 2. At population size of 200, the evolutionary trajectory of the generational distances (GDs) of the NSGA-III variants on the six FJSP instances (the average of the results of 30 independent runs; the initial population is a random population).

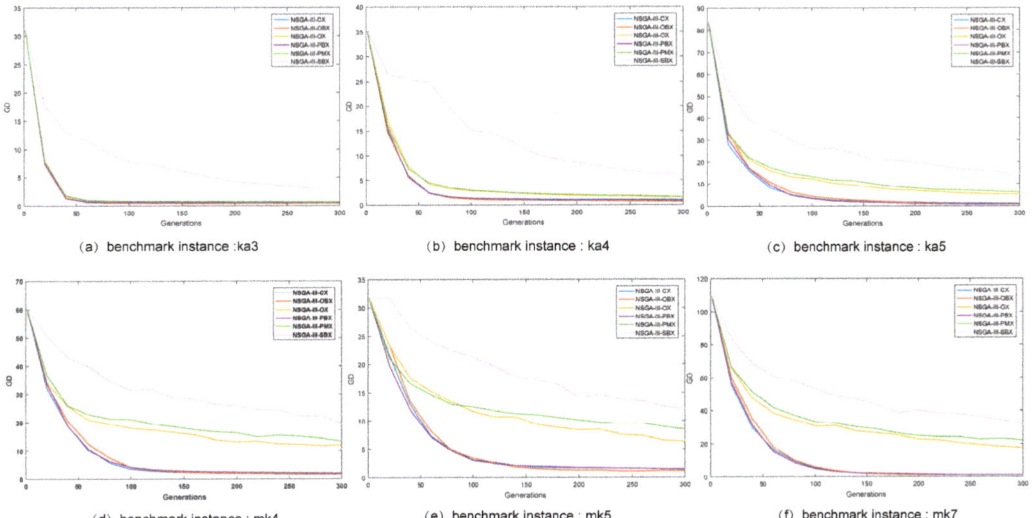

Figure 3. At population size of 300, the evolutionary trajectory of the GDs of the NSGA-III variants on the six FJSP instances (the average of the results of 30 independent runs; the initial population is a random population).

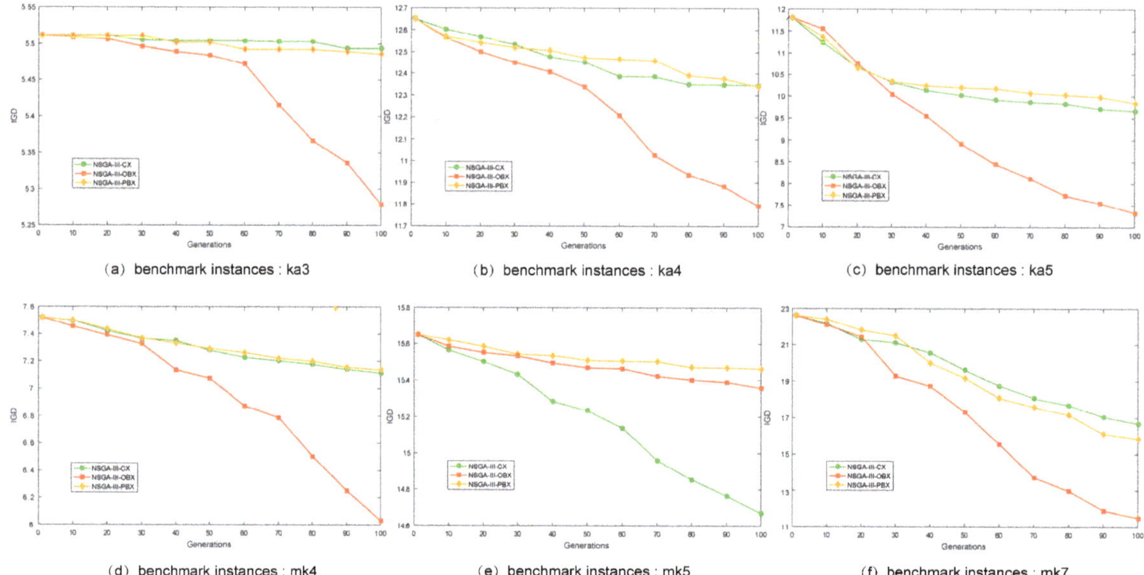

Figure 4. At population size of 200, the evolutionary trajectory of the inverted generational distance (IGD) for the NSGA-III variants on the six FJSP instances (the average of the results of 30 independent runs; the initial population is the population close to the Pareto front in the target space).

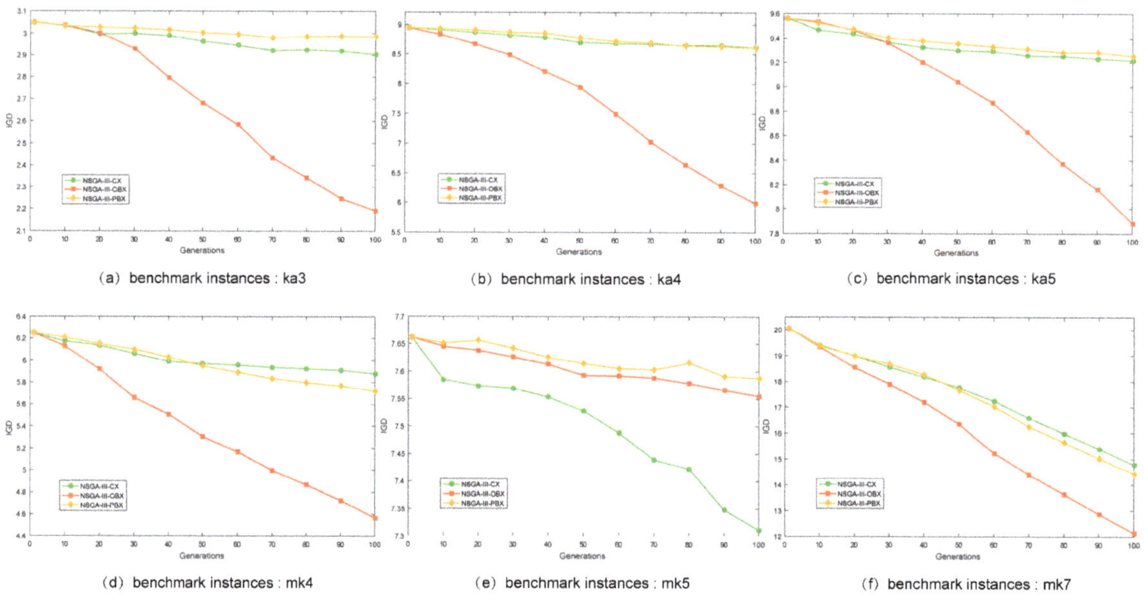

Figure 5. At population size of 300, the evolutionary trajectory of the IGD for the NSGA-III variants on the six FJSP instances (the average of the results of 30 independent runs; the initial population is the population close to the Pareto front in the target space).

3.3. The NSGA-III-COE Proposal

The research in the previous two sections shows that the NSGA-III variant using the three crossover operators CX, OBX, and PBX has better exploration capabilities than others in the decision space. When the initial population is close to the Pareto frontier, in most instances, NSGA-III-OBX has the best ability to develop better solutions in the decision space. However, in a few instances, NSGA-III-OBX does not have the best development capabilities. Therefore, which cross-operator has the best development capability is still uncertain. The above research aims to help us design a more effective evolutionary mechanism when solving the MO-FJSP.

Many studies have shown that exploring and developing strategies at the same time can find more useful information from the decision space in the process of finding a better solution. If we make full use of the three crossover operators of CX, OBX, and PBX, we can expect the algorithm to achieve better performance. This is the motivation for proposing the NSGA-III-COE algorithm. The effects of three different crossover operators are naturally integrated to improve the search ability of the decision space and maintain the diversity of the population. This is the main purpose of the NSGA-III-COE.

The NSGA-III-COE is the result of the combination of Pareto dominance and indicator-based thought. In order to achieve our purpose, we decided to coevolve three subpopulations using CX, OBX, and PBX crossover operators. In the process of evolution, natural selection is carried out by simulating the evolution of biological populations to achieve the purpose of survival of the fittest. To achieve natural selection, a certain parameter is necessary to guide the evolution of the population. Therefore, we combine the indicator-based idea with NSGA-III and add the concept of indicator into NSGA-III to guide the natural selection of the population.

In order to propose the NSGA-III-COE algorithm, we introduce the set coverage (SC) [30]. Assuming that both set A and set B are obtained approximate solution sets, the numerator of formula (10) represents the number of solutions in which the solution in B is dominated by at least one solution in A, and the denominator represents the total number of solutions contained in B. The SC is the probability that the solutions in B is dominated by at least one solution in A.

$$C(A,B) = \frac{|\{x \in B | \exists y \in A : y \text{ dominates } x\}|}{|B|}, \tag{10}$$

Each subpopulation in the initial population has the same number of individuals. In the evolution process, the evolution of biological populations is simulated, and a small number of individuals are randomly exchanged in each iteration to increase the diversity of chromosomes in the decision space and to increase the amount of useful information in the decision space.

When the evolution reaches half of the maximum number of iterations, the SC indicator intervenes. The subpopulation size is adjusted every 10 generations according to the SC indicator. The SC indicator makes natural selection of the subpopulation based on the exploration and the development ability of the subpopulation in the decision space. Natural selection in the evolutionary process means increasing the size of superior subpopulations and reducing the size of disadvantaged subpopulations in order to achieve the survival of the fittest. Algorithm 1 describes the evolutionary mechanism of the NSGA-III-COE.

Algorithm 1: Evolutionary mechanism of the NSGA-III-COE.

```
function Evolution(PopCX, PopOBX, PopPBX, genNow, gen)
if genNow > gen/2 and mod(genNow, gen/10) == 0 then
PopCX, PopOBX, PopPBX ← AdjustPopSzie(PopCX, PopOBX, PopPBX)
end if
PopCX, PopOBX, PopPBX ← RandomExchange(PopCX, PopOBX, PopPBX)
PopCX ← OperatorCX(PopCX)
PopCX ← OperatorCX(PopCX)
PopCX ← OperatorCX(PopCX)
return PopCX, PopOBX, PopPBX
end function
```

4. Experimental Results and Discussion

In order to verify the advantages of the NSGA-III-COE in the MO-FJSP decision space exploration and development capabilities, a large number of computational experiments were carried out. These experiments were implemented by MATLAB programming and were tested on three sets of well-known benchmark instances, including 5 Kacem instances (ka1, ka2, ka3, ka4, ka5) [22], 10 BRdata instances (mk1–mk10) [21], and 12 BRdata instances (01a–12a) [22]. These collections cover most of the problem instances used in the FJSP literature. In fact, most of the existing research only considers a small part of them. In our experiment, 27 instances are used to comprehensively evaluate the algorithm we propose.

Table 4 lists the parameter settings of the algorithm, and we use uniform parameter values for the algorithm. For all instances, the maximum number of iterations is set to 300, and it is the same for all implemented algorithms. When the number of iterations reaches the set maximum number of iterations, the algorithm terminates to ensure fair comparison. For each instance, all algorithms run independently 30 times starting with the same initial population.

Table 4. Experimental parameter settings for the co-evolutionary NSGA-III (NSGA-III-COE) performance evaluation.

Parameter	Value
Population size (N)	300
Initial size of each subpopulation (N_s)	100
Number of objectives (M)	3
Maximum number of iterations ($T_{\max}$)	300
Crossover probability (P_c)	0.95
Mutation probability (P_m)	0.05

We are not sure whether all the nondominant solutions of the 27 benchmark instances collected in this paper enable *IGD* to more accurately reflect the overall performance of the algorithm on all instances. Therefore, this paper uses the hypervolume (*HV*) [30] indicator to evaluate the overall performance of the algorithms for all 27 instances.

Suppose P is the solution set obtained by the algorithm, and $q = (q_1, q_2, \cdots, q_m)^T$ is a reference point in the target space, which is dominated by all the target vectors in the solution set P. Then, the HV for reference point q refers to the volume of the target space dominated by solution set P and bounded by reference point q.

$$HV(P,q) = \text{volume}\left(\cup_{p\in P}[p_1, q_1] \times [p_2, q_2] \cdots [p_m, q_m]\right), \tag{11}$$

Figure 6 illustrates the meaning of HV in a two-dimensional target space. The HV indicator calculation does not require a reference point set and can comprehensively reflect the convergence and the diversity of the solution set. The larger the HV is, the better the solution set obtained by the algorithm is and the better the overall performance of the algorithm will be.

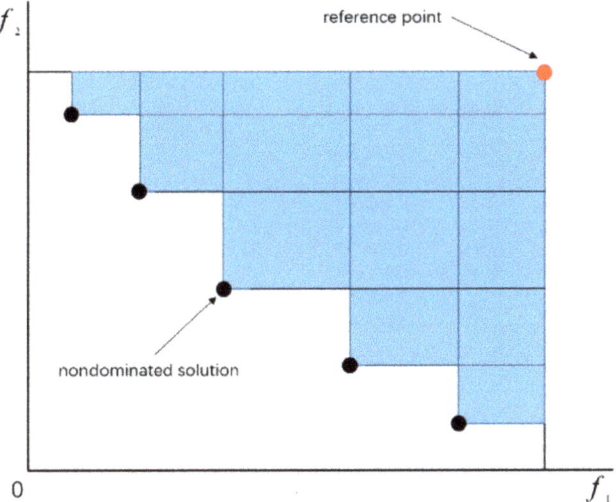

Figure 6. Diagram of the hypervolume (*HV*).

4.1. Comparison of NSGA-III-COE and NSGA-III Variants

In this section, we compare the NSGA-III-COE with NSGA-III-CX, NSGA-III-OBX, and NSGA-III-PBX to verify whether the population evolution mechanism proposed in this paper can integrate the effects of the three different crossover operators, enhance exploration and development capabilities in the decision space, and improve the performance of the algorithm. Table 5 shows the normalized average HVs obtained by running four algorithms 30 times independently on 27 benchmark instances. In addition, the features of the instance are also listed in the table. The first column indicates the name of the instance, and the second column indicates the size of the instance, where N_n indicates the number of processes, and N_m indicates the number of machines.

First, we analyze the three algorithms, NSGA-III-CX, NSGA-III-OBX, and NSGA-III-PBX. In 14 instances, the NSGA-III-OBX obtains the largest HV. In 11 instances, the NSGA-III-CX obtains the largest *HV*, and the NSGA-III-PBX obtains the largest *HV* in two instances. This phenomenon validates our conjecture when studying the developmental capabilities of NSGA-III variants. The same crossover operator shows different capabilities for developing better solutions when facing different MO-FJSPs.

Then, we add the NSGA-III-COE for analysis. Except for instances ka1 and mk2, the HVs of the NSGA-III-COE in the remaining 25 instances are all larger than those of the other NSGA-III variants. In instance ka1, the four algorithms have the same HVs. In instance mk2, NSGA-III-COE and NSGA-III-PBX both have the largest HVs. With the increase of instance complexity, the *HV* value of the NSGA-III-COE increases more and more obviously than other algorithms. This shows that the NSGA-III-COE performs best in all 27 instances, and as the complexity of the instance increases, the NSGA-III-COE shows better performance.

Table 5. Performance evaluation of NSGA-III-COE, NSGA-III-CX (cycle crossover), NSGA-III-OBX (order-based crossover), and NSGA-III-PBX (partially mapped crossover); the average HVs of 27 problem cases running independently 30 times. For each instance, the results which are better than the others are marked in bold (these have the largest *HV* value).

Instance	$N_n \times N_m$	NSGA-III-COE	NSGA-III-CX	NSGA-III-OBX	NSGA-III-PBX
ka1	3 × 4	0.038924	0.038924	0.038924	0.038924
ka2	4 × 5	**0.026456**	0.025696	0.026089	0.026372
ka3	10 × 7	**0.036638**	0.032826	0.036285	0.032998
ka4	10 × 10	**0.052563**	0.040711	0.043659	0.043500
ka5	15 × 10	**0.024589**	0.017029	0.022309	0.018040
mk1	10 × 6	**0.030030**	0.027031	0.029851	0.027547
mk2	10 × 6	**0.055304**	0.044472	0.049948	0.046155
mk3	15 × 8	**0.010112**	0.007421	0.007984	0.006651
mk4	15 × 8	**0.005349**	0.004029	0.003676	0.003089
mk5	15 × 4	**0.002695**	0.001993	0.001973	0.001895
mk6	10 × 15	**0.005967**	0.005599	0.004701	0.004495
mk7	20 × 5	**0.005084**	0.003309	0.004169	0.002962
mk8	20 × 10	**0.003186**	0.002989	0.002653	0.002296
mk9	20 × 10	**0.001387**	0.000730	0.000938	0.000414
mk10	20 × 15	**0.001743**	0.001098	0.001048	0.000477
01a	10 × 5	**0.008513**	0.007916	0.007997	0.008070
02a	10 × 5	0.003472	0.003414	0.003424	0.003472
03a	10 × 5	**0.007070**	0.005914	0.006099	0.005359
04a	10 × 5	**0.004745**	0.004028	0.004182	0.004116
05a	10 × 5	**0.005202**	0.004891	0.004659	0.004669
06a	10 × 5	**0.008040**	0.007375	0.007533	0.007330
07a	15 × 8	**0.003521**	0.003202	0.002860	0.002810
08a	15 × 8	**0.003402**	0.003080	0.002755	0.002622
09a	15 × 8	**0.007588**	0.007092	0.006975	0.006986
10a	15 × 8	**0.002659**	0.002087	0.001968	0.001715
11a	15 × 8	**0.003447**	0.003106	0.003053	0.002382
12a	15 × 8	**0.001557**	0.001222	0.001134	0.000959

In order to further verify whether the evolutionary mechanism proposed in this paper reaches our original design intention, Figure 7 shows the performances of the four algorithms of NSGA-III-COE, NSGA-III-CX, NSGA-III-OBX, and NSGA-III-PBX in instance mk2–mk10 at obtaining the non-dominated solution set in the same coordinate system obtained above. It can be seen from the figure that the non-dominated solution set obtained by the NSGA-III-COE is closer to the Pareto frontier than the other three algorithms and has good diversity. This indicates that the evolutionary mechanism proposed in this paper enhances search and development capabilities in the MO-FJSP decision space and speeds up the convergence speed while maintaining the diversity of the population, thus the algorithm obtains better performance.

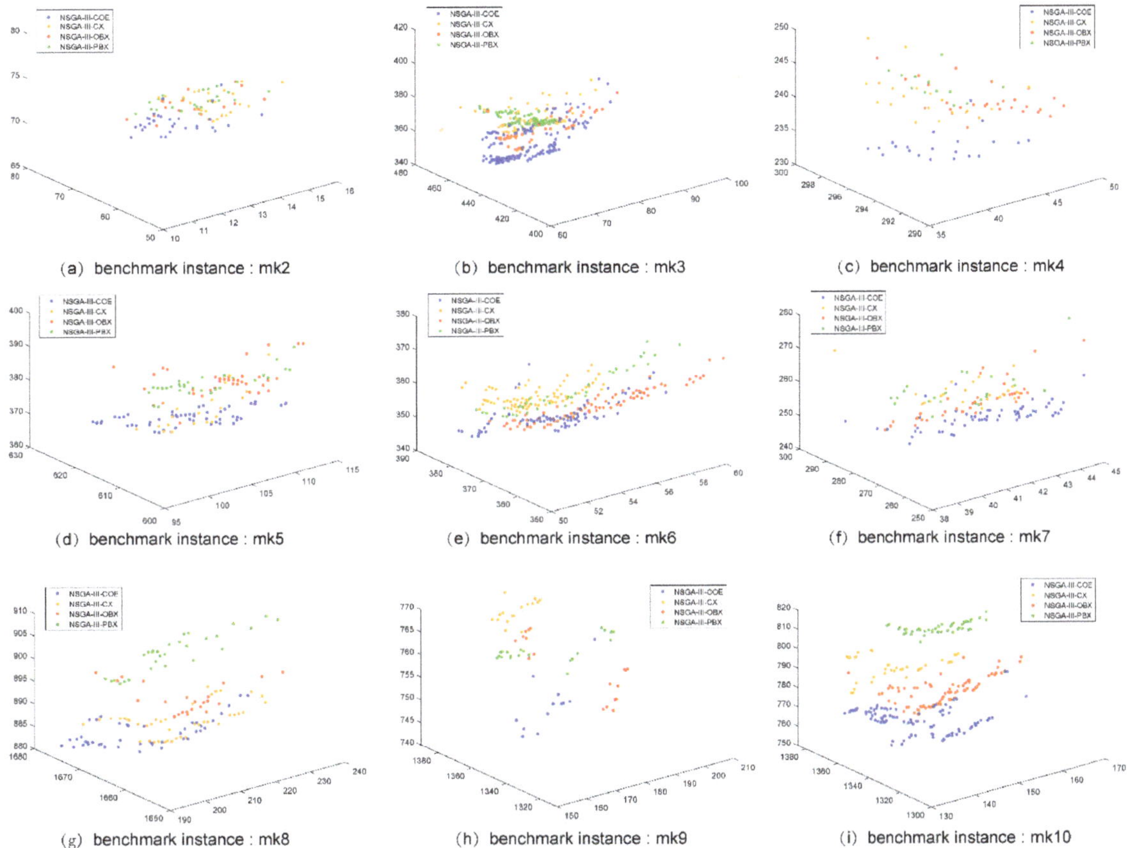

Figure 7. The non-dominated solutions of NSGA-III-COE, NSGA-III-CX, NSGA-III-OBX, and NSGA-III-PBX on instances mk2–mk10.

4.2. Comparison of the NSGA-III-COE with Widely Used MOEAs

This section compares the NSGA-III-COE with the existing widely used MOEAs. NSGA-III, NSGA-II, NSGA-II/strengthened dominance relation(NSGA-II/SDR) [31], improving the strength pareto evolutionary algorithm (SPEA2) [32] and hypervolume estimation algorithm for multi-objective optimization (HypE) [33] are chosen as the comparison algorithms because they are all currently widely used MOEAs that can be directly applied to solve the MO-FJSP after simple modification, and some of them have been used many times in the previous FJSP literature. Ahmadi et al. [34] used the NSGA-II to solve the FJSP with random failures. Bandyopadhyay et al. [10] compared the calculation results with NSGA-II and SPEA2 when solving a parallel machine scheduling problem. Yuan Y et al. [9] also used the NSGA-II for comparison when solving the FJSP. NSGA-II is used very frequently in solving scheduling problems. Therefore, this paper adds the newer modified NSGA-II algorithm NSGA-II/SDR to the comparison algorithms.

Table 6 shows the average HVs for 30 independent runs on 27 instances of NSGA-III-COE, NSGA-III, NSGA-II, NSGA-II/SDR, SPEA2, and HypE. The table shows that the HVs of the NSGA-III-COE on the three instances of ka1, ka2, and mk1 are slightly larger than those of the other algorithms. On the remaining 24 instances, the HVs of the NSGA-III-COE are much larger than those of the other algorithms. As the complexity of the benchmark instances increases, the advantages of the NSGA-III-COE become increasingly more obvious.

Table 6. Performance evaluation of the NSGA-III-COE and other comparison algorithms; the average HVs of 27 problem instances running 30 times independently. For each instance, the results which are significantly better than the others are marked in bold (these have much larger *HV* value than the other algorithms).

Instance	NSGA-III-COE	NSGA-III	NSGA-II/SDR	NSGA-II	SPEA2	HypE
ka1	0.038917	0.038763	0.038318	0.038678	0.038736	0.038609
ka2	**0.026269**	0.020522	0.020711	0.020292	0.023692	0.019619
ka3	**0.014032**	0.002858	0.002535	0.003239	0.002040	0.002653
ka4	**0.073033**	0.006564	0.007325	0.006934	0.005123	0.008142
ka5	**0.045579**	0.000723	0.001969	0.000670	0.000588	0.000775
mk1	**0.027291**	0.018659	0.018743	0.018591	0.017901	0.019043
mk2	**0.085445**	0.021139	0.023545	0.019240	0.015013	0.019672
mk3	**0.055143**	0.002332	0.002295	0.002367	0.002356	0.002455
mk4	**0.031763**	0.002991	0.003800	0.003022	0.003653	0.002835
mk5	**0.014355**	0.005375	0.005355	0.005334	0.006319	0.005436
mk6	**0.069930**	0.002379	0.003002	0.002544	0.004145	0.002644
mk7	**0.043737**	0.000740	0.001216	0.000716	0.000538	0.000610
mk8	**0.020138**	0.004515	0.003760	0.004417	0.004931	0.004724
mk9	**0.043676**	0.002392	0.002172	0.002298	0.003155	0.002507
mk10	**0.039108**	0.001319	0.001241	0.001250	0.001844	0.001281
01a	**0.018298**	0.003541	0.002632	0.003667	0.004244	0.003577
02a	**0.022695**	0.005291	0.004244	0.005345	0.006299	0.005265
03a	**0.056274**	0.008078	0.006250	0.008256	0.008777	0.008300
04a	**0.024167**	0.003110	0.002453	0.003039	0.003906	0.003125
05a	**0.031675**	0.004629	0.003988	0.004528	0.005552	0.004746
06a	**0.016381**	0.000991	0.000761	0.000875	0.001517	0.000941
07a	**0.018122**	0.002454	0.001985	0.002615	0.003106	0.002458
08a	**0.020696**	0.001245	0.001055	0.001240	0.001732	0.001313
09a	**0.019022**	0.001157	0.000955	0.001201	0.001597	0.001191
10a	**0.022589**	0.003136	0.002363	0.003140	0.003853	0.003125
11a	**0.032415**	0.002615	0.002055	0.002671	0.003192	0.002660
12a	**0.022939**	0.001854	0.001272	0.001954	0.002330	0.001810

Figures 8–10 show the non-dominated solution set in the same coordinate system obtained by the NSGA-III-COE and the comparison algorithm used in this section on almost all benchmark instances. It can be seen from the figure that the performance of the original NSGA-III is relatively close to that of other comparison algorithms. However, the non-dominated solution sets obtained by all the comparison algorithms are far away from the Pareto front, which obviously falls into the local optimum. The non-dominated solution obtained by the NSGA-III-COE is far superior to other comparison algorithms. This shows that the population evolution mechanism proposed in this paper is not only conducive to the improvement of convergence speed but also improves the ability to jump out of the local optimum on the MO-FJSP and greatly improves the performance of NSGA-III in solving the MO-FJSP.

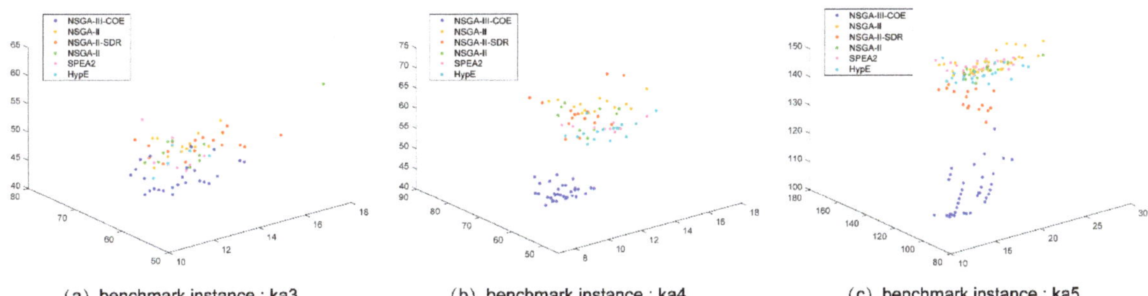

(a) benchmark instance : ka3 (b) benchmark instance : ka4 (c) benchmark instance : ka5

Figure 8. The non-dominated solutions of NSGA-III-COE, NSGA-III, NSGA-II, NSGA-II/SDR, SPEA2, and HypE on instances ka3, ka4, and ka5.

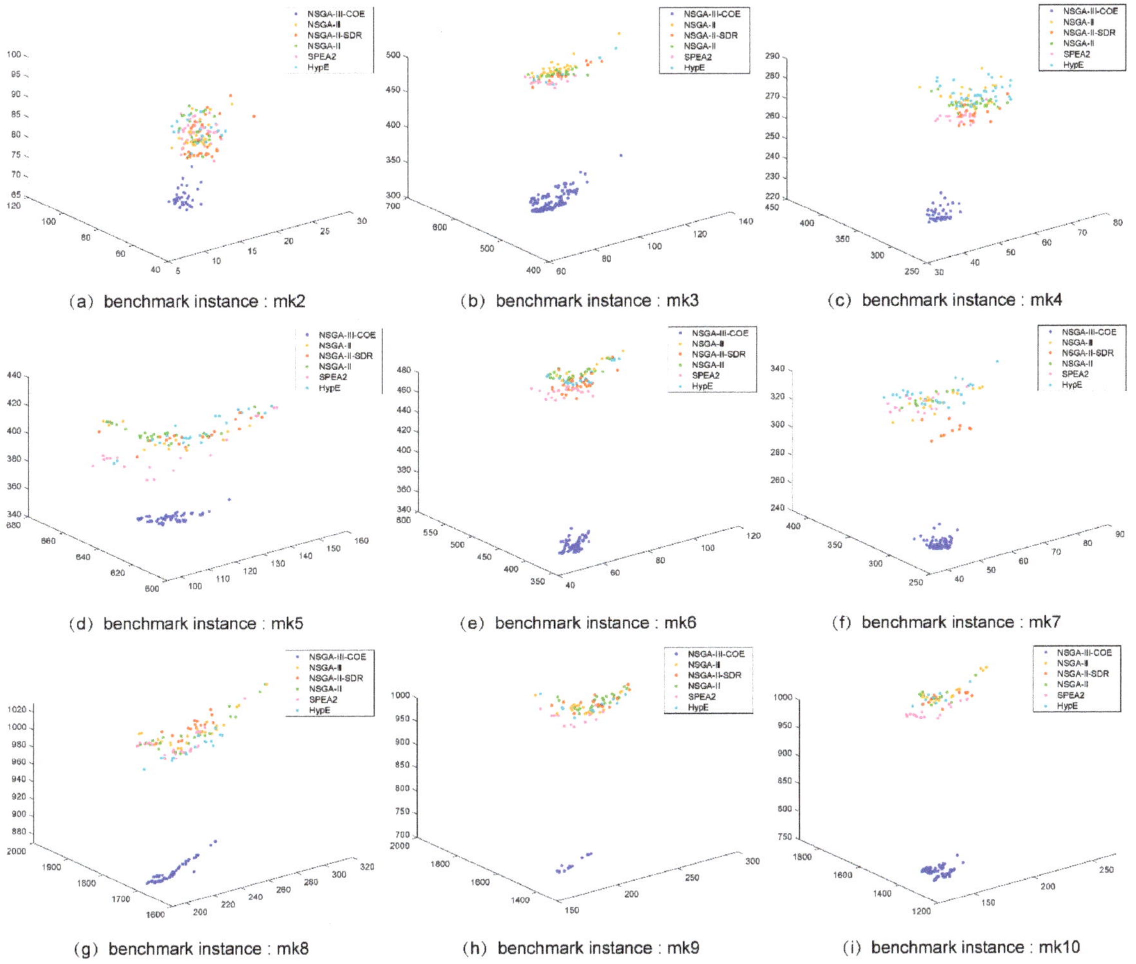

(a) benchmark instance : mk2 (b) benchmark instance : mk3 (c) benchmark instance : mk4

(d) benchmark instance : mk5 (e) benchmark instance : mk6 (f) benchmark instance : mk7

(g) benchmark instance : mk8 (h) benchmark instance : mk9 (i) benchmark instance : mk10

Figure 9. The non-dominated solutions of NSGA-III-COE, NSGA-III, NSGA-II, NSGA-II/SDR, SPEA2, and HypE on instances mk2–mk10.

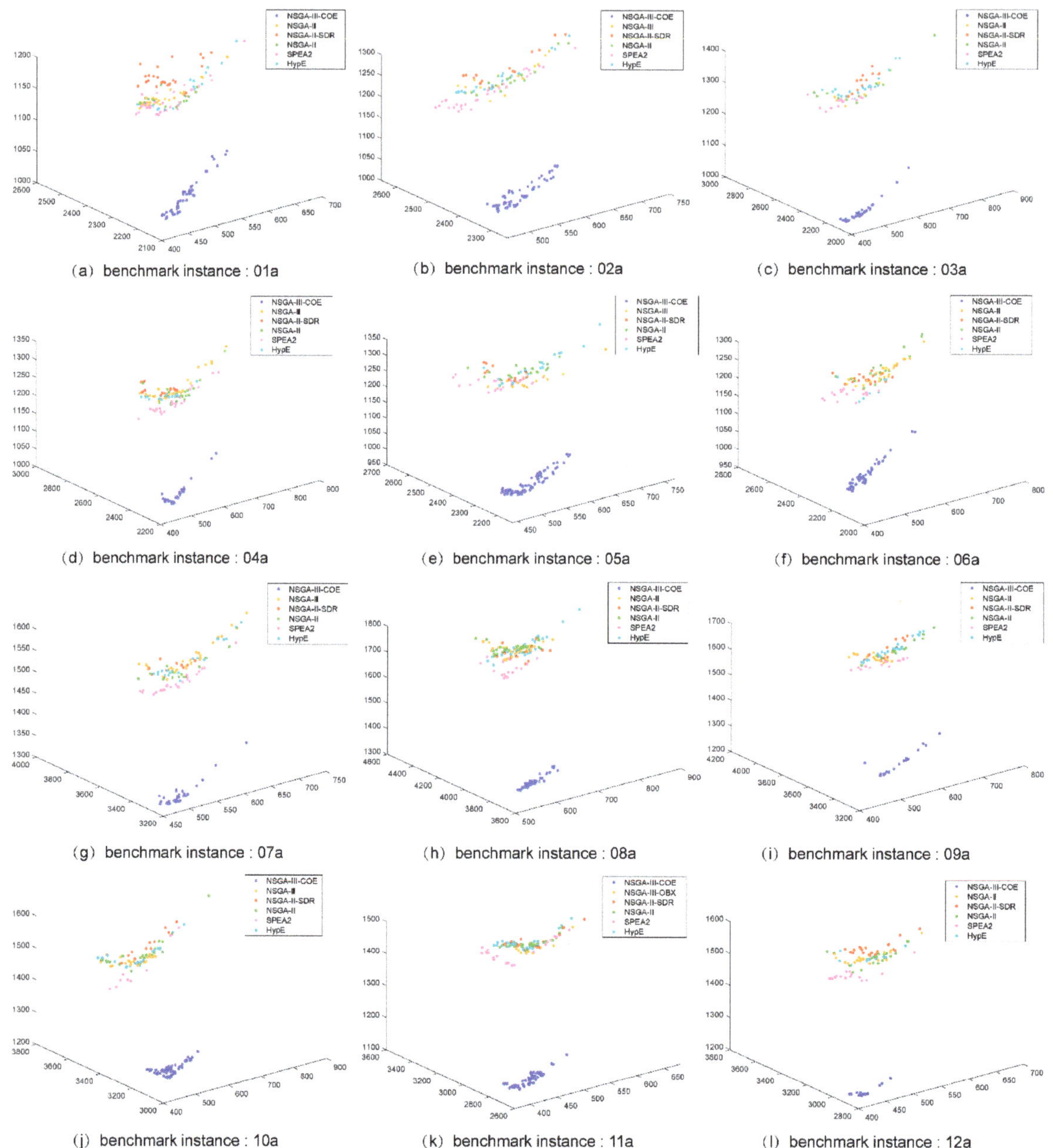

Figure 10. The non-dominated solutions of NSGA-III-COE, NSGA-III, NSGA-II, NSGA-II/SDR, SPEA2, and HypE on instances 01a–12a.

We count the running time of six algorithms on 27 instances, and the results are shown in Figure 11. The running time of the NSGA-III-COE is slightly longer than that of the NSGA-III. Compared with these widely used MOEAs, the computational cost of the NSGA-III-COE is at a moderate level.

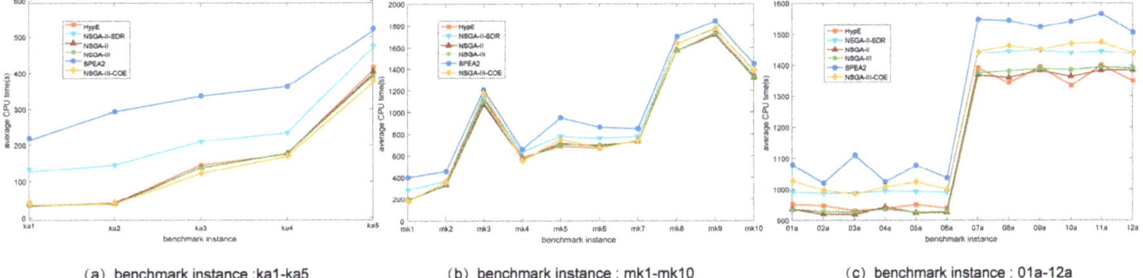

Figure 11. Average CPU time (in seconds) consumed by NSGA-III-COE and other comparison algorithms on 27 problem instances.

4.3. Sensitivity Analysis of Population Size

In this section, we associate the performance of the NSGA-III-COE and the other five MOEAs with the population size, which ranges from 30 to 500. We study the sensitivity of the NSGA-III-COE to population size on 27 FJSP benchmark instances.

We use *SR* (sum of ranks) to represent the performance of the algorithm:

$$SR = rank_{\text{ka1}} + \ldots + rank_{\text{ka5}} + rank_{\text{mk1}} + \ldots + rank_{\text{mk10}} + rank_{\text{01a}} + \ldots + rank_{\text{12a}}, \quad (12)$$

where *rank* represents the ranking of an algorithm among all algorithms for a benchmark instance. The ranking of the algorithm is based on the *HV* value of the non-dominated solution set. The larger the *HV* value is, the smaller the *SR* is, which means a higher ranking. For example, $rank_{\text{ka1}}$ represents the ranking of an algorithm on instance ka1. *SR* represents the cumulative sum rankings of an algorithm on all instances.

From Figure 12, the following experimental observation results can be obtained.

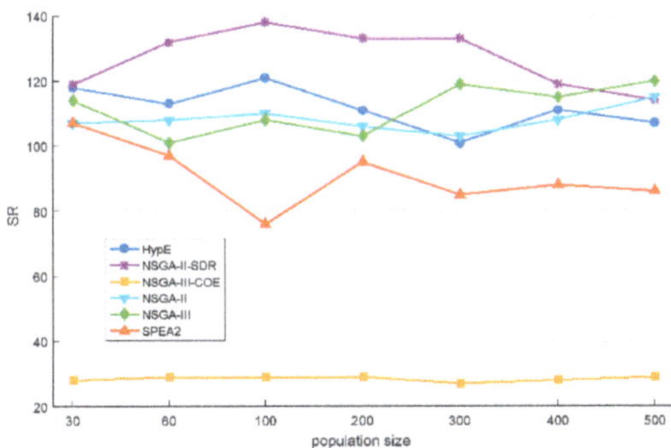

Figure 12. The impact of population size on the performance of the NSGA-III-COE and other multi-objective evolutionary algorithms (MOEAs) on 27 benchmark instances. The figure shows the sum of rankings of 27 instances obtained by each algorithm when the population size is 30, 60, 100, 200, 300, 400, and 500.

1. In experiments of different population sizes, the NSGA-III-COE achieves the highest ranking and is considerably ahead of other algorithms.
2. Some comparative MOEAs show sensitivity to population size. The most obvious is that the SPEA2 ranks best when the population size is 100.
3. In this set of comparative experiments, the NSGA-III-COE does not show obvious sensitivity to population size, and it performs well in the seven population sizes.

Considering experimental results, running time, and sensitivity to population size, the NSGA-III-COE is a very competitive algorithm for solving low carbon FJSP.

5. Conclusions

In this paper, a low carbon MO-FJSP mathematical model is established to minimize total completion time, total carbon emissions, and total machine load. This mathematical model is very close to the real production environment and conforms to the concepts of green manufacturing and sustainable development. By understanding the existing literature on the MO-FJSP research, this paper introduces five crossover operators into the NSGA-III to produce different NSGA-III variants. Then, the ability of different NSGA-III variants to explore and develop better solutions in the decision space on some FJSP benchmark instances is studied. Through research and analysis of the experimental results, the indicator-based thought is introduced into NSGA-III, and a new co-evolutionary mechanism incorporated with multi-crossover operator and natural selection is proposed, which combines the capabilities of different crossover operators to make the algorithm obtain better performance. Subsequently, we introduce the new evolutionary mechanism into the NSGA-III and propose the NSGA-III-COE. Using the NSGA-III-COE to solve low carbon MO-FJSP, multiple experiments are done. The NSGA-III-COE achieves good results in solving the MO-FJSP.

In the experiment, we compare the NSGA-III-COE with five existing widely used MOEAs on 27 benchmark instances in the three\ test sets of the FJSP. Experimental results show that the NSGA-III-COE has a strong ability to optimize the low carbon MO-FJSP, and the computational cost of solving the problem is similar to that of widely used MOEAs. Compared with other widely used algorithms, the NSGA-III-COE algorithm has obvious advantages in convergence speed and the ability to jump out of local optimum. Especially, it shows better performance when dealing with complex problem cases. Since the solving time of FJSP increases exponentially with the increase of the problem scale, our research is very meaningful for solving the low carbon MO-FJSP.

The work done in this paper only shows that the proposed the NSGA-III-COE is effective for solving the MO-FJSP. In the future, we will further study the production scheduling problem, continue to research and improve the algorithm, and apply the algorithm to solve other multi-objective production scheduling problems.

Author Contributions: Conceptualization, Y.W. and H.K.; Software, Y.W. and X.S.; Validation, Y.W. and X.S.; Formal Analysis, Y.W. and Y.S.; Investigation, H.K. and Q.C.; Data Curation, Y.W. and X.S.; Writing—Original Draft Preparation, Y.W.; Writing—Review & Editing, Y.W. and D.W.; Visualization, Y.W. and Y.S.; Funding Acquisition, X.S., H.K., Y.S. and Q.C. All authors have read and agreed to the published version of the manuscript.

Funding: This research was funded by National Natural Science Foundation of China, grant number 61663046, 1876166. This research was funded by Open Foundation of Key Laboratory of Software Engineering of Yunnan Province, grant number 2020SE308, 2020SE309. This research was funded by Science Research Foundation of Yunnan Education Committee of China, grant number 2020Y0004.

Institutional Review Board Statement: Not applicable.

Informed Consent Statement: Not applicable.

Data Availability Statement: Data sharing is not applicable to this article.

Conflicts of Interest: The authors declare no conflict of interest.

Abbreviations

p	Job number
q	Process number
h	Machine number
O_{pq}	qth process of job p
$J = \{J_1, J_1, \cdots J_n\}$	Collection of jobs
$M = \{M_1, M_1, \cdots M_m\}$	Collection of machines
M_{pq}	Collection of optional processing machines for process $O_{pq}(M_{pq} \in M)$
W_h	Load of machine M_h
N_n	Total number of jobs
N_{m}	Total number of machines
N_p	Number of processes contained in job p
S_{pq}	Start processing time of operation O_{pq}
σ_{pqh}	When the value is 1, it means that the qth process of job p is processed on machine M_h
T_{pqh}	Processing time of operation O_{pq} on machine M_h
T_p	Completion time of job J_p
T_{sh}	The time in standby state
C_{sh}	Carbon emission per unit time of machine M_h (M_h is in the standby state)
$C_{\text{pq}h}$	Carbon emission per unit time of machine M_h (M_h is performing operation O_{pq})

References

1. Cinar, D.; Oliveira, J.A.; Topcu, Y.I.; Pardalos, P.M. A priority-based genetic algorithm for a flexible job scheduling problem. *J. Ind. Manag. Optim.* **2017**, *12*, 1391–1415. [CrossRef]
2. Bozejko, W.; Uchronski, M.; Wodecki, M. Parallel hybrid metaheuristics for the flexible job problem. *Comput. Ind. Eng.* **2010**, *59*, 323–333. [CrossRef]
3. Jamrus, T.; Chien, C.F.; Gen, M.; Sethanan, K. Hybrid particle swarm optimization combined with genetic operators for flexible job-shop scheduling under uncertain processing time for semiconductor manufacturing. *IEEE Trans. Semicond. Manuf.* **2017**, *31*, 32–41. [CrossRef]
4. Yuan, Y.; Xu, H.; Yang, J. A hybrid harmony search algorithm for the flexible job scheduling problem. *Appl. Soft Comput. J.* **2013**, *13*, 3259–3272. [CrossRef]
5. Wu, X.; Wu, S. An elitist quantum-inspired evolutionary algorithm for the flexible job-shop scheduling problem. *J. Intell. Manuf.* **2017**, *28*, 1441–1457. [CrossRef]
6. Hmida, A.B.; Haouari, M.; Huguet, M.J.; Lopez, P. Discrepancy search for the flexible job scheduling problem. *Comput. Oper. Res.* **2010**, *37*, 2192–2201. [CrossRef]
7. Deb, K.; Pratap, A.; Agarwal, S.; Meyarivan, T. A fast and elitist multi-objective genetic algorithm: NSGA-II. *IEEE Trans. Evol. Comput.* **2002**, *6*, 182–197. [CrossRef]
8. Jiang, Z.Q.; Zuo, L. Multi-objective flexible job-shop scheduling based on low carbon strategy. *Comput. Integr. Manuf. Syst.* **2015**, *21*, 1023–1031.
9. Yuan, Y.; Xu, H. Multi-objective Flexible Job Scheduling Using Memetic Algorithms. *IEEE Trans. Autom. Sci. Eng.* **2014**, *12*, 336–353. [CrossRef]
10. Bhattacharya, B.R. Solving multi-objective parallel machine scheduling problem by a modified NSGA-II. *Appl. Math. Model.* **2013**, *37*, 6718–6729.
11. Zhao, B.; Gao, J.; Chen, K.; Guo, K. Two-generation Pareto ant colony algorithm for multi-objective job scheduling problem with alternative process plans and unrelated parallel machines. *J. Intell. Manuf.* **2018**, *29*, 93–108. [CrossRef]
12. Piroozfard, H.; Wong, K.Y.; Wong, W.P. Minimizing total carbon footprint and total late work criterion in flexible job scheduling by using an improved multi-objective genetic algorithm. *Resour. Conserv. Recycl.* **2018**, *128*, 267–283. [CrossRef]
13. Pomar, L.A.; Pulido, E.C.; Roa, J.D.T. A Hybrid Genetic Algorithm and Particle Swarm Optimization for Flow Shop Scheduling Problems. In Proceedings of the Workshop on Engineering Applications, Cartagena, CA, USA, 27–29 September 2017; pp. 601–612.
14. Burdett, R.L.; Kozan, E. An integrated approach for scheduling health care activities in a hospital. *Eur. J. Oper. Res.* **2018**, *264*, 756–773. [CrossRef]
15. Burdett, R.L.; Corry, P.; Yarlagadda, P.K.; Eustace, C.; Smith, S. A flexible job shop scheduling approach with operators for coal export terminals. *Comput. Oper. Res.* **2019**, *104*, 15–36. [CrossRef]
16. Mansouri, S.A.; Aktas, E.; Besikci, U. Green scheduling of a two-machine flowshop: Trade-off between makespan and energy consumption. *Eur. J. Oper. Res.* **2015**, *248*, 772–788. [CrossRef]
17. Mokhtari, H.; Hasani, A. An Energy-Efficient Multi-Objective Optimization for Flexible Job-Shop Scheduling Problem. *Comput. Chem. Eng.* **2017**, *104*, 339–352. [CrossRef]
18. Lu, C.; Li, X.; Gao, L.; Liao, W.; Yi, J. An effective multi-objective discrete virus optimization algorithm for flexible job-shop scheduling problem with controllable processing times. *Comput. Ind. Eng.* **2017**, *104*, 156–174. [CrossRef]

19. Ding, J.Y.; Song, S.; Wu, C. Carbon-efficient scheduling of flow shops by multi-objective optimization. *Eur. J. Oper. Res.* **2015**, *248*, 758–771. [CrossRef]
20. Deb, K.; Jain, H. An Evolutionary Many-Objective Optimization Algorithm Using Reference-Point-Based Non-dominated Sorting Approach, Part I: Solving Problems with Box Constraints. *IEEE Trans. Evol. Comput.* **2014**, *18*, 577–601. [CrossRef]
21. Brandimarte, P. Routing and scheduling in a flexible job by tabu search. *Ann. Oper. Res.* **1993**, *41*, 157–183. [CrossRef]
22. Kacem, I.; Hammadi, S.; Borne, P. Pareto-optimality approach for flexible job-shop scheduling problems: Hybridization of evolutionary algorithms and fuzzy logic. *Math. Comput. Simul.* **2014**, *60*, 245–276. [CrossRef]
23. Agrawal, R.B.; Deb, K.; Agrawal, R.B. Simulated Binary Crossover for Continuous Search Space. *Complex Syst.* **1994**, *9*, 115–148.
24. Oliver, I.M.; Smith, D.J.; Holland, J.R.C. A Study of Permutation Crossover Operators on the Traveling Salesman Problem. In Proceedings of the Second International Conference on Genetic Algorithms on Genetic Algorithms and Their application, Cambridge, MA, USA, 28–31 July 1987; pp. 224–230.
25. Syswerda, G. Schedule Optimization Using Genetic Algorithms. In Proceedings of the 4th International Conference on Genetic Algorithms, San Diego, CA, USA, 13–16 July 1991; pp. 94–101.
26. Davis, L. Applying adaptive algorithms to epistatic domains. In Proceedings of the International Joint Conference on Artificial Intelligence, Los Angeles, CA, USA, 18–24 August 1985; pp. 162–164.
27. Goldberg, D.; Lingle, R. Alleles, loci and the tsp en. In *Proceedings of the First International Conference on Genetic Algorithms*; Lawrence Erlbaum Associates: Hilldale, NJ, USA, 1985; pp. 154–159.
28. Veldhuizen, D.A.V. *Multi-Objective Evolutionary Algorithms: Classifications, Analyses, and New Innovations*; Air Force Institute of Technology: Wright-Patterson AFB, OH, USA, 1999; pp. 235–243.
29. Coello, C.A.C.; Cortes, N.C. Solving Multi-objective Optimization Problems Using an Artificial Immune System. *Genet. Program. Evolvable Mach.* **2005**, *6*, 163–190. [CrossRef]
30. Zitzler, E.; Thiele, L. Multi-objective evolutionary algorithms: A comparative case study and the strength Pareto approach. *IEEE Trans. Evol. Comput.* **1999**, *3*, 257–271. [CrossRef]
31. Tian, Y.; Cheng, R.; Zhang, X.; Su, Y.; Jin, Y. A strengthened dominance relation considering convergence and diversity for evolutionary many-objective optimization. *IEEE Trans. Evol. Comput.* **2019**, *23*, 331–345. [CrossRef]
32. Zitzler, E.; Laumanns, M.; Thiele, L. SPEA2: Improving the strength pareto evolutionary algorithm. *TIK-Report* **2002**, *103*, 742–751.
33. Bader, J.; Zitzler, E. HypE: An Algorithm for Fast Hypervolume-Based Many-Objective Optimization. *Evol. Comput.* **2011**, *19*, 45–76. [CrossRef]
34. Ahmadi, E.; Zandieh, M.; Farrokh, M.; Emami, S.M. A multi objective optimization approach for flexible job scheduling problem under random machine breakdown by evolutionary algorithms. *Comput. Oper. Res.* **2016**, *73*, 56–66. [CrossRef]

Article

Multi-Objective Parameter Optimization Dynamic Model of Grinding Processes for Promoting Low-Carbon and Low-Cost Production

Mingmao Hu [1], Yu Sun [1], Qingshan Gong [1,2,*], Shengyang Tian [1] and Yuemin Wu [1]

1 College of Mechanical Engineering, Hubei University of Automotive Technology, Shiyan 442002, China; hu@huat.edu.cn (M.H.); sunyu_huat@163.com (Y.S.); tsy679@126.com (S.T.); wuymcd@163.com (Y.W.)

2 College of Mechanical Engineering, Wuhan University of Science and Technology, Wuhan 430000, China

* Correspondence: gongqs_jx@huat.edu.cn; Tel.: +86-138-8681-5862

Received: 3 August 2019; Accepted: 4 December 2019; Published: 18 December 2019

Abstract: Grinding is widely used in mechanical manufacturing to obtain both precision and part requirements. In order to achieve carbon efficiency improvement and save costs, carbon emission and processing cost models of the grinding process are established in this study. In the modeling process, a speed-change-based adjustment function was introduced to dynamically derive the change of the target model. The carbon emission model was derived from the grinding force using regression. Considering the constraints of machine tool equipment performance and processing quality requirements, the grinding wheel's linear velocity, cutting feed rate, and the rotation speed of the workpiece were selected as the optimization variables, and the improved NSGA-II algorithm was applied to solve the optimization model. Finally, fuzzy matter element analysis was used to evaluate the most optimal processing plan.

Keywords: grinding optimization; low carbon; low cost; improved NSGA-II; fuzzy matter element

---

1. Introduction

In the production and processing of an enterprise, the grinding process induces a small amount of cutting, and the post-processing surface roughness of the parts are very low and is mainly used for precision and ultra-precision machining of parts. The high-speed rotation of the grinding wheel and the long processing time can lead to large carbon emissions from the machine tool, and this can be accompanied by a high processing cost and greater use of cutting fluid in the process. According to the analysis of the energy consumption of the grinding process [1–3] and the mathematical model of CNC (computer numerical control machine tools) [4–9], a grinding parameter optimization model based on carbon emissions and cutting costs is established.

Many scholars have carried out research on the optimization of the manufacturing process or system, and discussion on the connotation of energy efficiency of manufacturing systems [10–16]. Cai et al. [17,18] proposed a new concept of entitled, lean, energy-saving, emission reduction and fine energy consumption allowance. Greinacher S. et al. [19] focused on the identification of a cost-optimized combination of lean and green strategies with regard to green targets. Cai et al. [20] measured the eco-environment loss caused by industrial solid waste. Researchers have studied the model of green CNC machining. Feng Ma et al. [21] established the multi-objective, laser-sintering forming process optimization model, with minimum energy consumption and material cost. Jiang et al. [22] proposed a method to predict the remanufacturing cost based on dates. Lin et al. [23] proposed a method to directly quantify carbon emissions during the entire turning process and established a low-carbon, efficient turning model. Yan et al. [24] built a model to improve the thermal efficiency of the arc welding process, reducing energy consumption as a result. Bustillo et al. [25] took into account the optimization of the process by requiring calibration of the main input parameters in relation to the desired output values.

Grinding forces are key parameters in the grinding modeling process; however, most studies were based on the analysis of grinding motion and individual abrasive forces. Shen et al. [26] analyzed the characteristics of the non-circular grinding movement of special-shaped parts and established an empirical model of grinding force during the processing of special-shaped parts. Li et al. [27] considered the microscopic interaction between abrasive particles and workpieces at different processing stages and proposed a detailed cutting force model. Zhang et al. [28] proposed that the aggregate force was derived through the synthesis of each single-grain force, based on material-removal and plastic-stacking mechanisms.

For the analysis of the results of green manufacturing, many scholars have proposed many assessment and decision-making methods of energy efficiency [29–34]. Cai et al. [35] proposed energy performance certification to manage energy consumption and improve energy performance. Jia et al. [36] developed an energy consumption evaluation method for the activities related to machine tools and operators. Green manufacturing processing steps can also be evaluated by the general principles of fuzzy matter evaluation [37], and carbon emissions from it can be evaluated by aggregating the unit process to form a combined model [38].

In a comprehensive analysis of the above research, although the optimization of the machining process is discussed, the influence of the optimization variables on the dynamic changes of the multi-objective model is rarely considered. In order to improve the accuracy of the model, the dynamic modeling method needs to be studied. During the actual processing, the use of a single abrasive force to establish a theoretical model would cause errors, and few researchers dynamically fit the model at each stage of the process through experimental data. In the analysis of the results, each target of the multi-objective optimization model is incompatible. So, using the theory of fuzzy matter-element to evaluate Pareto front-end could improve the efficiency of evaluating incompatible problems in reality. Methods to improve the accuracy of the models and the results require further research.

2. Establishment of a Multi-Objective Optimization Model for the Grinding Process

2.1. Optimization Variable

Cutting speed v_c, feed rate f, and depth of cut a_p are three important variables in the machining process. For the outer circle cutting of the grinding process, the main motion is the rotary motion of the grinding wheel. The cutting speed is the linear speed v_s of the outer circle of the grinding wheel. The feed rate and the depth of cut are determined by the cutting feed rate v_r of the grinding wheel, and the rotation speed of the workpiece is v_w. The optimization variable is

$$U = (v_s, v_r, v_w)^T \tag{1}$$

2.2. Carbon Emission Model in the Grinding Process

The operation process of the CNC grinding machine is generally divided into start-up, standby, no-load, cutting, and retracting stages. The energy-consuming components of each stage are relatively fixed. Therefore, the power jump corresponding to each stage is relatively stable. The carbon emission during the grinding process is mainly composed of non-cutting and cutting. Non-cutting parts include carbon emissions from auxiliary systems such as standby, air cut, and cooling. The cutting part is mainly the overall carbon emissions of the machine tool during the material removal process. With the start-stop process of the CNC grinding machine as the aim, the energy consumption characteristics of each component are individually analyzed, and an energy-based carbon emission model is established.

With energy consumption as the basic input and greenhouse gas (GHG) as the output, the corresponding carbon emissions in this process are converted through the carbon emission coefficient of various energies [39]. ξ is the carbon emission coefficient of the energy type (e.g., fuel, electricity). δ is the environmental impact, such as the time of production and the area where the workshop is located. μ is the influence of the auxiliary process, such as the cooling of the processing environment, etc. The carbon emissions' factor W can be defined as

$$W = \xi(1 + \delta + \mu)E \tag{2}$$

According to the study of Li et al. [40], the relationship between cutting speed and carbon emissions should follow the curve shown in Figure 1. When the cutting speed is increased in Area 1, the cutting power is also increased, but as processing time decreases, the energy consumption reduced by the time reduction is greater than the energy consumption increased by the increase of the spindle load, that is, the carbon emission is reduced. In Area 3, when the cutting speed is increasing and the power is increased, the energy consumption of the spindle load is greater than the energy consumption reduced by the time reduction, so the carbon emissions increase. There is an optimum cutting speed in Area 2 that minimizes carbon emissions.

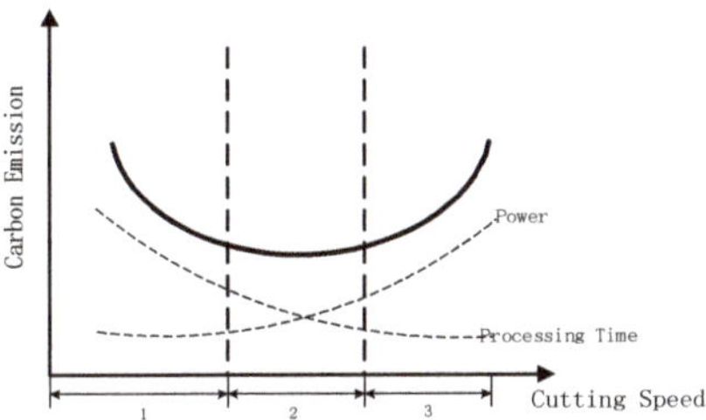

Figure 1. Cutting speed-carbon emission curve

When the grinding machine power is turned on, the lighting system, operation panel, machine tool frequency converter, servo driver, and other components are turned on. The time is short and the power fluctuation is large, so the start-up energy consumption is ignored. The standby power of the machine P_1 consists of the power of the auxiliary system, the motor and the servo power. The time for standby preparation and input of the program before processing is t_1. When the Z-axis starts to rotate, the spindle's no-load rotation power is P_{z_1}. No-load standby power is $P_s = P_1 + P_{Z_1}$, and no-load standby time is t_s. Then the energy consumption E_s after the machine is turned on can be expressed as

$$E_s = \int_0^{t_1} P_1 dt + \int_0^{t_s} P_s dt \tag{3}$$

The empty cutting energy consumption of the cylindrical grinding machine includes the movement of the X-axis of the head frame and the movement of the Z-axis of the grinding wheel. The X-axis rotation and the Z-axis movement power are respectively P_X, P_{Z_2}. Therefore, the energy consumption of air cut can be expressed as

$$E_{air} = \int_0^{t_X} P_X dt + \int_0^{t_{Z_2}} P_{Z_2} dt + \int_0^{t_X + t_{Z2}} P_s dt \tag{4}$$

In order to ensure the surface roughness of the parts and avoid the quality problems such as grinding burns, the flow rate and consumption of the cutting fluid change with different grinding wheel linear speeds and table feed speeds, and the wear amount of the grinding wheel also changes, resulting in dynamic changes of the model. For the change of the linear speed of the grinding wheel and the feed rate of the table, the adjustment function is defined as

$$\psi_i = \alpha_i(\Delta v_s)^2 + \beta_i(\Delta v_r)^2 + \gamma_i(\Delta v_w)^2 (i = 1, 2) \tag{5}$$

where α_i, β_i, γ_i are the adjustment factors, ψ_1 is the cooling adjustment factor, and ψ_2 is the grinding wheel wear adjustment function. The energy consumption of the auxiliary system includes the energy consumption generated during the cutting time t_m by the power of the filtration and cooling system P_c, and the energy consumption generated during the standby time t_{ch} of the machine tool in the process of loading and unloading parts, so the auxiliary energy consumption of each part processing can be expressed as

$$E_{\text{as}} = \psi_1 \int_0^{t_m} P_c dt + \int_0^{t_{\text{ch}}} P_1 dt \tag{6}$$

Non-cutting process carbon emissions can be expressed as

$$W_1 = \xi(1 + \delta + \mu) \cdot (E_{\text{s}} + E_{\text{air}} + E_{\text{as}}) \tag{7}$$

As shown in Figure 2, the grinding force of the grinding wheel is divided into the normal grinding force F_n and the tangential grinding force F_t. The machining power of the grinding P_{m} is mainly determined by the tangential grinding force and the linear velocity.

$$Pm = Ftvs75 \times 1.36 \times 9.81kW \tag{8}$$

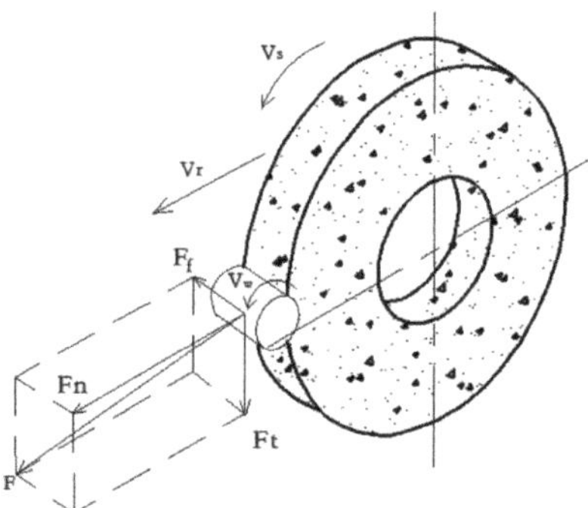

Figure 2. Grinding wheel force analysis

The speed of the wheel is recorded as follows:

$$v_{\text{s}} = \pi d_0 \cdot n_0 / (60 \times 1000) \tag{9}$$

where d_0 is the diameter of the grinding wheel, and n_0 is the grinding wheel speed.

The tangential grinding force is an important parameter for power calculation. The index in the empirical formula is calculated by experimental data to obtain the actual tangential force of this kind of grinding wheel. The mathematical formula of the cylindrical grinding force is $F_t = F_p v_s^x v_r^y v_w^z$, and F_p is an experimental variable based on different processing environments. The experimental value of the grinding force of the grinding wheel is taken as the natural logarithm, and the regression equation is shown as follows:

$$\ln F_t = \ln F_p + \text{x} \ln v_s + \text{y} \ln v_r + \text{z} \ln v_w \tag{10}$$

$$y = b_0 + b_1 \text{x}_1 + b_2 \text{x}_2 + b_3 \text{x}_3 \tag{11}$$

The data of grinding consumption obtained in the experiment were coded, the large value is +1, the small value is −1, and the four coefficients b_0, b_1, b_2, and b_3, of the regression equation are calculated according to the data of the grinding force. In the grinding, F_i is the i-th grinding variable, F_1 is the grinding wheel linear speed v_s, F_2 is the grinding wheel cutting feed amount v_r, F_3 is the workpiece rotation speed, and f_{ij} is the j-th experimental value of the i-th grinding amount. Then find the three values in the regression equation:

$$x_i = \frac{2(\ln F_i - \ln f_{ij\max})}{\ln f_{ij\max} - \ln f_{ij\min}} + 1 = \frac{2}{\ln f_{ij\max} - \ln f_{ij\min}} \ln F_i + \frac{-\ln f_{ij\max} - \ln f_{ij\min}}{\ln f_{ij\max} - \ln f_{ij\min}} \quad (12)$$

Assuming that $A_i = \frac{2}{\ln f_{ij\max} - \ln f_{ij\min}}$,$a_i = \frac{-\ln f_{ij\max} - \ln f_{ij\min}}{\ln f_{ij\max} - \ln f_{ij\min}}$, then $x_i = A_i \ln F_i + a_i$. Substitute it for x_i in the regression equation.

$$\begin{aligned} y &= b_0 + b_1(A_1 \ln F_1 + a_1) + b_2(A_2 \ln F_2 + a_3) + b_3(A_3 \ln F_3 + a_3) \\ &= (b_0 + b_1 a_1 + b_2 a_2 + b_3 a_3) + b_1 A_1 \ln F_1 + b_2 A_2 \ln F_2 + b_3 A_3 \ln F_3 \end{aligned} \quad (13)$$

$$F_t = e^{(b_0 + b_1 a_1 + b_2 a_2 + b_3 a_3)} v_s^{b_1 A_1} v_r^{b_2 A_2} v_w^{b_3 A_3} \quad (14)$$

Substitute b_0, b_1, b_2, and b_3 with their values in Equation (14) to calculate the F_t, and the actual tangential force is determined according to different depths of cut, wheel speeds, and table feed speeds. The processing energy consumption model takes the form of integral, and the power from 0 to t_m is integrated to obtain the energy consumption value. The E_m expression of grinding energy consumption is

$$E_m = \int_0^{t_m} \frac{F_t v_s}{75 \times 1.36 \times 9.81} dt \quad (15)$$

Then total carbon emissions from the grinding process are

$$W = \xi(1 + \delta + \mu) \cdot (E_s + E_{air} + E_{as} + E_m) \quad (16)$$

2.3. Cost Model in the Grinding Process

The processing cost of a single part increases with the increase of processing time. The grinding cost is mainly divided into two aspects: processing cost and loss cost. With the part affected by the optimization variable taken into consideration, the grinding process cost model is established. The processing cost includes standby, empty cutting, energy consumption cost of cutting operation, the labor cost during processing time and the use cost of auxiliary equipment, and thus the processing cost expression for each process is shown as follows:

$$C_m = (M_e + M_{as}) \times (\frac{t_1 + t_s}{Q} + t_{air} + \frac{a_p}{v_r} + t_{ch}) \quad (17)$$

where C_m is the grinding cost (yuan), M_e is the electricity cost (yuan/s), and M_{as} is the labor cost and the use cost of auxiliary equipment (yuan/s); t_{air} is the empty cut time (s); a_p is the depth of cut (m) at which the grinding load is generated and it can be known based on processing requirements; the number of processing batches is Q.

Loss costs include wheel loss and cutting fluid consumption, and multiple wheel dressings are required during machining of the part until the wheel is reduced to the minimum diameter. The cost of the grinding wheel C_{loss} is

$$C_{loss} = M_a \cdot \psi_2 \cdot \frac{\pi r^2 b - [\pi(r - a_p)^2 b]}{G} \quad (18)$$

where M_a is the grinding wheel cost (yuan/mm^3); b is the wheel width; r is the workpiece radius; G is the grinding ratio.

The consumption of cutting fluid consists of the portion of the machined surface that rises in temperature and evaporates into the air, the portion taken away by the chip, and the portion deposited on the surface of the part. Cutting fluid consumption cost C_{lub} (yuan) is

$$C_{\text{lub}} = M_l \times [\psi_1 \cdot (\frac{a_p}{v_r} \cdot l_{\text{lub}})] \tag{19}$$

where M_l is the unit cost of cutting fluid consumption (yuan/L); l_{lub} is the cutting fluid flow rate (L/s).

The total cost of grinding is

$$C = (M_e + M_{as}) \times (\frac{t_1 + t_s}{Q} + t_{air} + \frac{a_p}{v_r} + t_{ch}) + M_a \cdot \psi_2 \cdot \frac{\pi r^2 b - [\pi (r - a_p)^2 b]}{G} + M_l \times [\psi_1 \cdot (\frac{a_p}{v_r} \cdot l_{\text{lub}})] \tag{20}$$

2.4. Constraint and Optimization Model

With the grinding speed constrained, the grinding wheel speed v_s must meet the following requirement:

$$v_s \in \left[\frac{\pi d_0 n_{\min}}{1000} \le v_s \le \frac{\pi d_0 n_{\max}}{1000} \right] \tag{21}$$

where $n_{\min}$ and $n_{\max}$ represent the minimum and maximum speeds of the CNC grinding machine spindle, respectively.

The tangential feed amount v_r of the grinding wheel needs to be within the range of the maximum and minimum values of the spindle feed of the machine tool, that is

$$v_{r\min} \le v_r \le v_{r\max} \tag{22}$$

The workpiece rotation speed v_w needs to be within the range of the maximum and minimum values of the workpiece speed of the CNC grinding machine, that is

$$v_{w\min} \le v_w \le v_{w\max} \tag{23}$$

With the cutting force constrained, the grinding wheel's tangential force needs to be less than the maximum cutting force to protect the grinding wheel and the surface quality of the part, namely

$$F_t = F_p v_s^x v_r^y v_w^z \le F_{t\max} \tag{24}$$

Power constraint, the calculated power needs to be less than the maximum power of the CNC grinding machine, namely

$$P_m = \frac{F_t \cdot v_s}{75 \times 1.36 \times 9.81 \times \eta} \le P_{\max} \tag{25}$$

The surface quality of the machine is constrained. The surface roughness value needs to be greater than the minimum machining roughness value of the machine tool. It is also an important condition for restraining the linear speed v_s of the grinding wheel and the feed rate v_w of the table. According to the Ono theory [41], the surface roughness expression of the external grinding is

$$Ra = 0.975\gamma^{1.2} \times (\cot \varphi)^{0.1} \times (\frac{v_w}{v_s} \sqrt{\frac{1}{2r} + \frac{1}{d_s}})^{0.4} \ge Ra_{\min} \tag{26}$$

where γ is the cutting edge spacing considered by volume density; φ is half of the cutting edge angle; d_s is the wheel diameter.

In summary, the low-carbon and low-cost parameter multi-objective optimization model in the grinding process is

$$
\begin{cases} \min f(U) = (\min W, \min C)^T \\ U = (v_s, v_w)^T \end{cases}
$$
$$
\begin{aligned}
& s.t. \\
& v_s \in \left[\tfrac{\pi d_0 n_{\min}}{1000} \le v_s \le \tfrac{\pi d_0 n_{\max}}{1000}\right], \\
& v_{r\min} \le v_r \le v_{r\max}, \\
& v_{w\min} \le v_w \le v_{w\max}, \\
& F_t = F_p v_s^x v_r^y v_w^z \le F_{t\max}, \\
& P_m = \tfrac{F_t \cdot v_s}{75 \times 1.36 \times 9.81 \times \eta} \le P_{\max}, \\
& Ra = 0.975 \gamma^{1.2} \times (\cot \varphi)^{0.1} \times \left(\tfrac{v_w}{v_s} \sqrt{\tfrac{1}{2r} + \tfrac{1}{d_s}}\right)^{0.4} \le Ra_{\max}.
\end{aligned} \tag{27}
$$

3. Parameter Optimization Based on Improved NSGA-II Algorithm

3.1. Improved NSGA-II Algorithm

To solve the multi-objective optimization problems, the weighted summation method is used to assign weights to each target value, and the multi-objective problem is simplified to a single-objective problem. However, there is no standard for the assignment of target weights, and the objective functions usually have different dimensions. If the weight value cannot be determined between the carbon emissions and the cost cash, such as in this paper, it will have a greater impact on the calculation results. The algorithm used by a multi-objective function in the MATLAB program is an improved multi-objective optimization algorithm based on the non-dominated sorting genetic algorithm (NSGA-II) with an elite strategy, which can effectively solve the multi-objective optimization problem.

The characteristics of the genetic algorithm include determining the dominance and non-inferiority of the individual, and comparing and judging the better target individual. Based on the dominating judgment order value, the individuals in the population are assigned to different front-ends according to the size, and the higher the front, the stronger the dominance. The crowding distance is used to calculate the distance between a certain body in a front-end and other individuals in the front-end, and to characterize the degree of crowding between individuals—the greater the distance, the better the diversity of the population. The improved algorithm introduces the optimal front-end individual coefficients unique to the gamultiobj function, defines the proportion of individuals in the optimal front-end in the population, and also directly determines the number of individuals retained during the pruning process.

The algorithm flow is to first determine the constraint type of the optimization problem, generate the initial population, and judge whether the algorithm can be exited. And if it exits, get the Pareto optimal solution. If not, the population will evolve into the next generation. In the process of evolution, the gamultiobj function only uses the tournament selection method based on the order value and the crowded distance. The selected individual is assigned to several front-ends to generate the parent population. The parent population crosses, and the mutation produces the children. The gamultiobj function allows the elite to automatically retain, and the scaling function is no longer needed. The parent and the child are merged, and the individuals in the population are sorted by the non-dominated sorting function so that all the merged individuals are assigned to different front-ends. Then the crowded distance is used to calculate the distance between each body in a front-end and its neighbors. According to the optimal front-end individual coefficient, individuals equal to the size of the population are pruned in the population twice as large as the parent–child mergence to obtain a new parent population, and it is judged whether the iteration is terminated or whether the algorithm can be exited. The algorithm flow is shown in Figure 3.

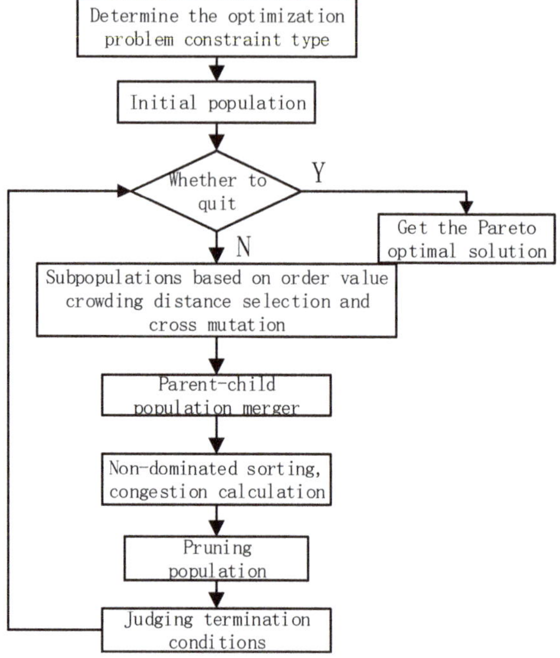

Figure 3. Algorithm flow chart.

3.2. Optimization Target Solving

The air compressor crankshaft blank of 45# steel was selected, and the second main journal of the crankshaft is finely ground by a face cylindrical grinding machine, M-181. The diameter of the main journal is φ 37.96–37.944, the root radius is R 1.4–1.7, the outer diameter of the main shaft is 0.015, and the roughness of the outer circle is Ra 0.8. The process drawing is shown in Figure 4, and the actual processing is shown in Figure 5. The grinding machine uses a 100# resin-bonded diamond grinding wheel to grind the outer end surface of the shaft parts, and the grinding precision and smoothness are high; the longitudinal movement of the grinding wheel frame is driven by the gear oil pump, and the movement is stable to ensure the uniform feeding speed of the grinding wheel; double-paired high-precision rolling bearings can make high rotation accuracy and rigidity. The relevant parameters of the machine tool are shown in Table 1.

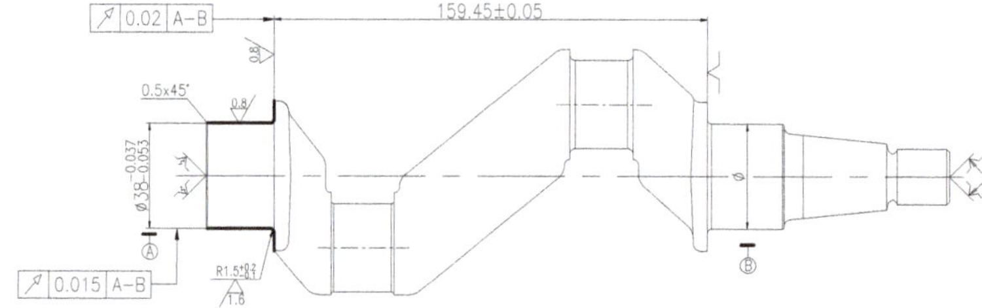

Figure 4. Fine-grinding the second main journal of the crankshaft.

Figure 5. Actual machining of the crankshaft.

Table 1. CNC cylindrical grinding machine parameters.

Parameters	Numerical Value
Wheel speed (r/min)	1500–3000
Head frame speed (r/min)	50–500
Grinding wheel feed rate (mm/min)	0.05–5000
Wheel size (outer diameter × width × inner diameter; mm)	Ø 500 × 80 × Ø 203
Motor power (kW)	7.5
Surface roughness (μm)	Ra 0.08

Control the grinding machine standby, air cutting, and other stages of operation. At each stage, use the three-phase four-wire wiring method to lap the power recorder and collect the power value of each stage of the machine tool. At the same time, the current-monitoring equipment is used to monitor the data reliability of the power meter. Finally, when the second main journal of the crankshaft is ground, a top force-measuring instrument is installed to detect and collect the grinding force at different cutting speeds in real-time. The experiment for the grinding force is carried out to obtain the parameters in Table 2, and the parameters are introduced into the model of the second part. The specific experimental instruments are shown in Figures 6 and 7, and the corresponding parameter values are shown in Table 3.

Table 2. Grinding force experiment.

No.	Grinding Variable			Coding				Measured Value of Tangential Grinding Force	
	ap/mm	vs/m·min^{-1}	vw/m·min^{-1}	b_0	b_1	b_2	b_3	Ft/N	lnFt
1	0.005	2.5	12	+1	−1	+1	−1	48	$y_1 = 3.87$
2	0.005	1.0	48	+1	−1	−1	+1	77	$y_2 = 4.34$
3	0.020	1.0	12	+1	+1	−1	−1	87	$y_3 = 4.46$
4	0.020	2.5	48	+1	+1	+1	+1	449	$y_4 = 6.11$

Figure 6. Grinding machine power recorder.

Figure 7. Grinding machine current monitor.

Table 3. Grinding parameters.

Parameters	Numerical Value
P_1	1.526 kW
P_s	2.772 kW
P_c	0.322 kW
t_1	1200 s
t_s	600 s
t_{ch}	120 s
a_p	5 μm
L	150 mm
Q	30 pieces
N	120 pieces
G	21

Set the optimal front-end individual coefficient to 0.3, the population size to 100, the maximum evolutionary algebra to 300, the stop algebra to 300, and the fitness function deviation to 0.001 to calculate the results. Since the initial population of the algorithm is randomly generated, the operation results obtained each time are different. The Pareto front-end values obtained from the result of a certain operation is shown in Figure 8. Table 4 shows the specific parameters of the 30 optimal front individuals, each row representing an individual speed of the grinding wheel outer circle, the cutting feed rate, the rotational speed of the workpiece, the individual's carbon emissions, and costs.

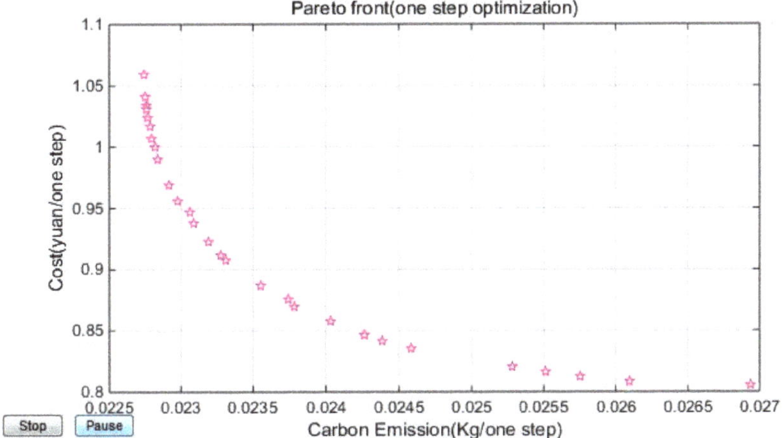

Figure 8. First front-end individual distribution map.

Table 4. Pareto front-end values obtained from a certain operation.

	v_s	v_r	v_w	W	C
1	24.994	1.602×10^{-4}	13.057	0.023	1.055
2	23.998	1.637×10^{-4}	14.339	0.027	0.808
3	25.000	1.602×10^{-4}	13.003	0.023	1.069
...	...	...	...	...	...
21	24.235	1.609×10^{-4}	14.028	0.025	0.825
...	...	...	...	...	...
30	24.975	1.611×10^{-4}	13.012	0.023	1.060

3.3. Fuzzy Matter Element-Based Decision-Making

Multi-objective decision-making schemes include use things, features, and fuzzy magnitudes to quantitatively analyze and calculate things. Assume that the thing is the plan K_i, and the feature is the evaluation index O_i (for the purposes of this paper, the evaluation index is the wheel speed, the table feed speed, the carbon emission, and the cost), and the fuzzy value gives the value of x_{ij}, thereby constituting the matter element. For the optimization model of this paper, take the smaller and better decision, and the metric is determined by the correlation coefficient λ_{ij}.

$$\lambda_{ij} = \frac{x_{ij} - \min x_{ij}}{\max x_{ij} - \min x_{ij}} \tag{28}$$

According to the scheme, the characteristics and the correlation coefficient, it can construct the complex fuzzy matter element of the dimension correlation coefficient of a thing.

$$R_\lambda == \begin{bmatrix} & K_1 & K_2 & K_3 & \cdots & K_{30} \\ O_1 & 0.994 & 0 & 1 & & 0.976 \\ O_2 & 0.741 & 1 & 0 & & 0.264 \\ O_3 & 0.040 & 1 & 0 & & 0.007 \\ O_4 & 0.001 & 1 & 0 & & 0.001 \\ O_5 & 0.946 & 0 & 1 & & 0.968 \end{bmatrix} \tag{29}$$

R_w is the weighted composite element for each decision-making indicator, and the weight of the i-th evaluation indicator for each scenario is $W_i = \sum\limits_{j=1}^{n} \lambda_{ji} / \sum\limits_{j=1}^{n}\sum\limits_{i=1}^{m} \lambda_{ji}$.

$$R_w = \begin{bmatrix} & O_1 & O_2 & O_3 & O_4 & O_5 \\ W_i & 0.285 & 0.249 & 0.153 & 0.092 & 0.222 \end{bmatrix} \tag{30}$$

The weighted average centralized processing is used to construct the correlation fuzzy matter element, namely

$$R_k = \begin{bmatrix} K_1 & K_2 & K_3 & \cdots & K_{30} \\ 0.4994 & 0.4933 & 0.5067 & \cdots & 0.5593 \end{bmatrix} \tag{31}$$

The minimum value is obtained by sorting the degree of association, and the optimal solution is K_{21}. The results in the single target case are obtained and compared, as shown in Table 5.

Table 5. Comparison of optimization results.

	v_s	v_r	v_w	W	C
Low cost	23.99770954	0.000163744	14.3388749	0.026951585	0.808222327
Low carbon and low cost	24.23538375	0.000160926	14.02766146	0.025212982	0.824788898
Low carbon	24.99970889	0.000160212	13.00277846	0.022746323	1.068775876

The best variable for the second main spindle refining of the crankshaft with low carbon and low cost is the grinding wheel linear velocity of 24.235 m/s, the spindle feed speed is 0.00016 m/s, the workpiece rotation speed is 14.0428 m/s, and the carbon emission is 0.0252 kg/piece. The processing cost is 0.825 yuan/piece. With low carbon as the goal, the grinding wheel's linear speed is high, the processing time is short, but the grinding wheel wear and cutting fluid usage are significant and the cost is high; With low cost as the target, the low linear speed of the grinding wheel reduces the amount of wear and the amount of cutting fluid, but the processing time is long and the carbon emission is high.

4. Conclusions

This paper systematically analyzes the energy consumption characteristics and parts-manufacturing costs of various stages of machine tools in grinding. An optimization model for the external grinding parameters with the minimum carbon emission and the optimal cost as the multi-objective is established. The use of auxiliary tools and the division of the whole process are considered in the modeling process. Considering the dynamic change of cutting fluid and the service life of the grinding wheel, an adjustment function is introduced based on the linear speed of the grinding wheel and the feed rate of the working table. The optimized grinding parameters are calculated by using the NSGA-II algorithm, and these parameters are evaluated through the fuzzy matter-element decision method. The machining process is fitted in a single grinding depth during the modeling process, but during the actual production, one part often requires multiple-time grinding. According to different grinding depth and surface roughness values, applying them in the multi-objective optimization dynamic model can realize step-by-step optimization in machining and be referred to for the selection of grinding process parameters.

Author Contributions: Conceptualisation, M.H. and Q.G.; Investigation and experiment, Y.S., S.T. and Y.W.; Visualisation, writing—review and editing, Y.S., M.H. and Q.G. All authors have read and agreed to the published version of the manuscript.

Funding: This research received no external funding.

Acknowledgments: The work described in this paper was supported by the 58th postdoctoral science foundation program of China (Grant No. 2015M581301), the National Natural Science Foundation of China (Grant No. 51775392 and 51675388), the Educational Commission of Hubei Province (Grant No. B2018069), the National Science Foundation of Hubei province (Grant No. 2019CFB384), and the Key Laboratory of Automotive Power Train and Electronics (Grant No. ZDK1201802). These financial contributions are gratefully acknowledged.

Conflicts of Interest: The authors declare no conflict of interest.

References

1. Priarone, P.C. Quality-conscious optimization of energy consumption in a grinding process applying sustainability indicators. *Int. J. Adv. Manuf. Technol.* **2016**, *86*, 2107–2117.
2. Arriandiaga, A.; Portillo, E.; Sánchez, J. A new approach for dynamic modelling of energy consumption in the grinding process using recurrent neural networks. *J. Neural Comput. Appl.* **2016**, *27*, 1577–1592.
3. Salonitis, K. Energy efficiency assessment of grinding strategy. *Int. J. Energy Sect. Manag.* **2015**, *9*, 20–37. [CrossRef]
4. Jiang, P.; Li, G.L.; Liu, P.X.; Jiang, L.; Li, X.Z. Energy consumption model and energy efficiency evaluation for CNC continuous generating grinding machine tools. *Int. J. Sustain. Eng.* **2017**, *10*, 226–232. [CrossRef]
5. Lv, J.; Peng, T.; Tang, R. Energy Modeling and a Method for Reducing Energy Loss Due to Cutting Load During Machining Operations. *J. Proc. Inst. Mech. Eng. Part B J. Eng. Manuf.* **2019**, *233*, 699–710. [CrossRef]
6. Calleja, A.; Bo, P.; González, H.; Bartoň, M.; de López Lacalle, L.N. Highly-accurate 5-axis flank CNC machining with conical tools. *Int. J. Adv. Manuf. Technol.* **2018**, *97*, 1605–1615.
7. Draganescu, F.; Gheorghe, M.; Doicin, C. Models of Machine Tool Efficiency and Specific Consumed Energy. *J. Mater. Process. Technol.* **2003**, *141*, 9–15. [CrossRef]
8. Aramcharoen, A.; Mativenga, P.T. Critical factors in energy demand modelling for CNC milling and impact of toolpath strategy. *J. Clean. Prod.* **2014**, *78*, 63–74. [CrossRef]
9. Gutowski, T.; Dahmus, J.; Thiriez, A. Electrical energy requirements for manufacturing processes. In *13th CIRP International Conference on Life Cycle Engineering*; CIRP International Leuven: Leuven, Belgium, 2006.
10. Liu, F.; Wang, Q.; Liu, G. Content Architecture and Future Trends of Energy Efficiency Research on Machining Systems. *J. Mech. Eng.* **2013**, *49*, 87. [CrossRef]
11. Shin, S.J.; Woo, J.; Rachuri, S. Energy efficiency of milling machining: Component modeling and online optimization of cutting parameters. *J. Clean. Prod.* **2017**, *161*, 12–29. [CrossRef]
12. Liu, F.; Liu, P.J.; Li, C.B.; Tuo, J.B.; Cai, W. The Statue and Difficult Problems of Research on Energy Efficiency of Manufacturing Systems. *J. Mech. Eng.* **2017**, *53*, 1–11. [CrossRef]

13. Park, C.-W.; Kwon, K.-S.; Kim, W.-B.; Min, B.-K.; Park, S.-J.; Sung, I.-H.; Yoon, Y.-S.; Lee, K.-S.; Lee, J.-H.; Seok, J. Energy consumption reduction technology in manufacturing—A selective review of policies, standards, and research. *Int. J. Precis. Eng. Manuf.* **2009**, *10*, 151–173. [CrossRef]
14. Trianni, A.; Cagno, E.; Farné, S. Barriers drivers and decision making process for indus-trial energy efficiency: A broad study among manufacturing small and medium-sized enterprises. *J. Appl. Energy* **2016**, *162*, 1537–1551. [CrossRef]
15. Abdelaziz, E.A.; Saidur, R.; Mekhilef, S. A review on energy saving strategies in industrial sector. *J. Renew. Sustain. Energy Rev.* **2011**, *15*, 150–168. [CrossRef]
16. Dietmair, A.; Alexander, V. A Generic Energy Consumption Model for Decision Making and Energy Efficiency Optimisation in Manufacturing. *Int. J. Sustain. Eng.* **2009**, *2*, 123–133. [CrossRef]
17. Cai, W.; Liu, C.; Zhang, C.; Ma, M.; Rao, W.; Li, W.; He, K.; Gao, M. Developing the ecological compensation criterion of industrial solid waste based on emergy for sustainable development. *J. Energy* **2018**, *157*, 940–948. [CrossRef]
18. Cai, W.; Liu, F.; Zhou, X.; Xie, J. Fine energy consumption allowance of workpieces in the mechanical manufacturing industry. *J. Energy* **2016**, *114*, 623–633. [CrossRef]
19. Greinacher, S.; Lanza, G. Optimisation of lean and green strategy deployment in manufacturing systems. *J. Appl. Mech. Mater.* **2015**, *794*, 748–785. [CrossRef]
20. Cai, W.; Liu, C.; Lai, K.-H.; Li, L.; Cunha, J.; Hu, L. Energy performance certification in mechanical manufacturing industry: A review and analysis. *J. Energy Convers. Manag.* **2019**, *186*, 415–432. [CrossRef]
21. Ma, F.; Zhang, H.; Hon, K.; Gong, Q.S. An optimization approach of selective laser sintering considering energy consumption and material cost. *J. Clean. Prod.* **2018**, *199*, 529–537. [CrossRef]
22. Jiang, Z.; Ding, Z.; Liu, Y.; Wang, Y.; Hu, X.; Yang, Y. A data-driven based decomposition–integration method for remanufacturing cost prediction of end-of-life products. *Robot. J. Comput. Integr. Manuf.* **2020**, *61*, 101838. [CrossRef]
23. Lin, W.; Yu, D.; Wang, S.; Zhang, C.; Zhang, S.; Tian, H.; Luo, M.; Liu, S. Multi-objective teaching–learning-based optimization algorithm for reducing carbon emissions and operation time in turning operations. *J. Eng. Optim.* **2015**, *47*, 994–1007. [CrossRef]
24. Wei, Y.; Hua, Z.; Zhi-gang, J.; Hon, K.K.B. A new multi-source and dynamic energy modeling method for machine tools. *Int. J. Adv. Manuf. Technol.* **2018**, *95*, 4485–4495. [CrossRef]
25. Bustillo, A.; Urbikain, G.; Perez, J.M.; Pereira, O.M.; Lopez de Lacalle, L.N. Smart optimization of a friction-drilling process based on boosting ensembles. *J. Manuf. Syst.* **2018**, *48*, 108–121. [CrossRef]
26. Shen, N.Y.; Wang, W.D.; Li, J.; Cao, Y.L.; Wang, Y. Modelling and Analysis of Grinding Energy Consumption in Non-circular Grinding Process. *J. Mech. Eng.* **2017**, *53*, 208–216. [CrossRef]
27. Li, H.N.; Yu, T.B.; Wang, Z.X.; Zhu, L.D.; Wang, W.S. Detailed modeling of cutting forces in grinding process considering variable stages of grain-workpiece micro interactions. *Int. J. Mech. Sci.* **2016**, *126*, 319–339. [CrossRef]
28. Zhang, Y.B.; Li, C.H.; Ji, H.J. Analysis of grinding mechanics and improved predictive force model based on material-removal and plastic-stacking mechanisms. *Int. J. Mach. Tools Manuf.* **2017**, *122*, 67–83. [CrossRef]
29. Liu, P.J.; Junbo, T.; Liu, F.; Li, C.B.; Zhang, X.C. A Novel Method for Energy Efficiency Evaluation to Support Efficient Machine Tool Selection. *J. Clean. Prod.* **2018**, *191*, 57–66. [CrossRef]
30. Paetzold, J.; Kolouch, M.; Wittstock, V.; Putz, M. Methodology for process-independent energetic assessment of machine tools. *J. Procedia Manuf.* **2017**, *8*, 254–261. [CrossRef]
31. Jiang, Z.; Ding, Z.; Zhang, H.; Cai, W.; Liu, Y. Data-driven ecological performance evaluation for remanufacturing process. *J. Energy Convers. Manag.* **2019**, *198*, 111844. [CrossRef]
32. Ding, Z.; Jiang, Z.; Zhang, H.; Cai, W.; Liu, Y. An integrated decision-making method for selecting machine tool guideways considering remanufacturability. *Int. J. Comput. Integr. Manuf.* **2018**, 1–12. [CrossRef]
33. Schudeleit, T.; Züst, S.; Wegener, K. Methods for evaluation of energy efficiency of machine tools. *J. Energy* **2015**, *93*, 1964–1970. [CrossRef]
34. Ma, F.; Zhang, H.; Gong, Q.S.; Hong, K.K.B. A Novel Energy Evaluation Approach of Machining Processes Based on Data Analysis. *J. Energy Sources Part A Recovery Util. Environ. Eff.* **2019**, 1–15. [CrossRef]
35. Cai, W.; Lai, K.; Liu, C.; Wei, F.F.; Ma, M.D.; Jia, S.; Jiang, Z.G.; Lv, L. Promoting sustainability of manufacturing industry through the lean energy-saving and emission-reduction strategy. *J. Sci. Total Environ.* **2019**, *665*, 23–32. [CrossRef]

36. Jia, S.; Yuan, Q.; Cai, W.; Li, M.; Li, Z. Energy modeling method of machine-operator system for sustainable machining. *J. Energy Convers. Manag.* **2018**, *172*, 265–276. [CrossRef]
37. Liu, Z.; Wang, S.; Wan, J.; Liu, G. Green Product Assessment Method Based on Fuzzy-Matter Element. *J. China Mech. Eng.* **2007**, *18*, 166–170.
38. Wang, Y.; Zhang, H.; Zhang, Z.; Wang, J. Development of an Evaluating Method for Carbon Emissions of Manufacturing Process Plans. *J. Discret. Dyn. Nat. Soc.* **2015**, *2015*. [CrossRef]
39. Zhang, L.; Ma, J.; Fu, Y.G. Carbon Emission Analysis for Product Assembly Process. *J. Mech. Eng.* **2016**, *52*, 151–160. [CrossRef]
40. Li, C.; Cui, L.; Liu, F.; Li, L. Multi-objective NC Machining Parameters Optimization Model for High Efficiency and Low Carbon. *J. Mech. Eng.* **2013**, *49*, 87–96. [CrossRef]
41. Zhuang, S.X. *Grinding Technology*; Mechanical Industry Press: Beijing, China, 2007; pp. 53–55.

Article

An Integrated Multicriteria Decision-Making Approach for Collection Modes Selection in Remanufacturing Reverse Logistics

Xumei Zhang [1], Zhizhao Li [1,*], Yan Wang [2] and Wei Yan [1]

[1] School of Automobile and Traffic Engineering, Wuhan University of Science and Technology, Wuhan 430065, China; zhangxumei@wust.edu.cn (X.Z.); yanwei81@wust.edu.cn (W.Y.)
[2] School of Computing, Engineering & Maths, University of Brighton, Brighton BN2 4GJ, UK; Y.Wang5@brighton.ac.uk
* Correspondence: lizhizhao98@wust.edu.cn; Tel.: +86-177-7130-0412

Abstract: Reverse logistics (RL) is closely related to remanufacturing and could have a profound impact on the remanufacturing industry. Different from sustainable development which is focused on economy, environment and society, circular economy (CE) puts forward more requirements on the circularity and resource efficiency of manufacturing industry. In order to select the best reverse logistics provider for remanufacturing, a multicriteria decision-making (MCDM) method considering the circular economy is proposed. In this article, a circularity dimension is included in the evaluation criteria. Then, analytic hierarchy process (AHP) is used to calculate the global weights of each criterion, which are used as the parameters in selecting RL providers. Finally, technique for order of preference by similarity to ideal solution (TOPSIS) is applied to rank reverse logistics providers with three different modes. A medium-sized engine manufacturer in China is taken as a case study to validate the applicability and effectiveness of the proposed framework.

Keywords: reverse logistics; multicriteria decision-making; circular economy; collection modes; remanufacturing

Citation: Zhang, X.; Li, Z.; Wang, Y.; Yan, W. An Integrated Multicriteria Decision-Making Approach for Collection Modes Selection in Remanufacturing Reverse Logistics. *Processes* **2021**, *9*, 631. https://doi.org/10.3390/pr9040631

Academic Editor: Peter Glavič

Received: 18 March 2021
Accepted: 2 April 2021
Published: 4 April 2021

1. Introduction

As the third source of profit next to resources and manpower [1], logistics has gained increasing attention in various industries globally. These activities include research on not only forward logistics but also reverse logistics (RL), which has been proved to have a positive impact on productivity and the natural environment [2]. The definition of RL was proposed by Rogers and Tibben-Lembke [3]: "RL is the process of planning, implementing, and controlling the efficient, cost effective flow of raw materials, in process inventory, finished goods and related information from the point of consumption to the point of origin for the purpose of recapturing value or proper disposal". Companies concentrate more on RL practices in order to gain more profits and be more socially responsible [4]. However, the implementation of RL requires a systematic and well-thought-out strategy; otherwise, it will increase the cost of the enterprise and will not achieve the desired objectives. On the other hand, along with green design and product recovery, RL is a new core aspect of supply chain management [5]. Traditionally, third-party reverse logistics providers are only responsible for transportation tasks. This can no longer meet the demand for sustainability and circularity. An RL that can meet the needs of enterprises in both economic and environmental aspects is needed.

The manufacturing industry is one of the industries that very closely related to RL. A good balance between economic benefits and environmental impacts may be achieved through RL via recycling and remanufacturing. Remanufactured products with high added value can reduce the overall cost of products by 50–60%, and their performance indexes are as good as those of new products, thus significantly improving enterprises' economic benefits [6]. In fact, most manufacturing enterprises cannot carry out remanufacturing activities well due to the lack of abundant capital and professional technical support, which requires significant investment and increases the economic burden of enterprises at the beginning [7]. In the current market situation, where the original equipment manufacturer (OEM) remanufacturing mode is difficult to implement due to the lack of a professional logistics system, the third-party remanufacturer (TPR) remanufacturing mode has become the mainstream mode due to its economic advantages provided by its efficient and mature logistics network [8]. A precondition of remanufacturing is the collection of waste products. Currently, there are three collection modes, depending on the operating party of the RL provider in the supply chain: third party take-back (TPT), manufacturing take-back (MT) and retailer take-back (RT) [9]. In MT mode, the manufacturer in the forward supply chain completes the process of collecting, disassembly, cleaning, renovation, testing, assembly, etc. and remanufactures the waste products into new products. In contrast, collecting is carried out by retailers/distributors in the forward supply chain in RT mode. Usually, this mode is used for collecting waste products with fewer varieties and small volumes. In TPT mode, the collection task is undertaken by a professional third-party logistics provider. Manufacturers outsource the collecting and part of remanufacturing of waste products to third-party logistics providers by paying outsourcing costs. In this article, a new kind of RL provider based on the concept of value-added service is proposed, in which the third-party logistics providers in TPT mode not only are responsible for the collection of waste products but also provide customized value-added services such as cleaning, damage degree identification and classification based on the OEMs' needs.

Normally, several alternative providers may be short-listed, the selection criteria are important for the identification of the optimal alternative. Although the evaluation index system for the selection of RL providers is formed by considering the needs of enterprises, most researchers build a complete index system based on an assumption that each criterion is dependent on the others, and their interactions are modeled and optimized to achieve a comprehensive evaluation and selection of alternatives [10]. In recent years, many articles have been published on the topic of "sustainability", and the circular economy, which is considered to be a subset of sustainability [11], has become a research hotspot. While sustainable development research is focused on the integration of economic and environmental aspects, the cores of circular economy are "reduction, reuse and recycling", which have more demands on the circularity and resource efficiency of the manufacturing industry [12]. Instead of focusing on the social, economic and environmental dimensions of sustainability, a new index system is proposed in this article taking "circularity" as one of the dimensions according to the circular economy theory.

Many studies have been conducted on RL provider selection, most of which have employed a hybrid approach of fuzzy number or grey theory and analytic hierarchy process (AHP). However, many researchers are skeptical of such an overly broad approach. For example, Saaty, the founder of AHP theory, questioned the scientific and logical nature of fuzzy AHP methods because AHP itself is generated from the fuzzy theory [13], which is difficult to verify. Thus, in this article, fuzzy logic is not adapted in the proposed decision method.

The rest of the article is organized as follows: In Section 2, a literature review and summary of the RL provider research are provided. In Section 3, the framework of the proposed RL provider selection problem is introduced and the composition of the criteria system is described. The framework is applied to a case study in Section 4, and the results are discussed in Section 5. Finally, in Section 6, the conclusions and limitations of this paper are summarized and views on future research are presented.

2. Literature Review

This section is divided into two parts. The first part describes the common methods used in the existing literature for RL provider selection. The second part introduces the criteria classification theory and criteria selection in relevant research.

2.1. RL Provider Selection Methods

RL provider selection is a multicriteria decision-making (MCDM) problem in which a number of alternative providers are evaluated according to a set of criteria [14]. Generally, a hybrid MCDM method, namely two (or more) different decision-making methods combined, is used to calculate the relative weight of evaluation criteria and rank the alternatives accordingly. There is a growing number of studies on RL provider evaluation and selection owning to the increasing recognition of RL in the circular economy. Zhang et al. (2020), in a systematic review of 41 articles on this topic, pointed out that fuzzy theory is useful for decision making since it can quantify the subjective criteria indicators. AHP, as a mature method used to deal with pairwise comparison, has become the most frequently used method in calculating the relative weights between criteria [10]. There are other methods that are also frequently used to calculate weights individually or in combination with other evaluation methods in different contexts, such as analysis network process (ANP) [15–18], data envelopment analysis (DEA) [19–21], interpretive structural modeling (ISM) [22–24] and best–worst method (BWM) [25–27]. The next step, when ranking alternatives, is to use a variety of ranking methods. The common ones are ViseKriterijumska Optimizacija I Kompromisno Resenje (VIKOR) [26,28–33], technique for order of preference by similarity to ideal solution (TOPSIS) [22,34–38] and multiobjective optimization by ratio analysis (MOORA) [39–41]. Zarbakhshnia et al. (2020) proposed a novel hybrid MCDM approach integrating fuzzy AHP and grey MOORA and used the MOORA method as the baseline to verify the reliability of the new method [39]. Santos et al. (2019) developed a hybrid Entropy-TOPSIS-F framework to weight the criteria and select the provider with the best environmental performance [34]. Garg et al. (2018) used an integrated BWM–VIKOR framework to evaluate and select outsourcing partners for an electronics company in India [26]. Li et al. (2018) developed an integrated cumulative prospect theory (CPT)-based MCDM approach to identify sustainable third-party reverse logistics providers (3PRLPs) [42]. Zarbakhshnia et al. (2018) proposed a fuzzy stepwise weight assessment ratio analysis (SWARA) to weight the evaluation criteria, based on which the approach of fuzzy complex proportional assessment of alternatives (COPRAS) was applied to rank and select the sustainable 3PRLPs for the automotive industry [43]. Sen et al. (2017) developed a decision support framework based on the dominance measure concept integrated with grey set theory, which can determine the dominance of two alternatives according to a specific criterion [44].

According to the above review, MCDM methods adapted by the combination of several techniques are widely utilized to deal with the problem of RL provider selection. Compared with other approaches, the ease of applicability in pairwise comparisons makes AHP a more useful tool. As the two most widely used ranking methods, both VIKOR and TOPSIS are based on the principle of compromise programming. The former focuses on sorting and selecting a set of alternatives in the presence of conflicting criteria, while the latter has no strict restrictions on data distribution and indicators.

2.2. RL Provider Selection Criteria

Establishing the selection criteria is the first step in RL provider selection [10]. Various evaluation criteria have been utilized depending on applications. Widely applied criteria include sustainability, green supply chain management (GSCM), circular economy (CE) and performance dimensions, as shown in Table 1.

Table 1. Summary of factors for reverse logistics (RL) provider selection.

Factors	Relevant Characteristics in the Literature	References
Sustainability	The evaluation criteria are determined based on the three dimensions of economy, environment and society	[20–25,27,29,37,38]
GSCM	Green recycling, green purchasing, green transportation, resource and environmental management capabilities, social responsibility benefits, green core competencies	[14,26,33]
CE	Air pollution, environmental standards, eco-friendly raw materials, eco-design, eco-friendly packaging, eco-friendly transportation, clean technology	[13]
Performance dimensions	Reverse logistics services, reverse logistics functions, organizational role, user satisfaction, RL activities, organizational performance criteria, IT application	[15,16,32,34]

Most articles in the field of sustainability focus on the three pillars of sustainability, namely society, economy and environment, to determine the evaluation criteria. In articles related to GSCM, more attention is paid to the reduction of the negative environmental impact in the transportation and manufacturing process when establishing evaluation criteria. Similarly, attention is also placed on the environment in the topic of CE, but when selecting evaluation criteria, greater importance is placed on the product being environmentally friendly in design for the reuse of end-of-life products. Another commonly used criterion is performance, for which evaluation criteria are established from the perspective of examining various aspects of RL provider capabilities. Some articles do not take the above four factors as the foundation but instead consider the application-specific requirements as the criteria. For instance, Liu et al. (2019) studied the recycling and disposal of used mobile phones. Considering that noises of different frequencies and intensity are emitted during the recycling process, noise pollution was taken as one of the evaluation criteria [25].

By summarizing and analyzing the existing research results, it can be concluded that sustainability is a popular research topic, and many integrated evaluation methods have been combined with fuzzy theory. Compared with existing works, this paper mainly makes the following contributions: (1) A new index system is proposed in this article taking circularity as one of the dimensions according to the circular economy theory. (2) Because the validity of the combination of fuzzy logic and various evaluation methods has not been proven, an MCDM approach consisting of AHP and TOPSIS was applied to select the best RL provider.

3. Methodology

Govindan et al. proposed a two-stage 3PRLP selection model that utilized AHP to identify factors and ANP to select providers [18]. Kumar and Dixit [28] and Liu et al. [29] combined this two-stage model with MCDM and upgraded it to a three-stage approach. This paper applies a three-stage process created through the synthesis of Govindan [18], Kumar and Dixit [28] and Liu et al.'s [29] frameworks for the RL provider selection as shown in Figure 1. In the first stage, the selection criteria for RL providers are determined based on an extensive literature review. In the second stage, AHP is used to calculate the relative weights of criteria. In this stage, the data of rating are collected from social groups through questionnaires via the Internet. In the last stage, the scores of RL providers in all three modes are collected, and then TOPSIS is used to calculate and rank the final scores of each RL provider. The best partner is the one with the highest final score.

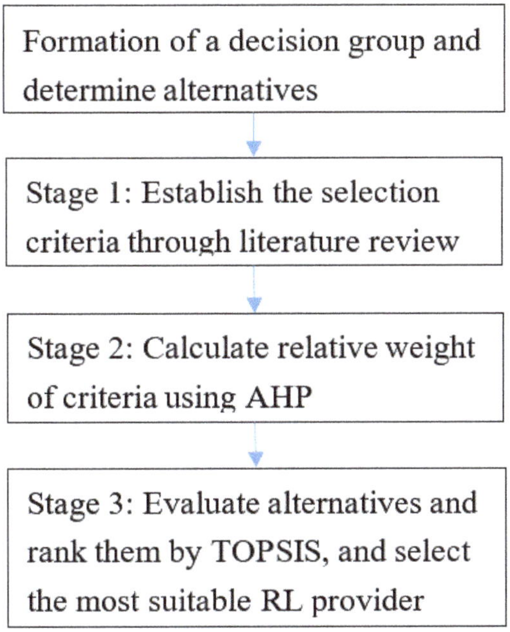

Formation of a decision group and determine alternatives

Stage 1: Establish the selection criteria through literature review

Stage 2: Calculate relative weight of criteria using AHP

Stage 3: Evaluate alternatives and rank them by TOPSIS, and select the most suitable RL provider

Figure 1. Proposed research framework for RL provider selection.

3.1. Evaluation Criteria of RL Provider Selection

Based on literature review and according to the Govindan's [15] application of CE theory in supplier selection and the operation mode of reverse logistics in the remanufacturing industry proposed by Tian et al. [45], a hierarchical structure of RL provider evaluation and selection is established as shown in Figure 2. The attributes contained in the first three dimensions of economy, society and technology are mainly derived from the improvement on the basis of the framework proposed by Tian et al. [45]. Tian et al.'s criteria focus on the evaluation of manufacturing enterprises [45]. By combining his framework with relevant research in the logistics industry [18,22,28,31,38,45–47], the criteria that need to be considered when evaluating and selecting RL providers can be determined. The attributes of the CE cluster are determined according to the theory of CE and the processes that may affect the environment when a product is moved in the supply chain. For example, the processes of collection, transportation and dismantling of waste products will all have an impact on the environment [15,45,48,49], which necessitates the consideration of these attributes.

Operating cost (E_1) is the expense paid by logistics providers to maintain their normal operation and development. It is related to the operation of enterprises and the operation of equipment, components and facilities. The operating cost composition of RL providers with different modes is different, but a sound operation can improve the efficiency of enterprises [18,45].

RL cost (E_2) determines whether remanufacturing activities can bring benefits to enterprises, and it includes transportation cost, inventory cost, inspection cost, packaging cost and material handling cost. The high cost of RL is one of the main obstacles to cost-effective remanufacturing production. Reducing RL cost is a common concern to both manufacturers and logistics companies [22,45].

Source of raw materials (E_3) refers to the sources and channels of collecting waste products of logistics companies. Waste products come from a variety of sources. One of the advantages of having a logistics provider for the collection of waste products is that they can benefit the existing logistics network, which makes the collection and transportation process more efficient and cost-effective, thus reducing time and capital costs.

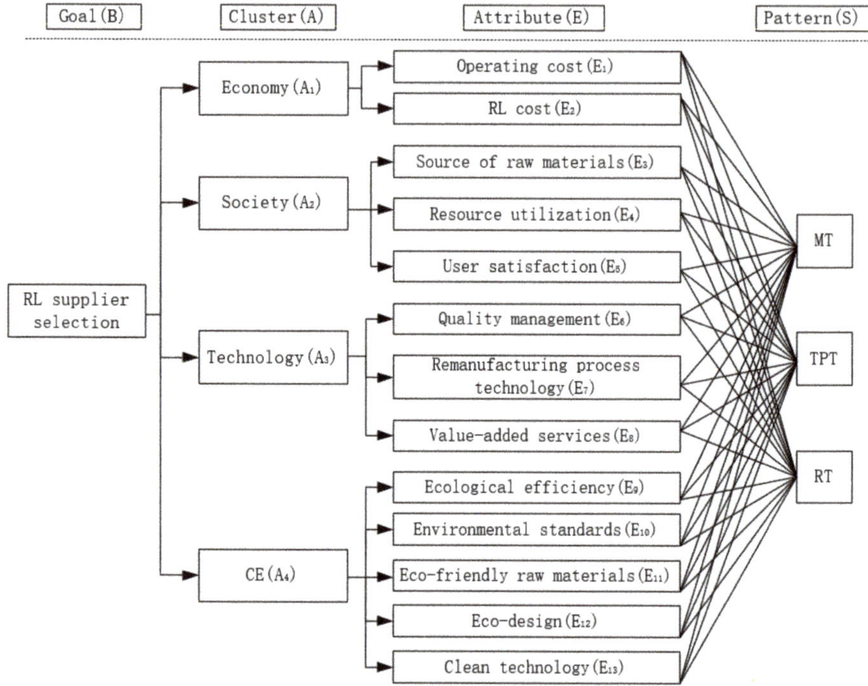

Figure 2. Structure of RL provider selection.

Resource utilization (E_4) refers to the ability to compare the value and cost of waste products after collection and to process them in different ways. Waste products are the essential resources of remanufacturing enterprises, and analyzing their effectiveness and utilization is an important step [45]. For RL providers, a preparation step is paramount to analyze and classify waste products and send them to different processing centers [28].

User satisfaction (E_5) includes effective communication with customers and partners, service improvement and overall working relationship. For an RL provider, whether communicating with partners or customers, efficient transmission of information is crucial as it can reduce unnecessary economic or time loss and improve the enterprise's reputation [42]. In addition, it is also necessary to pay attention to the timely upgrade and improvement of services [18].

Quality management (E_6) is used to measure integrity and quality after the processing of waste products. The quality of the remanufactured products is the final criterion for the entire remanufacturing process [45]. Although the quality of the remanufactured product should be comparable to that of the new product, more attention should be paid to quality management, such as providing a clear quality level after remanufacturing [31].

Remanufacturing process technology (E_7) is the key to product remanufacturing. For RL providers who are mainly responsible for the collection process, it mainly includes some remanufacturing pre-preparation technologies such as disassembly technology, cleaning technology and inspection technology [45].

Excellent RL providers can not only complete basic reverse logistics services according to the requirements of the entrusted enterprises but also provide some value-added services (E_8) [38]. Such services may include testing, classification of parts of waste products, remanufacturing of parts that only need to be simply refurbished before entering the secondary market, disposal of waste that is considered not to be worthy of remanufacturing and after-sales service [46,47].

Ecological efficiency (E_9) refers to the improvement of the resource transformation efficiency and environmental pollution reduction in the process of product collection through appropriate transportation and processing methods. Transportation activities are one of the main causes of ecological climate change [45]. Therefore, ecological efficiency is one of the evaluation criteria in the process of remanufacturing.

Environmental standards (E_{10}) are used to determine whether RL providers are complying with environmental protection standards for the collection of waste products [15].

Eco-friendly raw materials (E_{11}) refer to the utilization and reuse of recyclable materials in the product remanufacturing process. There is a trend to realize a recyclable product production process [48]. In this process, enterprises need to pay attention to what kind of raw materials are used in the manufacturing/remanufacturing process of products.

In the theory of circular economy, it is necessary to design a product with the minimum environmental degradation effect and the maximum recovery capacity while trying to reduce the cost in the manufacturing process [15]. This is the concept of eco-design (E_{12}). For the current situation where the scale of remanufacturing activities is about to be expanded, the design is one of the main factors affecting remanufacturing [49].

Clean technology (E_{13}) measures the ability of RL providers to process products using the appropriate technology. This criterion not only considers the process of collecting waste products but also includes the distribution processing of products, such as labeling, packaging and assembly procedures [15,50].

The framework consists of three parts: goal (B), cluster (A) and attribute (E). The goal is RL provider selection. The cluster includes economy (A_1), society (A_2), technology (A_3) and CE (A_4). The economic cluster includes attributes covering operating cost (E_1) and RL cost (E_2). The society cluster includes attributes of source of raw materials (E_3), resource utilization (E_4) and user satisfaction (E_5). The technology cluster includes attributes of quality management (E_6), remanufacturing process technology (E_7) and value-added services (E_8). The CE cluster includes attributes of ecological efficiency (E_9), environmental standards (E_{10}), eco-friendly raw materials (E_{11}), eco-design (E_{12}) and clean technology (E_{13}).

3.2. Selection Methodology for RL Provider

In order to determine the best RL provider for remanufacturing, an MCDM approach consisting of AHP and TOPSIS is applied. Saaty, the author of the AHP approach, questioned the rationality of the combination of fuzzy set theory and AHP [13] and believed that fuzzifying is a kind of interference without substantially changing the judgments. Therefore, the methods in this paper do not apply fuzzy numbers, and the results of AHP are compared with those of fuzzy AHP.

3.2.1. Analytic Hierarchy Process (AHP)

The basic idea of AHP is to hierarchize the problem to be analyzed (Saaty, 1980) [51]. According to the overall goal to be achieved, the problem is decomposed into different influence factors, which are combined in different levels according to their association and relationship to form a multilevel analysis structure model; finally, the alternatives are compared and ranked. The steps of AHP are as follows (Figure 3) [28,51]:

Step 1: Build a hierarchical structure. The top level indicates the problem to be solved. The second level consists of the decision criteria. The bottom level represents the alternatives.

Step 2: Construction of judgment matrix. In this step, data are usually collected through questionnaires or scores from a panel of experts. Decision-makers are required to make pairwise comparisons between decision dimensions and criteria according to the scale in Table 2. In order to obtain a more detailed attitude difference of the respondents and a more precise rating, this paper uses a 9-point scale.

Table 2. Scale of preference.

Scale	Explanation
1	Two factors are of equal importance compared to each other
3	One factor is slightly more important than the other
5	One factor is obviously more important than the other
7	One factor is strongly more important than the other
9	One factor is extremely more important than the other
2, 4, 6, 8	Intermediate values
Reciprocals	Reciprocals for inverse comparison

Step 3: Calculate the geometric mean of each row of the judgment matrix, and normalize it to obtain relative weights of dimensions and criteria, so as to obtain global weights and ranking of each criterion.

Step 4: Conduct the consistency test. Consistency ratio (CR) was calculated to evaluate the consistency of the comparisons. CR is the ratio of consistency index (CI) to random index (RI). CI is obtained as follows:

$$\mathrm{CI} = \frac{(\lambda_{\max} - N)}{(N-1)} \tag{1}$$

where $\lambda_{\max}$ is the maximum eigenvalue and N is the size of the matrix. Random index (RI) is obtained from Table 3. It is generally believed that when CR < 0.1, the inconsistency degree of the matrix is within the allowable range and indicates that the result has passed the consistency test. Otherwise, the judgment matrix should be reconstructed to improve the consistency.

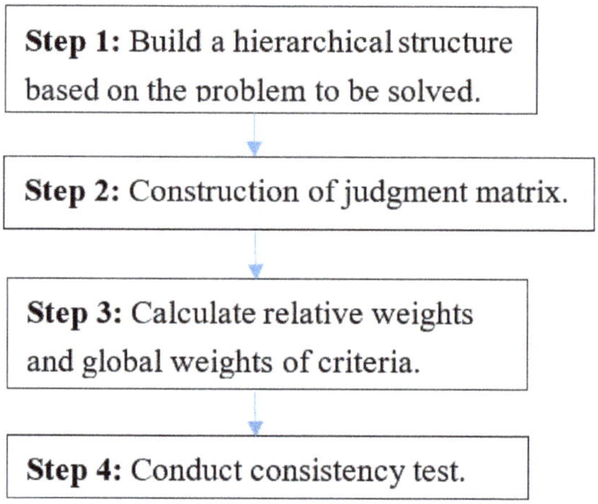

Step 1: Build a hierarchical structure based on the problem to be solved.

Step 2: Construction of judgment matrix.

Step 3: Calculate relative weights and global weights of criteria.

Step 4: Conduct consistency test.

Figure 3. Process of analytic hierarchy process (AHP).

Table 3. Random index (RI).

N	1	2	3	4	5	6	7	8	9	10	11
RI	0	0	0.58	0.90	1.12	1.24	1.32	1.41	1.45	1.49	1.51

3.2.2. TOPSIS

TOPSIS, an analysis method that compares and selects multiple alternatives based on multiple criteria, was firstly proposed by Hwang and Yoon (1981) [52]. Fundamental for TOPSIS is to determine the positive ideal solution and negative ideal solution of each attribute. The positive ideal solution is the optimal solution among the alternatives, and its attribute values reach the best value, while the negative ideal solution is the worst solution. After calculating the Euclidean distance between each scheme, the positive ideal solution, the distance between each scheme and the negative ideal solution, the approximate degree of each alternative to the optimal solution can be obtained, which can be used as the basis for evaluating the merits of the alternatives. The TOPSIS method has many advantages. It has no strict constraints on the data distribution and sample content. It is applicable to the analysis of small samples as well as large systems with multiple evaluation units and indexes. It is flexible and convenient to use and has universal applicability. The steps of TOPSIS are as follows (Figure 4):

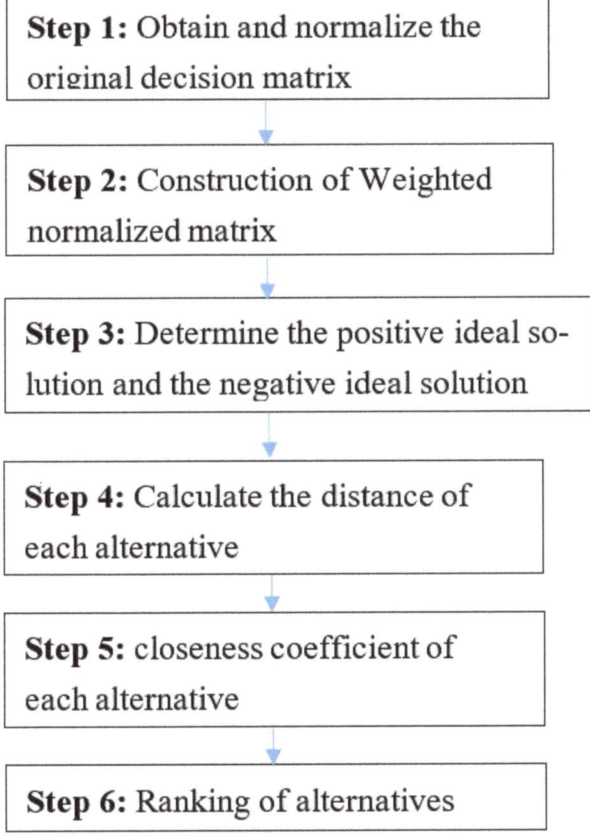

Figure 4. Process of technique for order of preference by similarity to ideal solution (TOPSIS).

Step 1: Obtain and normalize the original decision matrix.

Assuming that there are N criteria and M alternatives, the decision matrix formed by them is

$$X = \begin{bmatrix} x_{11} & \cdots & x_{1n} \\ \vdots & \ddots & \vdots \\ x_{m1} & \cdots & x_{mn} \end{bmatrix} \quad (2)$$

To eliminate the effects of various scales of criteria, the normalized decision matrix (B) is given by

$$b_{ij} = \frac{x_{ij}}{\sqrt{\sum_{i=1}^{m} x_{ij}^2}}, i = 1, 2, \ldots, m; j = 1, 2, \ldots, n \quad (3)$$

Step 2: Obtain the weighted normalized matrix.

The weighted normalized matrix (Z) is obtained by

$$z_{ij} = b_{ij} \cdot w_j, i = 1, 2, \ldots, m; j = 1, 2, \ldots, n \quad (4)$$

where w_j is the weight of the criterion.

Step 3: Determine the positive ideal solution (V^+) and the negative ideal solution (V^-) using the following equations:

$$V^+ = \left(z_1^+, z_2^+, \ldots, z_n^+\right) \text{ where } z_j^+ = \max\left\{z_{ij}\right\}, i = 1, 2, \ldots, m; j = 1, 2, \ldots, n \quad (5)$$

$$V^- = \left(z_1^-, z_2^-, \ldots, z_n^-\right) \text{ where } z_j^- = \min\left\{z_{ij}\right\}, i = 1, 2, \ldots, m; j = 1, 2, \ldots, n \quad (6)$$

Step 4: Calculate the distance of each alternative from V^+ and V^- using Equations (7) and (8):

$$D^+ = \sqrt{\sum_{j=1}^{n} \left(z_{ij} - z_j^+\right)^2}, i = 1, 2, \ldots, m \quad (7)$$

$$D^- = \sqrt{\sum_{j=1}^{n} \left(z_{ij} - z_j^-\right)^2}, i = 1, 2, \ldots, m \quad (8)$$

Step 5: Obtain the closeness coefficient (CC) of each alternative as follows:

$$CC_i = \frac{D_i^-}{D_i^+ + D_i^-}, i = 1, 2, \ldots, m \quad (9)$$

Step 6: Rank the alternatives.

The alternatives are ranked according to the value of CC. The best alternative has the highest closeness coefficient.

4. Application in Case Study

A case study with data from a Chinese engine manufacturer was conducted to verify the proposed approach. In recent years, increasing attention has been paid to the research and development of automobile remanufacturing in China. Compared with traditional manufacturing, remanufacturing has the characteristics of energy saving and material saving, which are important in the circular economy. With the introduction of relevant policies and the increase in public awareness of environmental issues, more and more automobile enterprises are actively engaging in remanufacturing activities. By June 2020, the motor vehicle population in China has reached 360 million, among which more than 270 million are automobiles [6]. The annual maintenance and replacement parts output value is nearly 1 trillion yuan, which represents a huge potential market for remanufacturing [6]. As an automobile part with high standardization and strong versatility, mainly made of metal materials, the engine is the automotive part that is commonly remanufactured.

For engine remanufacturing, the main process is the dismantling and cleaning of faulty or obsolete engine parts. This is followed by technical processing and transformation, according to the size modification requirements. Finally, after being reassembled and tested, it will be made into a new engine [53]. Before the remanufacturing processes, the reverse logistics of waste products should be completed first, which is a process requiring systematic operation. The company's production equipment level is above the average level of the same industry in China. At present, its products include remanufactured engines from various brands which have more than ten variations, including Steyr, Cummins and Chaochai 6102, which are used as maintenance parts for the after-sale market. This paper takes the problem of this company as an example and applies the proposed framework to select a suitable RL provider for it.

4.1. Application of AHP

First, an online AHP questionnaire was created according to the hierarchy shown in Figure 2. The questionnaire was sent to 340 respondents between September 14 and October 5, and 200 responses were collected after it was issued. Then, pairwise comparison matrices for each level were obtained according to the collected data. The weights of the four dimensions after data screening and processing are shown in Table 4 (this matrix passed the consistency test). "Technology" had the highest weight, followed by "circularity" and "society" at similar importance, and "economy" was the least important factor.

Table 4. The weights and ranking of the four dimensions.

Dimensions	Pairwise Comparisons				Weights	Rank
	A_1	A_2	A_3	A_4		
Economy (A_1)	1	0.5	0.2	0.25	0.0749	4
Society (A_2)	2	1	0.25	0.8	0.1419	3
Technology (A_3)	5	4	1	5	0.5950	1
Circularity (A_4)	4	1.25	0.2	1	0.1883	2

The weights of each criterion are listed in Table 5. Global weights were obtained by multiplying the relative weights between each criterion and the weights of each dimension. The CR of each matrix was less than 0.1 and could be used for evaluation and selection. Compared with "operating cost (E_1)", "RL cost (E_2)" is considered to be a more important indicator. "Resource utilization (E_4)" and "quality management (E_6)" are considered as the most important indicators in the social dimension and the technical dimension, respectively. "Eco-friendly raw materials (E_{11})" is the most important criterion and "environmental standards (E_{10})" is the least important criterion in the circularity dimension.

Table 5. The weights and ranking of criteria.

Dimensions	Criteria	Relative Weights	Relative Rank	Global Weights	Global Rank
Economy	E_1	0.1250	2	0.0094	13
	E_2	0.8750	1	0.0655	5
Society	E_3	0.2857	2	0.0405	8
	E_4	0.5714	1	0.0811	3
	E_5	0.1429	3	0.0203	10
Technology	E_6	0.5000	1	0.2975	1
	E_7	0.1000	3	0.0595	6
	E_8	0.4000	2	0.2380	2
Circularity	E_9	0.2034	3	0.0383	9
	E_{10}	0.0508	5	0.0096	12
	E_{11}	0.4068	1	0.0766	4
	E_{12}	0.0678	4	0.0128	11
	E_{13}	0.2712	2	0.0511	7

4.2. Application of TOPSIS

The decision-making team composed of eight experts scored three different modes (based on Figure 2, the three different modes were TPT, MT and RT) against the evaluation criteria, as shown in Table 6. For each mode, better performance for a criterion is reflected by a higher score. The range of the score is 1 to 9. Each final score was based on the sum of the average scores of the eight experts.

Table 6. Original decision matrix of alternatives.

	E_1	E_2	E_3	E_4	E_5	E_6	E_7	E_8	E_9	E_{10}	E_{11}	E_{12}	E_{13}
MT	2.875	2	6.25	6.375	6.125	7	7.5	6.375	7	7	7.125	7.5	7.375
TPT	3.75	4	6.375	6.625	6.5	6.625	6.375	6.875	6.875	6.5	6	5.625	6.375
RT	3.125	3.375	6.125	6.25	6.5	5.25	5	6.25	6.125	5.625	5.5	5.125	5.25

After normalization (Table 7), the weight of each index was calculated to obtain the weighted normalized matrix shown in Table 8. The positive ideal solutions V^+ and negative ideal solutions V^- are listed at the bottom of the table.

Table 7. Normalized decision matrix.

	E_1	E_2	E_3	E_4	E_5	E_6	E_7	E_8	E_9	E_{10}	E_{11}	E_{12}	E_{13}
MT	0.5075	0.3570	0.5773	0.5734	0.5545	0.6378	0.6793	0.5658	0.6052	0.6314	0.6587	0.7020	0.6661
TPT	0.6619	0.7139	0.5888	0.5959	0.5884	0.6036	0.5774	0.6101	0.5944	0.5863	0.5547	0.5265	0.5758
RT	0.5516	0.6024	0.5657	0.5622	0.5884	0.4784	0.4529	0.5547	0.5296	0.5074	0.5084	0.4797	0.4742

Table 8. Weighted normalized matrix.

	E_1	E_2	E_3	E_4	E_5	E_6	E_7	E_8	E_9	E_{10}	E_{11}	E_{12}	E_{13}
MT	0.0048	0.0234	0.0234	0.0465	0.0113	0.1897	0.0404	0.1347	0.0232	0.0061	0.0505	0.0090	0.0340
TPT	0.0062	0.0468	0.0238	0.0483	0.0119	0.1796	0.0344	0.1452	0.0228	0.0056	0.0425	0.0067	0.0294
RT	0.0052	0.0395	0.0229	0.0456	0.0119	0.1423	0.0269	0.1320	0.0203	0.0049	0.0389	0.0061	0.0242

V^+ = (0.0062, 0.0468, 0.0238, 0.0483, 0.0119, 0.1897, 0.0404, 0.1452, 0.0232, 0.0061, 0.0505, 0.0090, 0.0340)

V^- = (0.0048, 0.0234, 0.0229, 0.0456, 0.0113, 0.1423, 0.0269, 0.1320, 0.0203, 0.0049, 0.0389, 0.0061, 0.0242)

The distance of each alternative was then calculated using Equations (7) and (8). Finally, Equation (9) was used to calculate the closeness coefficient, and the results are shown in Table 9. The higher the closeness coefficient, the higher the alternative is ranked. The results show that TPT is the best solution, and its CC_i is 0.7565. MT has the furthest distance from the negative ideal solution, ranking the second.

Table 9. Distances to ideal solution and closeness coefficient.

Alternatives	D^+	D^-	CC_i	Rank
MT	0.0258	0.0518	0.6679	2nd
TPT	0.0152	0.0471	0.7565	1st
RT	0.0540	0.0161	0.2296	3rd

5. Discussions of Findings

As seen in Table 4, the technology (A_3) dimension is ranked the highest in the ranking of dimensions by a large margin. This result indicates that both the technology to be used in each specialized process of remanufacturing the engines and the general technology of processing other parts are considered as the most important criteria in the remanufacturing reverse logistics activities. As seen in Table 5, quality management (E_6) has the highest weight among the three criteria under this dimension. The final product of remanufacturing is the product whose quality and performance are the same as (or better than) the new product after the waste product has been processed. Therefore, it is particularly important to manage the quality of the output through professional technology and strict quality control; otherwise, core competitiveness will be lost. The criterion of value-added services (E_8) is ranked after E_6. Before remanufacturing, the waste products with excessive damage should be screened out. If the RL provider can complete a series of preparatory work steps before dispatching the good-quality waste products, it will save significant time and improve efficiency for manufacturers/remanufacturers. Therefore, this is also an important criterion. In contrast, remanufacturing process technology (E_7) is less important for RL providers who are mainly responsible for collection and logistics tasks.

Circularity (A_4) is slightly more important than society (A_2). The reason may be that the circularity dimension involves further requirements on both the technical and social dimensions, thus becoming more important to decision-makers. Eco-friendly raw materials (E_{11}) ranked first of the five criteria. It makes sense that the primary task of implementing the concept of circular economy is to use materials that have a less negative impact on the environment and to reuse them properly. Next is clean technology (E_{13}) and ecological efficiency (E_9). In addition to being environmentally friendly in terms of raw materials, proper methods and technologies should be used in transportation and processing to minimize unnecessary environmental pollution and energy consumption. It is also of significance to promote the concept of industry sustainability and the implementation of a circular economy [54]. Finally, eco-design (E_{12}) and environmental standards (E_{10}) rank last, both of which are less practical and more theoretical.

The purpose of the collection of waste products and remanufacturing is to adhere to the concept of sustainable development and improve the social prestige of companies [55]. For the immature engine remanufacturing industry in China, if a motor vehicle manufacturer is forced by policy or other reasons to carry out remanufacturing activities, it essentially trades the economic benefits for the benefits of other aspects, so in comparison, the impact of the economy (A_1) dimension is lower than that of the social dimension under this circumstance. The highest-ranking criterion for the society (A_2) dimension is resource utilization (E_4). Waste products are the resources of RL providers, and these products have varying availability. The ability to utilize these varying resources can reflect the adequacy of a company's basic business capacity in addition to transportation. Source of raw materials (E_3) and user satisfaction (E_5) mainly represent the business scope and operational level of RL providers, making them less important than resource utilization (E_4). Regarding the operating cost (E_1) and RL cost (E_2) under the economy (A_1) dimension, RL cost (E_2), which can better reflect the capabilities of the main business, is considered to be more important.

As seen in Table 9, the ranking of the alternatives in descending order is TPT, MT and RT. Since the first two have been applied to most cases, it is reasonable for RT to be ranked as last. MT is more suitable for large enterprises that normally have sufficient capital and can afford to invest in remanufacturing systems. For such medium-sized enterprises as the case in this paper, TPT can effectively make up for problems such as a lack of collecting channels, underdeveloped reverse logistics networks and systems with 3PRLPs' more specialized logistics systems. RL providers that provide strong value-added service capability can also share some of the manufacturing and remanufacturing processes.

6. Conclusions

Remanufacturing can be a risky decision for a manufacturer because it requires financial and technical support and affects the company's overall operating performance [56]. Choosing a good RL provider is particularly critical, starting with choosing the right take-back mode. This study proposes a new systematic index system and multicriteria decision-making method to select the best RL providers. First, evaluation criteria were established after an evaluation of the role of RL in manufacturing enterprises and a review of literature related to the selection of RL providers. Then, the decision method composed of AHP–TOPSIS was developed and the source data were collected in two different ways. The data used by AHP were obtained from questionnaires. The criteria weights obtained in this step were used as the input of TOPSIS, in which the environmental dimension obtained the highest weight. TOPSIS uses data provided by a panel of experts. The proposed framework was applied to a medium-sized automobile engine manufacturer in China, and the results show that TPT was the best RL provider mode.

The study presented in this paper has some limitations. The use of questionnaire data can compromise the results in certain ways; for example, the data may become irregular or too extreme because of a low level of professionalism of the respondents. This research also does not consider the uncertainty of expert scoring data and actual operation in the real situation due to the fuzzy theory not being introduced. Future research may discuss finding a new way to eliminate uncertainty. In addition, AHP and TOPSIS are methods that are widely applied due to their relatively simple calculation process, and they can easily be combined with other methods to form new approaches (such as connection-degree-based TOPSIS [57]). Future research may focus on the evaluation of various MCDM methods, including various alternative methods such as DEA, ISM, VIKOR and MOORA, to identify the optimal method that can be used to analyze the interrelationships between indicators and to rank criteria and alternatives.

Author Contributions: Conceptualization, Z.L. and X.Z.; writing—original draft preparation, Z.L.; writing—review and editing, Z.L., X.Z., Y.W. and W.Y.; supervision, X.Z. All authors have read and agreed to the published version of the manuscript.

Funding: This research was funded by the National Natural Science Foundation of China: Energy Efficiency Integrated Optimization of CNC Machining System Driven by Multi-source and On-line Energy Consumption Data Hybrid, grant number [51975432].

Institutional Review Board Statement: Not applicable.

Informed Consent Statement: Not applicable.

Data Availability Statement: Data is contained within the article.

Acknowledgments: The authors acknowledge the support and inspiration of Wuhan University of Science and Technology and the University of Brighton.

Conflicts of Interest: The authors declare no conflict of interest.

References

1. Nishizawa, O. *Circulation Expenses: An Unknown Third Source of Profits*; Koshinesha: Tokyo, Japan, 1970.
2. Kaviani, M.A.; Tavana, M.; Kumar, A.; Michnik, J.; Niknam, R.; de Campos, E.A.R. An integrated framework for evaluating the barriers to successful implementation of reverse logistics in the automotive industry. *J. Clean. Prod.* **2020**, *272*, 122714. [CrossRef]
3. Rogers, D.S.; Tibben-Lembke, R. *Going Backwards: Reverse Logistics Trends and Practices*; Reverse Logistics Executive Council: Reno, NV, USA, 1999.
4. Lai, K.-h.; Wu, S.J.; Wong, C.W.Y. Did reverse logistics practices hit the triple bottom line of Chinese manufacturers? *Int. J. Prod. Econ.* **2013**, *146*, 106–117. [CrossRef]
5. Jayaram, J.; Avittathur, B. Green supply chains: A perspective from an emerging economy. *Int. J. Prod. Econ.* **2015**, *164*, 234–244. [CrossRef]
6. National Development and Reform Commission. *Remanufacturing of Auto Parts: The "First Year" Has Arrived, and the Future Can Be Expected*; National Development and Reform Commission: Beijing, China, 2020.
7. Cao, J.; Chen, X.; Zhang, X.; Gao, Y.; Zhang, X.; Kumar, S. Overview of remanufacturing industry in China: Government policies, enterprise, and public awareness. *J. Clean. Prod.* **2020**, *242*, 118450. [CrossRef]
8. Zhang, Y.; Chen, W.; Mi, Y. Third-party remanufacturing mode selection for competitive closed-loop supply chain based on evolutionary game theory. *J. Clean. Prod.* **2020**, *263*, 121305. [CrossRef]
9. Wang, H.; Jiang, Z.; Zhang, H.; Wang, Y.; Yang, Y.; Li, Y. An integrated MCDM approach considering demands-matching for reverse logistics. *J. Clean. Prod.* **2019**, *208*, 199–210. [CrossRef]
10. Zhang, X.; Li, Z.; Wang, Y. A Review of the Criteria and Methods of Reverse Logistics Provider Selection. *Processes* **2020**, *8*, 705. [CrossRef]
11. Schöggl, J.-P.; Stumpf, L.; Baumgartner, R.J. The narrative of sustainability and circular economy—A longitudinal review of two decades of research. *Resour. Conserv. Recycl.* **2020**, *163*, 105073. [CrossRef]
12. Pieroni, M.P.P.; McAloone, T.C.; Pigosso, D.C.A. Circular economy business model innovation: Sectorial patterns within manufacturing companies. *J. Clean. Prod.* **2020**, *286*, 124921. [CrossRef]

13. Saaty, T.L.; Tran, L.T. On the invalidity of fuzzifying numerical judgments in the Analytic Hierarchy Process. *MComM* **2007**, *46*, 962–975. [CrossRef]
14. Haeri, S.A.S.; Rezaei, J. A grey-based green provider selection model for uncertain environments. *J. Clean. Prod.* **2019**, *221*, 768–784. [CrossRef]
15. Govindan, K.; Mina, H.; Esmaeili, A.; Gholami-Zanjani, S.M. An Integrated Hybrid Approach for Circular provider selection and Closed loop Supply Chain Network Design under Uncertainty. *J. Clean. Prod.* **2020**, *242*, 118317. [CrossRef]
16. Tosarkani, B.M.; Amin, S.H. A multi-objective model to configure an electronic reverse logistics network and third party selection. *J. Clean. Prod.* **2018**, *198*, 662–682. [CrossRef]
17. Tavana, M.; Zareinejad, M.; Santos-Arteaga, F.J.; Kaviani, M.A. A conceptual analytic network model for evaluating and selecting third-party reverse logistics providers. *Int. J. Adv. Manuf. Technol.* **2016**, *86*, 1705–1721. [CrossRef]
18. Govindan, K.; Sarkis, J.; Palaniappan, M. An analytic network process-based multicriteria decision making model for a reverse supply chain. *Int. J. Adv. Manuf. Technol.* **2013**, *68*, 863–880. [CrossRef]
19. Zarbakhshnia, N.; Jaghdani, T.J. Sustainable provider evaluation and selection with a novel two-stage DEA model in the presence of uncontrollable inputs and undesirable outputs: A plastic case study. *Int. J. Adv. Manuf. Technol.* **2018**, *97*, 2933–2945. [CrossRef]
20. Azadi, M.; Saen, R.F. A new chance-constrained data envelopment analysis for selecting third-party reverse logistics providers in the existence of dual-role factors. *Expert. Syst. Appl.* **2011**, *38*, 12231–12236. [CrossRef]
21. Momeni, E.; Azadi, M.; Saen, R. Measuring the efficiency of third party reverse logistics provider in supply chain by multi objective additive network DEA model. *Int. J. Ship. Transp. Log.* **2015**, *7*, 21–41. [CrossRef]
22. Kannan, G.; Pokharel, S.; Sasi Kumar, P. A hybrid approach using ISM and fuzzy TOPSIS for the selection of reverse logistics provider. *Resour. Conserv. Recycl.* **2009**, *54*, 28–36. [CrossRef]
23. Govindan, K.; Palaniappan, M.; Zhu, Q.; Kannan, D. Analysis of third party reverse logistics provider using interpretive structural modeling. *Int. J. Prod. Econ.* **2012**, *140*, 204–211. [CrossRef]
24. Gardas, B.B.; Raut, R.D.; Narkhede, B.E. Analysing the 3PL service provider's evaluation criteria through a sustainable approach. *Int. J. Prod. Perf. Manag.* **2019**, *68*, 958–980. [CrossRef]
25. Liu, A.; Ji, X.; Lu, H.; Liu, H. The selection of 3PRLs on self-service mobile recycling machine: Interval-valued pythagorean hesitant fuzzy best-worst multi-criteria group decision-making. *J. Clean. Prod.* **2019**, *230*, 734–750. [CrossRef]
26. Garg, C.P.; Sharma, A. Sustainable outsourcing partner selection and evaluation using an integrated BWM–VIKOR framework. *Environ. Dev. Sustain.* **2018**, *22*, 1529–1557. [CrossRef]
27. Govindan, K.; Jha, P.C.; Agarwal, V.; Darbari, J.D. Environmental management partner selection for reverse supply chain collaboration: A sustainable approach. *J. Environ. Manag.* **2019**, *236*, 784–797. [CrossRef]
28. Kumar, A.; Dixit, G. A novel hybrid MCDM framework for WEEE recycling partner evaluation on the basis of green competencies. *J. Clean. Prod.* **2019**, *241*, 118017. [CrossRef]
29. Liu, A.; Xiao, Y.; Lu, H.; Tsai, S.-B.; Song, W. A fuzzy three-stage multi-attribute decision-making approach based on customer needs for sustainable provider selection. *J. Clean. Prod.* **2019**, *239*, 118043. [CrossRef]
30. Zhou, F.; Wang, X.; Lim, M.K.; He, Y.; Li, L. Sustainable recycling partner selection using fuzzy DEMATEL-AEW-FVIKOR: A case study in small-and-medium enterprises (SMEs). *J. Clean. Prod.* **2018**, *196*, 489–504. [CrossRef]
31. Luthra, S.; Govindan, K.; Kannan, D.; Mangla, S.K.; Garg, C.P. An integrated framework for sustainable provider selection and evaluation in supply chains. *J. Clean. Prod.* **2017**, *140*, 1686–1698. [CrossRef]
32. Sasikumar, P.; Haq, A.N. Integration of closed loop distribution supply chain network and 3PRLP selection for the case of battery recycling. *IJPR* **2011**, *49*, 3363–3385. [CrossRef]
33. Prakash, C.; Barua, M.K. A combined MCDM approach for evaluation and selection of third-party reverse logistics partner for Indian electronics industry. *Sustain. Prod. Consump.* **2016**, *7*, 66–78. [CrossRef]
34. Dos Santos, B.M.; Godoy, L.P.; Campos, L.M.S. Performance evaluation of green providers using entropy-TOPSIS-F. *J. Clean. Prod.* **2019**, *207*, 498–509. [CrossRef]
35. Senthil, S.; Srirangacharyulu, B.; Ramesh, A. A robust hybrid multi-criteria decision making methodology for contractor evaluation and selection in third-party reverse logistics. *Expert Syst. Appl.* **2014**, *41*, 50–58. [CrossRef]
36. Prakash, C.; Barua, M.K. An analysis of integrated robust hybrid model for third-party reverse logistics partner selection under fuzzy environment. *Resour. Conserv. Recycl.* **2016**, *108*, 63–81. [CrossRef]
37. Govindan, K.; Agarwal, V.; Darbari, J.D.; Jha, P.C. An integrated decision making model for the selection of sustainable forward and reverse logistic providers. *AnOR* **2017**, *273*, 607–650. [CrossRef]
38. Chen, K.; Yu, X.; Yang, L. GI-TOPSIS Based on Combinational Weight Determination and its Application to Selection of Reverse Logistics Service Providers. *J. Grey Syst.* **2013**, *25*, 16–33.
39. Zarbakhshnia, N.; Wu, Y.; Govindan, K.; Soleimani, H. A novel hybrid multiple attribute decision-making approach for outsourcing sustainable reverse logistics. *J. Clean. Prod.* **2020**, *242*, 118461. [CrossRef]
40. Mavi, R.K.; Goh, M.; Zarbakhshnia, N. Sustainable third-party reverse logistic provider selection with fuzzy SWARA and fuzzy MOORA in plastic industry. *Int. J. Adv. Manuf. Technol.* **2017**, *91*, 2401–2418. [CrossRef]
41. Luo, Z.; Li, Z. A MAGDM Method Based on Possibility Distribution Hesitant Fuzzy Linguistic Term Set and Its Application. *Mathematics* **2019**, *7*, 1063. [CrossRef]

42. Li, Y.-L.; Ying, C.-S.; Chin, K.-S.; Yang, H.-T.; Xu, J. Third-party reverse logistics provider selection approach based on hybrid-information MCDM and cumulative prospect theory. *J. Clean. Prod.* **2018**, *195*, 573–584. [CrossRef]
43. Zarbakhshnia, N.; Soleimani, H.; Ghaderi, H. Sustainable third-party reverse logistics provider evaluation and selection using fuzzy SWARA and developed fuzzy COPRAS in the presence of risk criteria. *Appl. Soft. Comput.* **2018**, *65*, 307–319. [CrossRef]
44. Sen, D.K.; Datta, S.; Mahapatra, S.S. Decision Support Framework for Selection of 3PL Service Providers: Dominance-Based Approach in Combination with Grey Set Theory. *Int. J. Inf. Technol. Decis.* **2017**, *16*, 25–57. [CrossRef]
45. Tian, G.; Zhang, H.; Feng, Y.; Jia, H.; Zhang, C.; Jiang, Z.; Li, Z.; Li, P. Operation patterns analysis of automotive components remanufacturing industry development in China. *J. Clean. Prod.* **2017**, *164*, 1363–1375. [CrossRef]
46. Pourjavad, E.; Mayorga, R.V. A fuzzy rule-based approach to prioritize third-party reverse logistics based on sustainable development pillars. *J. Intell. Fuzzy Syst.* **2018**, *35*, 3125–3138. [CrossRef]
47. Guarnieri, P.; Sobreiro, V.A.; Nagano, M.S.; Marques Serrano, A.L. The challenge of selecting and evaluating third-party reverse logistics providers in a multicriteria perspective: A Brazilian case. *J. Clean. Prod.* **2015**, *96*, 209–219. [CrossRef]
48. Silva, A.L.E.; Moraes, J.A.R.; Machado, Ê.L. Proposta de produção mais limpa voltada às práticas de ecodesign e logística reversa. *Eng. Sanit. Ambient.* **2015**, *20*, 29–37. [CrossRef]
49. Singhal, D.; Tripathy, S.; Jena, S.K. Remanufacturing for the circular economy: Study and evaluation of critical factors. *Resour. Conserv. Recycl.* **2020**, *156*, 104681. [CrossRef]
50. Meherishi, L.; Narayana, S.A.; Ranjani, K.S. Sustainable packaging for supply chain management in the circular economy: A review. *J. Clean. Prod.* **2019**, *237*, 117582. [CrossRef]
51. Saaty, T.L. *The Analytic Hierarchy Process*; McGraw-Hill Book, Co.: New York, NY, USA, 1980.
52. Hwang, C.L.; Yoon, K. *Multiple Attributes Decision Making Methods and Applications*; Springer: Berlin, Germany, 1981.
53. Li, S. Research on remanufacturing technology and Development Trend of Automobile engine. *Intern. Combust. Engine Parts* **2019**, *18*, 203–204.
54. Jiang, Z.; Ding, Z.; Zhang, H.; Cai, W.; Liu, Y. Data-driven ecological performance evaluation for remanufacturing process. *Energ. Convers. Manag.* **2019**, *198*, 111844. [CrossRef]
55. Chu, J.W. *Automotive Recycling Engineering*, 2nd ed.; People Transportation Press: Beijing, China, 2012.
56. Jiang, Z.; Jiang, Y.; Wang, Y.; Zhang, H.; Cao, H.; Tian, G. A hybrid approach of rough set and case-based reasoning to remanufacturing process planning. *J. Intell. Manuf.* **2016**, *30*, 19–32. [CrossRef]
57. Ding, Z.; Jiang, Z.; Zhang, H.; Cai, W.; Liu, Y. An integrated decision-making method for selecting machine tool guideways considering remanufacturability. *Int. J. Comput. Integr. Manuf.* **2020**, *33*, 686–700. [CrossRef]

Review

A Review of the Criteria and Methods of Reverse Logistics Supplier Selection

Xumei Zhang [1,*], Zhizhao Li [1] and Yan Wang [2]

[1] School of Automobile and Traffic Engineering, Wuhan University of Science and Technology, Wuhan 430065, China; lizhizhao22@gmail.com

[2] School of Computing, Engineering & Maths, University of Brighton, Brighton BN2 4GJ, UK; Y.Wang5@brighton.ac.uk

* Correspondence: zhangxumei@wust.edu.cn; Tel.: +86-189-7101-2160

Received: 30 May 2020; Accepted: 15 June 2020; Published: 18 June 2020

Abstract: This article presents a literature review on reverse logistics (RL) supplier selection in terms of criteria and methods. A systematic view of past work published between 2008 and 2020 on Web of Science (WOS) databases is provided by reviewing, categorizing, and analyzing relevant papers. Based on the analyses of 41 articles, we propose a three-stage typology of decision-making frameworks to understanding RL supplier selection, including (a) establishment of the selection criteria; (b) calculation of the relative weights and ranking of the selection criteria; (c) ranking of alternatives (suppliers). The main discoveries of this review are as follows. (1) Attention to the field of RL supplier selection is increasing, as evidenced by the increasing number of papers in the field. With the adaption of circular economy legislation and the need resource and business resilience, it is expected that RL and RL supplier selection will be a hot topic in the near future. (2) A large number of papers take "sustainability" as the theoretical approach to carry out research and use it as the basis for determining the criteria. (3) Multi-criteria decision making (MCDM) methods have been widely used in RL supplier selection and have been constantly innovated. (4) Artificial intelligence methods are also gradually being applied. Finally, gaps in the literature are identified to provide directions for future research. (5) Value-added service is underrepresented in the current study and needs further attention.

Keywords: reverse logistics; supplier selection; MCDM; sustainability

---

1. Introduction

In recent years, reverse logistics (RL) has attracted increasing attention from researchers and industrialists due to the fact that it can recover the surplus value of end-of-life (EOL) products, meet environmental requirements and attach importance to customers' rights and interests [1]. The earliest widely accepted definition of reverse logistics was proposed by Rogers and Tibben-Lembke [2]: "RL is the process of planning, implementing, and controlling the efficient, cost effective flow of raw materials, in process inventory, finished goods and related information from the point of consumption to the point of origin for the purpose of recapturing value or proper disposal". Nowadays, environmental protection and development problems are a severe issue at the global scale; with the growth of population and the consumption of natural resources, implementing reverse logistics activities will soon become a "must" for many organizations. RL can achieve a viable and appropriate balance between the economy and the environment. In this way, it enables companies to generate profits and protect the environment through value retention of second-hand goods [3]. However, RL is a key problem in the closed-loop supply chain [4]. The supply chain management for RL mainly includes three aspects, namely, products, suppliers and raw materials. Among them, the evaluation and selection of suppliers is an important strategic decision to reduce operation costs, improve enterprise competitiveness and develop more business opportunities [5,6]. In recent years, the selection of suppliers has gradually become a problem with research value in reverse logistics practice. Issues include how to choose a third-party reverse logistics provider (3PRLP) when manufacturers want to implement remanufacturing activities and how electronics companies should choose recycling partners when recycling waste electrical and electronic equipment (WEEE).

There are many factors that prompt enterprises to implement RL, including stakeholders' concern about the environment, government's regulations on RL, cost effectiveness and sustainable development, etc., all of which contribute to the effective and efficient development of an enterprise [7]. In the current situation, the performance of an enterprise in a supply chain depends not only on its own performance but is also affected by the social performance and environmental performance of its partners. Any adverse effects produced by the other members of the supply chain could have negative effects on the enterprise [8]. Therefore, for enterprises involved in RL activities, whether they want to obtain better economic benefits or social prestige, there is a need to seriously evaluate and choose the suppliers they cooperate with. Under the buyer–supplier relationship, the primary problem that should be solved when choosing a supplier is often to achieve the ultimate goals of the buyer, such as maximizing economic benefit or achieving optimal environmental impact. Therefore, in RL practice, it is crucial to understand what theories and standards should be adopted and what methods should be employed to select partners to better achieve their goals. In this respect, the decisions mainly include supplier selection and supplier evaluation.

Few existing literature reviews on RL were focused on the analysis of supplier evaluation. For example, Govindan and Soleimani conducted a comprehensive review of the forefront areas of reverse logistics and closed-loop supply chains in clean production journals. The author did not take into account the aspect of suppliers in evaluation and decision; instead, their main focus was on manufacturing processes and remanufacturing processes. Govindan and Soleimani found that, between 2001 and 2014, RL research was mainly divided into eight categories, including general studies, remanufacturing, waste management, recycling, reuse, recovery, disassembling, and remanufacturing-recycling investigations [4]. Govindan et al. conducted a systematic review of 382 papers on RL and closed-loop supply chains published in scientific journals from January 2007 to March 2013 and found that 3PRLP selection accounted for only 2.9% of the total 382 papers. This means that, as the main supplier of RL, 3PRLP did not get much attention [9]. Prajapati et al. selected 449 related articles that respectively included the word "reverse logistics" in the title abstract and keywords and classified these articles into 11 different categories according to their structure and content. Although this category was not selected by 3PRLP, the author found that researchers often applied different multi-criteria decision-making (MCDM) technologies to help decision makers complete the decision when selecting suppliers. In all the literatures, there were 23 different MCDM technologies, among which the analytic hierarchy process (AHP) or fuzzy analytic hierarchy process (FAHP) was the most widely used method in the literature [7]. Aguezzoul reviewed the criteria and methods of third-party logistics (3PL) decision making. In terms of third-party logistics selection criteria, the author identified 11 key criteria, and found that cost was the most adopted criterion, followed by relationship service and quality. In terms of third-party logistics evaluation methods, this paper divides them into five categories: MCDM technical statistical methods, artificial intelligence mathematical programming and hybrid methods. In the conclusion, this paper pointed out that the study of third-party logistics choice is less theoretical and needs a more comprehensive conceptual framework, with a qualitative and quantitative combination [10]. Islam and Huda drew the conclusion that further research should be carried out in the area of 3PRLP selection [11].

In view of the existing research gaps in RL and third-party supplier selection, this paper makes a comprehensive review of the literature on two topics, including RL supplier evaluation and selection. This paper will also refer to the three stages of the decision-making process proposed in [8] to analyze and study the RL supplier question.

This article is structured as follows: Section 2 introduces the theoretical background, moving from logistics to RL and finally to RL supplier selection, as well as the steps of RL supplier selection which serve as the background of the literature review typology. Section 3 introduces research methodology and processes. Section 4 classifies and analyzes the selected sample of papers according to the selection process for RL suppliers. Finally, the conclusions of this paper and the limitations of this study are summarized, and suggestions for future research are given.

2. Theoretical Background

2.1. Background on RL

Logistics as an activity has a long history, but as a discipline it has only developed for a few decades. In 1918, an instant delivery corporation was established in the United Kingdom to deliver goods to wholesalers, retailers and consumers in a timely manner throughout the country. This activity is praised by some logistics scholars as an early document of logistics activities. During WWII, the United States first adopted the term "logistics management" in wartime arms supply from the perspective of military needs, and comprehensively managed the transportation and supply of arms. After WWII, the term "logistics" was borrowed by Americans into the management of enterprises, and was called "business logistics". Japan introduced the concept of "logistics" in the 1960s and interpreted it as "the circulation of goods". The concept of logistics varies from country to country, institution to institution, and period to period. In 1999, the United Nations logistics commission made a new definition of logistics. It pointed out that logistics is a process that realizes and controls the effective flow and storage of raw material inventory, final products and related information from the beginning to the end in order to meet the needs of consumers.

Since the 1990s, many countries in the world began to pay attention to the environment and raw material resources, which means RL has gradually received more attention [12]. RL has been proved to be beneficial to the economic situation and social image of enterprises; Jayaraman and Luo note that many of the things that companies dispose of, such as product waste resources, are actually valuable [13]. In a closed-loop supply chain, reverse flow can bring competitive advantage to enterprises. If enterprises fail to realize the importance of reverse logistics strategy, customer relations might be in jeopardy, which put them at a disadvantaged position in market competition. The value of reverse logistics has been reflected in many industries, such as the carpet industry [14], retail industry [15], bottling sector [16], paper industry [17], packaging industry [18], cell phone industry [19], pharmaceutical industry [20], and battery recycling industry [21]. Thus, it can be seen that reverse logistics has research significance in various degrees, whether it is the study of theory, such as what benefits it can bring to the enterprise, or the study of practice, such as how it should be applied in various fields. As for the manufacturing industry, studies on machining systems [22] and performance evaluation [23] are relatively mature, while there are relatively fewer studies on the combination of remanufacturing with RL.

According to the research conducted by Rachih et al., the process of RL can be described as collection, inspection and sorting, and the last step is re-processing [24]. Product recovery is the first step and also a key step in establishing an effective and profitable reverse logistics system [12]. 3PRLP has a better ability to respond to the complexity and randomness of the supply chain as well as the professionalism of reverse logistics activities [25], which makes the third-party collection model widely used. However, there are still three main collection modes—third-party take-back (TPT), manufacturer take-back (MT), and retailer take-back (RT)—which means manufacturers, retailers and 3PRLP can all act as reverse logistics suppliers. Moreover, the objectives of different types of supply chains are not the same, and different companies in different industries hope to get different benefits from reverse logistics activities [26]; the business scope and capabilities of each RL service provider are also different, which means that most enterprises will face a decision to choose the best RL supplier. Govindan et al. indicate that, if the reverse logistics system is operated effectively, it could be the main source of benefits [27]. Therefore, the selection of reverse logistics suppliers becomes one of the most important activities in the entire supply chain system, due to the fact that enterprises tend to outsource reverse logistics activities in order to focus on core business operation.

2.2. Literature Review Typology

In this literature review, the articles are categorized according to a three-stage process of supplier selection, created through the synthesis of Govindan [27], Kumar and Dixit [28] and Liu et al.'s [29] frameworks (see Figure 1). The three stages are: (A) establishment of the selection criteria; (B) calculation of the relative weights and ranking of the selection criteria; (C) ranking of alternatives (suppliers).

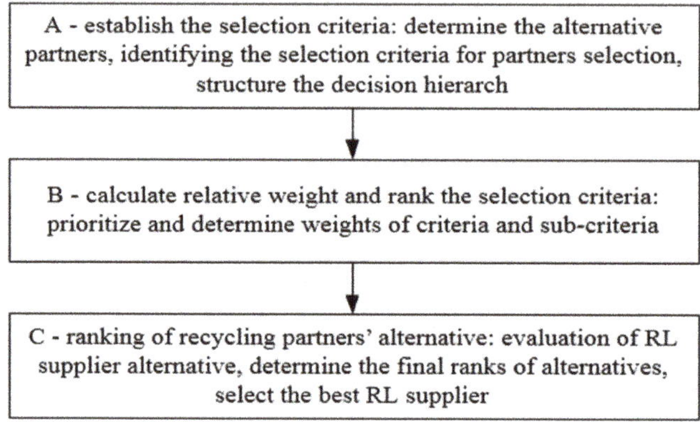

Figure 1. RL supplier selection process.

The starting point for this research is that the awareness of environmental protection and the concept of sustainable development in society are constantly improving, and understanding of the importance of reverse logistics in various industries is increasing year by year. In particular, with the stringent government environmental legislation implemented worldwide, the choice of suppliers has become an important issue for enterprises. In order to meet the demand of logistics activities in the supply chain and realize the requirements of environmental friendliness, enterprises tend to choose RL suppliers that can best match their own demand and maximize economic and environmental benefits. To understand the selection and evaluation of suppliers in RL activities, Kumar and Dixit [28] and Liu et al. [29]'s frameworks were studied, all of which started with the establishment of selection criteria (the Stage A). The completion of this stage usually requires a certain amount of literature review for support and expert opinion from the field of expertise as a reference [28]. As for the methods, these include, but are not limited to, Delphi and brainstorming methods. Determining the selection criteria is a complex and deliberate process, as each enterprise has different expectations. For example, Liu et al. [29] applied the triple bottom line (TBL) to establish sustainable supplier criteria. For the social aspect, it is necessary to meet the needs of customers as far as possible. For the economic aspect, it is necessary to make plans to reduce costs and maximize profits. For the environment aspect, it is required to consider resource consumption and waste disposal in actual activities, so as to minimize negative impact on the environment. These three aspects have their own sub-criteria, and there will be trade-offs when determining the indicator system under different circumstances.

At Stage B, which is to calculate the relative weight of the criteria, Kumar and Dixit explain that, due to the uncertainty in practical problems and the fuzziness of the criteria, it is usually necessary to combine the common evaluation method with the fuzzy theory to get a more reliable ranking of the criteria [28]. Liu et al. argued that a mixed model of the two evaluation methods can achieve roughly the same effect [29]. Govindan et al. used a systems analysis approach to deal with the criteria, but it is clear that the contextual relationships between these variables (criteria) are subject to the personal preferences of the decision makers, and their personal biases may affect the final result [27]. Therefore, Stage B will analyze the applicability of various approaches to the relationship between criteria with consideration of the subjectivism of the criteria.

At Stage C, ranking methods are used to rank alternatives to make the final choice. In many cases, the hybrid method formed by combining the two phases B and C is called the MCDM method, or it forms a new hybrid MCDM framework. The MCDM method has been widely used, especially in research in the manufacturing industry which is closely related to reverse logistics [30,31]. At this stage, various ranking methods are applied to evaluate and sort the alternatives according to the processed criteria in order to select the best RL supplier.

3. Research Methodology

A valid literature review can improve knowledge of related fields by identifying patterns, themes and issues as well as identifying key concepts that work as a path to new theoretical developments and new directions of research [32,33]. Islam and Huda define a literature review as having four steps: material collection, descriptive analysis, category selection and, finally, material evaluation to support the topic [11]. A valuable literature review can summarize past research in the field and point out potential research directions in the future.

In this study, material collection is accomplished by a two-phase process. For the first phase, refer to the search method in [8]; articles in the Web of Science (WOS) database were searched using the terms ("reverse logistics" AND "evaluation" AND "supplier") in the title, abstract and keywords (at least one of the three elements should include the search terms). Although the range of dates searched is not limited, no papers were found from before 2005. The reason may be that the logistics industry was immature at that time, which led to little research on reverse logistics. Papers published after April 2020 are not included, as the final updated data were completed on this date.

In the second phase, based on the work of Alexander et al. [34], two exclusion criteria—semantic relevance and relevance to the research problem—were used to filter articles. Semantic relevance refers to the different meanings a word may have in different contexts. For example, "reverse logistics" and "selection" can capture articles that refer to the selection of reverse logistics recovery methods or modes [35]; "reverse logistics" and "evaluation" can include articles that refer to evaluation of reverse logistics barriers or performance. Such articles had to be excluded. Relevance to the research problem means that, in order to determine the relevance of an article's topic, the entire content of the article has to be reviewed, as the title or abstract may not present these contents clearly. After reviewing these papers, 41 of the 107 articles were ultimately selected for the review using the exclusion criteria.

4. Analysis

This section conducts a descriptive analysis of the selected papers and then categorizes these articles according to the three stages described in Section 2.2 and Figure 1.

4.1. Descriptive Analysis

The number of articles published each year is shown in Figure 2. From Figure 2, we can see that the first article about RL supplier selection was published in 2008; thus, the articles selected for this review were published between 2008 and 2020. Starting in 2015, the number of relevant articles has increased, and the growth rate has increased rapidly in the past two years. There were 11 articles published in 2019 and 7 in 2018, but no articles related to the topic were found among the articles published in 2010. It should be noted that the number of articles for 2020 is not complete, and there may be further growth after data collection.

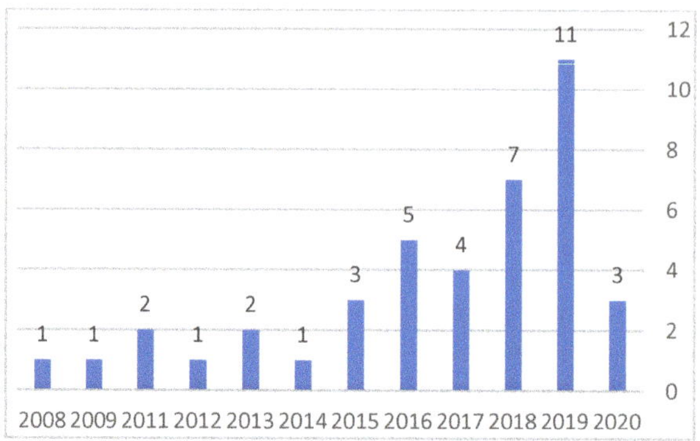

Figure 2. Number of papers per year.

Among the 41 articles, 12 were from the *Journal of Cleaner Production* (Table 1), which is the journal with the most published articles, accounting for 29%, followed by the *International Journal of Advanced Manufacturing Technology* (10%). The top contributing countries are shown in Table 2. (Since many articles are jointly published by scholars from multiple countries, only the countries of the first author of the articles are listed, and Table 2 only shows countries with more than three published articles.) India has the largest number of publications on the subject (15), followed by China (10). In addition, some articles are based on studies of enterprises in particular countries (Table 3). Overall, 14 articles chose Indian companies for case studies, 7 articles chose Iranian companies and 6 chose Chinese companies. The logistics industry in India has been developing rapidly due to its rapid economic growth, which has prompted more research on RL subjects in India. The implementation of circular economy promotion laws in China might be a driving force for the interest and the number of papers in China.

Table 1. List of journals and number of papers.

Journals	References	N.
Journal of Cleaner Production	[1,3,28,29,36–43]	12
International Journal of Advanced Manufacturing Technology	[44–47]	4
Applied Soft Computing	[48,49]	2
Expert Systems with Applications	[50,51]	2
Resources Conservation and Recycling	[27,52]	2
Annals of Operations Research	[53]	1
Computers Industrial Engineering	[54]	1
Computers Operations Research	[55]	1
Environment Development and Sustainability	[56]	1
International Journal of Information Technology Decision Making	[57]	1
International Journal of Production Economics	[58]	1
International Journal of Production Research	[59]	1
International Journal of Productivity and Performance Management	[60]	1
International Journal of Shipping and Transport Logistics	[61]	1
Journal of Environmental Management	[62]	1
Journal of Grey System	[63]	1
Journal of Intelligent Fuzzy Systems	[64]	1
Jurnal Teknologi	[65]	1
Mathematics	[66]	1
Omega International Journal of Management Science	[67]	1
Opsearch	[68]	1
Production Planning Control	[69]	1
Sustainability	[70]	1
Sustainable Production and Consumption	[71]	1

Table 2. Contributions by country.

Country	Number of Papers
India	15
China	10
Denmark	9
Iran	9
USA	5
Australia	3
Canada	3

Table 3. Number of articles for particular countries.

Country	Number of Papers
India	14
Iran	7
China	6
Brazil	2
USA	1
Korea	1
France	1

The top 10 most-cited articles were identified among the 41 articles analyzed (See Table 4), three of which were published in *the Journal of Cleaner Production*. *Resources Conservation and Recycling* published the most-cited articles, while the article with the highest average annual citation rate [42] was published in the *Journal of Cleaner Production*.

Table 4. Most cited articles.

References	Journal	Citations	Year of Publication	Average Citations/Year
[27]	Resources Conservation and Recycling	259	2009	21.58
[42]	Journal of Cleaner Production	176	2017	44.00
[58]	International Journal of Production Economics	171	2012	19.00
[54]	Computers & Industrial Engineering	138	2008	10.62
[50]	Expert Systems with Applications	85	2014	12.14
[41]	Journal of Cleaner Production	65	2018	21.67
[49]	Applied Soft Computing	64	2016	12.80
[51]	Expert Systems with Applications	60	2011	6.00
[43]	Journal of Cleaner Production	54	2015	9.00
[59]	International Journal of Production Research	50	2011	5.00

Each article is categorized based on the three stages of the RL supplier selection process, as described in Figure 1 (see Table 5). The following section describes the findings and analysis of these stages; the theories and methods used in each stage in the papers are discussed in detail in Section 4.2 to Section 4.4. It is important to note that [42] is discussed in a particular section as its content is a conceptual framework.

Table 5. Articles according to the three stages of the process.

References	Stages	No.
[1,3,27–29,36–45,47,48,50–54,56,57,59,61,63,64,66–71]	A, B and C	34
[46,49,58,62,65]	A and B	5
[55]	A and C	1
[60]	A	1

4.2. Stage A: Establish the Selection Criteria

In terms of the theoretical approaches used in the 41 articles selected, researchers mainly focus on "sustainability", which is utilized in 18 papers. Six papers use green supply chain management (GSCM) to investigate how to select RL suppliers. "Green" focuses more on the environment aspect than "sustainability". In addition, only one article uses circular economy (CE), which focuses more on the economic aspect than "sustainability". The remaining half mainly apply two theories to complete the classification of criteria. Of these, 11 papers summarize several "performance dimensions" [57] based on the business capabilities of the RL supplier. Organizational performance, RL Functions, IT application, service quality and user satisfaction are the most commonly used main groups. Four papers consider firms' requirements and operating strategies [36] to complete the classification of criteria. One article completes the criteria classification through strengths–weaknesses–opportunities–threats (SWOT) analysis [49] (Table 6).

Table 6. Theory of criteria classification.

Theory	References	No.
Sustainability	[3,29,38,42–45,48,53,55,56,60,62,64,66–68,70]	18
Performance dimensions	[27,40,46,47,50,52,57–59,61,65]	11
GSCM	[28,37,39,41,54,69]	6
Firm's requirements and operating strategy	[36,51,63,71]	4
SWOT	[49]	1
CE	[1]	1

Generally identifying the most-used criteria to select the 3PRLP was based on a systematic literature review (SLR), and several papers proposed a similar set of criteria: forward logistics; reverse logistics; financial; capacity; environmental; alliances [43,50,57,65]. In addition, different main parameters determine the selection process for an appropriate 3PRLP in different situations. Some of these parameters are common, such as the use of third-party logistics [46], service quality and service ability [63], and application of information technology [28,47,58], and some are used in special cases, like noise pollution [36]. Both Li et al. [40] and Prakash and Barua [52,71] consider firms' requirements and operating strategies. Momeni et al. [61], P.Sasikumar and A.Noorul Haq [59] determine the criteria according to the process of reverse logistics.

Sustainability is a popular research topic; many articles divide criteria into three categories, reflecting their impacts on society, economy and environment, respectively. The specific criteria, from economic, social and environmental perspectives according to previous sustainable/green practices and literature reviews, are explored [38,62,67]. Silva et al. define EES (economic, environmental and social) considerations as the three dimensions of sustainable development [8]. It should be noted that ISO14000 is often used as a reference for formulation of environmental criteria. Some authors add the consideration of risk, which includes operational risk and financial risk [48,66,68]; sometimes safety practices and health, along with compliance with the International Labour Organization (ILO) code [53] or respect for local rules and policies [45] are grouped together into the category of social factors. In the following research, Zarbakhshnia et al. [3] improve "risk and safety" by using the Delphi method to update the content of the criteria, and add two sub-criteria—organizational risk and safety— into the new dimension of "risk and safety". Garg et al. [56] take all sustainable outsourcing partners (SOPs) as alternative objects; reverse logistics and waste minimization are regarded as sub-criteria under the aspect of environmental factors, in which practices recovering materials, minimization of waste and disposal in an environmental friendly way are emphasized. Liu et al. [29] transform the criteria of the three sustainability dimensions based on customer needs (CNs) into engineering characteristics (ECs), which have nine criteria, including reverse logistics. Bai and Sarkis summarize cost, quality, time, flexibility, and innovativeness as five economic/business attributes that can be considered from a sustainability perspective [55].

The papers also link a range of other criteria to RL supplier selection. Pourjavad et al. divides the crucial criteria of 3PRLP selection into environmental, social and cost [64]. The cost criteria consists of the quality of products, value-added services, transport capacity and level of advanced equipment. Jung [70] set social sustainability as the main aspect, with focuses on three social criteria: philanthropy, average salary and management policy.

When selecting suppliers in the sustainable supply chain, both the collection capability as well as green technology capability are considered [44]; others employ waste management and pollution prevention as evaluation criteria [42]. Gardas et al. add the "sustainable eco-friendly process/recycling" to the 3PLP selection criteria of the pharmaceutical industry [60]. Govindan et al. developed a model for circular supplier selection, in which three criteria are identified: quality, on-time delivery and circularity, among which the sub-criterion "environmental standards" requires suppliers to utilize environmental standards to recycle products [1].

Chatterjee et al. present the dimensions and corresponding criteria of the GSCM, including green design, green production, green warehousing, green transportation and green purchasing [41]. Tosarkani and Amin added "social-cultural enablers" to various criteria, representing green logistics practices [39]. EES assessment is also cited by some authors to select reverse logistics centers [69]. Efendigil et al. proposed that RL is an important part of the effective operation of "green supply chains". However, their twelve performance indicators for 3PRLP are biased towards logistics practices and neglect of green practices [54]. In contrast, the criteria for selection and evaluation of green suppliers summarized by Santos et al. [37] are more oriented towards green practices. A set of more comprehensive criteria is proposed in further research; Kumar and Dixit [28] studied recycling partner evaluation for the disposal of waste electrical and electronic equipment (WEEE) using green competencies (GC), and proposed a mode for GC and recycling partner selection, using opportunism, service and delivery performance, resource and environmental management capabilities, social responsibility benefits, green core competencies, management and organizational competencies, regulatory obligations and risk compliances as the criteria.

4.3. Stage B: Calculate the Relative Weights

In stage B, in order to determine the degree of importance of the criteria, the weights for each dimension and then the weights of the sub-criteria are calculated in their separate dimensions. For articles that use the concept of sustainability to establish criteria, the most heavily weighted dimensions are generally environment [28,42,56] and economy [45,48,67,68]. Some articles are specific to cost or price [64,70]. Only one of the selected articles places the highest weight on the society dimension [3]. Often, the most critical criteria are tailored to different situations and objectives. Tavana et al. conclude that the third-party logistics services (3PLS) criterion has the most important impact on the selection of 3PRLPs [46]. That is, the highest priority should be given to the 3PLS criterion if a decision maker aims at improving the performance when selecting 3PRLPs [47,52,71]. Articles with green practices as the purpose place the most emphasis on various environmental criteria, such as "circular" [1], environmental practice [37,39,40], environmental expenditure [54] and green design [41]. Others focus on technical/engineering capability [27,36,58].

According to decision making techniques, the majority of articles employ fuzzy theory to assist in this process. These articles use fuzzy logic to deal with the uncertainty of qualitative criteria. FAHP was used in 13 articles; four articles used FANP (fuzzy analysis network process). BWM (best–worst method), DEA (data envelopment analysis) and ISM (interpretive structural modeling) were employed in three papers each. SWARA (stepwise weight assessment ratio analysis) was employed in two papers each. CPT (cumulative prospect theory) and AEW (anti-entropy weighting) are used in one article each. More details are listed in Table 7.

Table 7. Decision making techniques.

Decision Making Technique	References	No.
FAHP	[3,28,42,49,50,52–54,63,67,68,70,71]	13
FANP	[1,39,46,47]	4
DEA	[44,51,61]	3
ISM	[27,58,60]	3
BWM	[36,56,62]	3
SWARA	[45,48]	2
Variable weight theory	[66]	1
Entropy	[37]	1
CPT	[40]	1
DEMATEL	[64]	1
Rasch Model	[65]	1
QFD and IVITFNs	[29]	1
Fuzzy DEMATEL-AEW	[38]	1
Rough DEMATEL-ANP	[41]	1
SWOT	[49]	1
VIKOR	[51]	1
Linguistic data	[57]	1

Although most articles do not mention it, there is some basis for the method selection. Kumar and Dixit [28] consider that, compared with ANP, the ease of applicability in pairwise comparisons made AHP become a more useful tool. However, the weakness of AHP is that the process of AHP is time consuming because there will be a large number of required pairwise comparisons [67]. Chatterjee et al. find that, in the R'DEMATEL method (rough Decision Making Trial and Evaluation Laboratory Model), the relationship between the criteria is closer to the real systems compared with ANP, because the levels of interdependence of criteria and dimensions do not have any reciprocal values [41].

4.4. Stage C: Ranking of Alternatives

In stage C, the articles determine the final results using appropriate ranking methods. As can be seen from the Table 8, VIKOR (ViseKriterijumska Optimizacija I Kompromisno Resenje) and TOPSIS (technique for order of preference by similarity to ideal solution) are the two commonly used methods; MOORA (multi-objective optimization by ratio analysis) and DEA were applied in three papers each, two papers used the FANP method, and the rest of the articles used a variety of novel approaches including IPHFS (interval Pythagoras hesitant fuzzy set), MAIRCA (multi-attribute ideal-real comparative analysis), COPRAS (complex proportional assessment), NRS (neighborhood rough set), PROMETHEE (preference ranking organization method for enrichment evaluation), F-AIO (fuzzy artificial immune optimization).

Luo et al. [66] compared the results of PDHFLT (possibility distribution based hesitant fuzzy linguistic term sets)-MOORA with those of PDHFLT-TOPSIS and PDHFLT-VIKOR. Although there was no difference in the final results, the overall ranking was still different. TOPSIS ignores individual regret, while the VIKOR-based method considers individual regret, but the result is very sensitive to individual regret proportion, which is difficult to determine. The PDHFL-MULTIMOORA method considers both the group utility values and individual regret values and can overcome this shortcoming. Bai and Sarkis [55] mainly make innovations in ranking methods. The NRS theory is used to reduce the number of 3PRLPs to be selected at first, and then a preferred 3PRLP is eventually selected by using VIKOR combined with the TOPSIS. The solution with the shortest distance from the ideal solution and the longest distance from the negative ideal solution can be determined by TOPSIS, but the relative importance and conflict criterion of distance could not be taken into account. The proposed new method integrated the advantages of the TOPSIS and VIKOR methods and can take into account factors such as the ideal solution, the negative ideal solution and conflicting criteria, which can make results more reliable [55].

Table 8. Decision making techniques.

Decision Making Technique	References	No.
VIKOR	[28,29,38,42,56,59,71]	7
TOPSIS	[27,37,50,52,53,63]	6
MOORA	[3,45,66]	3
DEA	[44,51,61]	3
FANP	[39,47]	2
IPHFS	[36]	1
MAIRCA	[41]	1
COPRAS	[48]	1
Fuzzy inference system	[64]	1
NRS-TOPSIS-VIKOR	[55]	1
TOPSIS-PROMETHEE	[68]	1
Dominance-based decision-making approach	[57]	1
Artificial neural networks	[54]	1
DEMATEL	[1]	1
CPT	[40]	1
FAHP	[70]	1
F-AIO	[69]	1
ELECTRE I	[67]	1

The IPHFS adopted by Liu et al. [36] can more accurately show the attitude of decision makers than traditional ANP, TOPSIS and BWM methods because it has more liberal restrictions on membership and non-affiliation. Govindan et al. consider that widely used approaches such as AHP, TOPSIS, and DEA, which require decision makers to make great cognitive effort, may lead decision makers to make some arbitrary transformation of the performances scales, or provide poor recognition ability [67]. The advantage of the new proposed ELECTRE and SMAA approaches is that they allow the qualitative nature of some criteria and heterogeneous criteria scales, and they provide the possibility of modeling the effects of high strengths or critical weaknesses in the comparison of a pair of alternatives. Li et al. [40] proposed the CPT-based hybrid-information MCDM method, which takes into consideration the bounded rationality of DM. This approach offers better consistency with reality, as it addresses the shortcomings of FTOPSIS. Chatterjee et al. compared the ranking results of rough TOPSIS, rough COPRAS, rough MAIRCA and rough VIKOR; the conclusion is that the R'VIKOR method focuses on seeking a balance between total and individual satisfaction, while the R'MAIRCA method emphasizes ranking and selecting from a set of alternatives in the presence of conflicting criteria. R'TOPSIS introduces the ranking index, including the distances from the ideal point and from the negative ideal point, but these distances are simply summed as numerical values, without taking into account their relative importance; similar problems may appear in the R'COPRAS method [41].

5. Conclusions and Discussions

5.1. Main Findings and Implications

In this paper, 41 articles on RL supplier selection are analyzed in depth to identify the criteria and methods most used in this selection. The following conclusions can be drawn:

(a) The number of papers on RL supplier selection has increased in recent years; this result demonstrates that RL supplier selection has attracted organizations and research centers from around the globe. Table 2 shows that research on this topic is mainly popular in developing countries (India, China, Iran) or developed countries where agriculture is a traditional industry (such as Denmark). These countries are focusing on developing industry and manufacturing to become pillar industries, and their emphasis on RL supplier selection shows that RL is an important part of this process and indicates that, from the perspective of world development, this topic still has space for further development in the future. With the recent adoption of the Circular Economy Action Plan by the

European Commission, one of the main building blocks of the European Green Deal in 2020 will be a driving force for the adaption of RL in Europe.

The interruption of global supply chains by the COVID–19 pandemic is forcing businesses to convert to RL and to the circular economy to enable the resource and business resilience. This will further promote the attention to RL-related research globally.

(b) Although a small number of authors use the "green" and circular economy, or other theories, most of the existing articles focus on the perspective of sustainable development. Among them, environment and economy are the dominant two criteria, and few articles focus on the social aspect. Although such a trend conforms to the trend in the development of the logistics industry, too much concentration on a single subject may have a negative impact on further research, and more new theories should be developed as the core of subsequent research. Those articles that do not employ any theory mainly focus on RL capability in terms of criteria, so, in further research, this criterion can be refined to consider what theories and indicators should be applied from the needs of different industries or to determine new attributes.

(c) The RL supplier selection process is based on a large number of attributes which are not fully considered in some criteria. Value-added services are mentioned in some articles but not taken seriously. It is conceivable that an RL supplier would be competitive if it met the needs of different customers for various types of value-added services. Value-added services can be used to reflect the strength of an RL supplier's professional capabilities and adaptability to different markets, from which researchers can discuss the subject of evaluation of RL supplier performance.

(d) Almost all studies use more than one decision method, with MCDM being the most frequently used decision technique, and fuzzy logic is the most common method used to eliminate uncertainty. AHP and ANP are favored by researchers for their simplicity and intuitiveness, and they are often combined with traditional ranking methods such as VIKOR or TOPSIS. Such a hybrid method has broad applicability, but it does not satisfy all decision requirements. In addition, many articles do not elaborate on the rationality of the hybrid method used. More attention should be paid to the choice of methods for solving specific problems.

(e) The artificial intelligence method is used in only two articles. This research method, which is closely related to the currently important field of big data analysis, should be developed for a wider range of applications. The strong learning ability and data processing ability of AI technology can help RL activities improve efficiency in the process of collecting data and decision making.

According to the results of the literature review, the concept of RL supplier selection can be concluded as a process of first identifying criteria based on industry characteristics and the topic in the frontier of the research field of the research object, then determining the appropriate decision method based on a comparison of the advantages and disadvantages of various methods and their applicability to a specific problem, and finally selecting the best RL supplier through specific criteria and methods. The results of the presented review can serve as a basis for selecting criteria and methods in future research. It is suggested for researchers and practitioners that in future that they should not overlook those criteria, and methods that are not widely used, novel criteria and better decision-making methods can be studied. Finally, from a more macro perspective, research on reverse logistics needs to be combined with research in many other fields. For example, the application and implementation of reverse logistics in various industries can also become a further research direction.

5.2. Research Limitations

This study's main limitation lies in its focus on the selection and evaluation of reverse logistics suppliers. Other decision-making processes or model building methods in the process of reverse logistics are not included in it. Since most of the references are based on the establishment of a decision-making framework and a case study, the research methods of these articles are not classified and summarized. When analyzing the criteria, this paper stops at the criteria and does not delve into the detailed composition and relation of the sub-criteria. A more extensive and comprehensive study can be carried out by systematically summarizing relevant papers according to the whole reverse logistics process from consumer to manufacturer. In terms of the selection of references, due to the influence of various restrictions, this paper only contains the English papers that can be retrieved in WOS. If the search scope of is further expanded, a better analysis of the research on this topic may be obtained.

Author Contributions: Conceptualization, Z.L. and X.Z.; writing—original draft preparation, Z.L.; writing—review and editing, Z.L., X.Z. and Y.W.; supervision, X.Z. All authors have read and agreed to the published version of the manuscript.

Funding: This research was funded by the first batch of industry-school cooperative education projects of the Ministry of Education in 2018, grant number [201801287002].

Acknowledgments: The authors acknowledge the support and inspiration of Wuhan University of Science and Technology and the University of Brighton.

Conflicts of Interest: The authors declare no conflict of interest.

References

1. Govindan, K.; Mina, H.; Esmaeili, A.; Gholami-Zanjani, S.M. An Integrated Hybrid Approach for Circular supplier selection and Closed loop Supply Chain Network Design under Uncertainty. *J. Clean. Prod.* **2020**, *242*, 118317. [CrossRef]
2. Rogers, D.S.; Tibben-Lembke, R. *Going backwards: Reverse Logistics Trends and Practices*; Reverse Logistics Executive Council: Reno, NV, USA, 1999.
3. Zarbakhshnia, N.; Wu, Y.; Govindan, K.; Soleimani, H. A novel hybrid multiple attribute decision-making approach for outsourcing sustainable reverse logistics. *J. Clean. Prod.* **2020**, *242*, 118461. [CrossRef]
4. Govindan, K.; Soleimani, H. A review of reverse logistics and closed-loop supply chains: A Journal of Cleaner Production focus. *J. Clean. Prod.* **2017**, *142*, 371–384. [CrossRef]
5. Yazdani, M.; Chatterjee, P.; Zavadskas, E.K.; Zolfani, S.H. Integrated QFD-MCDM framework for green supplier selection. *J. Clean. Prod.* **2017**, *142*, 3728–3740. [CrossRef]
6. Mardani, A.; Kannan, D.; Hooke, R.E.; Ozkul, S.; Alrasheedi, M.; Tirkolaee, E.B. Evaluation of green and sustainable supply chain management using structural equation modelling: A systematic review of the state of the art literature and recommendations for future research. *J. Clean. Prod.* **2020**, *249*, 119383. [CrossRef]
7. Prajapati, H.; Kant, R.; Shankar, R. Bequeath life to death: State-of-art review on reverse logistics. *J. Clean. Prod.* **2019**, *211*, 503–520. [CrossRef]
8. Da Silva, E.M.; Ramos, M.O.; Alexander, A.; Jabbour, C.J.C. A systematic review of empirical and normative decision analysis of sustainability-related supplier risk management. *J. Clean. Prod.* **2020**, *244*, 118808. [CrossRef]
9. Govindan, K.; Soleimani, H.; Kannan, D. Reverse logistics and closed-loop supply chain: A comprehensive review to explore the future. *Eur. J. Oper. Res.* **2015**, *240*, 603–626. [CrossRef]
10. Aguezzoul, A. Third-party logistics selection problem: A literature review on criteria and methods. *Omega* **2014**, *49*, 69–78. [CrossRef]
11. Islam, M.T.; Huda, N. Reverse logistics and closed-loop supply chain of Waste Electrical and Electronic Equipment (WEEE)/E-waste: A comprehensive literature review. *Resour. Conserv. Recycl.* **2018**, *137*, 48–75. [CrossRef]
12. Agrawal, S.; Singh, R.K.; Murtaza, Q. A literature review and perspectives in reverse logistics. *Resour. Conserv. Recycl.* **2015**, *97*, 76–92. [CrossRef]
13. Jayaraman, V.; Luo, Y. Creating competitive advantages through new value creation: A reverse logistics perspective. *Acad. Manag. Perspect.* **2007**, *21*, 56–73. [CrossRef]
14. Biehl, M.; Prater, E.; Realff, M.J. Assessing performance and uncertainty in developing carpet reverse logistics systems. *Comput. Oper. Res.* **2007**, *34*, 443–463. [CrossRef]
15. Bernon, M.; Rossi, S.; Cullen, J. Retail reverse logistics: A call and grounding framework for research. *Int. J. Phys. Distrib. Logist. Manag.* **2011**, *41*, 484–510. [CrossRef]
16. González-Torre, P.L.; Adenso-Diaz, B.; Artiba, H. Environmental and reverse logistics policies in European bottling and packaging firms. *Int. J. Prod. Econ.* **2004**, *88*, 95–104. [CrossRef]

17. Ravi, V.; Shankar, R. Reverse logistics operations in paper industry: A case study. *J. Adv. Manag. Res.* **2006**, *3*, 88–94. [CrossRef]
18. González-Torre, P.L.; Adenso-Diaz, B. Reverse logistics practices in the glass sector in Spain and Belgium. *Int. Bus. Rev.* **2006**, *15*, 527–546. [CrossRef]
19. Rathore, P.; Kota, S.; Chakrabarti, A. Sustainability through remanufacturing in India: A case study on mobile handsets. *J. Clean. Prod.* **2011**, *19*, 1709–1722. [CrossRef]
20. Narayana, S.A.; Elias, A.A.; Pati, R.K. Reverse logistics in the pharmaceuticals industry: A systemic analysis. *Int. J. Logist. Manag.* **2014**, *25*, 379–398. [CrossRef]
21. Wang, X.; Gaustad, G.; Babbitt, C.W.; Richa, K. Economies of scale for future lithium-ion battery recycling infrastructure. *Resour. Conserv. Recycl.* **2014**, *83*, 53–62. [CrossRef]
22. Cai, W.; Li, L.; Jia, S.; Liu, C.; Xie, J.; Hu, L. Task-Oriented Energy Benchmark of Machining Systems for Energy-Efficient Production. *Int. J. Precis. Eng. Manuf.-Green Technol.* **2019**, *7*, 205–218. [CrossRef]
23. Jiang, Z.; Ding, Z.; Zhang, H.; Cai, W.; Liu, Y. Data-driven ecological performance evaluation for remanufacturing process. *Energ. Convers. Manag.* **2019**, *198*, 111844. [CrossRef]
24. Rachih, H.; Mhada, F.Z.; Chiheb, R. Meta-heuristics for reverse logistics: A literature review and perspectives. *Comput. Ind. Eng.* **2019**, *127*, 45–62. [CrossRef]
25. Kumar, S.; Putnam, V. Cradle to cradle: Reverse logistics strategies and opportunities across three industry sectors. *Int. J. Prod. Econ.* **2008**, *115*, 305–315. [CrossRef]
26. Darbari, J.D.; Kannan, D.; Agarwal, V.; Jha, P.C. Fuzzy criteria programming approach for optimizing the TBL performance of closed loop supply chain network design problem. *Ann. Oper. Res.* **2019**, *273*, 693–738. [CrossRef]
27. Govindan, K.; Pokharel, S.; Sasi Kumar, P. A hybrid approach using ISM and fuzzy TOPSIS for the selection of reverse logistics provider. *Resour. Conserv. Recycl.* **2009**, *54*, 28–36.
28. Kumar, A.; Dixit, G. A novel hybrid MCDM framework for WEEE recycling partner evaluation on the basis of green competencies. *J. Clean. Prod.* **2019**, *241*, 118017. [CrossRef]
29. Liu, A.; Xiao, Y.; Lu, H.; Tsai, S.B.; Song, W. A fuzzy three-stage multi-attribute decision-making approach based on customer needs for sustainable supplier selection. *J. Clean. Prod.* **2019**, *239*, 118043. [CrossRef]
30. Jiang, Z.; Jiang, Y.; Wang, Y.; Zhang, H.; Cao, H.; Tian, G. A hybrid approach of rough set and case-based reasoning to remanufacturing process planning. *J. Intell. Manuf.* **2016**, *30*, 19–32. [CrossRef]
31. Ding, Z.; Jiang, Z.; Zhang, H.; Cai, W.; Liu, Y. An integrated decision-making method for selecting machine tool guideways considering remanufacturability. *Int. J. Comput. Integr. Manuf.* **2018**, 1–15. [CrossRef]
32. French, S.; Geldermann, J. The varied contexts of environmental decision problems and their implications for decision support. *Environ. Sci. Policy* **2005**, *8*, 378–391. [CrossRef]
33. Meredith, J. Theory building through conceptual methods. *Int. J. Oper. Prod. Manag.* **1993**, *13*, 3–11. [CrossRef]
34. Alexander, A.; Walker, H.; Naim, M. Decision theory in sustainable supply chain management: A literature review. *Int. J. Supply Chain Manag.* **2014**, *19*, 504–522. [CrossRef]
35. Wang, H.; Jiang, Z.; Zhang, H.; Wang, Y.; Yang, Y.; Li, Y. An integrated MCDM approach considering demands-matching for reverse logistics. *J. Clean. Prod.* **2019**, *208*, 199–210. [CrossRef]
36. Liu, A.; Ji, X.; Lu, H.; Liu, H. The selection of 3PRLs on self-service mobile recycling machine: Interval-valued pythagorean hesitant fuzzy best-worst multi-criteria group decision-making. *J. Clean. Prod.* **2019**, *230*, 734–750. [CrossRef]
37. Dos Santos, B.M.; Godoy, L.P.; Campos, L.M.S. Performance evaluation of green suppliers using entropy-TOPSIS-F. *J. Clean. Prod.* **2019**, *207*, 498–509. [CrossRef]
38. Zhou, F.; Wang, X.; Lim, M.K.; He, Y.; Li, L. Sustainable recycling partner selection using fuzzy DEMATEL-AEW-FVIKOR: A case study in small-and-medium enterprises (SMEs). *J. Clean. Prod.* **2018**, *196*, 489–504. [CrossRef]
39. Tosarkani, B.M.; Amin, S.H. A multi-objective model to configure an electronic reverse logistics network and third party selection. *J. Clean. Prod.* **2018**, *198*, 662–682. [CrossRef]
40. Li, Y.L.; Ying, C.S.; Chin, K.S.; Yang, H.T.; Xu, J. Third-party reverse logistics provider selection approach based on hybrid-information MCDM and cumulative prospect theory. *J. Clean. Prod.* **2018**, *195*, 573–584. [CrossRef]

41. Chatterjee, K.; Pamucar, D.; Zavadskas, E.K. Evaluating the performance of suppliers based on using the R'AMATEL-MAIRCA method for green supply chain implementation in electronics industry. *J. Clean. Prod.* **2018**, *184*, 101–129. [CrossRef]
42. Luthra, S.; Govindan, K.; Kannan, D.; Mangla, S.K.; Garg, C.P. An integrated framework for sustainable supplier selection and evaluation in supply chains. *J. Clean. Prod.* **2017**, *140*, 1686–1698. [CrossRef]
43. Guarnieri, P.; Sobreiro, V.A.; Nagano, M.S.; Marques Serrano, A.L. The challenge of selecting and evaluating third-party reverse logistics providers in a multicriteria perspective: A Brazilian case. *J. Clean. Prod.* **2015**, *96*, 209–219. [CrossRef]
44. Zarbakhshnia, N.; Jaghdani, T.J. Sustainable supplier evaluation and selection with a novel two-stage DEA model in the presence of uncontrollable inputs and undesirable outputs: A plastic case study. *Int. J. Adv. Manuf. Technol.* **2018**, *97*, 2933–2945. [CrossRef]
45. Mavi, R.K.; Goh, M.; Zarbakhshnia, N. Sustainable third-party reverse logistic provider selection with fuzzy SWARA and fuzzy MOORA in plastic industry. *Int. J. Adv. Manuf. Technol.* **2017**, *91*, 2401–2418. [CrossRef]
46. Tavana, M.; Zareinejad, M.; Santos-Arteaga, F.J.; Kaviani, M.A. A conceptual analytic network model for evaluating and selecting third-party reverse logistics providers. *Int. J. Adv. Manuf. Technol.* **2016**, *86*, 1705–1721. [CrossRef]
47. Govindan, K.; Sarkis, J.; Palaniappan, M. An analytic network process-based multicriteria decision making model for a reverse supply chain. *Int. J. Adv. Manuf. Technol.* **2013**, *68*, 863–880. [CrossRef]
48. Zarbakhshnia, N.; Soleimani, H.; Ghaderi, H. Sustainable third-party reverse logistics provider evaluation and selection using fuzzy SWARA and developed fuzzy COPRAS in the presence of risk criteria. *Appl. Soft Comput.* **2018**, *65*, 307–319. [CrossRef]
49. Tavana, M.; Zareinejad, M.; Di Caprio, D.; Kaviani, M.A. An integrated intuitionistic fuzzy AHP and SWOT method for outsourcing reverse logistics. *Appl. Soft Comput.* **2016**, *40*, 544–557. [CrossRef]
50. Senthil, S.; Srirangacharyulu, B.; Ramesh, A. A robust hybrid multi-criteria decision making methodology for contractor evaluation and selection in third-party reverse logistics. *Expert Syst. Appl.* **2014**, *41*, 50–58. [CrossRef]
51. Azadi, M.; Saen, R.F. A new chance-constrained data envelopment analysis for selecting third-party reverse logistics providers in the existence of dual-role factors. *Expert Syst. Appl.* **2011**, *38*, 12231–12236. [CrossRef]
52. Prakash, C.; Barua, M.K. An analysis of integrated robust hybrid model for third-party reverse logistics partner selection under fuzzy environment. *Resour. Conserv. Recycl.* **2016**, *108*, 63–81. [CrossRef]
53. Govindan, K.; Agarwal, V.; Darbari, J.D.; Jha, P.C. An integrated decision making model for the selection of sustainable forward and reverse logistic providers. *Ann. Oper. Res.* **2017**, *273*, 607–650. [CrossRef]
54. Efendigil, T.; Önüt, S.; Kongar, E. A holistic approach for selecting a third-party reverse logistics provider in the presence of vagueness. *Comput. Ind. Eng.* **2008**, *54*, 269–287. [CrossRef]
55. Bai, C.; Sarkis, J. Integrating and extending data and decision tools for sustainable third-party reverse logistics provider selection. *Comput. Oper. Res.* **2019**, *110*, 188–207. [CrossRef]
56. Garg, C.P.; Sharma, A. Sustainable outsourcing partner selection and evaluation using an integrated BWM–VIKOR framework. *Environ. Dev. Sustain.* **2018**, *22*, 1529–1557. [CrossRef]
57. Sen, D.K.; Datta, S.; Mahapatra, S.S. Decision Support Framework for Selection of 3PL Service Providers: Dominance-Based Approach in Combination with Grey Set Theory. *Int. J. Inf. Tech. Decis.* **2017**, *16*, 25–57. [CrossRef]
58. Govindan, K.; Palaniappan, M.; Zhu, Q.; Kannan, D. Analysis of third party reverse logistics provider using interpretive structural modeling. *Int. J. Prod. Econ.* **2012**, *140*, 204–211. [CrossRef]
59. Sasikumar, P.; Haq, A.N. Integration of closed loop distribution supply chain network and 3PRLP selection for the case of battery recycling. *Int. J. Prod. Res.* **2011**, *49*, 3363–3385. [CrossRef]
60. Gardas, B.B.; Raut, R.D.; Narkhede, B.E. Analysing the 3PL service provider's evaluation criteria through a sustainable approach. *Int. J. Prod. Perform. Manag.* **2019**, *68*, 958–980. [CrossRef]
61. Momeni, E.; Azadi, M.; Saen, R. Measuring the efficiency of third party reverse logistics provider in supply chain by multi objective additive network DEA model. *Int. J. Ship. Transp. Log.* **2015**, *7*, 21–41. [CrossRef]
62. Govindan, K.; Jha, P.C.; Agarwal, V.; Darbari, J.D. Environmental management partner selection for reverse supply chain collaboration: A sustainable approach. *J. Environ. Manag.* **2019**, *236*, 784–797. [CrossRef] [PubMed]

63. Chen, K.; Yu, X.; Yang, L. GI-TOPSIS Based on Combinational Weight Determination and its Application to Selection of Reverse Logistics Service Providers. *J. Grey Syst.* **2013**, *25*, 16–33.
64. Pourjavad, E.; Mayorga, R.V. A fuzzy rule-based approach to prioritize third-party reverse logistics based on sustainable development pillars. *J. Intell. Fuzzy Syst.* **2018**, *35*, 3125–3138. [CrossRef]
65. Sabtu, M.; Saibani, N.; Ramli, R.; Rahman, M.N.A. Multi-Criteria Decision Making for Reverse Logistic Contractor Selection in E-Waste Recycling Industry Using Polytomous Rasch Model. *J. Teknol.* **2015**, *77*, 119–125. [CrossRef]
66. Luo, Z.; Li, Z. A MAGDM Method Based on Possibility Distribution Hesitant Fuzzy Linguistic Term Set and Its Application. *Mathematics* **2019**, *7*, 1063. [CrossRef]
67. Govindan, K.; Kadziński, M.; Ehling, R.; Miebs, G. Selection of a sustainable third-party reverse logistics provider based on the robustness analysis of an outranking graph kernel conducted with ELECTRE I and SMAA. *Omega* **2019**, *85*, 1–15. [CrossRef]
68. Kafa, N.; Hani, Y.; El Mhamedi, A. Evaluating and selecting partners in sustainable supply chain network: A comparative analysis of combined fuzzy multi-criteria approaches. *Opsearch* **2017**, *55*, 14–49. [CrossRef]
69. Wu, C.; Barnes, D. Partner selection for reverse logistics centres in green supply chains: A fuzzy artificial immune optimisation approach. *Prod. Plan. Control* **2016**, *27*, 1356–1372. [CrossRef]
70. Jung, H. Evaluation of Third Party Logistics Providers Considering Social Sustainability. *Sustainability* **2017**, *9*, 777. [CrossRef]
71. Prakash, C.; Barua, M.K. A combined MCDM approach for evaluation and selection of third-party reverse logistics partner for Indian electronics industry. *Sustain. Prod. Consump.* **2016**, *7*, 66–78. [CrossRef]

Article

Study on the Sustainability Evaluation Method of Logistics Parks Based on Emergy

Cui Wang [1,2], Hongjun Liu [1,*], Li'e Yu [1] and Hongyan Wang [1]

1 Business School, Suzhou University, Suzhou 234000, China; wangcui@ahszu.edu.cn (C.W.); yulie@ahszu.edu.cn (L.Y.); sxywhy@ahszu.edu.cn (H.W.)
2 Center for International Education, Philippine Christian University, Manila 1004, Philippines
* Correspondence: sxylhj@ahszu.edu.cn

Received: 8 August 2020; Accepted: 29 September 2020; Published: 2 October 2020

Abstract: To improve the sustainable development ability of logistics parks, this study constructs a sustainability evaluation method of logistics parks based on emergy; analyzes the input (energy, land, investment, equipment, information technology, and human resources) and output (income and waste) of logistics parks from the perspective of emergy; studies the characteristics of the emergy flow of logistics parks; and constructs the function, structure, ecological efficiency, and sustainable development indexes of logistics parks. The basic situation, resource efficiency, and environmental friendliness of the logistics parks are comprehensively evaluated from the emergy point of view. On this basis, targeted decision suggestions are provided for the sustainable development of logistics parks. Finally, the feasibility and effectiveness of the method are verified by an example. This study reveals the internal relationship among economic, environmental, and social benefits of logistics parks through emergy and provides theoretical and methodological support for the sustainable development of logistics parks.

Keywords: emergy; logistics parks; sustainability

1. Introduction

With the continuous development of the global economy and e-commerce, the demand for the logistics industry shows a steady growth trend. As an important form of the logistics industry, logistics parks have an agglomeration effect, which can improve logistics efficiency, reduce logistics cost, and promote the linkage development of the logistics industry, the manufacturing industry, and modern commerce. However, various problems emerge in the development process of logistics parks. In the early stage, the market demand was not scientifically analyzed, and the function orientation was not clear, which made the construction of logistics parks inconsistent with the actual needs of enterprises. This inconsistency led to the vacancy of parks, the low utilization rate of land and equipment in parks, the low efficiency of parks, the lack of relevant facilities, the low level of informatization, the lack of added value of logistics services provided for enterprises, the operation mode remaining in the primary stage, which results in the low-profit margin of logistics parks, and the lack of reuse of waste generated in the process of logistics in parks. Therefore, how to evaluate the basic situation, resource efficiency, and environmental friendliness of logistics parks; improve the economic and social benefits of logistics parks; and realize sustainable development have become urgent research topics.

Sustainable logistics is a trendy research theme that addresses societal, environmental, and industrial challenges. It has attracted the interest of many experts and scholars. The entropy method and the coupling coordination degree model are used to study how to coordinate economic, logistics, and ecological environment [1]. The coordinated development between the metropolitan economy and logistics for sustainability is analyzed by Lan [2]. The association between green logistics performance and sustainability reporting is investigated drawing on signaling theory [3]. The sustainability of national logistics performance is evaluated by using data envelopment analysis [4]. Logistics themes and challenges that are environmentally sustainable are explored from a logistics service provider's view [5]. Trends in sustainable logistics in major cities in China are studied by Lan [6]. The multi-method approach, including literature review, text analysis, text mining, and statistical analysis is implemented to study environmental sustainability in city logistics measures [7–9]. The best–worst method is adopted to evaluate and rank the challenges of implementing eco-innovation practices for freight logistics sustainability [10]. The network design and planning of sustainable multi-period reverse logistics under uncertainty utilizing conditional value at risk for recycling construction and demolition waste is studied by Rahimi [11]. Operational and environmental sustainability tradeoffs in the planning of multimodal freight transportation are evaluated by Kelle [12]. A fuzzy multi-criteria model is structured by Bandeira for evaluating sustainable urban freight transportation operations [13]. Melkonyan assesses the sustainability of last-mile logistics and distribution strategies using the case of local food networks [14]. Pourhejazy integrates sustainability into the optimization of fuel logistics networks [15]. Sustainability challenges in the maritime transport and logistics industry and its way ahead are studied by Lee [16]. Some scholars directed their studies toward the methods of logistics sustainable development. The development of innovative green infrastructure solutions is mentioned to improve the sustainability of ports logistics [17]. Customer satisfaction, sufficient security and privacy, affordability, and competitive pressure are indicated as the highest-ranked critical success factors to achieve supply chain social sustainability using social media [18]. The sharing economy is used as a pathway to sustainable development in organic food supply chains [19]. A novel taxonomy of green initiatives and to investigate their diffusion among logistics service providers is provided [20]. The main challenges and opportunities for the development of river logistics as a sustainable alternative are studied by Ademar [21].

Environmental and social issues are an important part of sustainable development. The research on the sustainable development of logistics makes logistics pay attention to the impact on society and the natural environment while improving economic benefits. However, the current research on the quantitative evaluation of logistics sustainability is limited, and no unified standard exists between logistics economic and ecological systems. Starting from the ecological environment, the emergy analysis method integrates energy flow, material flow, and currency flow, and calculates a series of comprehensive index systems reflecting ecological and economic efficiency. This method also studies the interaction between human society and nature. It is a bridge connecting economic and ecological systems [22]. Emergy theory was initially used in the analysis of eco-economic systems or industrial park evaluation analysis [23–25]; now, its application scope is increasingly extensive. Emergy based sustainability evaluation of remanufacturing machining systems is studied by Liu [26], and proposes an integrated optimization control method for remanufacturing assembly system [27]. Two calculation methods of emergy indices are used to compare 10 power generation systems [28]. The comprehensive evaluation of the environmental sustainability of the case study hydropower projects on the Tibetan Plateau in 2016 using an emergy analysis approach is done by Chen [29]. An introduction and background are provided on how emergy accounting analysis can be adjusted and applied at the supply chain level [30]. Emergy analysis and combined emergy and life cycle assessment (EM-LCA) were used to evaluate the sustainability of an open-pit gold mine and an alluvial gold mine in Colombia [31]. Mohammad uses emergy to evaluate the sustainability of greenhouse systems, leading to management recommendations to increase the sustainability of production in these systems [32]. Cai proposes an emergy-based sustainability evaluation method for the outsourcing machining resources for improving resource utilization efficiency [33]. Logistics parks have a similar organization and operation mode with industrial parks. Therefore, the application of emergy theory to the eco-economic system analysis of logistics parks and the quantitative analysis and evaluation of the sustainability of logistics parks can further enrich the application scope of emergy theory.

Based on the theory and method of emergy analysis, this study first defines the boundary and scope of logistics park activities and puts forward the emergy measurement model and the evaluation index system of logistics parks. This study then selects a Logistics Park in Chuzhou City, Anhui Province, as an example to verify the feasibility of the index system in practice and finally forms a conclusion.

This study uses the theory and method of emergy analysis to transform the input of land, equipment, human resource, energy, and other resources and the output of incomes and waste into a unified emergy unit. The economic and social benefits of logistics parks are quantitatively analyzed and comprehensively evaluated. This study reveals the internal mechanism of the sustainability of logistics parks and provides a new perspective for the benefit evaluation of logistics parks.

In practice, the use of emergy theory to evaluate the benefits of logistics parks can help park managers to identify the problems in operation and the causes of low efficiency and provide directions for logistics parks to improve efficiency. The related suggestions on sustainable development provide a theoretical reference for other logistics parks. It can also help decision-makers understand the operation status of logistics parks and provide the decision-making basis for the overall planning and resource allocation of logistics parks.

2. Model

2.1. Boundary and Scope

A lot of resources need to be invested in the construction of logistics parks. They include natural resources, such as land, sunshine, and rain; other renewable resources, such as coal and oil; other non-renewable resources; and non-natural resources, such as logistics infrastructure equipment, human resources, and information technology, which all need a lot of funds. However, due to various reasons, the resource efficiency of logistics parks is low and needs to be improved. Therefore, logistics parks should analyze the various resource elements closely related to their operation to make full use of existing resources, make it play a role as much as possible, and ensure the long-term and sustainable service of logistics parks.

According to the activity flow and characteristics of logistics parks, the boundary is determined, and the emergy flow diagram of logistics parks is drawn.

From the above figure, we can see that the emergy elements are closely related to the activities of a logistics park (Figure 1). The energy input of the logistics park is mainly renewable energy, including solar, wind, geothermal, rainwater potential, and rainwater chemical energies. The emergy input outside the park includes renewable resource input and non-renewable resource input. It includes the water, electricity, coal and fuel, land resources, facilities and equipment costs, capital investment, labor, and information costs required by the operation. The emergy output of the logistics park is the business income and net profit brought by logistics and related services, as well as various wastes generated in the operation process.

In view of the availability and accuracy of production data, the model mainly considers the input and output of the logistics park. For the convenience and feasibility of the model calculation, the extension problem is not considered, such as the social benefits and risks brought by the logistics park. From another level, this is actually included in the logistics park management and service, but their impact on the emergy flow is implicit.

2.2. Emergy Measurement Model

The operation of logistics parks needs to import a variety of resources, including renewable resources, such as solar energy, rainwater, and wind from nature, and non-renewable resources, such as water, electricity, fuel oil, and coal, as well as land, facilities and equipment, human resources, and information technology. The output is mainly waste (waste water, waste gas, and waste residue) and services through the provision of logistics-related services to achieve business income and make profits. Energy input and output have many kinds, the quantity is large, and the dimension is not unified. Based on related literature, this study uses emergy to unify measuring energy [34]. Whose basic expression is as follows:

$$EM = UEV \times N \quad (1)$$

where EM is emergy, UEV is the emergy conversion rate of different substances, and N is the different units of input flow (mass g or energy J).

Based on the above formula, we construct the input and output emergy measurement model of logistics parks as follows.

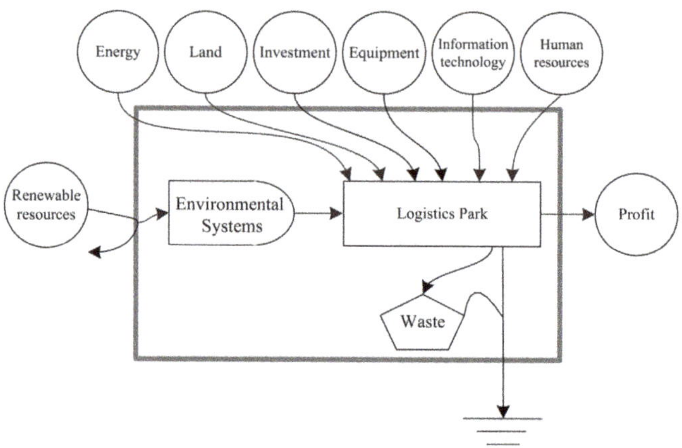

Figure 1. Emergy system diagram of a logistics park.

2.2.1. Input Emergy Value

(1) Emergy of energy. When logistics parks provide logistics services, they are less dependent on the resources provided by the natural environment and involve relatively less energy. The input energy mainly includes solar, rainwater, and wind energy provided by nature, as well as water, electricity, coal, fuel oil, and other resources used in logistics and other value-added services [35].

$$M_e = \sum_{i=1}^{I} e^i \times UEV_e^i \quad (2)$$

where M_e is the emergy provided by the nature of the logistics system, e^i is the i-th energy provided by nature, and I is the total number of types of natural energy used in the logistics system.

(2) Emergy of management. It mainly refers to all kinds of management expenses for organizing and managing business activities in the operation process of logistics parks, including the rent of warehouse and logistics sites, maintenance and depreciation costs of logistics infrastructure and equipment, business entertainment expenses generated by foreign economic exchanges, material consumption, and expenses incurred by various departments due to office needs.

$$M_m = \sum_{i=1}^{I} m^i \times UEV_m^i \quad (3)$$

where M_m is the emergy of various management costs in the logistics system, m^i is the i-th management cost in the logistics system, and I is the total number of types.

(3) Emergy of capital costs. Logistics parks occupy much land and need corresponding logistics facilities and equipment. Thus, the demand for funds is large. Capital costs refer to the cost

paid by logistics parks for raising and using various funds, including fund-raising expenses and capital occupation expenses, such as printing, notarization, and guarantee fees generated from issuing stocks and bonds, shareholders' dividends, and bank loan interest paid for occupying others' funds.

$$M_c = \sum_{i=1}^{I} c^i \times UEV_c^i \tag{4}$$

where M_c is the emergy of various capital costs in the logistics system, c^i is the i-th capital cost in the logistics system, and I is the total number of types of capital costs used in the logistics system.

(4) Emergy of human resource. It refers to the salary, bonus, performance, welfare, allowance, and other expenses of all kinds of managers and employees in logistics parks, as well as the training and recruitment expenses of employees.

$$M_h = \sum_{i=1}^{I} h^i \times UEV_h^i \tag{5}$$

where M_h is the emergy of human resource in the logistics system, h^i is the i-th human resource cost in the logistics system, and I is the total number of types of human resource costs used in the logistics system.

(5) Emergy of information. Information is an important part of logistics parks. As an indispensable part of logistics parks, a logistics information platform is the bond and support of logistics park operation and an important means to provide logistics services for other enterprises. Information costs include the cost of building an information platform and the cost of management and maintenance during the operation of the platform.

$$M_f = \sum_{i=1}^{I} f^i \times UEV_f^i \tag{6}$$

where M_f is the emergy of information cost in the logistics system, f^i is the i-th information cost in the logistics system, and I is the total number of types of information costs used in the logistics system.

2.2.2. Output Emergy Value

(1) Emergy of waste. It mainly refers to waste gas, wastewater, and solid waste generated in the operation of logistics parks.

$$M_w = \sum_{i=1}^{I} w^i \times UEV_w^i \tag{7}$$

where M_w is the emergy of waste in the logistics system, w^i is the i-th waste in the logistics system, and I is the total number of types of waste produced by the logistics system.

(2) Emergy of income. Logistics parks provide warehousing, transportation, circulation processing, and other logistics services and various value-added services. By providing these services, it can obtain various incomes, such as storage fees, equipment rentals, processing costs, and value-added services, and form the operating income of the park.

$$M_b = \sum_{i=1}^{I} b^i \times UEV_b^i \tag{8}$$

where M_b is the emergy of incomes in the logistics system, b^i is the i-th income in the logistics system, and I is the total number of types of incomes of the logistics system.

2.3. Sustainability Evaluation Index System of Logistics Parks

Based on the analysis of the emergy elements and the emergy measurement model of logistics parks, combined with their characteristics, the corresponding indicators were constructed to evaluate the ecological efficiency and sustainable development of logistics parks.

2.3.1. Structural Indexes

(1) Emergy self-sufficiency ratio (ESR). It is the ratio of the emergy input of local resources and total emergy. For the logistics park, the emergy input of local resources is mainly renewable resources in the system. Its formula is as follows:

$$ESR = EMR/EMU \tag{9}$$

(2) Purchasing emergy ratio. It is the ratio of input emergy from the outside of logistics parks and the total emergy the system:

$$Purchasing\ emergy\ ratio = EMI/EMU \tag{10}$$

Logistics parks focus on providing a variety of logistics services and value-added services. These two indexes reflect the dependence of logistics parks on external resources. The lower the ESR and the higher the purchasing emergy ratio, the stronger dependence of the system is on the outside.

(3) Emergy ratio of equipment and labor. It is the ratio between the emergy of equipment, facilities, labor, and information purchased by logistics parks and the total emergy. The calculation formula is as follows:

$$Emergy\ of\ equipment\ and\ labor\ ratio = EMel/EMU \tag{11}$$

This index reflects the emergy ratio of the facilities and equipment, labor, and information resources purchased by logistics parks. The higher the index is, the more developed the system is, and the less dependent the system is on the environment.

2.3.2. Functional Indexes

(1) Emergy exchange ratio (EER). It refers to the ratio of the commodity emergy (the emergy obtained by the buyer) to the emergy equivalent to the buyer's payment currency. Its expression can be described as follows:

$$EER = EMI/EMO \tag{12}$$

For logistics parks, the buyer can obtain warehousing, transportation, circulation processing, and information services by paying currency. The index is used to study the exchange proportion of emergy in the process of inflow and outflow. The larger the index is, the more emergy exchange the system has, and the more advantageous it is in foreign exchange.

(2) Emergy yield ratio (EYR). It is the ratio of the output emergy to input emergy of parks [36]. The calculation formula is as follows:

$$EYR = EMJ/EMU \tag{13}$$

Through this index, the production efficiency and development degree of the system can be comprehensively evaluated. For logistics parks, the emergy of its output is the net profit and waste, whereas the emergy input is mainly the emergy of various materials consumed in the process of logistics activities, including energy, human resources, and information. The higher the EYR is, the higher the emergy output of logistics park under a certain emergy input, which means that the utilization rate of various resources in parks is higher and can bring greater economic benefits to the park, and the competitiveness of the logistics park is stronger.

(3) Emergy density of land. It is the ratio of emergy output to land area.

$$Emergy\ land\ density = EMO/A \tag{14}$$

where A refers to the land area of the logistics park. This index reflects the land-use efficiency of logistics parks. The higher the index, the greater the degree of land use is, and the higher the output is.

2.3.3. Ecological Efficiency Indexes

(1) Emergy investment ratio (EIR). It refers to the ratio of feedback emergy from the economy to free input emergy from the environment. The former includes all kinds of emergy purchased by logistics parks, such as water, electricity, labor, and information, whereas the latter is the emergy of renewable and non-renewable resources provided by the system free of charge.

$$EIR = EMI/EMR \tag{15}$$

The index is an important indicator to measure the degree of economic development and environmental load of the system. It reflects the corresponding emergy input required by the developing unit of natural resources. The larger the ratio is, the higher the degree of economic development is, and the less dependence on the environment is.

(2) Environmental loading ratio (ELR). It is the ratio of non-renewable emergy to renewable emergy [37]. Its calculation formula is as follows:

$$ELR = (EMn + EMel)/(EMr + EMR) \tag{16}$$

This index mainly evaluates the impact of system activities on the environment and is a warning to the economic system. The larger the index is, the higher the degree of science and technology of the system, high-intensity emergy utilization is observed, and the pressure of the environmental loading is greater in parks. If the system's environmental load ratio is high for a long time, its self-organizing ability will decline, which may cause irreparable strucltural changes or loss of function.

(3) Emergy ratio of waste (EWR). It is the ratio of waste emergy to the total emergy of the system. Its formula is as follows:

$$EWR = EMW/EMU \tag{17}$$

It reflects the resource utilization degree and the recycling level of the system. The lower the index is, the higher the resource recycling degree of logistics parks is.

2.3.4. Sustainable Development Index

Considering the dual attributes of logistics parks, which realizes its economic benefits and considers some government public welfare, the emergy index for sustainable development (EISD) is adopted to evaluate the sustainable development of logistics parks. The calculation method of this index is as follows:

$$ELSD = EYR \times EER/ELR \tag{18}$$

EER reflects the gain and loss of the system in the external exchange, and *EYR* reflects the output efficiency of the system. The product of the two is the emergy output benefit of the system. *ELR* reflects the impact of various activities of the system on the environment. Therefore, *EISD* considers the environmental and socio-economic benefits of the system [38]. The higher the ratio is, the better the comprehensive benefits of the system are under unit environmental pressure, and the more competitive advantages the system has in sustainable development.

Based on the above emergy indexes, the emergy evaluation index system of logistics parks is constructed, as shown (Table 1).

Table 1. Emergy evaluation index system of logistics parks.

Emergy Index	Expression	Remark
Renewable energy in the system (EMR)		System owned resource
Input emergy outside the system (EMI)	EMI = EMr + EMn + EMel	
Emergy of renewable resources (EMr)		Input of renewable resources
Emergy of non-renewable resources (EMn)		Input of non-renewable resources from outside of the system, such as electricity, gas
Emergy of equipment and labor (EMel)		input of capital, information, labor resources from outside of the system
Total emergy (EMU)	EMU = EMR + EMI	
Output emergy (EMO)	EMO = EMW + EMJ	
Emergy of waste (EMW)		Output of wastes
Emergy of profits (EMJ)		Output of net profits
Structural index		
Emergy self-sufficiency ratio (ESR)	EMR/EMU	Evaluating the natural environment support capacity
Purchasing emergy ratio	EMI/EMU	Reflecting the dependence of logistics park on external resources
Emergy ratio of equipment and labor	EMr/EMU	Reflecting the degree of system development and technology dependence
Functional indexes		
Emergy exchange ratio (EER)	EMI/EM0	Evaluating the gains and losses in the external exchange
Emergy yield ratio (EYR)	EMJ/EMU	Measuring the contribution of system output to the economy
Emergy density of land	EMO/A	Reflecting the land use efficiency
Ecological efficiency indexes		
Emergy investment ratio (EIR)	EMI/EMR	Measuring the degree of economic development and environmental load
Environmental loading ratio (ELR)	(EMn + EMel)/(EMr + EMR)	Evaluating the impact of activities on the environment
Emergy ratio of waste (EWR)	EMW/EMU	Reflecting the resource utilization degree and recycling level
Sustainable development index		
Emergy index for sustainable development (EISD)	EYR × EER/ELR	Measuring the status and level of sustainable development

3. Case Study

3.1. Background

A Logistics Park is located in Chuzhou City, Anhui Province. It was established in 2011, covering an area of 35 hectares, with a registered capital of 1,300,000,000 CNY (Unit of Chinese currency). Based on the surrounding industrial parks, the logistics park serves the surrounding economic areas and provides an integrated platform for logistics operation for the modern manufacturing industry, third-party logistics enterprises, and the business circulation industry. It integrates basic logistics services, such as multimodal transport, warehousing and distribution, transit distribution, container transportation, and other extended logistics services, such as raw material procurement, warehouse receipt pledge, supply chain management, and information services. It is a modern logistics park that is multi-functional, regional, and service-oriented.

3.2. Results

Through consulting the company's public annual report, combined with the investigation of the logistics park, relevant data from 2014 to 2018 were collected with the help of relevant personnel of the park. The specific data and emergy flow are shown in Tables 2 and 3.

According to the emergy flow table, the emergy evaluation index of the logistics park is calculated, as shown (Table 4).

Table 2. The energy data of the logistics park in 2014–2018.

Collection Object	Original Data					Emergy Conversion Rate (Sej/Unit)
	2014	2015	2016	2017	2018	
Renewable Resources in the System						
Solar energy (J)	7.05×10^{15}	7.11×10^{15}	6.99×10^{15}	7.26×10^{15}	7.17×10^{15}	1.00
Wind energy (J)	2.47×10^{11}	2.98×10^{11}	2.65×10^{11}	2.72×10^{11}	3.06×10^{11}	6.63×10^{2}
Rainwater potential energy (J)	3.42×10^{8}	3.79×10^{8}	2.93×10^{8}	4.55×10^{8}	4.21×10^{8}	8.89×10^{3}
Rainwater chemical energy (J)	8.92×10^{7}	9.34×10^{7}	8.17×10^{7}	9.86×10^{7}	9.53×10^{7}	1.54×10^{4}
Geothermal energy (J)	4.58×10^{12}	5.13×10^{12}	5.66×10^{12}	4.91×10^{12}	5.27×10^{12}	2.90×10^{4}
Input Resources						
Water (G)	2.23×10^{9}	2.89×10^{9}	3.32×10^{9}	3.72×10^{9}	3.37×10^{9}	4.10×10^{4}
Electric (J)	1.44×10^{12}	1.51×10^{12}	1.80×10^{12}	1.98×10^{12}	2.16×10^{12}	1.60×10^{5}
Fuel (J)	1.01×10^{12}	1.15×10^{12}	1.38×10^{12}	1.47×10^{12}	1.61×10^{13}	6.60×10^{4}
Management cost (CNY)	3.90×10^{6}	4.85×10^{6}	5.40×10^{6}	6.15×10^{6}	7.00×10^{6}	8.61×10^{11}
Capital cost (CNY)	5.85×10^{6}	6.84×10^{6}	6.84×10^{6}	7.38×10^{6}	7.38×10^{6}	8.61×10^{11}
Information (CNY)	7.50×10^{5}	8.80×10^{5}	8.00×10^{5}	1.20×10^{6}	5.50×10^{5}	8.61×10^{11}
Human resource (CNY)	4.60×10^{6}	4.80×10^{6}	5.40×10^{6}	6.80×10^{6}	7.50×10^{6}	8.61×10^{11}
Output Resources						
Incomes (CNY)	3.42×10^{9}	3.68×10^{9}	5.34×10^{9}	4.82×10^{9}	4.10×10^{9}	8.61×10^{11}
Net profit (CNY)	6.90×10^{6}	6.15×10^{6}	6.02×10^{6}	7.78×10^{6}	1.26×10^{7}	8.61×10^{11}
Waste water (G)	1.78×10^{8}	2.31×10^{8}	2.66×10^{8}	2.98×10^{8}	2.70×10^{8}	1.24×10^{9}
Waste gas (G)	1.66×10^{5}	1.77×10^{5}	2.35×10^{5}	2.31×10^{5}	6.3×10^{5}	1.84×10^{8}
Solid waste (G)	4.88×10^{6}	5.12×10^{6}	5.57×10^{6}	6.01×10^{6}	5.69×10^{6}	2.52×10^{8}

Table 3. The emergy data of the logistics park in 2014–2018.

Collection Object	Emergy Data				
	2014	2015	2016	2017	2018
Renewable resources in the system					
Solar energy (J)	7.05×10^{15}	7.11×10^{15}	6.99×10^{15}	7.26×10^{15}	7.17×10^{15}
Wind energy (J)	1.64×10^{14}	1.98×10^{14}	1.76×10^{14}	1.80×10^{14}	2.03×10^{14}
Rainwater potential energy (J)	3.04×10^{12}	3.37×10^{12}	2.60×10^{12}	4.04×10^{12}	3.74×10^{12}
Rainwater chemical energy (J)	1.37×10^{12}	1.44×10^{12}	1.26×10^{12}	1.52×10^{12}	1.47×10^{12}
Geothermal energy (J)	1.33×10^{17}	1.49×10^{17}	1.64×10^{17}	1.42×10^{17}	1.53×10^{17}
Input Resources					
Water (G)	9.14×10^{13}	1.18×10^{14}	1.35×10^{14}	1.53×10^{14}	1.38×10^{14}
Electric (J)	2.31×10^{17}	2.42×10^{17}	2.88×10^{17}	3.17×10^{17}	3.46×10^{17}
Fuel (J)	6.67×10^{16}	7.59×10^{16}	9.11×10^{16}	9.70×10^{16}	1.06×10^{18}
Management cost (CNY)	3.36×10^{18}	4.18×10^{18}	4.65×10^{18}	5.30×10^{18}	6.03×10^{18}
Capital cost (CNY)	5.04×10^{18}	5.89×10^{18}	5.89×10^{18}	6.35×10^{18}	6.35×10^{18}
Information (CNY)	6.46×10^{17}	7.58×10^{17}	6.89×10^{17}	1.03×10^{18}	4.74×10^{17}
Human resource (CNY)	3.96×10^{18}	4.13×10^{18}	4.65×10^{18}	5.85×10^{18}	6.46×10^{18}
Output Resources					
Incomes (CNY)	2.94×10^{21}	3.16×10^{21}	4.60×10^{21}	4.15×10^{21}	3.53×10^{21}
Net profit (CNY)	5.94×10^{18}	5.30×10^{18}	5.18×10^{18}	6.70×10^{18}	1.08×10^{19}
Waste water (G)	2.21×10^{17}	2.86×10^{17}	3.30×10^{17}	3.70×10^{17}	3.35×10^{17}
Waste gas (g)	3.05×10^{13}	1.77×10^{13}	3.26×10^{13}	4.25×10^{13}	1.16×10^{14}
Solid waste (g)	1.23×10^{15}	1.29×10^{15}	1.40×10^{15}	1.51×10^{15}	1.43×10^{15}

Table 4. The sustainable development evaluation index data of the logistics park.

Emergy Indexes	Data				
	2014	2015	2016	2017	2018
Renewable energy in the system (EMR)	1.40×10^{17}	1.56×10^{17}	1.71×10^{17}	1.49×10^{17}	1.60×10^{17}
Input emergy outside the system (EMI)	1.33×10^{19}	1.53×10^{19}	1.63×10^{19}	1.89×10^{19}	2.07×10^{19}
Emergy of equipment, labor, information, and other resources (Emr)	9.14×10^{13}	1.18×10^{14}	1.35×10^{14}	1.53×10^{14}	1.38×10^{14}
Emergy of non-renewable resources (Emn)	1.33×10^{19}	1.53×10^{19}	1.63×10^{19}	1.89×10^{19}	1.98×10^{19}
Total emergy (EMU)	1.47×10^{19}	1.68×10^{19}	1.80×10^{19}	2.04×10^{19}	2.14×10^{19}
Output emergy (EMO)	5.72×10^{18}	5.01×10^{18}	4.85×10^{18}	6.33×10^{18}	1.05×10^{19}
Emergy of waste (EMW)	2.22×10^{17}	2.87×10^{17}	3.90×10^{17}	3.72×10^{17}	3.37×10^{17}
Emergy of profits (EMJ)	5.94×10^{18}	5.30×10^{18}	5.18×10^{18}	6.70×10^{18}	1.08×10^{19}
Structural indexes					
Emergy self-sufficiency ratio (ESR)	1.04×10^{-2}	1.01×10^{-2}	1.04×10^{-2}	7.80×10^{-3}	8.03×10^{-3}
Purchasing emergy ratio	9.90×10^{-1}	9.90×10^{-1}	9.90×10^{-1}	9.92×10^{-1}	9.92×10^{-1}
.Emergy ratio of equipment and labor	9.67×10^{-1}	9.69×10^{-1}	9.66×10^{-1}	9.70×10^{-1}	9.25×10^{-1}
Functional indexes					
Emergy exchange ratio (EER)	2.33	3.05	3.35	2.99	1.98
Emergy yield ratio (EYR)	4.25×10^{-1}	3.25×10^{-1}	2.95×10^{-1}	3.31×10^{-1}	5.01×10^{-1}
Emergy density of land	1.64×10^{13}	1.43×10^{13}	1.39×10^{13}	1.81×10^{13}	3.00×10^{13}
Ecological efficiency indexes					
Emergy investment ratio (EIR)	9.49×10	9.77×10	9.50×10	1.27×10^{2}	1.29×10^{2}
Environmental loading ratio (ELR)	9.69×10	9.97×10	9.71×10	1.29×10^{2}	1.38×10^{2}
Emergy ratio of waste (EWR)	1.65×10^{-2}	1.86×10^{-2}	2.02×10^{-2}	1.95×10^{-2}	1.61×10^{-2}
Sustainable development index					
Emergy index for sustainable development (EISD)	1.02×10^{-2}	9.93×10^{-3}	1.02×10^{-2}	7.67×10^{-3}	7.20×10^{-3}

3.3. Index Evaluation and Analysis of the Logistics Park

According to the above emergy data, the emergy structure, system function, ecological efficiency, and sustainable development of the logistics park are analyzed as follows.

3.3.1. Analysis of Emergy Structure Indexes

The three structural indexes show that the logistics park is less dependent on the natural environment, and has a low degree of local resource development. ESR reflects the dependence of the system on natural resources. The higher the index is, the greater the contribution of local emergy resources is to the logistics system, and the higher the degree of resource development is. The ESR of the logistics park is low, ranging from 0.7% to 1%, and has a significant decline in 2017 and 2018, while the purchasing emergy ratio, which is related to the ESR, is stable and maintained at approximately 99% (Figure 2). The low ESR and high purchasing emergy ratio of the logistics park are determined by the characteristics of the logistics park itself. Different from the agricultural ecosystem or industrial products, the logistics park is a production and service-oriented logistics park, which mainly provides comprehensive services, such as raw material procurement, transportation, warehousing, circulation processing, and supply chain management for the surrounding industrial parks. Its production process is limited to the simple circulation and processing of products, and mainly rely on labor and mechanized operation, the direct use of natural environmental resources, such as sunlight, and rain, geothermal is less. The dependence on external energy is particularly large. Most of the emergy resources in logistics park need to be purchased from outside. The emergy ratio of equipment and labor is also high, ranging from 92.5% to 97.05%, which conforms to the emergy structure of the logistics park; logistics parks mainly depend on the inputs of equipment, technology, human resources, and so on.

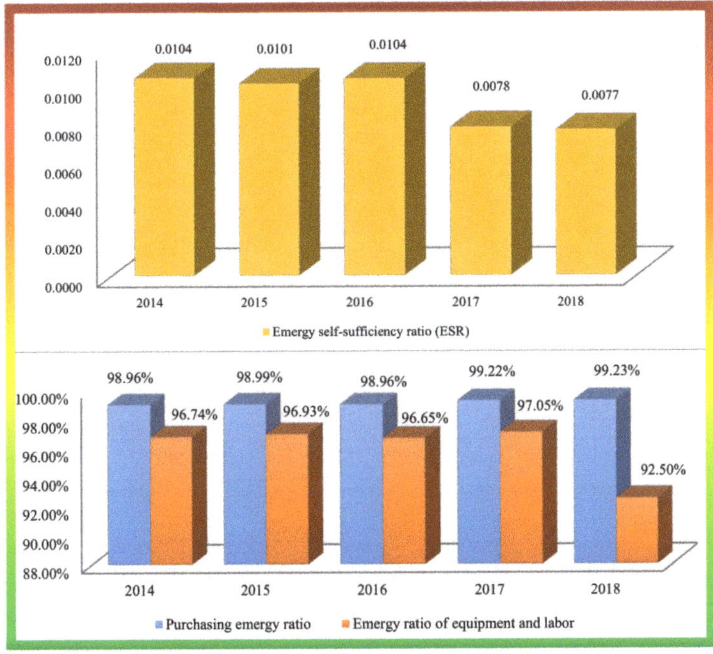

Figure 2. Analysis of emergy structure indexes.

3.3.2. Analysis of System Function Indexes

EER can be used to evaluate the gain and loss benefits of the system, whereas EYR reflects the resource utilization efficiency of the system. The higher the two ratios are, the higher the environment sustainable development is. If EER = 1, it means that the system is in a break-even state, EER > 1 indicates that the system is in a profit state, otherwise, it is in a loss. When EYR < 2, the emergy of the product is not increased due to production activities, but is only a process of consumption or transformation. The system operation is not a sustainable production and development process [39]. It shows that the EER of the logistics park is between 1.9 and 3.5 (Figure 3), which is always greater than 1. It indicates that, in the process of foreign economic activities, the logistics park has been in a steady state of profit and is in a favorable position. The highest EYR of the park is 0.5, which is far lower than 2 (Figure 3). EYR shows an upward trend after the previous decline. On the one hand, the logistics park mainly uses the corresponding facilities and equipment to provide logistics services for other enterprises, and the products involved are often only a transformation process from upstream suppliers to downstream customers. Production is limited to a simple process of circulation processing. Its product emergy is only consumption or transformation, and production efficiency is low. On the other hand, the logistics park is in the initial development stage, the resource system is not effectively configured, the resource structure is not reasonable, the utilization of machinery and equipment is not sufficient, and the workers have more idle time, which leads to low efficiency. Moreover, the emergy density of land in the logistics park is average because the park is still under development. Although it has been fully planned, part of the land is idle in the early stage, and the land resources are not fully utilized. With the development of the park, the ratio is gradually increasing.

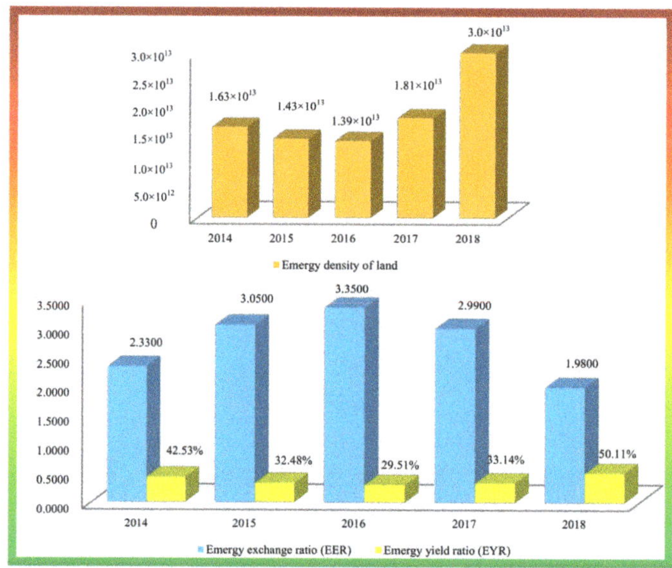

Figure 3. Analysis of system function indexes.

3.3.3. Analysis of Ecological Efficiency Evaluation Indexes

EIR is used to measure the degree of economic development and environmental load. ELR reflects the environmental pressure brought by the system [40]. EWR reflects the utilization degree and circulation of the system resources. The higher the ELR and EWR are, the lower the sustainable development level is. The above table shows that the EIR of the logistics park is distributed between 90 and 140, indicating that the logistics park has a high level of economic development and less dependence on the environment. The ELR of the park is also between 90 and 140, which is a medium load state (Figure 4). Therefore, the park absorbs more non-renewable resources, product emergy, and financial resources outside with a small range of land and a small amount of natural resources. The two indexes showed an upward trend, indicating that the pressure on the environment is increasing and needs to be controlled. In the process of logistics activities, part of the waste is formed, and EWR fluctuates around 1.8%. Therefore, more waste is produced in the logistics process of the park, and the recycling degree is poor, which bring certain pressure to the environment. The waste in the park mainly includes domestic garbage and garbage generated in the process of loading, unloading, and warehousing. The source of garbage is little. Thus, the relevant personnel in the park do not pay enough attention to waste management and lack the corresponding management of waste recycling. Therefore, EWR is high.

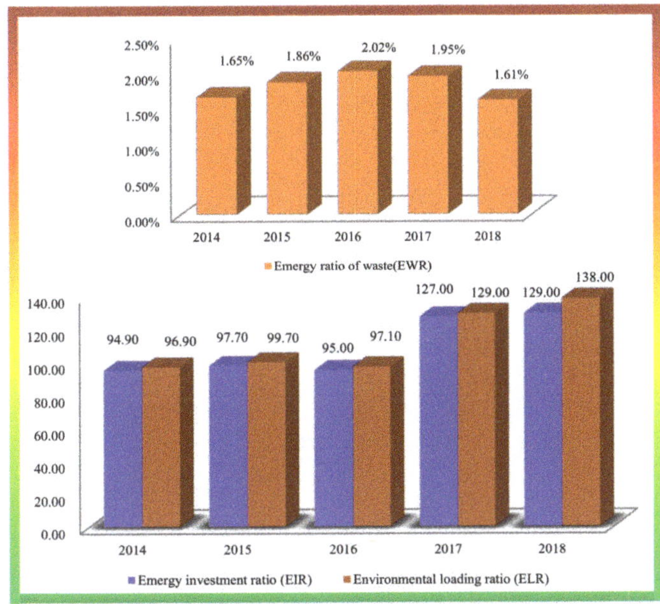

Figure 4. Analysis of ecological efficiency evaluation indexes.

3.3.4. Analysis of Sustainable Development Index

The final data shows that the improved EISD is used to measure the sustainable development of the logistics park. The sustainable development level is low. It shows a downward trend from 1.02×10^{-2} to 7.20×10^{-3}, which shows that the logistics park activities are inefficient and have great pressure on the environment and that it is in an unsustainable stage (Figure 5). The main reasons are the low EYR of the park, the decreased dependence on the natural environment in the process of logistics activities, and the increased use of non-renewable resources, resulting in high environmental load ratio and poor sustainable development level. This ratio decreased significantly in 2017 and 2018 mainly because the logistics park began to carry out container business in 2017. To load and unload containers, some fuel oil frontal cranes were introduced, which increased the investment and the use of non-renewable resources of the park. Moreover, the extensive management, the lack of enough attention from leadership, and the blind pursuit of profits ignored the input–output ratio and the recycling of waste, leading to the poor resource utilization ratio of the logistics park, which further reduced the sustainable development level of the park.

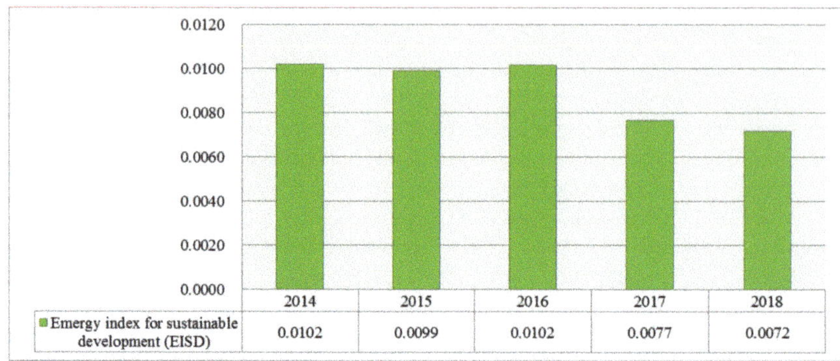

Figure 5. Analysis of the sustainable development index.

3.4. Discussion

Compared with similar research results [4,8–10], our research has the following advantages: The methods, such as data envelopment analysis, text mining, statistical analysis and best–worst method, evaluate the sustainability of logistics park, which are pay more attention to economy, technology and pollution emissions, but ignore the contribution of ecosystem to the activity process of logistics park. Emergy analysis method makes up for this defect. Moreover, it is better than other methods on measuring, evaluating, and optimizing the sustainability of logistics parks for quantitative measurement.

Combined with the above research and conclusions, we can get the following three management implications:

- First, the research on the sustainable development of logistics parks can promote the logistics park to strengthen the effective use of resources, and reduce the discharge of waste, so as to reduce the negative impact on the environment.
- Second, the sustainable development of the logistics park is the embodiment of the social responsibility of the logistics park. Improving the sustainable development ability of the logistics park can help the logistics park reduce the cost, improve the efficiency, establish a good image, and avoid the administrative punishment of the government.
- Third, the improvement of sustainable development ability of logistics parks involves a wide range, including EYR, EER, and ELR of the park. These indicators are determined by the input resources, total emergy, output resources, and waste of the system. The identification and analysis of these influencing factors and the evaluation of indicators related to sustainable development. It can improve the sustainable development ability of logistics parks.

4. Suggestions for Sustainable Development of the Logistics Park

4.1. Improving the ESR and Emergy Structure

The above analysis shows that the direct use of various resources from the natural environment of the logistics park is less, ESR is less than 1%, and its emergy mainly comes from the purchasing emergy outside the system, the purchasing emergy ratio is always maintained at approximately 99%. To improve the emergy structure, the logistics park should focus on reducing the emergy ratio of the external purchase of the system.

First, resource sharing should be strengthened, and the duplication of investment should be reduced. At present, the logistics park is developing rapidly. Before its construction, a comprehensive investigation must be conducted on the regional logistics demand and logistics development, and the radiation scope and location of the logistics park must be determined based on the local industrial planning, economic development planning, and transportation planning to reduce the repeated investment in warehouses, facilities, and equipment and improve the utilization rate of resources.

Second, the construction of logistics standardization should be promoted further. Standardization can improve the efficiency of logistics, reduce the cost of logistics operation, speed up the turnover of logistics equipment, and reduce the investment of logistics enterprises on some equipment. At present, the emergy of capital costs for the park accounts for about 35% of the total emergy, and the emergy related to equipment accounts for about 20%. Through these two methods, the logistics park can reduce the investment in logistics facilities and equipment. However, the implementation of these two methods is a long-term process, and its benefits cannot be seen temporarily.

Third, the utilization of natural resources should be improved by, for example, making full use of sunlight and rainwater to provide lighting power and domestic water for the park. Through these aspects, the ESR can be increased, and the emergy structure can be optimized. The ESR of the logistics park is less than 1% now. By strengthening the utilization of natural resources, the ESR can reach 1.5–2%.

4.2. Improving the Operation Efficiency of the Logistics Park and the Emergy Yield Ratio

The evaluation index system shows that although the EER of the logistics park is good, but the EYR is only approximately 0.5, which has a certain gap with the average level, and it directly leads to the low level of sustainable development of the park. Improving EYR is one of the important methods to improve the sustainable development level of the park, which requires the park to improve its operation efficiency. First, the integration of existing resources should be strengthened, and the optimal allocation of resources should be realized. Logistics resources include capital, land, technology, equipment, human resources, and other factors. The Emergy density of land of the park is between 1.39×10^{13} and 3.00×10^{13}, which is in the middle level. It is necessary to plan the park reasonably, improve the utilization efficiency of land and capital, and avoid idle land. Advanced logistics technology and information technology can improve work efficiency, reduce emergy loss in the logistics process, and improve the effective use of emergy. However, the emergy of information only accounts for about 4% of the total emergy, the investment in information technology can be appropriately increased. In addition, the logistics park should also improve the ability of the logistics park to integrate resources such as capital and equipment, strengthen infrastructure construction, reduce the repeated construction of a single enterprise, and realize the scale benefit of logistics facilities utilization.

Second, on the basis of the existing business, the basic service functions must be improved, and more value-added services should be provided. Through diversified and perfect logistics services, more profits can be obtained, and the EYR of the logistics park can be improved.

4.3. Implementing Green Logistics to Reduce Waste Discharge

In the park logistics system, the more waste generated by various activities, the less net output, which will reduce the EYR index of the system, thus, affecting the sustainable development level of the system.

Compared with industrial parks, logistics parks' discharge has less waste. Thus, many parks pay little attention to waste emissions. However, from the emergy view, the data shows that the emergy of waste in the logistics park accounts for approximately 5% of the output, which has a great impact on the EYR of the park, and reduces the sustainable development index of the park. The logistics park has a lot of waste, such as waste from loading and unloading, warehousing, circulation processing, and supporting catering accommodation.

On the one hand, the park should reasonably plan the logistics park and logistics activities, reduce the discharge of waste in the process of operation, form a waste recycling system, and strengthen the supervision and management of waste. For example, we can find that waste water discharge accounts for the largest proportion of waste discharge. The park can strengthen the management of waste water discharge, realize the recycling of waste water and reduce the discharge.

On the other hand, the use of clean energy should be increased, and the environmental pollution caused by harmful gas emissions should be reduced [41–43]. For example, in the selection of handling equipment, the logistics park may tend to choose new energy equipment. When cooperating with transportation companies, the park can adopt certain ways to encourage transportation companies to use new energy vehicles.

5. Conclusions

Based on emergy theory, this study establishes the sustainable development indexes of logistics parks, including EER, ELR, EYR, and EISD. Taking a Logistics Park as an example, this study analyzes and evaluates the data of the logistics park and proposes that the park should reduce a proportion of external purchasing emergy, improve the emergy structure, enhance the operational efficiency of the park, improve EYR, implement green logistics, and reduce the environmental load ratio and other sustainable development suggestions. The study provides decision-making suggestions for park managers.

Emergy theory can be applied to the comprehensive evaluation of sustainable development of logistics parks, which can better meet the requirements of evaluation and analysis, and has operability. By establishing the emergy evaluation index system of logistics parks, the system composition and functional structure of logistics parks can be analyzed, the relationship between logistics parks and external environment can be reflected, and the comprehensive evaluation of logistics parks from economic and ecological perspectives can be realized.

Our model evaluates the sustainability of the logistics park based on the input and output resources of the system. Social problems, threats, and risks faced by investors, and so on, are difficult to quantify, which is the next direction we are trying to explore. In the future, we will conduct further research on the correlation between various elements of the system and the economic and ecological benefits of the logistics park, so as to improve the sustainability of logistics parks. Emergy theory can be further applied to all aspects of the logistics system, which is of great significance to the sustainable development of the logistics industry.

Author Contributions: Conceptualization, C.W. and H.L.; methodology, C.W.; investigation, L.Y.; resources, H.W.; data curation, C.W. and H.L.; writing—original draft preparation, C.W.; writing—review and editing, C.W. and H.L. All authors have read and agreed to the published version of the manuscript.

Funding: This research is supported by Humanities and Social Sciences Research Projects of Anhui Universities (No.SK2019A0538), Key project of natural science research in Universities of Anhui Province (No. kj2019a0673), Teaching and research project of quality engineering in Anhui Province (No.2018jyxm1035), Key scientific research projects of Suzhou University (No.2016yzd11), Key Support Program of Anhui Provincial University Excellent Talent (No.gxyqZD2019078), Research project of Suzhou University (No.2019jb12), Program for Innovative Research Team in Suzhou University(No.2018kytd01, No.2018kytd02), Key research base project of Humanities and Social Sciences in universities of Anhui Province(No.SK2018A0474) and Open research platform of Suzhou University(No.2019ykf19).

Conflicts of Interest: The authors declare no conflict of interest.

References

1. Zhang, W.; Zhang, X.; Zhang, M.; Li, W. How to Coordinate Economic, Logistics and Ecological Environment? Evidences from 30 Provinces and Cities in China. *Sustainability* **2020**, *12*, 1058. [CrossRef]
2. Lan, S.L.; Zhong, R.Y. Coordinated development between metropolitan economy and logistics for sustainability. Resources. *Conserv. Recycl.* **2018**, *128*, 345–354. [CrossRef]
3. Karaman, A.S.; Kilic, M.; Uyar, A.G. Logistics performance and sustainability reporting practices of the logistics sector: The moderating effect of corporate governance. *J. Clean. Prod.* **2020**, *258*, 120718. [CrossRef]
4. Rashidi, K.; Cullinane, K. Evaluating the sustainability of national logistics performance using Data Envelopment Analysis. *Transp. Policy* **2019**, *74*, 35–46. [CrossRef]

5. Abbasi, M.; Nilsson, F. Developing environmentally sustainable logistics. Exploring themes and challenges from a logistics service providers' perspective Transportation Research Part D. *Transp. Environ.* **2016**, *46*, 273–283.
6. Lan, S.; Tseng, M.-L.; Yang, C.; Huisingh, D. Trends in sustainable logistics in major cities in China. *Sci. Total Environ.* **2020**, *712*, 136381. [CrossRef] [PubMed]
7. Jagienka, R.-C.; Agnieszka, S.-J. Environmental sustainability in city logistics measures. *Energies* **2020**, *13*, 1303.
8. Nathanail, E.; Adamos, G.; Gogas, M.C. A novel approach for assessing sustainable city logistics Transportation Research. *Procedia* **2017**, *25*, 1036–1045.
9. Zhan, C.; Zhao, R.; Hu, S. Emergy-based sustainability assessment of forest ecosystem with the aid of mountain eco-hydrological model in Huanjiang County, China. *J. Clean. Prod.* **2020**, *2511*, 119638. [CrossRef]
10. Orji, I.J.; Simonov, K.S.; Himanshu, G.; Modestus, O. Evaluating challenges to implementing eco-innovation for freight logistics sustainability in Nigeria Transportation Research Part A. *Policy Pract.* **2019**, *129*, 288–305.
11. Rahimi, M.; Ghezavati, V. Sustainable multi-period reverse logistics network design and planning under uncertainty utilizing conditional value at risk (CVaR) for recycling construction and demolition waste. *J. Clean. Prod.* **2018**, *172*, 1567–1581. [CrossRef]
12. Kelle, P.; Song, J.; Jin, M.; Schneider, H.; Claypool, C. Evaluation of operational and environmental sustainability tradeoffs in multimodal freight transportation planning. *Int. J. Prod. Econ.* **2019**, *209*, 411–420. [CrossRef]
13. Bandeira, R.A.M.; D'Agosto, M.A.; Ribeiro, S.K.; Bandeira, A.P.F.; Goes, G.V.A. Fuzzy multi-criteria model for evaluating sustainable urban freight transportation operations. *J. Clean. Prod.* **2018**, *184*, 727–739. [CrossRef]
14. Melkonyan, A.; Gruchmann, T.; Lohmar, F.; Kamath, V.; Spinler, S. Sustainability assessment of last-mile logistics and distribution strategies: The case of local food networks. *Int. J. Prod. Econ.* **2020**, *228*, 107746. [CrossRef]
15. Pourhejazy, P.; Kwon, O.K.; Lim, H. Integrating Sustainability into the Optimization of Fuel Logistics Networks. *KSCE J. Civ. Eng.* **2019**, *23*, 1369–1383. [CrossRef]
16. Lee, P.T.-W.; Kwon, O.K.; Ruan, X. Sustainability challenges in maritime transport and logistics industry and its way ahead. *Sustainability* **2019**, *11*, 1331. [CrossRef]
17. Twrdy, E.; Zanne, M. Improvement of the sustainability of ports logistics by the development of innovative green infrastructure solutions. *Transp. Res. Procedia* **2020**, *45*, 539–546. [CrossRef]
18. Orji, I.J.; Simonov, K.-S.; Himanshu, G. The critical success factors of using social media for supply chain social sustainability in the freight logistics industry. *Int. J. Prod. Res.* **2020**, *58*, 1522–1539. [CrossRef]
19. Sobhan, A.; Ashkan, H.; Jacob, J.J. Sharing economy in organic food supply chains: A pathway to sustainable development. *Int. J. Prod. Econ.* **2019**, *218*, 322–338.
20. Piera, C.; Roberto, C.; Emilio, E. Developing the WH2 framework for environmental sustainability in logistics service providers: A taxonomy of green initiatives. *J. Clean. Prod.* **2017**, *165*, 1063–1077. [CrossRef]
21. Vilarinho, A.; Liboni, L.B.; Siegler, J. Challenges and opportunities for the development of river logistics as a sustainable alternative: A systematic review. *Transp. Res. Procedia* **2019**, *39*, 576–586. [CrossRef]
22. Odum, T. *Environmental Accounting: Emergy and Environ-Mental Decisionmaking*; Wiley: Hoboken, NJ, USA, 1996; pp. 15–163.
23. He, Z.; Jiang, L.; Wang, Z.; Zeng, R.; Liu, J. The emergy analysis of southern China agro-ecosystem and its relation with its regional sustainable development. *Glob. Ecol. Conserv.* **2019**, *20*, e00721. [CrossRef]
24. Liu, X.; Guo, P.; Nie, L. Applying emergy and decoupling analysis to assess the sustainability of China's coal mining area. *J. Clean. Prod.* **2020**, *243*, 118577. [CrossRef]
25. Yang, Q.; Liu, G.; Biagio, F.; Agostinho, G.F.; Casazza, M. Emergy-based ecosystem services valuation and classification management applied to China's grasslands Ecosystem. *Services* **2020**, *42*, 101073. [CrossRef]
26. Liu, C.; Cai, W.; Dinolov, O. Emergy based sustainability evaluation of remanufacturing machining systems. *Energy* **2018**, *150*, 670–680. [CrossRef]
27. Liu, C.; Zhu, Q.; Wei, F.; Rao, W.; Liu, J.; Hu, J.; Cai, W. An integrated optimization control method for remanufacturing assembly system. *J. Clean. Prod.* **2019**, *248*, 119261. [CrossRef]
28. Ren, S.; Feng, X.; Yang, M. Emergy evaluation of power generation systems. *Energy Convers. Manag.* **2020**, *2111*, 112749. [CrossRef]

29. Chen, J.; Mei, Y.; Ben, Y.; Hu, T. Emergy-based sustainability evaluation of two hydropower projects on the Tibetan Plateau. *Ecol. Eng.* **2020**, *1501*, 105838. [CrossRef]
30. Tian, X.; Sarkis, J. Expanding green supply chain performance measurement through emergy accounting and analysis. *Int. J. Prod. Econ.* **2020**, *225*, 107576. [CrossRef]
31. Natalia, A.; Londoño, C.; Velásquez, H.I.; McIntyre, N. Comparing the environmental sustainability of two gold production methods using integrated Emergy and Life Cycle Assessment. *Ecol. Indic.* **2019**, *107*, 105600.
32. Asgharipour, M.R.; Amiri, Z.; Daniel, E. Campbell Evaluation of the sustainability of four greenhouse vegetable production ecosystems based on an analysis of emergy and social characteristics. *Ecol. Model.* **2020**, *42415*, 109021. [CrossRef]
33. Cai, W.; Liu, C.; Jia, S.; Felix, T.; Chan, S.; Ma, X. An emergy-based sustainability evaluation method for outsourcing machining resources. *J. Clean. Prod.* **2020**, *2451*, 118849. [CrossRef]
34. Odum, H.T.; Arding, J.E. *Emergy Analysis of Shrimp Mariculture in Ecuador. Narragansett*; CRC Press: Boca Raton, FL, USA, 1991; Volume 61–64, pp. 17–21.
35. Sun, H.; Liu, C.; Chen, J.; Gao, M.; Shen, X. A Novel Method of Sustainability Evaluation in Machining Processes. *Processes* **2019**, *7*, 275. [CrossRef]
36. Pan, H.; Zhang, X.; Wu, J.; Zhang, Y.; Lin, L.; Yang, G.; Deng, S.; Li, L.; Yu, X.; Qi, H.; et al. Sustainability evaluation of a steel production system in China based on emergy. *J. Clean. Prod.* **2016**, *112*, 1498–1509. [CrossRef]
37. Brown, M.T.; Ulgiati, S. Assessing the global environmental sources driving the geobiosphere: A revised emergy baseline. *Ecol. Model.* **2016**, *339*, 126–132. [CrossRef]
38. Cuadra, M.; Rydberg, T. Emergy evaluation on the production, processing and export of coffeein Nicaragua. *Ecol. Modeling* **2006**, *196*, 421–433. [CrossRef]
39. Geng, Y.; Zhang, R.; Ulgiati, S. Emergy analysis of an industrial park: The case of Dalian, China. *Sci. Total Environ.* **2010**, *408*, 5273–5283. [CrossRef]
40. Zhang, X.; Zhang, R.; Wu, j.; Zhang, Y.-Z.; Lin, L.-L.; Deng, S.-H.; Li, L.; Yang, G.; Yu, X.-Y.; Qi, H.; et al. An emergy evaluation of the sustainability of Chinese crop production system during 2000–2010. *Ecolindic* **2016**, *60*, 622–633. [CrossRef]
41. Liu, C.; Zhu, Q.; Wei, F.; Rao, W.; Liu, J.; Hu, J.; Cai, W. A review on remanufacturing assembly management and technology. *Int. J. Adv. Manuf. Technol.* **2019**, *105*, 4797–4808. [CrossRef]
42. Liu, J.; Feng, Y.; Zhu, Q.; Sarkis, J. Green supply chain management and the circular economy: Reviewing theory for advancement of both fields. *Int. J. Phys. Distrib. Logist. Manag.* **2018**, *48*, 794–817. [CrossRef]
43. Tang, J.; Li, B.Y.; Li, K.W.; Liu, Z.; Huang, J. Pricing and warranty decisions in a two-period closed-loop supply chain. *Int. J. Prod. Res.* **2020**, *58*, 1688–1704. [CrossRef]

Article

Multi-Objective Optimization of Workshop Scheduling with Multiprocess Route Considering Logistics Intensity

Yu Sun [1,2], Qingshan Gong [1], Mingmao Hu [1,*] and Ning Yang [1]

1 College of Mechanical Engineering, Hubei University of Automotive Technology, Shiyan 442002, China; sunyu_huat@163.com (Y.S.); gongqingshan@163.com (Q.G.); 15971919591@163.com (N.Y.)

2 China International Telecommunication Construction Corporation, Beijing 100071, China

* Correspondence: hu@huat.edu.cn; Tel.: +86-137-3353-8300

Received: 6 May 2020; Accepted: 10 July 2020; Published: 15 July 2020

Abstract: In order to solve the problems of flexible process route and workshop scheduling scheme changes frequently in the multi-variety small batch production mode, a multiprocess route scheduling optimization model with carbon emissions and cost as the multi-objective was established. At the same time, it is considered to optimize under the existing machine tool conditions in the workshop, then the theory of logistics intensity between equipment is introduced into the model. By designing efficient constraints to ensure reasonable processing logic, and then applying multilayer coding genetic algorithm to solve the case. The optimization results under single-target and multi-target conditions are contrasted and analyzed, so as to guide enterprises to choose a reasonable scheduling plan, improve the carbon efficiency of the production line, and save costs.

Keywords: multiprocess route; workshop scheduling; multi-objective optimization; logistics intensity

1. Introduction

Small and medium-sized enterprises often design different products in response to the changing needs of customers, so that a variety of small batch production models are formed from the limited production cycle and manufacturing resources. Many scholars have used a variety of methods to solve the optimization problems of energy saving, emission reduction, cost reduction, and time saving for the existing traditional manufacturing system, so that the green upgrade optimization of the system can be achieved without a large amount of investment [1–9]. It is worth mentioning that Li Congbo et al. [10] used the concept of feature elements and processing elements to establish a multi-objective machining route optimization model. Oleh et al. [11] built a process route planning model for complex and flexible workshop based on BOM (Bill of Material) tables. Liu et al. [12] planned route planning problem is transformed into a constrained traveling salesman problem, which was calculated using the ant colony algorithm. Fang et al. [13] proposed a mathematical programming model for the pipeline workshop scheduling problem considering peak power load, energy consumption, carbon emissions, and cycle time. Gonzalez et al. [14] proposed an effective neighborhood structure for flexible job shop scheduling problems, and realized the reassociation of discrete search process routes. Li Congbo et al. [15] considered job batch optimization and established multi-objective multiprocess flexible operation workshop scheduling model. Huang Xuewen et al. [16] established a multiprocess route scheduling model based on OR subgraphs, and proposed a new four-tuple mathematical description method to describe the process path and machine tool flexibility. Liu Qiong et al. [17] optimized the integrated process route a feature of the model with multiple parameters and mutual influence, a NSGA-II algorithm with four-stage coding was proposed.

Based on the existing research on such problems, this paper proposes a workshop scheduling optimization model with low carbon and low cost as the multi-objective. At the same time, in the optimization process of the existing research, the emergency task insertion, machine tools failure, and workpiece features are considered [18–21], but their combination with the actual situation of the equipment in the workshop needs to be supplemented. For this reason, this paper considers the impact of logistics strength between machine tools on optimization to guide the existing equipment, so as to help enterprises achieve the goal of effectively utilizing resources, reducing costs, reducing logistics movement, and improving operational efficiency.

In order to ensure reasonable processing logic, many scholars also made innovative research on the establishment of constraints. Zheng Yongqian et al. [22] introduced the feature constraint matrix and processing priority coefficient, and established the process ordering model of the machining center. Huang Yuyue et al. [23] considered the scheduling optimization of the shift time and other constraints in actual production. Yan Jungang et al. [24] proposed waiting dual time window constraints caused by limited time and equipment capabilities. An Xianghua [25] proposed the establishment of processing element constraints based on intuitionistic fuzzy numbers, and using the cellular automaton-SPEA2 multi-objective optimization algorithm. Chang Zhiyong et al. [26] established constraints of the geometric positional relationship between the features and used adaptive ant colony optimization algorithms to solve practical problems. Chunlong Li et al. [27] proposed a SEEHS constraint treatment method for hydropower plants based on feasible space and adopted a contrary adjustment method to deal with power balance constraints.

In order to solve problems, scholars also have their own ideas and innovations on algorithms. Nikhil Padhye et al. [28] propose a unified approach to improving different evolution. Kalyanmoy Deb et al. [29] proposed to improve the performance of particle swarm optimization by linking with the algorithm of genetic algorithm. De Jong et al. [30,31] introduced a new algorithm, an evolutionary algorithm, in detail. R. Čapek, P. et al. [32] proposed a heuristic algorithm based on priority construction with unscheduled steps and applied it to the case study of wire harness production. Gökan May et al. [33] reduced total energy consumption and processing time through a green genetic algorithm. Dunbing Tang et al. [34] used an improved particle swarm optimization algorithm to find the Pareto optimal solution of the dynamic flexible flow shop scheduling problem. Xiao-Ning Shen et al. [35] proposed an active response method based on multi-objective evolutionary algorithm (MOEA) for multiple targets such as efficiency and stability.

Based on the existing research, most of them take into account the influencing factors in the optimization process or consider the process route order to establish constraints. In this paper, some 0–1 variables are defined to constrain the processing sequence between processes and features, and ensure reasonable optimization and improve calculation efficiency.

2. Establishment of Multi-Objective Optimization Model for Process

Assumptions: (1) Features can select multiple process routes. (2) Processes in the process route are processed only once on one machine. (3) The machine tool will not malfunction regardless of inventory accumulation and part quality and volume.

The definitions i', j', k', l' are the previous part of the ith part, the previous machine of the jth machine tool, the previous feature of the kth feature, and the previous process route of the lth process route. The total processing time of a process of a feature of a part on the machine tool is t_{ijk}, the starting time is st_{ijk}, and the part delivery time is $\mathrm{ed_i}$.

Collection of parts $P = \{\mathrm{p}_1, \cdots, \mathrm{p_i}, \cdots, \mathrm{p_a}\}$, p_i indicates the ith of part, $i = 1, 2, \cdots, a$.

Collection of machine tools $M = \{\mathrm{m}_1, \cdots, \mathrm{m_j}, \cdots, \mathrm{m_b}\}$, m_j indicates the jth of machine tool, $j = 1, 2, \cdots, b$.

Collection of features for all parts $OP = \{\mathrm{op}_1, \cdots, \mathrm{op_i}, \cdots, \mathrm{op_a}\}$, op_i indicates the ith of part's all features.

The collection of features for each part $\mathrm{op_i} = \{\mathrm{op_{i1}}, \cdots, \mathrm{op_{ik}}, \cdots, \mathrm{op_{ic}}\}$, op_{ik} indicates the ith of part's the kth of feature, $k = 1, 2, \cdots, c$. Defining variables $\chi_{\mathrm{ik}} \in \{0, 1\}$, which means whether the feature $op_{ik'}$ of the ith of part needs to be machined before the feature op_{ik}, if necessary $\chi_{ik} = 1$; otherwise, it is zero.

The collection of process routes selectable for each feature is $OPR = \{\mathrm{opr_{ik1}}, \cdots, \mathrm{opr_{ikl}}, \cdots, \mathrm{opr_{ikd}}\}$, opr_{ikl} indicates that the kth feature of the ith part selects the lth process route for processing, $l = 1, 2, \cdots, d$.

The collection $OPM = \{\mathrm{opm_{ikl1}}, \cdots, \mathrm{opr_{iklj}}, \cdots, \mathrm{opr_{iklb}}\}$ represents the process included in the routing (corresponding machine tool), and opm_{iklj} represents the kth feature of the ith part selected the lth process route and processed by the jth machine tool. The variable $\iota_{\mathrm{iklj}} \in \{0, 1\}$ is defined, which means whether the previous process $opm_{iklj'}$ of process route l needs to be completed before opm_{iklj}, if necessary $\iota_{iklj} = 1$; otherwise, it is zero.

2.1. The Model of Carbon Emission

When the CNC machine is in the standby or preparation stage, the standby power of the machine P_1 consists of the power-related auxiliary system, the motor, and the servo. The waiting time before the machining and the input program is t_1; when a machine tool is processing different parts, the tool needs to be adjusted more; the tool setting time is t_2, the movement time of the X, Y and Z axes of the machine tool are $t_{X_1}, t_{Y_1}, t_{Z_1}$, the moving power of each axis is $P_{X_1}, P_{Y_1}, P_{Z_1}$, and the power is 0 if no movement occurs (such as the spindle of the lathe); when the spindle Z starts to rotate, the no-load rotary power is P_{Z_2}, the no-load standby power is $P_s = P_1 + P_{Z_2}$, and the no-load standby time is t_s, then the kth feature of the ith part before the jth machine tool starts processing and can be expressed as

$$E^{s}_{ijk} = P_1 \cdot t_1 + P_1 \cdot t_2 + P_{X_1} \cdot \mathrm{t}_{X_1} + P_{Y_1} \cdot \mathrm{t}_{Y_1} + P_{Z_1} \cdot \mathrm{t}_{Z_1} + P_{\mathrm{s}} \cdot t_{\mathrm{s}} \tag{1}$$

$$t_2 = \mathrm{t}_{X_1} + \mathrm{t}_{Y_1} + \mathrm{t}_{Z_1} \tag{2}$$

The cutting process of each procedure includes the air-cut energy E^{air}_{ijk} and cutting energy E^{m}_{ijk}. The air-cut energy consumption of the machine tool includes the power of the X and Y axes of the machine tool. The power of each axis is P_X, P_Y, and the energy consumption generated during the moving time is t_{X_2}, t_{Y_2}. If the spindle has movement (such as milling machine, grinding machine, etc.), the moving power P_{Z_2} of the spindle Z generates energy consumption in time t_{Z_2}. And the energy consumption of the no-load and standby parts of the machine during the air-cut time need to be considered. The cutting energy consumption is the energy consumed by the machine tool in the cutting time t_m. For different machine tools and different cutting three-factor machine power, the actual machining power P_{ijk} of each process is measured. Therefore, the kth feature of the ith part is machined on the jth machine tool can be expressed as

$$E^{air}_{ijk} + E^{m}_{ijk} = P_X \cdot \mathrm{t}_{X_2} + P_Y \cdot \mathrm{t}_{Y_2} + P_{Z_2} \cdot \mathrm{t}_{Z_2} + P_{\mathrm{s}} \cdot (t_{X_2} + t_{Y_2} + t_{Z_2}) + P_{ijk} \cdot t_m \tag{3}$$

The auxiliary system energy consumption includes the energy consumed by the power P_c such as the filtration and cooling system during the cutting time t_m; the energy consumption generated during the standby time t_{chp} of the machine during the loading and unloading of the parts; after machining N times on the jth machine tool, the tool needs to be replaced. The energy consumption of the machine tool standby during the tool change time t_{cht}, the auxiliary energy consumption of the the kth feature of the ith part is machined on the jth machine tool can be expressed as

$$E^{\mathrm{as}}_{\mathrm{ijk}} = P_c \cdot t_m + P_1 \cdot t_{\mathrm{chp}} + \frac{P_1 \cdot t_{\mathrm{cht}}}{N} \tag{4}$$

Taking energy consumption as the basic input and greenhouse gas (GHG) as the output, the corresponding carbon emissions in this process are converted through the carbon emission coefficient of various energies [36]. ξ is the carbon emission coefficient of energy (fuel, electricity, etc.), and taking into account the environmental (regional and time) impact factors δ and auxiliary process influence factors μ, the carbon emissions can be defined as

$$W = \xi(1 + \delta + \mu)E \tag{5}$$

The carbon emissions for cutting in each process can be expressed as

$$W_{ijk} = \xi(1 + \delta + \mu) \cdot (E^{s}_{ijk} + E^{air}_{ijk} + E^{m}_{ijk} + E^{as}_{ijk}) \tag{6}$$

It is assumed that no more than one machine is used for each process of the part; the collection of carbon emissions from part processing can be expressed as $W = \{W_{111} \cdots W_{ijk} \cdots W_{abc}\}$, where W_{ijk} indicates the processing carbon emission of the kth feature of the ith part that was machined on the jth machine tool.

2.2. The Model of Cost

The processing cost of parts increases with the increase of processing time. It is divided into three aspects: machine tool cost, loss cost, and ancillary costs. Machine tool costs include energy consumption costs during standby, air cutting, cutting operations, etc. The expression for each process is

$$\begin{aligned} &C^{m}_{ijk} = M_e \times t_{ijk}, \\ &t_{ijk} = t_1 + t_2 + t_{X_1} + t_{Y_1} + t_{Z_1} + t_{X_2} + t_{Y_2} + t_{Z_2} + t_s + t_m + t_{chp} + \tfrac{t_{cht}}{N}, \end{aligned} \tag{7}$$

where C^{m}_{ijk} is the processing cost (US$) for each process and M_e is the unit cost (US$/s). t_{ijk} (s) is the total processing time of the part, which includes all the time associated with processing.

The loss cost includes the grinding wheel loss and the cutting fluid consumption. During the processing of N procedures, multiple tool dressings are required until the tool is worn to the minimum size and scrapped. The consumption of the cutting fluid is composed of the part where the temperature of the surface rises to evaporate the cutting fluid into the air, the part taken away by the chips, and the part deposited on the surface of the part. The loss cost C^{loss}_{ijk} (US$) generated by each process is

$$C^{loss}_{ijk} = \frac{M_a + M_l \times L}{N} \tag{8}$$

where M_a is the tool cost (US$), L is the amount of cutting fluid (L) added during N times processes, and M_l is the cutting fluid cost (US$/L).

The ancillary cost includes labor costs, which is the labor remuneration paid by workers during the process, and also includes the operating costs of each process, which include the costs of lubricants, water, equipment maintenance, and repair. Then, each process ancillary cost C^{as}_{ijk} (US$) is

$$C^{as}_{ijk} = (M_p + M_u) \times t_{ijk} \tag{9}$$

where M_p is labor cost (US$/s) and M_u is operating cost (US$/s).

Then, the cost of each processing operation is

$$C_{ijk} = C^{m}_{ijk} + C^{loss}_{ijk} + C^{as}_{ijk} \tag{10}$$

Collection of parts cost can be obtained as $C = \{C_{111} \cdots C_{ijk} \cdots C_{abc}\}$; C_{ijk} indicates the processing cost of the kth feature of the ith part that was machined on the jth machine tool.

2.3. Optimization Model Considering Logistics Strength

The system layout design of System Layout Planning (SLP) is the method used by the manufacturing enterprise in equipment planning [37]. The functions that the production system should perform include the system design and planning of the equipment, personnel, logistics, and investment within the system. The core content is based on the analysis of factory production logistics, combined with the degree of interconnection of operating units, and then the design and plan of the location of the production equipment.

To carry out workshop scheduling via the multiprocess route in manufacturing enterprises, the machine tool equipment has been relatively fixed, so it should be considered to carry out multi-objective optimization in the existing equipment planning and logistics system. We analyzed the logistics and other elements of the production line:

(1) Under the cluster layout, a temporary storage area is set between two areas or near the processing machine to facilitate the overall transportation of parts. The evaluation coefficient of the storage area is defined as $A_{j'j}$. When it is necessary to use the storage area to store parts, the value of the coefficient is bigger, and vice versa.

(2) The transportation of parts will occur between the production areas or the machine tools. The material flow evaluation coefficient between the two processes connected is defined as $B_{j'j}$. The more numerous and heavier the parts that need to be transported between the processes are, the greater the value of the coefficient, and vice versa.

(3) The layout position of the machine tool determines the length of the logistics route. The distance evaluation coefficient between two processes connected is defined as $C_{j'j}$. The longer the distance between the processes, the greater the value of the coefficient, and vice versa.

(4) There is a difference in the logistics frequency between processes. We define the logistics frequency evaluation coefficient between the two processes as $D_{j'j}$. The higher the frequency of transportation between machine tools, the greater the value of the coefficient, and vice versa.

(5) We analyzed the environmental conditions between the areas or the machine tools, defining the conversion evaluation coefficient as $I_{j'j}$. When the logistics environment is more complicated, the value of the coefficient is greater, and vice versa.

In summary, the logistics intensity evaluation function between two machine tools is established, and the expression is

$$LO_{j'j} = A_{j'j}B_{j'j}C_{j'j}D_{j'j}I_{j'j} \tag{11}$$

According to the actual situation of the enterprise's production line, define the evaluation indicators of each coefficient. The logistics information in the factory can be collected, and substituted into Formula (11) to be calculated, to then obtain the logistics strength evaluation function $LO_{j'j}$. The range is divided into five levels, so that the logistics intensity between machine tools is divided into intensity levels (A, E, I, O, U). Based on score, the logistics priority coefficient $G_{j'j}$ can be established to define the logistics strength of the machine tool j' to the machine tool j. The intensity level can converted into a natural number of 0–4; the smaller the value, the higher the logistics intensity between processes. The unit logistics related table can be drawn, as shown in Figure 1.

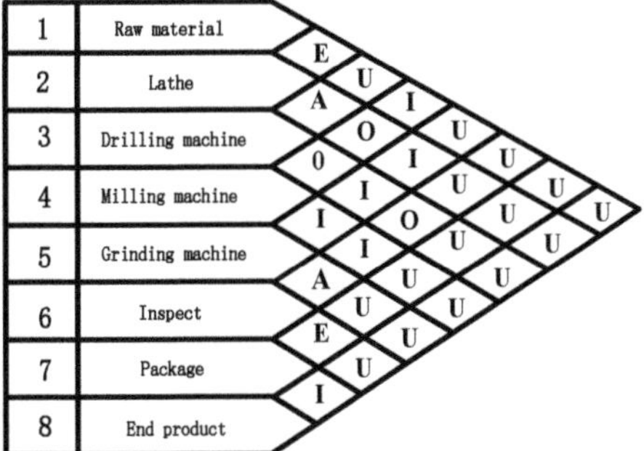

Figure 1. Logistics-related table.

Combined with the analysis of logistics, we can get a collection of logistics strength between machine tools $G = \{G_{11} \cdots G_{j'j} \cdots C_{b'b}\}$.

The definition of the independent variable $\gamma_{ikl} \in \{0,1\}$ indicates whether the routing opr_{ikl} is selected, and if the lth route of the kth feature is selected $\gamma_{ikl} = 1$; otherwise, it is zero.

The objective function can be obtained according to the carbon emission and cost of the machine tool processing corresponding to the selected process route, and the logistics strength between the machine tools used in the process of the preceding and following process routes,

$$\min(\sum G_{j'j} \cdot \gamma_{ikl} \cdot W_{ijk}, \sum G_{j'j} \cdot \gamma_{ikl} \cdot C_{ijk})$$

s.t.

$$st_{ij'k} + t_{ij'k} + Q(1-\gamma_{ikl}) < \chi_{ik} t_{iklj} st_{ijk}, \quad \forall i, j, j', k \tag{12}$$

$$st_{ijk} - st_{i'jk'} > t_{i'jk'} + Q(1-\gamma_{ikl}), \quad \forall i, i', j, k, k' \tag{13}$$

$$\sum_{l}^{d} \gamma_{ikl} = 1, \quad \forall i, k, l \tag{14}$$

$$st_{ij'k} \geq 0, \quad \forall i, j', k \tag{15}$$

$$\max(st_{ijk} + t_{ijk}) \leq ed_i, \quad \forall i, j', k \tag{16}$$

where Q is a maximum value, and Formula (12) indicates that the process route satisfies the processing order between the features and the order of processing between processes; Formula (13) means that one machine can only process one part at a time; Formula (14) means that one feature of a part can only select one process route; Formula (15) indicates that the part processing should start from time 0; Formula (16) indicates that the last process of the part needs to be completed before the part delivery date.

3. Multilayer Coding Genetic Algorithm

The multilayer coding genetic algorithm divides the individual coding into multiple layers; each layer of coding contains different meanings, and jointly expresses the complete solution of the problem, thereby using a chromosome to accurately express the solution of the complex problem.

First, the constraint type of the optimization problem is determined, the solution set of the population composition problem is initialized, and an extended process-based coding is designed. Each chromosome represents a feasible solution under the target optimization. There are multiple kinds of products to be processed in this paper. When the product n_i has m_j processes, the individual lengths of the chromosomes have a total of $2\sum_{i=1}^{k} n_i m_j$ integers, where in the first half is the processing order of the product features on the machine, and the second half represents the machine tool corresponding to the selected machining method. As shown in Figure 2, if the second feature can be processed by a third or second machining method, the corresponding machine tools are 5 and 8.

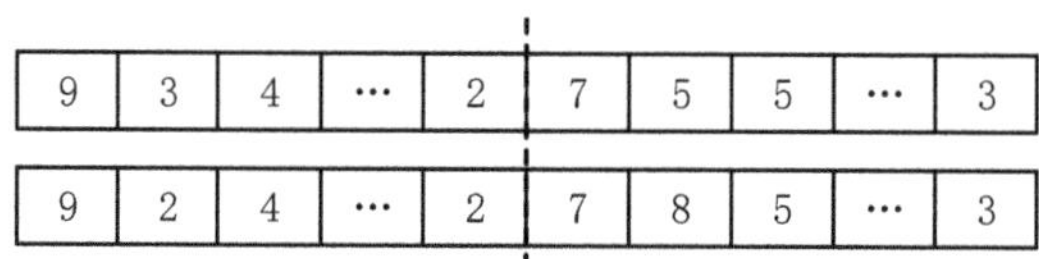

Figure 2. Coding method.

To determine the fitness value of a chromosome, it is necessary to reduce the chromosome into a process. It is also necessary to consider the logistics strength of the machine tool used in the current process and the machine tool used in the previous process, and also to consider how the target allocates the ratio of the processed carbon emissions and the processing cost in the multi-objective optimization. Forming a fitness calculation function, the calculation formula is

$$fitness(i) = k_1 \cdot W_{ijk} \cdot G_{j'j} + k_2 \cdot C_{ijk} \cdot G_{j'j} \tag{17}$$

where k_1, k_2 are the coefficients, the smaller the total carbon emissions and cost generated by the process route of all parts, the better the chromosome.

The roulette method is used to select chromosomes with good fitness. The probability that chromosome i is selected every time is $p(i)$. The better the chromosome is, the smaller the fitness value is. If the value is larger after derivative, the selected probability will be greater.

$$p(i) = Fitness(i) / \sum_{i=1}^{n} Fitness(i) \tag{18}$$

$$Fitness(i) = 1/fitness(i) \tag{19}$$

The crossover of the integer crossover method is used to obtain new chromosomes, and the population is continuously evolved. Two chromosomes are selected from the population by roulette method, the $\sum_{i=1}^{k} n_i m_j$ bits are selected from the individual codes, and the intersection positions are randomly selected for intersection. After the intersection, the process in each process route will be redundant or missing, and the excess part of the process will be added to the missing part. It is judged whether the process conforms to the processing logic according to the constraint conditions. Then, the processing machine corresponding to each process is adjusted to a machine tool of $\sum_{i=1}^{k} n_i m_j + 1$ to $2 \sum_{i=1}^{k} n_i m_j$ position.

Integer variation is used to obtain excellent individuals and promote the evolution of the entire population. The chromosomes are randomly selected from the population, the first half of the individual is selected to perform the mutation operation, and the machine tool corresponding to the second half is exchanged. The entire algorithm process is shown in Figure 3.

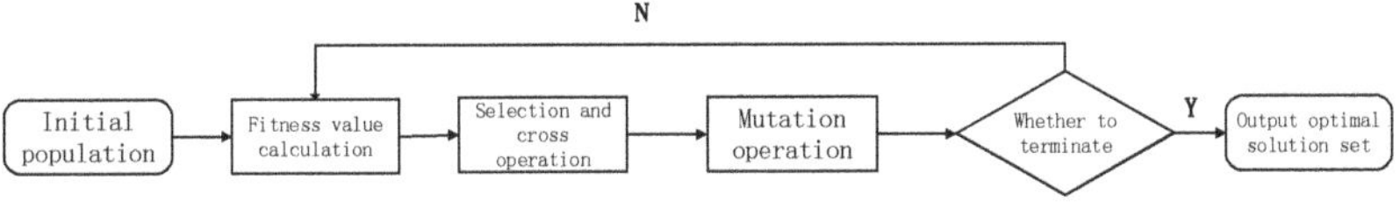

Figure 3. Algorithm flow.

4. Examples and Analysis

According to the production situation of an air compressor crankshaft of a commercial vehicle brake system company, a processing example can be fitted. In this example, set the part set P, that is, process eight kinds of parts at the same time on a production line. Any part contains three major features and the available feature set OP. Any feature has multiple process routes that can be selected and the available process route set OPR. The number of the machine tools in the workshop form a set M, that is, the production line has a total of 15 selectable processing machine tools, including machine tools with the same processing method but different numbers. Comprehensively considering the machine tool set and the process route set, the process set OPM can be obtained.

The machine tool can be selected according to the features' process route. Table 1 expresses the optional machine matrix J_m, which longitudinally represent eight parts, horizontally represent the machine tool corresponding to the features of the parts, and the brackets indicate the various machine tools that can be selected; 0 means no such process. For example, the processing route of part

1 includes 5-12-4-2-3-5-9, 5-12-4-2-8-5-9, 5-12-4-2-8-5-11, 5-12-4-2-3-5-11, 5-12-4-15-3-5-9, 5-12-4-15-8-5-9, 5-12-4-15-8-5-9, 5-12-4-15-8-5-11, and 5-12-4-15-3-5-11; there are a total of eight process routes.

Table 1. Parts' features selectable machine tools.

Part	Feature1			Feature2			Feature3		
1	5	12	4	[2,15]	[3,8]	5	[9,11]	0	
2	4	[2,15]	8	[6,10]	5	[1,12]	3	[2,15]	
3	1	[6,10]	13	[2,15]	[7,14]	5	[9,11]	[3,8]	
4	5	15	[7,14]	10	[2,15]	[3,8]	0	0	
5	[7,14]	5	[9,11]	11	2	[3,8]	[1,12]	[2,15]	
6	[6,10]	4	[11,9]	14	8	[3,8]	7	[7,14]	
7	1	[6,10]	10	[2,15]	[2,15]	5	[1,12]	0	
8	11	[7,14]	15	[2,15]	5	7	[1,12]	9	

Similar to the optional machine tool matrix J_m, a processing time matrix T is generated based on the machining time of the part features on the machine, as shown in Table 2.

Table 2. Parts' features processing time.

Part	Feature1			Feature2			Feature3	
1	3	10	9	[5,4]	[5,3]	10	[5,7]	0
2	6	[8,6]	4	[5,6]	3	[3,3]	6	[2,3]
3	4	[5,7]	7	[5,5]	[9,11]	5	[8,5]	[4,3]
4	7	3	[4,6]	3	[5,7]	[3,6]	0	0
5	[6,4]	10	[7,9]	8	5	[4,7]	[4,6]	[5,6]
6	[3,5]	10	[8,7]	9	4	[9,4]	5	[4,6]
7	6	7	9	[8,7]	[4,5]	3	[5,5]	0
8	8	[6,7]	5	[10,8]	6	4	[7,5]	7

Set the logistics strength matrix G_{Jm} of 15 machine tools according to the collection of logistics strength G,

$$G_{Jm} = \begin{bmatrix} & 1 & 2 & 3 & \cdots & 13 & 14 & 15 \\ 1 & 0 & 4 & 2 & & 4 & 4 & 2 \\ 2 & 4 & 0 & 1 & \cdots & 3 & 4 & 3 \\ 3 & 2 & 1 & 0 & & 4 & 3 & 2 \\ \vdots & & \vdots & & \ddots & & \vdots & \\ 13 & 4 & 3 & 4 & & 0 & 2 & 1 \\ 14 & 4 & 4 & 3 & \cdots & 2 & 0 & 1 \\ 15 & 2 & 3 & 2 & & 1 & 1 & 0 \end{bmatrix}$$

In order to complete the calculation of the model, the power value of each machine tool in the carbon emission model needs to be collected, and the control equipment enters the processing stages of standby and empty cutting. We used the three-phase four-wire connection method to connect the power recorder to the motor used in each stage. The control equipment works stably and collects the corresponding power value. The data is substituted into the expressions of the model. At the same time, the current monitoring device can be connected to the device to monitor the current situation at various stages of stable operation, and the accuracy of the power value can be reversed to improve the accuracy of the data [5]. The connection of the power recorder is shown in Figure 4a, and the connection of the current monitoring device is shown in Figure 4b.

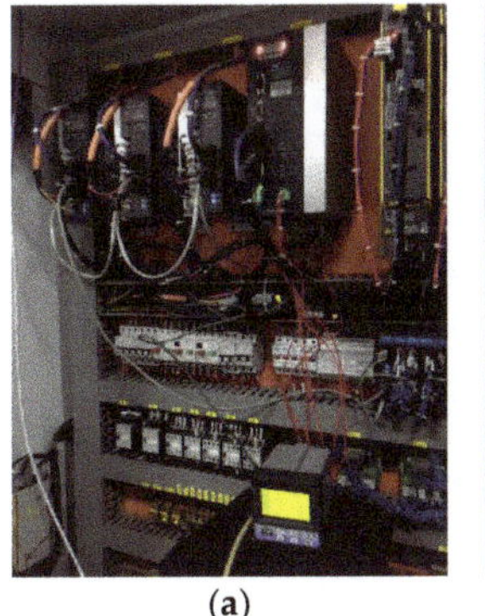
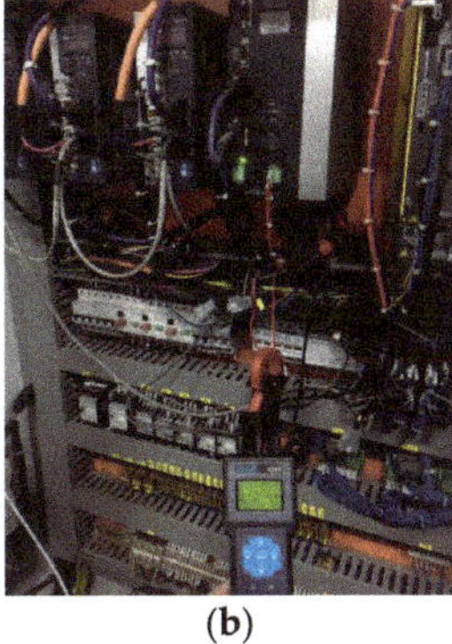

(a) (b)

Figure 4. Power detection equipment. (**a**) shows the connection of the power recorder, (**b**) shows the connection of the current monitoring device.

Using the power detection experimental method and the timing of the processing process, the carbon emissions W_{ijk} and processing costs C_{ijk} of each machine tool during processing can be calculated. Then, the carbon emission matrix W and processing cost matrix C can be determined, which are similar to the optional machine tool matrix J_{m} and processing time matrix T.

Substituting five matrices into the algorithm calculation, the number of populations is 40, the maximum number of iterations is 50, the crossover probability is 0.8, and the mutation probability is 0.6.

When the cost is optimized separately, the algorithm is iterated for 24 generations. The results are shown in Figures 5 and 6. The completion time is 61 min and the carbon emission is 28.8 Kg. Each part feature mainly considers selecting a machine tool with a short machining time. The process route selects a route with less machine conflict and expands the parallel machining, which reduces the waiting time of the workpiece and reduces the maximum completion time.

When increasing carbon emissions for multi-objective optimization, the fitness coefficients k1 and k2 are taken as 0.5 and 0.5, respectively, and the algorithm is iterated for 27 generations. The results are shown in Figures 7 and 8. The total carbon emissions of the workshop are 23.7 Kg, which is 17.7% lower than the cost optimization alone. Most of the characteristics of each part will choose the processing route with less carbon emission and its corresponding machine tool for processing, which reduces the total carbon emission. However, the concentration of processing machine tools leads to the increase of waiting time in the process. The time has increased by 15 min and the cost has also increased.

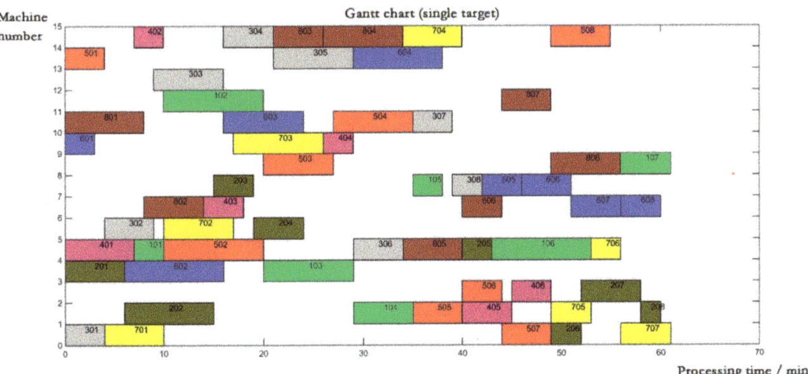

Figure 5. Single-target optimization result Gantt chart.

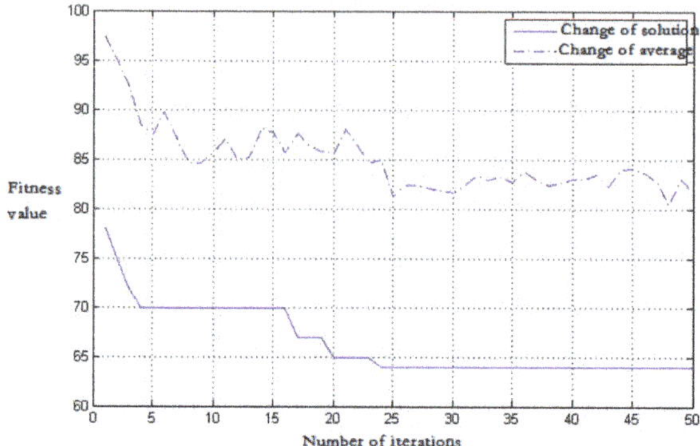

Figure 6. Single target algorithm search process.

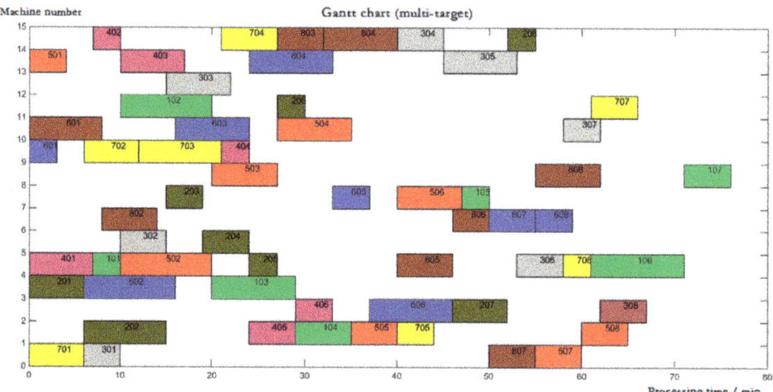

Figure 7. Multi-objective optimization results Gantt chart.

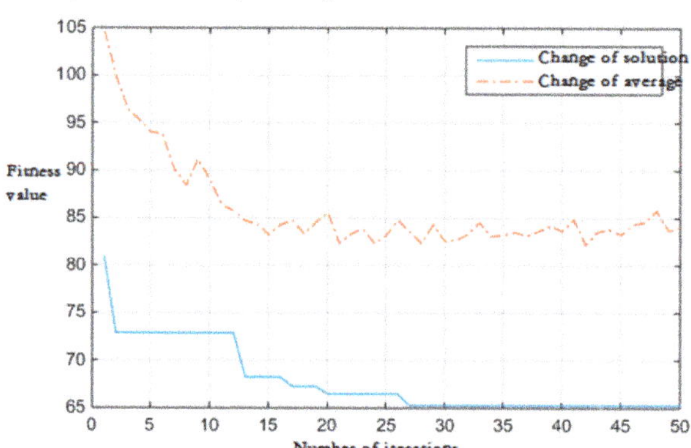

Figure 8. Multi-objective algorithm search process.

5. Conclusions

This paper establishes a multiprocess route workshop scheduling optimization model with the lowest carbon emissions and the best cost by defining a set of factors that affect machining production. At the same time, in order to make the model accurately fit the actual facility conditions, the existing equipment and the factors that affect the logistics intensity are also analyzed. The evaluation coefficients and functions are defined, and the logistics intensity model is fitted and substituted into the objective function. The scheduling modeling and optimization process of the production line has certain reference value. Considering the actual conditions, such as the processing order between the parts' features and the processing order between processes, 0–1 variables χ, γ, and τ are introduced to ensure that the optimization process meets the actual processing order. Finally, through the fitting processing example, the multilayer coding genetic algorithm was used to calculate the Gantt chart under single-objective and multi-objective optimization, and the conclusion was drawn through comparative analysis, which verified the practicability and accuracy of the model, thus guiding the multi-objective. The process of workshop scheduling optimization for small and large batch parts manufacturing enterprises is defined. However, the modeling does not take into account the production line dynamics such as machine tool equipment sudden failures and order changes, and ignores the impact of peak power and processing quality, so optimization research needs to be further in-depth.

Author Contributions: Visualisation, writing, review and editing, and data curation: Y.S. and N.Y.; conceptualization, investigation, and experiment: Q.G. and M.H. All authors have read and agreed to the published version of the manuscript.

Funding: This research received no external funding.

Acknowledgments: This work was supported by the 58th postdoctoral science foundation program of China (Grant No. 2015M581301), the National Natural Science Foundation of China (Grant No. 51775392), the Doctoral Research Startup Foundation of Hubei University of Automotive Technology (Grant No. BK202001), and the Key Laboratory of Automotive Power Train and Electronics (Grant No. ZDK1201802). These financial contributions are gratefully acknowledged.

Conflicts of Interest: The authors declare no conflict of interest.

References

1. Liu, Y.; Dong, H.; Lohse, N.; Petrovic, S.; Gindy, N. An investigation into minimising total energy consumption and total weighted tardiness in job shops. *J. Clean. Prod.* **2014**, *65*, 87–96. [CrossRef]
2. Ding, J.-Y.; Song, S.; Wu, C. Carbon-efficient scheduling of flow shops by multi-objective optimization. *Eur. J. Oper. Res.* **2016**, *248*, 758–771. [CrossRef]
3. Cai, W.; Lai, K.-H.; Liu, C.; Wei, F.; Ma, M.; Jia, S.; Jiang, Z.; Lv, L. Promoting sustainability of manufacturing industry through the lean energy-saving and emission-reduction strategy. *Sci. Total. Environ.* **2019**, *665*, 23–32. [CrossRef]
4. Gong, Q.-S.; Zhang, H.; Jiang, Z.; Wang, H.; Wang, Y.; Hu, X.-L. Nonempirical hybrid multi-attribute decision-making method for design for remanufacturing. *Adv. Manuf.* **2019**, *7*, 423–437. [CrossRef]
5. Hu, M.; Sun, Y.; Gong, Q.; Tian, S.; Wu, Y. Multi-Objective Parameter Optimization Dynamic Model of Grinding Processes for Promoting Low-Carbon and Low-Cost Production. *Processes* **2019**, *8*, 3. [CrossRef]
6. Cai, W.; Liu, C.; Zhang, C.; Ma, M.; Rao, W.; Li, W.; He, K.; Gao, M. Developing the ecological compensation criterion of industrial solid waste based on emergy for sustainable development. *Energy* **2018**, *157*, 940–948. [CrossRef]
7. Jiang, Z.; Ding, Z.; Zhang, H.; Cai, W.; Liu, Y. Data-driven ecological performance evaluation for remanufacturing process. *Energy Convers. Manag.* **2019**, *198*, 111844. [CrossRef]
8. Cai, W.; Liu, F.; Zhou, X.; Xie, J. Fine energy consumption allowance of workpieces in the mechanical manufacturing industry. *Energy* **2016**, *114*, 623–633. [CrossRef]
9. Jiang, Z.; Ding, Z.; Liu, Y.; Wang, Y.; Hu, X.; Yang, Y. A data-driven based decomposition–integration method for remanufacturing cost prediction of end-of-life products. *Robot. Comput. Manuf.* **2020**, *61*, 101838. [CrossRef]
10. Li, C. Multi-objective Machining Process Route Optimization Model for High Efficiency and Low Carbon. *Chin. J. Mech. Eng.* **2014**, *50*, 133–141. [CrossRef]
11. Sobeyko, O.; Mönch, L. Integrated process planning and scheduling for large-scale flexible job shops using metaheuristics. *Int. J. Prod. Res.* **2016**, *55*, 1–18. [CrossRef]
12. Liu, X.-J.; Yi, H.; Ni, Z. Application of ant colony optimization algorithm in process planning optimization. *J. Intell. Manuf.* **2010**, *24*, 1–13. [CrossRef]
13. Fang, K.; Uhan, N.; Zhao, F.; Sutherland, J.W. A new approach to scheduling in manufacturing for power consumption and carbon footprint reduction. *J. Manuf. Syst.* **2011**, *30*, 234–240. [CrossRef]

14. González, M.A.; Vela, C.R.; Varela, R. Scatter search with path relinking for the flexible job shop scheduling problem. *Eur. J. Oper. Res.* **2015**, *245*, 35–45. [CrossRef]
15. Li, C. A Batch Splitting Flexible Job Shop Scheduling Model for Energy Saving under Alternative Process Plans. *Chin. J. Mech. Eng.* **2017**, *53*, 12–23. [CrossRef]
16. Huang, X.; Zhang, X.; Ai, Y. ACO integrated approach for solving flexible job-shop scheduling with multiple process plan. *Comput. Integr. Manuf. Syst.* **2018**, *24*, 558–569.
17. Liu, Q. Integrated Optimization of Process Planning and Shop Scheduling For Reducing Manufacturing Carbon Emissions. *Chin. J. Mech. Eng.* **2017**, *53*, 164–174. [CrossRef]
18. Li, C.; Kou, Y.; Lei, Y.; Xiao, Q.; Li, L. Flexible job shop rescheduling optimization method for energy-saving based on dynamic events. *Comput. Integr. Manuf. Syst.* **2020**, *26*, 288–299.
19. Tang, Q.; Chen, S.; Zhao, M.; Zhang, L. Prediction of Optimal Rescheduling Mode under Machine Failures with in Job Shops. *China Mech. Eng.* **2019**, *30*, 188–195.
20. Kong, L.; Wang, L.; Li, F.; Liu, X.; Wang, G. Sustainable Scheduling for Hybrid Flow-Shop Based on Performance Matching of Machine Tools. *Comput. Integr. Manuf. Syst.* **2019**, *25*, 1075–1085.
21. Du, Y.; Zhang, F. Flexible Job Shop Scheduling with Degenerate and Preventive Maintenance. *Digit. Manuf. Sci.* **2018**, *16*, 133–138.
22. Zheng, Y.; Wang, Y. Optimization of Process Selection and Sequencing Based on Genetic Algorithm. *China Mech. Eng.* **2012**, *23*, 59–65.
23. Huang, Y.; Li, K.; Zheng, X. A Scheduling Algorithm to Equal Amount of Batches for Job Shop Considering the Constraint of Work-shifts. *Sci. Technol. Eng.* **2013**, *13*, 2.
24. Yan, J.; Xing, L.; Zhang, Z.; He, R. Dual Time Window Constrained Job-shop Scheduling Algorithm. *Sci. Technol. Eng.* **2016**, *16*, 85–92.
25. An, X. Optimization ofprocess route based on intuitionistic fuzzy number and multi-objective optimization algorithm. *Comput. Integr. Manuf. Syst.* **2019**, *25*, 1180–1191.
26. Chang, Z. Optimization of Process Based on Adaptive Ant Colony Algorithm. *Chin. J. Mech. Eng.* **2012**, *48*, 163. [CrossRef]
27. Li, C.; Zhou, J.; Lu, P.; Wang, C. Short-term economic environmental hydrothermal scheduling using improved multi-objective gravitational search algorithm. *Energy Convers. Manag.* **2015**, *89*, 127–136. [CrossRef]
28. Padhye, N.; Bhardawaj, P.; Deb, K. Improving differential evolution through a unified approach. *J. Glob. Optim.* **2012**, *55*, 771–799. [CrossRef]
29. Deb, K.; Padhye, N. Enhancing performance of particle swarm optimization through an algorithmic link with genetic algorithms. *Comput. Optim. Appl.* **2013**, *57*, 761–794. [CrossRef]
30. De Jong, K. Evolutionary Computation: A unified approach. In *Proceedings of the 2016 on Genetic and Evolutionary Computation Conference—GECCO '16*; Association for Computing Machinery (ACM): New York, NY, USA, 2016; pp. 185–199.
31. De Jong, K. Evolutionary computation: A unified approach. In *Proceedings of the Genetic and Evolutionary Computation Conference Companion*; Association for Computing Machinery (ACM): New York, NY, USA, 2019.
32. Čapek, R.; Šůcha, P.; Hanzalek, Z. Production scheduling with alternative process plans. *Eur. J. Oper. Res.* **2011**, *217*. [CrossRef]
33. May, G.; Stahl, B.; Taisch, M.; Prabhu, V. Multi-objective genetic algorithm for energy-efficient job shop scheduling. *Int. J. Prod. Res.* **2015**, *53*, 7071–7089. [CrossRef]
34. Tang, D.; Dai, M.; Salido, M.A.; Giret, A. Energy-efficient dynamic scheduling for a flexible flow shop using an improved particle swarm optimization. *Comput. Ind.* **2016**, *81*, 82–95. [CrossRef]
35. Xiao-Ning, S.; Xin, Y. Mathematical modeling and multi-objective evolutionary algorithms applied to dynamic flexible job shop scheduling problems. *Inf. Sci.* **2015**, *298*, 198–224.
36. Zhang, L. Carbon Emission Analysis for Product Assembly Process. *Chin. J. Mech. Eng.* **2016**, *52*, 151–160. [CrossRef]
37. Hanwu, M. *The Planning and Designing of Logistics and Facilities*; High Education Press: Beijing, China, 2005; pp. 78–98.

Article

Multiple Scenarios Forecast of Electric Power Substitution Potential in China: From Perspective of Green and Sustainable Development

Jing Wu [1,2,*], Zhongfu Tan [1,2,3], Gejirifu De [1,2], Lei Pu [1,2], Keke Wang [1,2], Qingkun Tan [4] and Liwei Ju [1,2]

1 School of Economics and Management, North China Electric Power University, Beijing 102206, China
2 Beijing Key Laboratory of New Energy and Low-Carbon Development, North China Electric Power University, Beijing 102206, China
3 School of Economics and Management, Yan'an University, Yan'an 716000, China
4 State Grid Energy Research Institute Co., Ltd., Changping District, Beijing 102209, China
* Correspondence: 1182106009@ncepu.edu.cn; Tel.: +86-178-0113-7856

Received: 1 August 2019; Accepted: 27 August 2019; Published: 2 September 2019

Abstract: To achieve sustainable social development, the Chinese government conducts electric power substitution strategy as a green move. Traditional fuels such as coal and oil could be replaced by electric power to achieve fundamental transformation of energy consumption structure. In order to forecast and analyze the developing potential of electric power substitution, a forecasting model based on a correlation test, the cuckoo search optimization (CSO) algorithm and extreme learning machine (ELM) method is constructed. Besides, China's present situation of electric power substitution is analyzed as well and important influencing factors are selected and transmitted to the CSO-ELM model to carry out the fitting analysis. The results showed that the CSO-ELM model has great forecasting accuracy. Finally, combining with the cost, policy supports, subsidy mechanism and China's power consumption data in the past 21 years, four forecasting scenarios are designed and the forecasting results of 2019–2030 are calculated, respectively. Results under multiple scenarios may give suggestions for future sustainable development.

Keywords: electric power substitution; potential forecasting; CSO-ELM; green sustainable development

---

1. Introduction

With the rapid growth of economy and the increasing consumption of fossil resources, China is facing problems of resource shortage, climate change and environmental governance [1], showing an increasing contradiction between social development and unsustainable energy structure. The overdose of coal combustion, large number of automotive exhaust emissions and improper treatment of pollutants are all leading to serious atmospheric pollution, which seriously threatens the quality of people's lives and has an irreversible impact overall on the ecological environment [2]. In order to mitigate the effects of pollution emissions, it is urgent to develop and promote highly efficient and green energy technologies in order to reach social sustainability. According to China Energy Statistical Yearbook (2017), 50% to 60% of particulate matter 2.5 (PM 2.5) air pollution comes from coal combustion and 20% to 30% from oil combustion. Meanwhile, the National Development and Reform Commission and the National Energy Administration of China has proposed that, by 2020, the total amount of electricity replacing coal and oil combustion is estimated to reach 130 million tons of standard coal, and the proportion of electric energy in the end-stage energy consumption should be 27%, increasing by about 1.5%. The additional consumption of electricity in the "13th Five-Year Plan" is set to be

450 billion kWh [3]. Under the global trend of low-carbon green development, "Two substitutions", which includes both clean-energy substitution and electric power substitution, is meant to guide the energy structure optimal reform [4]. Therefore, studies on electric power substitution potential will give suggestions and guidance for its further sustainable development.

As an energy consumption pattern [5], electric power substitution can make further use of the environmental capacity in different regions, in order to reach the balance of pollutants emissions and optimal resources allocation [6,7]. By replacing coal and fuel with electric power, pollutions can be effectively cut down and energy efficiency improved [8]. With the development of energy technology revolution, electric power substitution can be applied into transportations, electric boilers, electric kilns, electric heating and electric cookers, replacing fossil resources such as oil and coal. At present, the industrial energy efficiency is relatively low. Clean energy substitution would effectively improve the efficiency of energy utilization [9]. The market potential for promoting electric power substitution in China is about 22 trillion kWh. The potential of substituting coal and oil with electricity is prospectively 18 trillion and 400 billion kWh, showing huge potential of the substitution market. Presently, electric power substitution is gradually becoming a research focus both in China and globally. Scholars have made achievements regarding electric power substitution with two different approaches: one is the technical and economic research of the power substitution, and the other is related to methods used for forecasting.

By analyzing the technical and economic efficiency of energy substitutions, Wu et al. [10] proposed that the sustainable use of energy would be the main direction of future development. Based on the system dynamics model, Song et al. [11] analyzed the emission reduction effect of the renewable energy substitution in China. Besides, while researching energy substitution, many scholars achieved energy substitutions on the power supply side by combining multiple renewable energy resources [12]. He et al. [13] introduced the environmental utility in the environmental consumption (EUEC) model to discuss the relationship between urban energy consumption and environmental utility changes. Barreto [14] demonstrated the dynamic substitution effect of renewable energy replacing fossil fuel by building a theoretical framework that incorporates alternative energy and traditional fossil energy into the endogenous growth model. Liu et al. [15] established a dynamic system model combining multiple renewable energy sources and made an empirical analysis. By changing the proportion of electric vehicles in the power system, André et al. [16] compared and analyzed the emission reduction benefits of the system under multiple scenarios. Kumar et al. [17] discussed the innovative capabilities that enterprises need to adopt, such as pollution prevention and clean technology strategies, in order to achieve sustainable development.

The development of electric power substitution is influenced by many factors, such as technology, economy, environmental protection requirements, policy measures, demand response, etc. Wu et al. [18] pointed out that the initial investment was high; therefore, promoting the projects is facing greater resistance. Similarly, Shaligram [19] argued that the current barriers to substitution mainly contained the high cost of substitute technology. By summarizing the practical experience in the Jiangsu Province, Li [20] found that there were some problems in the promotion work, such as insufficient policy support, less response from users and lower technology level. Combined with various factors, Liang et al. [21] constructed the evaluation index model of power substitution scheme and analyzed its substitution potential. Lu and Xie [22] empirically analyzed the intensity of enterprises conducting clean substitution under the pressure of carbon emission reduction. Faced with the carbon emission reduction, the sense of conducting cleaner production will increasingly rise.

For forecasting methods, many scholars have made many achievements. Common methods include single forecasting models and combined forecasting models, such as Support Vector Machine (SVM), artificial neural network, genetic algorithm, grey forecasting model, and a combination of forecasting methods. Wei et al. [23] combined a neural network and statistical linear model to predict wind power output. Michael et al. [24] incorporated the forecasting model based on machine learning to predict the household energy consumption. Zhang et al. [25] proposed a forecasting model for building demand response with random forest and ensemble learning method. Wu et al. [26] forecasted the short-term load of a power system based on generalized regression neural network method. Based on artificial neural network method, Xia et al. [27] combined a virtual instrument and radial basis function neural network and created long-term, medium-term and short-term load forecasting. Shan et al. [28] used the extremum learning machine (ELM) method based on Back Propagatio (BP) and SVM to forecast photovoltaic (PV) generation. Chen and Yu [29] used SVM on wind signal prediction. Lee and Tong [30] combined grey model and incorporating genetic algorithm together to forecast and analyze the demand for electric power. A hybrid model composed by SVM and a Seasonal Auto Regressive Integrated Moving Average was proposed for short-term PV generation forecasting [31]. Yu and Xu [32] used an optimized genetic algorithm and improved BP neural network (BPNN) to forecast the load of natural gas. In the actual implementation process of power substitution work, an accurate forecast on power consumption and the changing trend can provide data support and policy guidance for further power substitution work promoting [33,34]. Sun et al. [35] forecasted the potential of power substitution using particle swarm optimization (PSO)-SVM. Yin [36] established a grey energy demand-forecasting model to forecast the terminal energy demand in Beijing. Zheng [37] constructed the potential forecasting model on rural electric power substitution, and made middle and long-term analysis of electricity consumption. Li [38] used the improved TOPSIS method to analyze the potential of regional power substitution.

Above all, scholars have made achievements in presenting electric power substitution, promoting methods and digging related influencing factors [39]. However, few have mentioned the developing potential forecast of the substitution works. Most existing works on potential analysis used the comprehensive evaluation method, which cannot show the future developing trend of the power substitutions. Therefore, in order to make up for the deficiencies of the existing research on the potential analysis of power substitution, a CSO-ELM model based on the Pearson correlation test is constructed to forecast the market potential of electric power substitution projects. The main contributions of this work are summarized as follows.

(1). To overcome the limitation of single algorithms, a forecasting algorithm combining Cuckoo Search Optimization (CSO) and Extreme Learning Machine (ELM) is constructed, which can make full use of the superior global search ability of the Cuckoo algorithm and the learning efficiency and generalization ability ELM. The proposed algorithm is highly sensitive to the market potential of power substitution projects and can accurately reflect the development potential.
(2). Starting from the current electric power substitution in China, factors that affect the promotion of electric power substitution projects, such as economy, policy and technology are analyzed systematically, so as to make up for the deficiency of the existing research on the influencing factors analysis of electric power substitution potential.
(3). A relevance test is conducted to choose effective influencing factors. Factors with a significant influencing level will be taken as input factors of the forecasting model, and factors with insignificant or general influencing degree are excluded.
(4). The validity and superiority of the forecasting model are verified by the actual situation of China's electric power substitution. some feasible suggestions are put forward based on the forecasting results under multiple scenarios to promote the orderly progress of electric power substitution work.

The paper is organized as follows: Section 2 presents the structures and features of the forecasting method. Section 3 introduces the current situation of the electric power substitution in China and gives summary of all influencing factors. In Section 4, specific data is used to verify the effectiveness of the proposed method, and four scenarios are given to address further discussion of the future development of the substitution work. Finally, the conclusion is given in Section 5.

2. Current Situation and Influencing Factors Selection of Electric Power Substitution in China

2.1. Current Situation of Electric Power Substitution

The rapid growth of China's economy is accompanied by excessive energy consumption. In order to narrow the gap of energy utility between China and developed countries, the concept of energy conservation should be penetrated through energy exploitation, transportation and utilization. Proper use of energy and improving energy efficiency are the main goals of energy development in the future. Electric power substitution projects have great development potential in the future.

Electric power substitution uses electric power to replace coal-fired heating. Through the large-scale centralized conversion of power, electric power substitution can improve the efficiency of fuel use and reduce pollutant emissions to optimize the terminal energy structure and promote environmental protection. Electric power substitutions include coal substitution by electricity, oil substitution by electricity, and electrification of agricultural production, etc. Different substitution methods for each field are shown in the Table 1 below.

Table 1. Electric power substitution methods.

Substitution Fields	Substitute Methods
Substitute Coal by Electricity	Electric heating replacing coal-fired heating Electric boiler replacing industry coal-fired boiler Electrification in family and catering industry
Substitute Oil by Electricity	Electric vehicles replacing fossil-fueled vehicles Electric pump replacing agricultural oil pump Shore power supply replacing diesel engines
Substitute Gas by Electricity	Electricity replacing natural gas
Power Trans-Regional Transmission	Long-distance transmission of clean power

Despite various methods to promote electric power substitution, scientific and reasonable policies are necessary in promoting the substitution work. Since 2015, the Chinese government has promulgated 226 supporting policies to encourage electric power substitution, guiding the society to choose electric energy actively, and gradually eliminate the high pollution and low efficiency of energy use. Government support is not only subjective to propaganda and guidance, but also to strengthen the construction of electric power to improve the competitiveness of electricity continuously in power market. The relevant electricity alternative development policies are shown in the Table 2 below.

As can be seen from Table 2, a good policy environment has been provided to develop energy substitutions. The government issued guidance on electric power substitution promotion, using the substitution work as a national strategy. Then, supporting policies came to support the pilot work of clean energy heating in winter in northern areas, to promote the prevention and control of air pollution in Beijing-Tianjin-Hebei Region and the surrounding areas. Electric power substitution has been regarded as an important part of the national "13th Five-Year Plan" in the electricity industry and modern comprehensive transportation system.

Table 2. Policies supporting electric power substitution work.

Policies	Year	Contents
Guidelines for Improving Power System Regulation Capability	2018	Comprehensively promote electric power substitution. By 2020, the aim is to achieve 450 billion kWh of the substituting electricity, and the electricity consumption accounts for 27% of the total energy consumption.
Guidelines on Promoting Clean Energy Heating in Cities and Towns in Northern China	2017	Promote the work of "coal to gas", "coal to electricity" and renewable energy heating in "2 + 26" cities in Beijing-Tianjin-Hebei region and the surrounding areas to improve the efficiency of clean energy heating
Demand Side Management in Electricity Industry (Revised)	2017	Promote the electricity work in the demand side to help control air pollution and expand electricity consumption. Bring innovations to electric power substitution work and create new methods and the substituting fields to expand the promoting scale.
Work Plan for Air Pollution Control and Prevention in Beijing-Tianjin-Hebei and Surrounding Areas in 2017	2017	Support the heating and gas substituted projects that provide services for residents' livelihood, and give capital supports for equipment purchasing.
"13th Five-Year Plan" of Comprehensive Work on Energy Conservation and Emission Reduction"	2017	During the "13th Five-Year Plan" period, the proportion of electric energy in the consumption of terminal energy should be continuously increased. The construction of charging facilities should be pushed forward. Promote central electric heating to replace small coal-fired boilers gradually.
Intensive Plan for Air Pollution Prevention and Control in Beijing-Tianjin-Hebei Region	2016	Promote the clean energy substitution work in rural areas of Beijing-Tianjin-Hebei Region.
Guidelines on Promoting Electric Power Substitution	2016	Electric power substitution will be promoted in the four key areas, such as heating in northern areas, production and manufacturing, transportation, power supply and consumption.
Notice on Further Pilot Work of Energy Conservation and New Energy Vehicle Demonstration and Promotion	2016	Encourage the development of new energy vehicles and electric vehicles, and mobilize the enthusiasm of all units to use energy-saving and new energy vehicles.

With the creation of new policies, the scale of electric power substitution has been expanding. Implementation plans of electric power substitution have come out as well. Presently, China's electric power substitution projects are mainly carried out in the field of substituting coal and oil. Electric heating and electric vehicles—as the main alternative methods—have achieved remarkable results. In 2017, 101,807 electric power substitution projects were implemented nationwide, with 128.6 billion kWh of electricity in all substituting fields. Among them, 8.8 billion kWh comes from residential heating, 77.4 billion kWh from industrial (agricultural) production and manufacturing, 12.8 billion kWh from transportation, 23.9 billion kWh from electric power supply and consumption, and 5.8 billion kWh from household electrification and other fields, which is equivalent to a reduction of 64.4 million tons of coal-fired burning. The emission reduction is about 110 million tons of carbon dioxide, 5.2 million tons of sulfur dioxide and nitrogen oxides.

2.2. Analysis on Influencing Factors of Electric Power Substitution Potential

With the development of global energy internet, electric power substitution is facing new opportunities. The substitution work is affected by energy consumption, GDP, energy prices, investment in renewable power assets and average concentration of P.M. 2.5 and other factors.

(1). Electricity consumption

Electricity consumption is an important index used to measure the level of national electrification. Electrification represents the proportion of electric power, and reflects the changes of social energy consumption structure. The increase of electricity consumption directly leads to the improvement of social electrification. Meanwhile, the improvement of electrification shows an increase in social energy-use technology, which can effectively reduce the cost of electric power substitution projects and reach further promotion of electric power substitution.

(2). GDP per capita

GDP is an important indicator reflecting the economic development in China, which is an important factor affecting the demand of electricity. The economy of a region will have impacts on electricity and other energies' consumption. Research on power and energy has been made, and scholars have regarded GDP per capita as the decisive factor affecting electricity demand, which means that China's electricity demand and GDP growth are endogenous with a significant and stable relationship. In addition, the rise in GDP can promote residents' living standards. As people are becoming richer, they may pay more attention to the energy structure. Thus, the promotion of electric power substitution can be further improved. Therefore, GDP per capita is chosen as an indicator to show the impact of economic development on power substitution in China.

(3). Annual investment increment in electric power industry

The relationship between the investment in electric power industry and social electricity consumption is positive. Investing in electric power assets shows the attention society attaches to the development of the electric power industry. The investment in the electric power industry includes investment in power grid construction and investment in generators. Both will bring an increase in electricity consumption, which can indirectly improve the replacing effects of electricity to other energy resources. Two indicators—the annual investment increment in electric power industry and in power grid construction—are chosen in the following analysis.

(4). Electric power installed capacity

The electric power installed capacity is proportional to the total generating capacity, which helps promote electric power substitution work. In addition, renewable power generation has lower operating costs and less pollution emissions. Substituting fossil resources with renewable power will further improve the social benefits and achieve pollution reduction from power supply side. Therefore, as an important indicator, electric power installed capacity of renewable energy is considered in forecasting the market potential of electric power substitution projects.

(5). Renewable power generation

Renewable energy utilization is significant to energy structure adjustment under low-carbon mechanism. Using clean energies to generate will effectively reduce the proportion of coal-fired thermal power, thereby reducing the environmental burden. Besides, integrating more renewable power into the substitution work will help with the power curtailment problem, and large-scale utilization of renewable energy achieves substitutions for traditional fuels from the generation side. Thus, the implementation scope of electric power substitutions will be further expanded, and the environmental benefits will be significantly improved.

(6). Carbon emissions

The rising carbon emissions has forced the government and all sectors of society to pay more attention to the energy consumption structure. According to China's "National Independent Contribution" in 2030—compared with the situation in 2005—carbon dioxide emissions should achieve the peak value, carbon dioxide emissions per unit GDP will decrease by 60% to 65% and the proportion of non-fossil energy in primary energy consumption should reach about 20%. Facing the double pressures of international emission reduction commitments and domestic resources and environment, promoting clean energy usage is an important means to achieve carbon emission reduction. Constrained by carbon emission targets, enterprises' awareness of environmental protection will raise. With the development of clean technology, the electric power substitution work shall be promoted in large scale [22].

3. Methodology

3.1. Pearson Test

Pearson correlation coefficient is a method used to measure the degree of correlation between two variables. The correlation value is between 1 and −1, where 1 means that variables are completely positive, 0 means irrelevant, and −1 means completely negative. The correlation value between (X, Y) is calculated as follows.

$$\begin{aligned} P_{x,y} &= \frac{\mathrm{cov}(X,Y)}{\sigma_X \sigma_Y} = \frac{E((X-\mu_X)(Y-\mu_Y))}{\sigma_X \sigma_Y} \\ &= \frac{E(XY)-E(X)E(Y)}{\sqrt{E(X^2)-E^2(X)}\sqrt{E(Y^2)-E^2(Y)}} \end{aligned} \tag{1}$$

The numerical value of $P_{x,y}$ reflects the linear correlation degree of Y and X, which is between $[-1, 1]$. The conclusion is as follows.

(1). When $|P_{x,y}| \to 1$, the correlation between Y and X is stronger.

(2). When $|P_{x,y}| \to 0$, the correlation between Y and X is weaker, or non-linear related, even when not related.

3.2. Cuckoo Search Optimization

Enlighted by the brood parasitism behaviors of cuckoo birds, the Cuckoo Search optimization (CSO) algorithm, which is a natural heuristic algorithm developed by Yang in 2009 [40]. The cuckoo bird lays eggs in the nest of the host-bird's nest and removes the eggs of the host. Occasionally, some cuckoo eggs look similar to host eggs and get the opportunity to be raised. In other cases, these eggs may be found by the host birds, who will throw them away or leave the nests to find other places to build new nests. Each egg in a nest represents a solution, and a cuckoo egg stands for a new solution. CSO uses new and potentially better solutions to replace not-so-good solutions in the nests.

The CSO method operates as follows. The cuckoo lays only one egg at a time and randomly places the egg in a nest. The nest with the highest quality egg (the solution to the problem) will remain to the next turn. The number of the nests which can be laid eggs in is fixed, and the probability that the host bird can select the cuckoo eggs is $p_a \in [0, 1]$.

Under this situation, the host bird can choose to throw the cuckoo eggs or find a new nest to replace the old one. If not detected, the cuckoo eggs will be successfully hatched up and find new hatching sites through Lévy flight. Considering the Lévy flight behavior of the cuckoo birds' nest-seeking feature, assume that there are N cuckoo eggs in the D-dimensional search space. The location of the number i egg under the kth iteration is x_i^k, and the new solution x_i^{k+1} can be expressed as follows.

$$x_i^{k+1} = x_i^k + \delta_i \tag{2}$$

$$\delta_i = \alpha \times s_i \oplus (x_i^k - x^{best}), \tag{3}$$

where, $\alpha > 0$ is the step size, which relates to the scale of the problem. δ_i is the changing amount of position that needs to be taken, and $\oplus$ is the entry wise multiplications.

The random step is produced by the symmetric Lévy distribution.

$$s_i = \frac{u}{|v|^{1/\beta}}, \tag{4}$$

where, $u(u_1, u_2, \cdots, u_d)$, $v(v_1, v_2, \cdots, v_d)$ are vectors in the D-dimensional space. $\beta = 3/2$. The sub-vectors of u and v obey normal distribution.

$$u \sim N(0, \sigma_u^2), v \sim N(0, \sigma_v^2) \tag{5}$$

$$\sigma_u \sim \left(\frac{\Gamma(1+\beta) \cdot \sin(\pi \cdot \beta/2)}{\Gamma((1+\beta)/2) \cdot \beta \cdot 2^{(\beta-1)/2}}\right)^{1/\beta}, \sigma_v = 1. \tag{6}$$

Lévy flight includes a random directional linear motion sequence with no characteristic scale, and the step size of each sequence satisfies the heavy-tailed distribution. The relatively short straight-line move with a larger frequency will be intermittently replaced by the longer-step move with less frequency. Lévy flight ensures the comprehensiveness of searching, so that the search efficiency of CSO is higher than that of the standard Gauss random processes.

3.3. Extreme Learning Machine

ELM, proposed by Professor Huang in 2004, is a fast and efficient single-layer feedforward neural network algorithm. ELM is essentially a linear-in-the-parameter model, so its learning process is easy to converge to the global minimum. Due to the random input weights and hidden layer thresholds, the number of hidden layer nodes significantly influence the performance of the model. For a Single-hidden Layer Feedforward Neural network (SLFN), ELM uses the number of hidden layers to train—which greatly reduces the training time and computational complexity. The main idea of the ELM model is to randomly set the network weights and then obtain the inverse output matrix of the hidden layer. Compared with other learning models, the ELM model operates extremely fast and maintains a better accuracy and has therefore been widely used in many fields. In the ELM training process, the number of neurons in the hidden layer is the only need. Therefore, the output weight matrix of the hidden layer can be calculated without adjusting the connection weight between the input layer neurons and the hidden layer neurons and the deviation of the hidden layer neurons.

The network-training model of extreme learning machine adopts the former single hidden layer structure. Assuming there are N sets of initial training set (x_i, t_t), the input layer includes $x_i = [x_{i1}, x_{i2}, \ldots x_{in}]^{\mathrm{T}} \in R^n$, and the target output layer is $t_i = [t_{1i}, t_{2i}, \ldots t_{mi}]^{\mathrm{T}} \in R^m$. The hidden layer contains L nodes. The activation function $g(x)$ is expressed as follows.

$$\sum_{i=1}^{L} \beta_i g_i(x_i) = \sum_{i=1}^{L} \beta_i g(w_i \cdot x_j + b_i) = y_j \quad j = 1, 2, \ldots N, \tag{7}$$

where, y_j represents the output vectors of ELM. β_i represents the weight vectors connecting the hidden layer and the output layer. w_i represents the weight vectors that connects the hidden layer and the input layer. b_i and $g(w_i \cdot x_j + b_i)$ are threshold value and the output value of the hidden node i.

The objective of an ELM is to search for a suitable set of β, ω, and b to approximate all training sample pairs with zero error.

$$\sum_{j=1}^{N} ||t_j - y_j|| = \sum_{j=1}^{N} ||t_j - \sum_{i=1}^{L} \beta_i g(w_i x_j + b_i)|| = 0. \tag{8}$$

Formula (7) can be expressed as:

$$H\beta = T \tag{9}$$

$$H = \begin{bmatrix} g(w_1 x_1 b_1) g(w_2 x_1 b_2) \cdots g(w_L x_1 b_L) \\ g(w_1 x_2 b_1) g(w_2 x_2 b_2) \cdots g(w_L x_2 b_L) \\ \cdots \\ g(w_1 x_N b_1) g(w_2 x_N b_2) \cdots g(w_L x_N b_L) \end{bmatrix}_{N\times L}, \tag{10}$$

$$\beta = [\beta_1, \beta_2, \ldots \beta_L]_{L\times 1}^{-1}, \text{ and } T = [t_1, t_2, \ldots t_L]_{L\times 1}^{-1}, \tag{11}$$

where H is the output matrix of the hidden layer; β is the weights vector connecting the hidden layer nodes with the output layer neurons; and T represents the target output.

When the activation function is infinite differentiable, ELM analytically calculates the hidden-output weights by searching for a minimal norm least square solution of the following linear equation.

$$||H\hat{\beta} - T|| = \min_{\beta} ||H\beta - T||\,, \tag{12}$$

$$H\beta = T\,, \tag{13}$$

$$\hat{\beta} = H^{\mathrm{T}} T, \tag{14}$$

where H^{T} denotes the Moore-Penrose generalized inverse of the hidden-layer output matrix H.

3.4. CSO-ELM Model

In the basic ELM model, when the number of hidden layer nodes is fixed, the input weights and hidden layer deviations of the network structure are randomly determined, which affects the stability of the model itself. After using CSO to improve ELM algorithm, the number of hidden layer nodes, input weights and thresholds value of ELM model are adaptively selected. The flow chart is shown in Figure 1. In the CS algorithm, N nests are set. Then, round the position value as the number of the hidden nodes M. After inputting the sample data, a set of input weight matrices W for ELM networks will be generated. Then, the threshold value of the hidden layer b will be obtained, and the model output the weight matrices β. The prediction error calculated by the sample data will be the fitness value of the present nest. At the end of each iteration, the optimal nest location is preserved, and the values of W, b, and β are outputted. After satisfying the iteration termination conditions, the fitness values of nests are compared. Finally, the optimal position of the bird's nest are obtained, and the corresponding values of W, b, β will be exported. The improved ELM method operates as follows.

1. Set the parameters of the CSO algorithm. The probability parameter of the nest being discovered is set as p_a. The initial positions of N nests are generated as $nest_0 = [x_1^0, x_2^0, \ldots, x_N^0]$. Round the position values to N different values as the number of nodes of the ELM hidden layer. Input the sample and calculate the root mean square error (RMSE) as the corresponding optimal fitness value F_0. The maximum number of iterations is max_it.
2. Choose the last optimum nest position x_i. Search for the nest position j through the Lévy flight, and round the position value as the number of ELM hidden nodes. Then, calculate F_j and compare it with F_i to keep the better solution.
3. Generate random number p_r and compare it with p_a. If $p_r > p_a$, choose the nest position randomly and replace the worst one. Otherwise, nothing changes.

4. Stop searching after the number of iterations is satisfied.
5. Select the position with minimum fitness as the number of hidden layer nodes of ELM and output the corresponding values of W, b and β. The electric power substitution market potential prediction model cased on CSO-ELM is established.

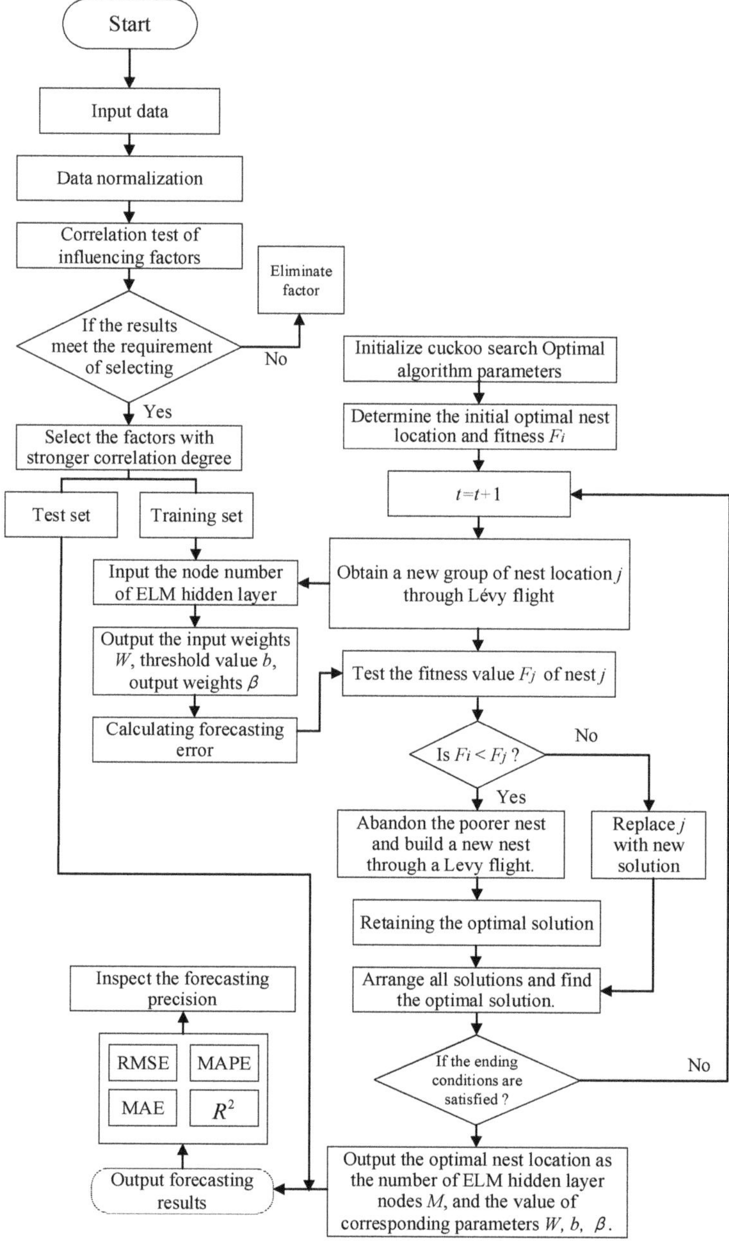

Figure 1. Outline of procedure of proposed method.

4. Case Study and Results

4.1. Validation Test of Forecasting Model

The proposed forecasting method is tested by historical data inputting. The data used in this research is collected from both the "Statistical Yearbook of China" and the "Annual Report of Electric Power Industry" for the latest 20 years. Given the development of electric power substitution projects being affected by many factors, Pearson correlation test is carried out for test the correlations of each factor. Factors with high correlation are used as input data of the forecasting model to accurately predict the market potential of electric power substitution projects. The historical data are shown in the Table 3. The correlation test of each factor is carried out by Statistical Product and Service Solutions (SPSS) software and the results are shown in Table 3 and Figure 2.

Table 3. Correlation degree of influence factors and market potential of electric power substitution.

	Factors	Correlation Degree		Factors	Correlation Degree
1	GDP per capita	0.966	5	Renewable Energy Generation	0.931
2	Electricity Consumption	0.987	6	Renewable Energy Installed Capacity	0.915
3	Electric Power Installed Capacity	0.968	7	Annual Increment Investment in Electric Power Industry	0.963
4	Carbon Emissions	0.985	8	Annual Increment Investment in Power Grids	0.975

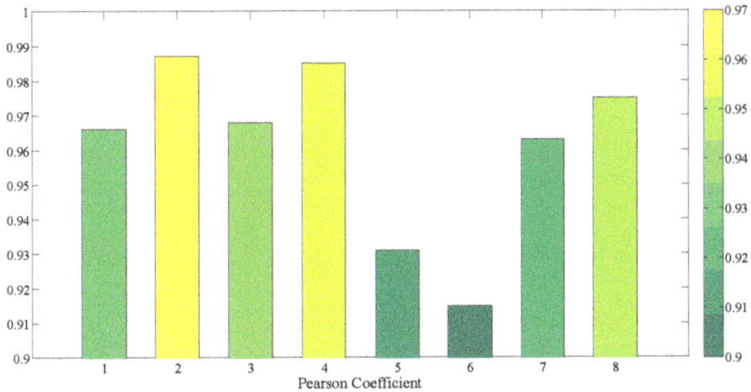

Figure 2. Pearson coefficient.

The research is used to forecast the market potential of power substitution projects in 2018–2030. A forecasting model of extreme learning machine based on the cuckoo search optimization is constructed. By taking historical data of 1998–2017 as an example, the validity of this model is tested. Due to the substitution work being a gradual process—when the CSO-ELM model is used to forecast the market potential of electric power substitution projects—the electric power substitution amount in every year in the future is forecasted in turn. When forecasting, the data of the target year is added to the training set of the model to give out more accurate results.

Data from 1998 to 2012 is chosen as the training set, and data from 2013 to 2017 is used as the test set. The BP neural network, ELM and CSO-ELM forecasting models are all used to forecast the electric power substitution from 2013 to 2017. ELM and BP neural network are introduced as the contrast algorithm, and the parameters are set to be at the optimal values after comparison. In setting the parameters settings of BPNN, the number of neurons in hidden layer is set as 6. In addition, the tansig function is used as the transfer function. The output layer is set as the purelin function. The training times are set to 1000, and the precision target is set as 0.001. In the ELM forecasting model, the number of nodes in hidden layer is 30, and the activation function is sig function. The initial population number of CSO-ELM is 20 and the maximum iteration number is 200. Historical data is listed in Table 4.

Table 4. Historical Data.

Year	GDP Per Capita/Yuan	Electric Power Consumption/ Billion kWh	Renewable Energy Generation / Billion kWh	Electric Power Installed Capacity/ Million kW	Renewable Energy Installed Capacity/Million kW	Annual Increment Investment in Electric Power Grid/Billion Yuan	Annual Increment Investment in Electric Power Industry/Billion Yuan	Carbon Emission /Million Tons	Total Substitution Amount/Million Tons of Standard coal
2018	63,186	6844.9	1867	18,994.8	72,869	537.3	793	10,000	36,946.3
2017	59,660	6307.7	1832.4	17,770.3	65,000	535.6	801.5	9232.6	34,133.5
2016	53,935	5919.8	1866.6	16,457.5	56,400	542.6	883.6	9113.6	32,658.7
2015	50,251	5702	1509.2	15,252.7	48,000	464	857.6	9163.2	29,340.4
2014	47,203	5638.37	1378.4	13,701.8	44,037	411.9	780.5	9204.2	29,544.1
2013	43,852	5420.34	1151	12,576.8	38,700	385.6	772.7	8966.3	28,047.9
2012	40,007	4976.26	1066.6	11,467.6	31,300	369.3	746.5	8792.3	24,957.9
2011	36,403	4700.1	886.3	10,625.3	28,100	368.7	739.3	8104.9	24,957.9
2010	30,876	4193.45	811.1	9664.1	25,648.5	344.8	705.1	7680.7	23,060.9
2009	26,222	3703.21	703.5	8741.0	22,302.1	389.8	770.2	7174.6	18,640.4
2008	24,121	3454.13	674.8	7929.3	18,987.3	288.4	576.3	6710	17,530.9
2007	20,505	3271.18	591.1	7132.9	16,128	245.1	549.3	6515.26	16,638.6
2006	16,738	2858.8	533.8	6222	14,317	209.3	528.7	6002.7	14,271.9
2005	14,368	2494.03	476	5171.8	12,428	152.6	475.4	5447.8	8791
2004	12,487	2197.14	389.8	4423.9	11,294.7	123.8	328.6	4782.7	1873.6
2003	10,666	1903.16	326.2	3914.1	10,163.7	128.4	318.1	4110.4	843.4
2002	9506	1646.54	302	3565.7	9102.4	154.9	229.6	3551.3	312.6
2001	8717	1472.35	279.4	3384.9	8547.5	120.3	238.4	3296.8	123.7
2000	7942	1347.24	260.6	3193.2	8178.1	101.8	204.3	3139.4	38.9
1999	7229	1230.52	228.4	2987.7	7533.4	91	197.8	2908.6	8.1
1998	6860	1159.84	196.3	2772.8	6739.6	77.2	175.4	2991.3	0

The fitting results of the three methods are shown in Figure 3. The overall fitting effect of ELM is better than that of BPNN, which is mainly because the forecasting effect of BPNN largely depends on the quality of historical data and the training of a large number of data. The ELM method is an improved BPNN, which maintains the advantages of the fast learning speed and strong generalization ability, and can accurately analyze in the case of small amount of data. The CSO-ELM method combines the global optimization ability of CSO and optimizing transmission weight of ELM, contributing to a better forecasting effect.

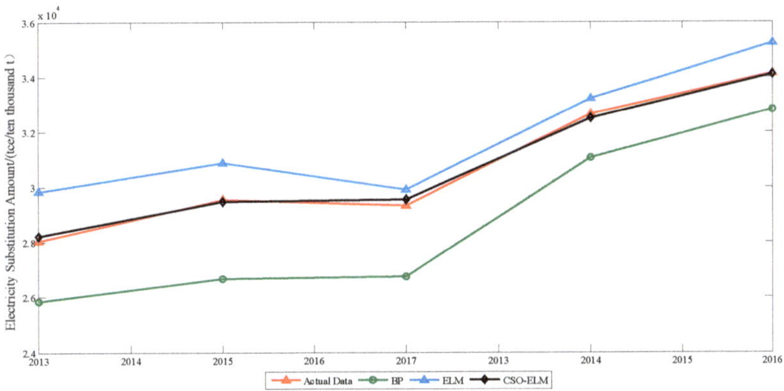

Figure 3. Results of three forecasting models.

To quantify the effect of the forecasting model, Root Mean Square Error (RMSE), Mean Absolute Percentage Error (MAPE), Mean Absolute Error (MAE) and Determining Factor R^2 are selected as indicators to evaluate the effect of the forecasting model. The calculation of the four error indicators are as follows.

$$RMSE = \sqrt{\frac{1}{n}\sum_{i=1}^{n}(p_i - p'_i)^2}, \tag{15}$$

$$MAPE = \frac{1}{n}\sum_{i=1}^{N}\left|\frac{p'_i - p_i}{p_i}\right| \times 100\%, \tag{16}$$

$$MAE = \frac{1}{n}\sum_{i=1}^{N}|p'_i - p_i|, \tag{17}$$

$$R^2 = \frac{\sum_{i=1}^{n}\left(p'_i - \overline{p}_i\right)^2}{\sum_{i=1}^{n}\left(p_i - \overline{p}_i\right)^2}, \tag{18}$$

where, p_i represents the actual electric power substitution amount. p'_i is the forecasting amount. $\overline{p}_i$ is the average of the actual electric power substitution amount, and n is the data size. The calculation results of four indicators are shown in Table 5 and the contrast is shown in Figure 4.

Table 5. Calculation results of error indicators.

Models	RSME	MAPE	MAE	R^2
BPNN	2186.66	7.00	2106.62	2.35
ELM	1168.72	3.56	1071.04	1.083
CSO-ELM	148.10	0.44	132.08	0.92

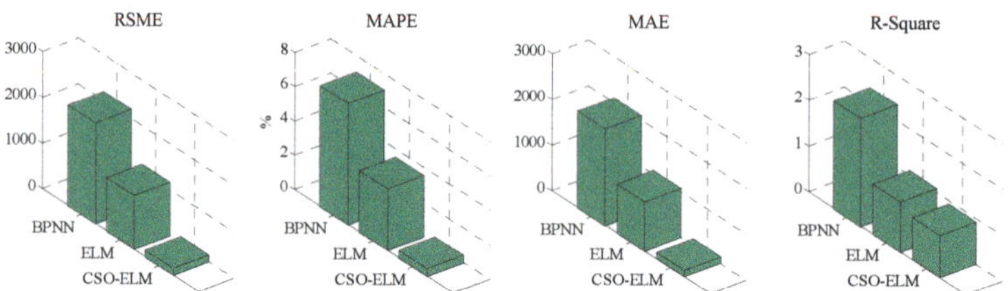

Figure 4. Errors of three methods.

From Table 5 and Figure 4, the constructed CSO-ELM model has a better fitting effect, higher prediction accuracy and less error in the market potential forecasting of electric power substitution. Therefore, using CSO-ELM model to forecast the substitution potential in the next 13 years can reflect the future market potential of electric power substitution projects.

4.2. Scenarios Setting

In forecasting the developing potential of future electric power substitution projects, factors such as the costs, policy support and subsidy mechanism are fully considered. Thus, combined with the political objectives of energy-consumption structure optimization, energy conservation and emission reduction, and renewable energy development, four scenarios are designed: basic scenario, high-cost restraint scenario, policies-supported scenario and subsidy weakening scenario.

In basic scenario, according to the development trend of power substitution related factors in 1998–2018, the market development potential of electric power substitution in 2019–2030 is forecasted. In the high-cost restraint scenario, the substitution project will be slowed down by the purchase of expensive equipment and high operating costs. In the policies-support scenario, in order to reduce the cost resistance and alleviate the environmental pressure caused by fossil energy combustion, the government encourages the substitution work by policy mechanism. China's "13th Five-Year Plan for Electric Power Industry" proposed to encourage the developing of electric power substitution and clean energy substitution to expand the proportion of electricity in energy consumption. In addition, the substitution work is supported with appropriate subsidies mechanisms. Relevant subsidy mechanisms can stimulate the rapid development of electric power substitution projects. With the popularization of electric power substitution, the cost would be gradually recovered, and the projects may be profitable, which is when the subsidy mechanism can be weakened or cancelled accordingly, to achieve an independent development of the substitution projects.

In addition, considering the stage characteristics of social and economic development, the forecasting period is divided into three stages: the first stage (2019–2020), the second stage (2021–2025) and the third stage (2026–2030). The parameter settings for each scenario and time period are shown in the Table 6 below.

Table 6. Scenarios and primary parameter settings.

Primary Parameters	Basic Scenario	High-Cost Restraint Scenario	Policies-Supported Scenario	Subsidy Weakening Scenario
Per Capita GDP	China's future economy will be at the stage of medium-high speed growth, with GDP per capita growing about 6.4% to 7.2% each year. GDP per capita is an objective condition and may barely change in all four scenarios.			
Electricity Consumption	In view of the trend of China's power consumption, the electricity consumption is in an increasing trend with a rate of 4%.	At first, electricity consumption will continue to grow at a rate of about 4% in the short term. Then, due to the high cost of the substitution projects, the growth rate will be slightly lower than that in the basic scenario. So, the increase rate is about 3.8%.	Relevant industrial policies and subsidy mechanism will bring positive effects. Electricity consumption will steadily increase with a growth trend of about 4.2–4.4%.	The weakening of subsidies only occurs in the later stage, so the impact on electricity consumption is not significant. The growth rate of electricity consumption can still be maintained at 4.2% to 4.4%.
Renewable Energy Generation	The growth of renewable energy consumption may bring an increase in electricity consumption. Based on the increasing trend, the growth rate is about 10% to 12%.	The high cost of electric power substitution projects brings constraints. Without political supports, the development of renewable energy power is slower, with an annual growth rate of about 9.5%.	Chinese government proposed to achieve the target of 15% and 20% of substituting fossil energy by non-fossil energy in 2020 and 2030 respectively. With proper subsidies, the growth rate can reach 12% to 13%.	The goal of developing renewable energy remains unchanged, but it is assumed that the subsidy coefficient of renewable energy will gradually decrease in the next 13 years. The growth rate is slightly lower, about 11.8% to 12.5%.
Electric Power Installed Capacity	With the gradual saturation of power-installed capacity, the installed capacity shows the trend of deceleration increase, at a rate of 4% to 4.6%.	Due to the high cost of electric power substitution projects and the risk of loss (no subsidy given), the growth rate of electric energy consumption is slower. The construction of newly-built electric assets will be affected, and the growth rate will slow down to about 3.2% to 4.3%.	Renewable energy may increase its installed capacity. The growth rate of installed capacity is about 4.1% to 4.8%.	Similar to the policies-supported scenario.

Table 6. *Cont.*

Primary Parameters	Basic Scenario	High-Cost Restraint Scenario	Policies-Supported Scenario	Subsidy Weakening Scenario
Per Capita GDP	China's future economy will be at the stage of medium-high speed growth, with GDP per capita growing about 6.4% to 7.2% each year. GDP per capita is an objective condition and may barely change in all four scenarios.			
Renewable Energy Installed Capacity	Electric power substitution strategy brings positive effects on developing renewable energy However, the growth rate, which is about 8.3% to 9%, is limited by the social demand.	If technology improvement is the only driving force and no subsidy incentive is given, the development of renewable energy will be restrained. The growth rate of renewable energy installed capacity will be slow, about 5.2%.	Supported by policies and subsidies, the renewable energy installed capacity will increase. The increasing rate is about 8.4% to 12.54%.	The renewable energy installed capacity will be slightly lower than that in the policies-supported scenario, and the increasing rate is about 8% to 12%.
Annual Increment Investment in Power Grid	The increase in renewable energy power will bring investments in power grid construction. The growth rate of power-grid construction investment is about 1.7% to 2.3%.	Compared with the basic scenario, less investment will be made to construct the power grid for accessing renewable power, so the growth rate is about 1.6% to 2%.	Same as the basic scenario.	Same as the basic scenario.
Annual Increment Investment in Electric Power Industry	According to historical data from 1998 to 2017, the growth rate of the annual investment in power industry will gradually decrease in the future, about 5.2% to 6.1%.	The volume ratio of the annual investment under this scenario is slightly lower than that of the basic scenario, which is about 4.1% to 4.4%.	Due to subsidy mechanism, more investment enters into power industry. The growth rate of electricity investment will slow down in the future, which is about 6.2% to 7.3%.	The investment will increase progressively at a rate about 6.2% to 7.1%, which is slightly lower than that of the policies-supported scenario.
Carbon Emissions	According to the carbon emission situation in the past 20 years, assume that the trend of carbon emission reduction in the future is about 2.1% to 2.5%.	While high cost brings resistance to the development of renewable energy, it will also weaken the environmental benefits and reduce the efficiency of carbon emission reduction indirectly. The trend of emission reduction is about 1.4% to 1.7%.	The development of renewable energy will effectively reduce carbon emissions. Therefore, in this scenario, the reduction rate of carbon emissions is about 3.6% to 4%.	The decreasing rate is slightly lower than that in policies-supported scenario, at about 3.6% to 3.8%

4.3. Results of Scenarios and Discussion

In the four different scenarios, the amount of electric power substitution shows significant growth trends before 2030. Affected by electricity consumption, renewable energy generation and other factors, the market potential of electric power substitution has broad market prospects. Electric energy can effectively reduce carbon emissions to a minimum through scale-effect and technological means (such as smart grid) in power production and transmission section. Therefore, the realization of electricity substitution is a low-carbon energy development and utilization strategy as a whole, which will inevitably have a positive impact on China's low-carbon economy. Restricted by high construction costs and operation fees, people's subjective acceptance of electric power substitution project is relatively low, which brings difficulties in popularizing. The forecasting results are shown in Table 7.

Table 7. Multi-scenarios forecasting result.

Year	Substituting Amount			
	Basic Scenario	High-Cost Restraint Scenario	Policies-Supported Scenario	Subsidy Weakening Scenario
2019	382.7618	373.5561	385.4891	385.618
2020	396.7616	378.3581	412.536	415.7296
2021	415.1382	389.6555	429.1873	428.0477
2022	422.5893	401.019	440.9279	441.7451
2023	442.4443	417.0991	463.5306	463.9663
2024	472.7527	431.8294	491.1256	492.1716
2025	493.6642	445.6533	514.5227	515.4268
2026	523.7452	458.8113	543.0128	546.7548
2027	554.6676	473.256	576.6824	587.3033
2028	584.2509	489.9646	608.059	618.2991
2029	604.934	524.5843	640.7256	652.0776
2030	628.2181	538.1725	680.3083	693.8293

In all four scenarios, substituting electricity shows significant increase. In the first stage, the growth rate is relatively small. During this period, the subsidies of the subsidy weakening scenario have not been weakened, so there is little difference with other scenarios. In the second stage, due to social development and technological progress, electric power substitution projects begin to have a certain scale of promotion. The substitute electricity in the high-cost restraint scenario shows the slowest increase. In the policies-supported scenario, the substitute electricity amount shows accelerated growth, while the subsidy mechanism in the subsidy weakening scenario begins to fade down slightly. In the third stage, after the promotion of the first two stages, the electric power substitution project has a certain scale effect, and the electric power substitution quantity shows a trend of accelerating growth.

As can be seen in Figure 5, the process of electric power substitution in the policies-support scenario and subsidy weakening scenario is significantly higher than that in the basic scenario, which shows that the government support is very important in substitution promoting work. In addition, the substitution amount in the high-cost restraint scenario is the least one in all four scenarios, about 538.17 million tons of standard coal in 2030. High costs lower the residents' acceptance of projects, and the promotion power of manufacturers is insufficient without any subsidies given. In the policies-support scenario, the government supports by giving subsidies in equipment purchasing, installation, operation and maintenance process—so that the projects are promoted—with the most increment in electricity amount, reaching 693.8 million tons of standard coal in 2030. Differing from the policies-support scenario, changes after power substitution reaching certain scale effects are considered in the subsidy weakening scenario. With the popularization of electric power substitution projects and technical progress, government support gradually reduced to raise independence of the substitution industry, as can be seen from the curve of the subsidy weakening scenario in Figure 5. The increasing trend of substitute electricity under the subsidy weakening scenario is similar to that of the policies-support scenario. With the reduction of subsidies, the amount of electric power substitution become less than that in the policies-support scenario after 2023, but still higher than that in the basic scenario, which shows the effectiveness of subsidy weakening.

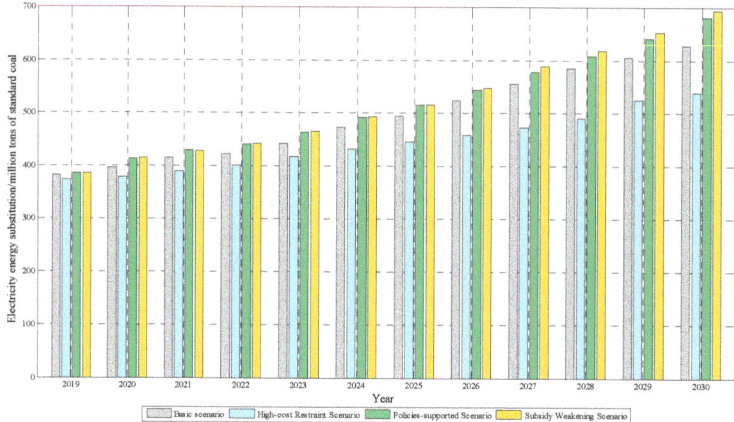

Figure 5. Forecasting results of electric power substitution under all scenarios (million tons of standard coal).

In the early stage of developing electric power substitution, the main fields for promotion are concentrated on substitutions of coal, coal-fired boilers and coal-fired heating—where technology is relatively mature. In the middle and later stages, electric power substitution work turns to household electrification, substitution to internal-combustion engines by electricity and substitution to oil-fired vehicle by electric vehicle. Implementing electric power substitution will bring significant changes in China's energy consumption structure, significantly reducing the proportion of oil and natural gas consumption and significantly reduce carbon emissions.

5. Conclusions

Electric power substitution is important in environmental management and for optimizing energy structure. Market forecasting on electric power substitution projects is conducive to the government and society to develop and promote the substituting work. Therefore, a market potential forecast model of electric power substitution based CSO-ELM method is proposed. Firstly, the validity of the influencing factors on the potential of power substitution development are verified through correlation test. Secondly, combined with CSO algorithm, the ELM method is improved, which overcomes the limitation of the search ability of the original ELM and improves the forecasting accuracy. In addition, compared with BP, ELM and other forecasting models, the proposed CSO-ELM model has lower error and deviation, showing better forecasting effect. Due to government support and the limitations of industrial development, four scenarios were designed to give specific forecasting results to show how different factors influence the development of electric power substitution projects. In addition, the forecasting time period was divided into three stages to take on further analysis. Through comparative analysis of the forecasting results under different scenarios, some suggestions for promoting electric power substitution are proposed below.

(1). Compared with the direct utilization of coal, oil and other traditional energy sources on the user side, electricity production has lower carbon emissions. Therefore, it is of great significance to promote social energy conservation and emission reduction, and to solve the Urban Haze problem as well.

(2). Subsidies would help reduce the actual costs for users in applying electric power substitution projects, which would arouse users and producers' initiative. For improving the economic benefits, proper electricity price mechanism and corresponding subsidies should be made to expand the market scale of substitution projects.

(3). The forecasting results in the subsidy weakening scenario show that the substituting electricity amount keeps increasing with weakened subsidies due to the scale effect built in the early stage with full subsidy mechanism. Therefore, while promoting, the government can change the subsidy coefficient flexibly according to the economic benefits of electric power substitution projects. Adjustable subsidy coefficient can also reduce the financial burden of the government, and encourage the industry to achieve independent development.
(4). Instead of applying renewable energy power into electricity energy substitution, renewable power can be used in the generation side to substitute the thermal units as well. The application of cleaner power will help optimize the structure of the whole power industry chain, and realize high proportion and high efficiency utilization of renewable energy.
(5). Another way to promote the development of renewable energy is to improve the generation cost of traditional generators. By improving the standard of pollutant emission payments and applying Renewable Portfolio Standards (RPS) and green power certifications, the utilization of renewable energy will be further improved.

Author Contributions: Conceptualization, J.W. and G.D.; methodology, K.W.; software, K.W.; validation, G.D., Z.T. and L.P.; formal analysis, J.W.; investigation, Q.T.; resources, L.J.; data curation, K.W.; writing—original draft preparation, J.W.; writing—review and editing, G.D. and Q.T.; K.W.; supervision, Z.T.; project administration, G.D. and L.J.; funding acquisition, Z.T. and L.J.

Funding: This work was partially supported by Supported by the Project funded by China Postdoctoral Science Foundation (2019M650024), the National Nature Science Foundation of China (Grant Nos. 71904049,71874053, 71573084), the Beijing Social Science Fund(18GLC058) and the 2018 Key Projects of Philosophy and Social Science Research, Ministry of Education, China (18JZD032).

Acknowledgments: Teachers and classmates helped complete this paper. We would like to express our gratitude to them for their help and guidance.

Conflicts of Interest: The authors declare no conflict of interest.

References

1. Zhihua, W.; Peng, A.; Fangting, X. The coordination of three institutions to promote electric power replacement. *Power Demand Side Manag.* **2016**, *18*, 103–104.
2. Qing, W.; Feng, W.; Yan, M. The green power supply office construction based on clean alternative and electric power alternative. *Power Demand Side Manag.* **2016**, *18*, 36–39.
3. Pu, L.; Wang, X.; Tan, Z.; Wu, J.; Long, C.; Kong, W. Feasible electricity price calculation and environmental benefits analysis of the regional nighttime wind power utilization in electric heating in Beijing. *J. Clean. Prod.* **2019**, *212*, 1434–1445. [CrossRef]
4. Shou-bo, X.U. Review of reformation of China's energy development strategy. *Power Syst. Clean Energy* **2008**, *24*, 1–5.
5. Ming, L.; Diangang, H.; Youxue, Z. Research and practice of renewable energy local consumption mode in Gansu Province based on "double alternative" strategy. *Power Grid Technol.* **2016**, *40*, 2991–2997.
6. Song, X.; Kening, C.; Li, M.; Menghua, F. Key Issues in China's Demand Side Resources Promoting Non-Hydro Renewable Power Intermittent. *J. Tech. Econ. Manag.* **2018**, *4*, 87–92.
7. Haibo, P.; Dezhi, L.; Wanjiao, H.; Wangzhu, G. Electric power replacement technologies facing new energy accommodation. *Power Demand Side Manag.* **2016**, *18*, 45–48.
8. Cai, W.; Liu, C.; Lai, K.H.; Li, L.; Cunha, J.; Hu, L. Energy performance certification in mechanical manufacturing industry: A review and analysis. *Energy Convers. Manag.* **2019**, *186*, 415–432. [CrossRef]
9. Cai, W.; Li, L.; Jia, S.; Liu, C.; Xie, J.; Hu, L. Task-Oriented Energy Benchmark of Machining Systems for Energy-Efficient Production. *Int. J. Precis. Eng. Manuf. Green Tech.* **2019**, 1–14. [CrossRef]
10. Wu, J.; Yan, J.; Jia, H.; Nikos, H.; Ned, D.; Hongbin, S. Integrated Energy Systems. *Appl. Energy* **2016**, *167*, 155–157. [CrossRef]
11. Hui, S.; Xiaoping, W. Research on Renewable Energy Substitute Model Based on System Dynamics in China. *Math. Pract. Theory* **2013**, *10*, 58–70.

12. Qian, S.; Xiaodong, L. Identifying the underpin of green and low carbon technology innovation research: A literature review from 1994 to 2010. *Technol. Forecast. Soc. Chang.* **2013**, *80*, 839–864.
13. He, H.; Jim, C.Y. Coupling model of energy consumption with changes in environmental utility. *Energy Policy* **2012**, *43*, 235–243. [CrossRef]
14. Raul, A.B. Fossil fuels, alternative energy and economic growth. *Econ. Model.* **2018**, *75*, 196–220.
15. Yan, L.; Bo, Y.; Fuyan, H. Study on renewable energy dynamic state growth model on sustainable development theory. *China Soft Sci.* **2011**, *S1*, 240–246.
16. Pina, A.; Baptista, P.; Silva, C.; Ferrão, P. Energy reduction potential from the shift to electric vehicles: The Flores island case study. *Energy Policy* **2014**, *67*, 37–47. [CrossRef]
17. Bhupendra, K.V.; Sangle, S. What drives successful implementation of pollution prevention and cleaner technology strategy? The role of innovative capability. *J. Environ. Manag.* **2015**, *155*, 184–192. [CrossRef]
18. Zhong, W.; Qiang, L.; Hongtao, X. Efficiency evaluation of alternative energy in heat supply sustem. *J. Zhejiang Univ. Technol.* **2015**, *43*, 508–511.
19. Pokharel, S. Promotional issues on alternative energy technologies in Nepal. *Energy Policy* **2003**, *31*, 307–318. [CrossRef]
20. Zuofeng, L. The research and practice of electric energy alternative in Jiangsu. *Power Demand Side Manag.* **2016**, *18*, 1–3.
21. Xiaoli, L.; Wenbing, L.; Haiming, Z. Alternative energy in the energy network. *Smart Grid* **2015**, *3*, 1192–1196.
22. Wencong, L.; Changcai, X. The effect of carbon emissions trading market on the clean production in various regions in China [J/OL]. *Acta Ecol. Sin.* **2019**, *18*, 1–9. Available online: http://kns.cnki.net/kcms/detail/11.2031.Q.20190704.1606.068.html (accessed on 21 August 2019).
23. Wei, Y.; Xia, L.; Pan, S.; Wu, J.; Zhang, X.; Han, M.; Zhang, W.; Xie, J.; Li, Q. Prediction of occupancy level and energy consumption in office building using blind system identification and neural networks. *Appl. Energy* **2019**, *240*, 276–294. [CrossRef]
24. Kostmann, M.; Härdle, W.K. Forecasting in Blockchain-Based Local Energy Markets. *Energies* **2019**, *12*, 2718. [CrossRef]
25. Li, Y.; Han, Y.; Wang, J.; Zhao, Q. A MBCRF Algorithm Based on Ensemble Learning for Building Demand Response Considering the Thermal Comfort. *Energies* **2018**, *11*, 3495. [CrossRef]
26. Zhouchun, W.; Xiaochen, Z.; Yuqing, M.; Xinyan, Z. A hybrid model based on modified multi-objective cuckoo search algorithm for short-term load forecasting. *Appl. Energy* **2019**, *237*, 896–909.
27. Changhao, X.; Jian, W.; McMenemy, K. Short, medium and long term load forecasting model and virtual load forecaster based on radial basis function neural networks. *Electr. Power Energy Syst.* **2010**, *32*, 743–750.
28. Shan, Y.; Fu, Q.; Geng, X.; Zhu, C. Combined forecasting of photovoltaic power generation in microgrid based on the improved BP-SVM-ELM and SOM-LSF with particlization. *Proc. CSEE* **2016**, *36*, 3334–3343.
29. Chen, K.; Yu, J. Short-term wind speed prediction using an unscented Kalman filter based state-space support vector regression approach. *Appl. Energy* **2014**, *113*, 690–705. [CrossRef]
30. Lee, Y.; Tong, L. Forecasting energy consumption using a grey model improved by incorporating genetic programming. *Energy Convers. Manag.* **2011**, *52*, 147–152. [CrossRef]
31. Ferlito, S.; Adinolfi, G.; Graditi, G. Comparative analysis of data-driven methods online and offline trained to the forecasting of grid-connected photovoltaic plant production. *Appl. Energy* **2017**, *205*, 116–129. [CrossRef]
32. Yu, F.; Xu, X. A short-term load forecasting model of natural gas based on optimized genetic algorithm and improved BP neural network. *Appl. Energy* **2014**, *134*, 102–113. [CrossRef]
33. Guochang, F.; Lixin, T.; Min, F.; Mei, S. Impacts of new energy on energy intensity and economic growth. *Syst. Eng. Theory Pract.* **2013**, *33*, 2795–2803.
34. Qingyou, Y.; Mingliang, Z.; Xinfa, T. Electrical energy alternative research based on the cost utility analysis. *Oper. Res. Manag. Sci.* **2015**, *24*, 176–183.
35. Yi, S.; Mo, S.; Baoguo, S.; Fang, C. Electric energy substitution potential analysis method based on particle swarm optimization support vector machine. *Power Syst. Technol.* **2017**, *41*, 1767–1771.
36. Hang, Y. *Energy Conservation and Emission Reduction Environment Power to Substitute Other Energy Evaluation Method Research*; North China Electric Power University: Beijing, China, 2013.
37. Jincheng, Z. *Analysis on Potential and Environmental Benefit of Rural Power Substitute*; North China Electric Power University: Beijing, China, 2015.

38. Yanmei, L.; Zeng, C. Study on Regional Electric Energy Substitution Potential Evaluation Based on the TOPSIS Method Improved by Connection Degree. *Power Syst. Technol.* **2019**, *43*, 687–695.
39. Dongli, C.; Yue, Y.; Zhixiang, L. Application and Efficiency Evaluation of Alternative Energy. *Power Syst. Clean Energy* **2011**, *27*, 30–34.
40. Yang, X.S.; Deb, S. Cuckoo search via Lévy flights. In Proceedings of the 2009 World Congress on Nature & Biologically Inspired Computing (NaBIC 2009), Coimbatore, India, 9–11 December 2009; pp. 210–214.

 processes

MDPI

Article

Rapid and Green Preparation of Multi-Branched Gold Nanoparticles Using Surfactant-Free, Combined Ultrasound-Assisted Method

Phat Trong Huynh [1,2], Giang Dang Nguyen [1], Khanh Thi Le Tran [1], Thu Minh Ho [1], Vinh Quang Lam [2] and Thanh Vo Ke Ngo [1,*]

1 Research Laboratories of Saigon Hi-tech Park, Ho Chi Minh City 700000, Vietnam; phat.huynhtrong@shtplabs.org (P.T.H.); giang.nguyendang@shtplabs.org (G.D.N.); khanh.tranthile@shtplabs.org (K.T.L.T.); thu.hominh@shtplabs.org (T.M.H.)

2 Faculty of Physics and Engineering Physics, VNU.HCM, University of Science, Ho Chi Minh City 700000, Vietnam; lqvinh@vnuhcm.edu.vn

* Correspondence: thanh.ngovoke@shtplabs.org; Tel.: +(84)-373608890

Abstract: The conventional seed-mediated preparation of multi-branched gold nanoparticles uses either cetyltrimethylammonium bromide or sodium dodecyl sulfate. However, both surfactants are toxic to cells so they have to be removed before the multi-branched gold nanoparticles can be used in biomedical applications. This study describes a green and facile method for the preparation of multi-branched gold nanoparticles using hydroquinone as a reducing agent and chitosan as a stabilizer, through ultrasound irradiation to improve the multi-branched shape and stability. The influence of pH, mass concentration of chitosan, hydroquinone concentration, as well as sonication conditions such as amplitude and time of US on the growth of multi-branched gold nanoparticles, were also investigated. The spectra showed a broad band from 500 to over 1100 nm, an indication of the effects of both aggregation and contribution of multi-branches to the surface plasmon resonance signal. Transmission electron microscopy measurements of GNS under optimum conditions showed an average core diameter of 64.85 ± 6.79 nm and 76.11 ± 14.23 nm of the branches of multi-branched particles. Fourier Transfer Infrared Spectroscopy was employed to characterize the interaction between colloidal gold nanoparticles and chitosan, and the results showed the presence of the latter on the surface of the GNS. The cytotoxicity of chitosan capped GNS was tested on normal rat fibroblast NIH/3T3 and normal human fibroblast BJ-5ta using MTT assay concentrations from 50–125 μg/mL, with no adverse effect on cell viability.

Keywords: green synthesis; multi-branched gold nanoparticles; ultrasound; hydroquinone; chitosan

Citation: Huynh, P.T.; Nguyen, G.D.; Thi Le Tran, K.; Minh Ho, T.; Lam, V.Q.; Ngo, T.V.K. Rapid and Green Preparation of Multi-Branched Gold Nanoparticles Using Surfactant-Free, Combined Ultrasound-Assisted Method. *Processes* **2021**, *9*, 112. https://doi.org/10.3390/pr9010112

Received: 5 November 2020
Accepted: 5 January 2021
Published: 7 January 2021

1. Introduction

Anisotropic gold nanoparticles are of interest in physics, chemistry, and optics, due to their unique chemical and physical properties [1–5]. In addition, they are biocompatible and less toxic [6–9]. Recently, multi-branched gold nanoparticles (GNS) attracted much attention because of their absorption in the NIR region of the electromagnetic spectrum [10]. These applications are based on the LSPR phenomenon, caused by excitation with electromagnetic radiation of the NPs [11]. The characteristics of the LSPR band, including the width, peak position, and intensity, are directly related to particles size, morphology, and properties of capping agents [12]. Depending on the nature of the GNS, the absorption spectrum is composed of a weak band at 520–550 nm and a broad band between 700–1100 nm [10]. The band at the shorter wavelengths is attributed to the transverse plasmon resonance, whereas the longitudinal component is at the longer wavelengths. Due to the shape of the anisotropic particles, there are contributions to the plasmon resonance spectrum from both the transverse and longitudinal directions, with the latter being predominant, as the core size decreased and the length of the arms increased. However, in circumstances where the core diameters of the particles are longer than the arms, the band broadening could be due to particle aggregation, where the neighbor core plasmon resonance interact. Highly branched gold nanoparticles with small cores are characterized by broad and red-shifted LSPR maximum, with very weak or absent bands between 520–550 nm.

Multi-branched gold nanoparticles or gold nanostars are usually synthesized by the seed-mediated method, in the presence of anionic or cationic surfactants [13,14]. They take the role of shape-directing and capping agents to drive structural growth and prevent the aggregation of gold nanoparticles. The presence of surfactants result in cellular toxicity and they are also difficult to remove before the particles are used [15,16]. Many reports of surfactants-free GNS in recent years [17,18] are mainly based on seed-mediated protocols. The use of sodium borohydride, hydrazine, or trisodium citrate in seed-mediated synthesis of GNS, results in cytotoxicity and they are harmful to the environment.

Recently, green chemistry has become a popular trend in a variety of fields, as it offers a number of advantages, including safety, energy efficiency, and the production of less toxic waste [19–21]. In green synthesis of nanoparticles, people evaluate and select new nontoxic reductants, protecting agents, innocuous solvents from nature (leaf extract, microorganism, or nature polymer) to replace toxic materials [22,23], and develop advanced and energy-efficient techniques such as microwave and sonochemical methods [24,25].

Sonochemical effects are caused by acoustic cavitation in liquids with the creation and collapse of bubbles. The collapse of the cavitation bubbles is more rapid than thermal transport and generates "localized hot spots", which have a temperature of 5000 K, pressures of about 2000 atm and cooling rates of more than 109 K/s [26]. The advantages of using the sonochemical approach include the production of high purity, uniform shape, high yields, and cost effective synthesis of nanoparticles.

Several one-pot synthesis techniques of multi-branched gold nanoparticles were studied [27–29]. The advantages of these procedures are facile and free-surfactant synthesis but just short-branches gold nanostars were formed (less than 10 nm branches). Sonochemical synthesis is a popular method for spherical nanoparticles [30]. However, it is rarely studied using ultrasound assist to synthesize gold nanostars, except reports of Badilescu et al. [31]. In this work, we introduce a rapid and green preparation of multi-branched gold nanoparticles using surfactant-free and seedless combined ultrasound (US) assisted protocol. Chitosan (CS) was used as a stabilizer while hydroquinone (HQ) was used as the reducing agent. CS is a polysaccharide derived from shrimps, crabs, and other crustaceans. Due to its many advantageous properties such as biocompatibility, nontoxicity, low-cost, biodegradability, and antimicrobial agent, CS has a number of commercial and biomedical uses. Additionally, HQ has a variety of uses as a reducing agent that is soluble in water. Furthermore, HA was used to replace a traditional reducing agent, ascorbic acid, which could tune the adsorption of anisotropic gold nanoparticles towards the far NIR region [32]. Additionally, sonochemical or ultrasound assistance was applied to control the size and shape, and enhance the stability of GNS.

2. Materials and Methods

2.1. Materials

Chloroauric acid ($HAuCl_4xH_2O$, ~52% Au basis), sodium hydroxide 97%; chitosan (low molecular weight), hydroquinone 99%, phosphate buffer saline (tablet, pH 7.2–7.6), acid acetic 99% were purchased from Sigma-Aldrich, St. Louis, Missouri, US. The materials for cytotoxicity include Dulbecco's Modified Eagle Medium (DMEM), Fetal Bovine Serum (FBS), Bovine Calf Serum (BCS) were supplied by Sigma-Aldrich, Darmstadt Germany. Thiazolyl blue tetrazolium bromide (MTT), and antibiotic solution (penicillin-streptomycin) were supplied by Sigma-Aldrich, St. Louis, Missouri, US. Lysis buffer solution was purchased from Biobasic, Markham, Ontario, Canada. Normal mouse fibroblast (NIH/3T3,

CRL-1658) and normal human fibroblast (BJ-5ta, CRL-4001) were ordered from ATCC, Manassas, Virgina, US. Deionized (DI) water 17.8 MΩ was used throughout the experiments.

2.2. Methods

2.2.1. Surfactant-Free Preparation of Multi-Branched Gold Nanoparticles (GNS) by the One-Step Method

First, 2 g chitosan (CS) powder was homogenized into a 98 g acid acetic 1% solution to make CS 2% solution. Next, 1 mL of $HAuCl_4$ 2×10^{-2} M solution was added to 9 mL chitosan solution, under stirring. This mixture was pH-adjusted by acetic acid. Finally, hydroquinone (HQ) 0.1 M was mixed immediately into this mixture. The reaction solution was kept constant for 30 min at room temperature. Throughout the period of reaction, the color of the solution changed from colorless to cobalt blue, indicating that multi-branched particles had formed. The influences of conditions including pH, mass concentration of CS, hydroquinone concentration, and morphology of GNS were studied.

2.2.2. Surfactant-Free Preparation of GNS Combined Ultrasound

The procedure of preparation for the combined ultrasound (US) was similar to the process above, except that the reaction solution was sonicated instead of being kept at room temperature without stirring. Sonication was carried out using the Q2000 sonicator—Qsonica, Newtown, Connecticut, US (power 1.375 watts, frequency 20 kHz). Chitosan-coated GNS was synthesized at a constant frequency of 20 kHz and six different levels of amplitude (0, 20, 40, 60, 80, and 100 μm) and six time-periods (0, 2, 4, 6, 8, and 10 min), in order to investigate the effect of amplitude and sonication time on size, shape, and stability of GNS.

2.2.3. Characterization

UV–Vis spectrophotometer Dynamica Halo RB-10 (Dynamica, Livingston, UK) was used to record the surface plasmon resonance (SPR) of GNS in the wavelength range of 400–1100 nm, at a scanning rate of 200 nm/min. The interaction between GNS and CS was shown by FT-IR analysis (Bruker Tensor 27, Bruker Optics, Ettlingen, Germany). All FT-IR results were obtained from the powder samples and smoothing or correction baseline was not applied. The crystal structure of GNS was determined by employing X-ray diffraction (XRD). Scanning was carried out in the 2 theta range of 20–100°, using the X-ray diffractometer Bruker D5005 (Bruker AXS, Karlsruhe, Germany). Transmission electron microscope (TEM) analysis was examined by JEM1010-JEOL (Jeol, Tokyo, Japan). The J-Image software (NIH Image) was used to calculate the average length of branches as well as diameter of cores of GNS, based on thirty particles of each three sample from the TEM images. All analyses including UV-Vis, FT-IR, XRD, and TEM, as well as the cytotoxicity test below were carried out through separation from the same sample. The GNS solutions were sonicated before examinations and measurements.

2.2.4. Cytotoxicity of Chitosan-Capped GNS

NIH/3T3 were cultured in DMEM (Dulbecco's Modified Eagle Medium) supplemented with 1% penicillin–streptomycin (Pen–Strep) and 10% bovine calf serum (BCS), while the BJ-5ta cells were in DMEM with 1% Pen–Strep and 10% FBS (Fetal Bovine Serum. Both cell lines were grown in a humidified incubator with 5% CO_2. The cell lines were detached from the culture flasks using a trypsin (0.25%)—EDTA (0.53 mM) solution. Cell viability was evaluated using thiazolyl blue tetrazolium bromide (MTT). Both cell lines were seeded onto 96 well plates at 10^4 cells/well. CS-coated GNS was added to each well so that the final concentration samples were 50, 75, 100 and 125 μg/mL. The negative control subject was Lysis buffer. Cells were incubated for 24 h in an incubator (37 °C and 5% CO_2). Then, MTT solution (0.5 mg/mL) was added to each well. The plates were incubated at 37 °C for 4 h to form MTT formazan. Lysis buffer (4 mM HCl) was added to each well to dissolve the crystallized MTT formazan.

The plate absorbance (OD) was read at wavelength λ = 570 nm, using a Microreader. Cell viability could be calculated and compared to the control samples, as follows:

$$\% \text{ cell viability} = \frac{OD\ of\ sample}{OD\ of\ negative\ control} \times 100\% \quad (1)$$

3. Results and Discussion

3.1. One-Step Surfactant-Free Preparation of GNS

3.1.1. Effect of Mass Concentration of CS

Figure 1 shows that the UV–Vis result of GNS prepared in different mass concentration of CS. According to previous reports, the absorption spectrum was a plasmon resonance peak (SPR) ranging from 500 to over 1100 nm, indicating multi-branched particle presence. The SPR absorbance of GNS shifted, depending on the size and shape [33]. An increase in length of branches led to SPR shift toward the NIR region [34], while SPR absorbance moved to blue shift, due to the short branched formations [35]. On the other hand, broadening and increasing the maximum absorbance wavelength of SPR toward the NIR region suggested a form of aggregation, due to the interaction of the spherical core of the particles with each other [36]. It was clear that the intensity of SPR increased when the mass concentration of CS was increased from the beginning, evaluating the concentration to 1%. However, the SPR absorbance moved to the blue shift and the intensity of SPR decreased when mass concentration increased to 2%. This could be related to an increase in the size of the core particles. Table 1 describes the influence of % CS to SPR and intensity of SPR of GNS. At the beginning, 0.25% CS both of the SPR and the intensity of SPR rose from 845 nm and 0.15, to 875 nm and 0.18, respectively. With the increasing % CS, although the SPR increased slightly to 878, the intensity of SPR reached the highest value at 0.33. However, both SPR dropped to 667 nm and 877 nm, when % CS increased to 2. TEM images of GNS are shown in Figure 2. The prepared GNS at 1% CS (Figure 2a) had an average core of 57.33 ± 5.91 nm, and long, sharp branches with an average length of 44.32 ± 9.27 nm. Meanwhile at 1.5% CS, the average length of the branches decreased to 20.71 ± 8.57 nm (Table 2).

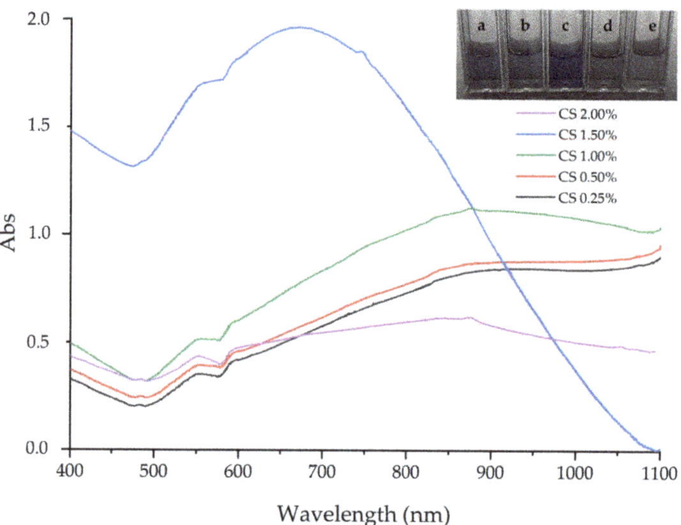

Figure 1. Absorption spectra of GNS prepared in various mass concentrations of CS: (**a**) 0.25%, (**b**) 0.50%, (**c**) 1.0%, (**d**) 1.5% and (**e**) 2.0%.

Table 1. The influence of mass concentration of chitosan to surface plasmon resonance and intensity absorbance.

% CS (*w*/*v*)	SPR (nm)	Int. of SPR
0.25	845	0.15
0.50	875	0.18
1.00	880	0.33
1.50	667	1.04
2.00	877	0.21

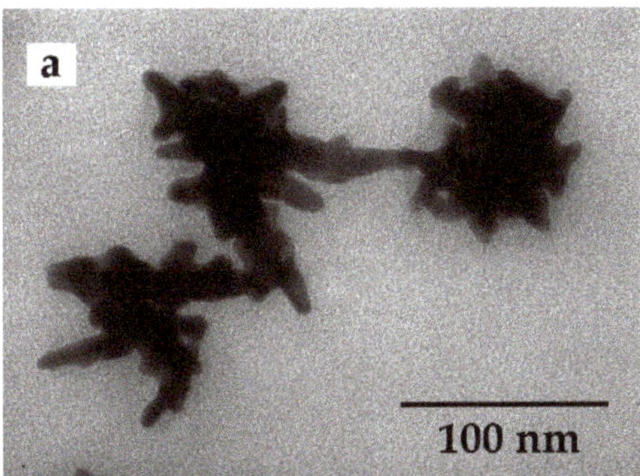

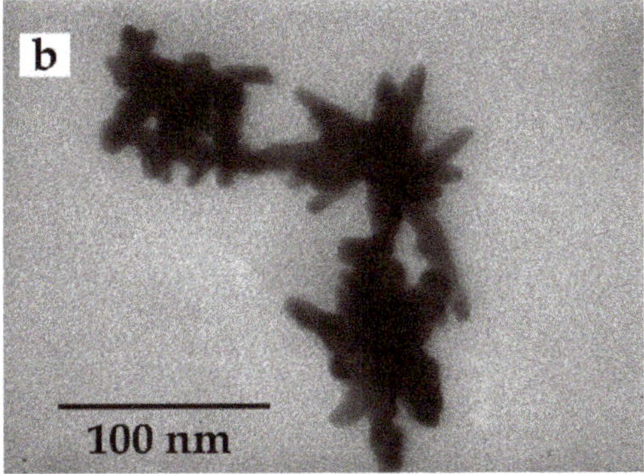

Figure 2. TEM images of GNS prepared in 1.0% (**a**) and 1.5% CS (**b**).

Table 2. The influence of mass concentration of chitosan on the morphology, average core and branches of GNS.

% CS (*w*/*v*)	Morphology	Core (nm)	Branches (nm)
1.00	Long multi-branches	57.33 ± 5.91	44.32 ± 9.27
1.50	Short multi-branches	51.38 ± 9.14	20.71 ± 8.57

3.1.2. Effect of pH

The role of pH on the formation of GNS was illustrated by UV–Vis absorption spectra (Figure 3) and TEM images (Figure 4). There was a rise of SPR absorbance and intensity of SPR, as pH 1.0 was adjusted to pH 1.5 and reached the highest value, 865 nm of absorbance and 0.49 of intensity. The SPR and intensity of SPR significantly declined when pH was adjusted to 2.0, 2.5, and finally 3.0. The SPR dropped to the lowest peak 833 nm, while intensity went down to 0.04, following pH adjustment (Table 3). As per the UV–Vis spectra and TEM examinations, the effect of both size of core and branches of multi-branched particles on SPR absorbance was shown. Adjusting pH from 1.0 to 2.0, the SPR absorbance spectra shifted to the NIR region and SPR intensity declined from 0.54 to 0.16. These were contributed by an increase of size of core and prolongated branches (Table 4). At pH 1.0, the GNS formed were short, multi-branch particles with 53.44 ± 8.30 nm average core and 36.23 ± 8.84 nm branches (Figure 4a). At pH adjusted to 1.5, the prepared GNS particles had elongated branches of 45.23 ± 10.03 nm (Figure 4a) and a core diameter of 59.32 ± 8.08 nm. Meanwhile, when the pH increased to 2.0, and the branches of GNS were short and unsharp (Figure 4b). At pH 3.0, a just, non-star shape were formed (few branches particles) and aggregation of the core of nanoparticles was observed from Figure 4c. Both TEM images and decrease in SPR intensity indicated that there was a greater contribution to the adsorption spectrum because of the core aggregation than the surface plasmon resonance from multibranches, at basic pH. The influence of pH to formation mechanism of gold nanoparticles could be explained by the relation between pH and HQ, as a reducing agent [37]. HQ can reduce Au^{3+} ions to Au^0 atoms by electrons produced in the oxidation/reduction process. Oxidation/reduction is a reversible process at equilibrium. At high acidic conditions (pH below 1.0), HQ could not reduce Au^{3+} to Au atoms, in the absence of gold seeds, because there were not any produced electrons. Adjusted to pH 1.5, few gold seeds appeared due to rapid reduced Au^{3+} ions. Furthermore, the reversible reaction promoted some electrons that reduced Au^{3+} to Au^+ ions. Under protection of CS, Au^+ attached to the gold seeds, and grew in anisotropic directions to form multi-branched particles. However, at pH towards basic condition (pH upper 2.0), the process became irreversible and was driven to the right. This promoted numerous electrons and uncontrolled gold nanoparticles were synthesized.

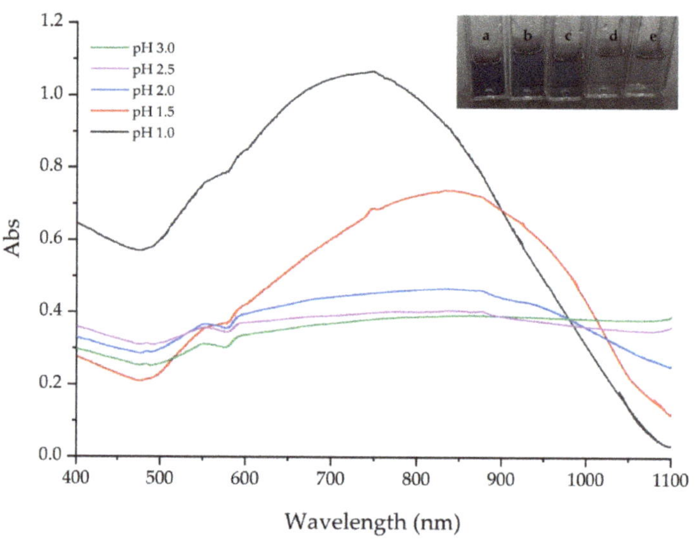

Figure 3. Absorption spectra of GNS prepared in different pH: (**a**) 1.0, (**b**) 1.5, (**c**) 2.0, (**d**) 2.5 and (**e**) 3.0.

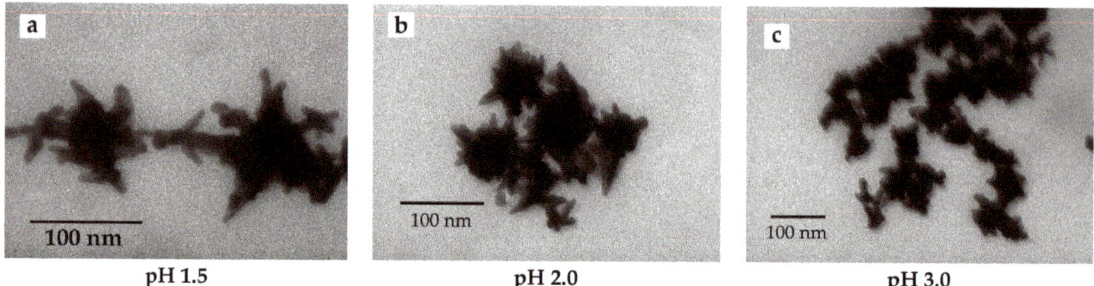

Figure 4. TEM images of GNS prepared in different pH: (**a**) 1.5, (**b**) 2.0, and (**c**) 3.0.

Table 3. The influence of pH to surface plasmon resonance and intensity absorbance.

pH	SPR (nm)	Int. of SPR
3.0	833	0.04
2.5	834	0.06
2.0	835	0.16
1.5	865	0.49
1.0	746	0.54

Table 4. The influence of pH to morphology, average core and branches of GNS.

pH	Morphology	Core (nm)	Branches (nm)
1.0	Long multi-branches	53.44 ± 8.30	36.23 ± 8.84
1.5	Long multi-branches	59.32 ± 8.08	45.23 ± 10.03
2.0	Short multi-branches	55.66 ± 12.28	19.55 ± 9.60
3.0	Few branches	99.89 ± 16.84	34.83 ±13.22

3.1.3. Effect of Hydroquinone

Different volumes of HQ 0.1 M were adjusted to investigate the effect of HQ for preparation of GNS. The absorption spectra of GNS prepared in a variety of HQ volume are displayed in Figure 5. There was an increase of intensity of SPR, from 0.40 to 1.16, and the SPR absorbance also shifted to 708 nm, as 1.0 to 2.0 mL of HQ volume was added (Table 5). The intensity of SPR declined rapidly to 0.67 and 611 nm of SPR absorbance, when amounts of 2.0 mL to 3.0 mL HQ 0.1 M were added, respectively. TEM micrographs in Figure 6 revealed morphology of three samples of GNS. At low volume of HQ (2.0 mL or less), GNS exhibited short multi-branched particles of 19.58 ± 5.19 nm, with an average core of 51.22 ± 6.67 nm. The 2.0 mL HQ volume resulted in long, sharp branches of GNS. The average branched length was 76.11 ± 14.23 nm and the core was 64.85 ± 6.79 nm. However, the branches of GNS were shorter when the HQ volume increased to 3.0 mL. This resulted in 55.15 ± 10.68 nm average core and 49.85 ± 7.42 nm in branched length (Table 6). Additionally, TEM revealed to the interaction of core particles. The influence of volume of HQ 0.1 M could be explained as follows. Gold seeds were first formed, then Au^{3+} ions were reduced to Au^{+}. Then, the Au^{+} ions attached to the gold seeds and grew in anisotropic directions to form multi-branched particles. At low HQ, the electrons produced by the reversibility of HQ was enough to promote seed-particles, meanwhile small amounts of Au^{+} ions appeared. Therefore, the short multi -branched particles were synthesized. At high HQ, not only seed particles formed but also all Au^{3+} changed to Au^{+} ions, resulting in the formation of long and sharp multi-branched particles. However, if the HQ was too high, many electrons were promoted, which led to an uncontrolled reaction [38].

Table 5. The influence of HQ to surface plasmon resonance and intensity absorbance.

Volume HQ 0.1 M (mL)	SPR (nm)	Absorbance
1.0	630	0.40
1.5	746	0.64
2.0	708	1.16
2.5	652	0.77
3.0	611	0.67

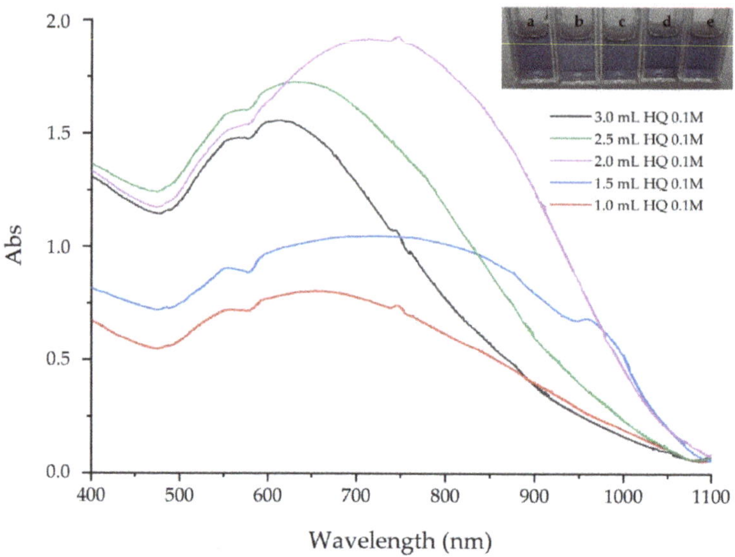

Figure 5. Absorption spectra of GNS prepared in various HQ 0.1 M volume—(**a**) 1.0 mL, (**b**) 1.5 mL, (**c**) 2.0 mL, (**d**) 2.5 mL and (**e**) 3.0 mL.

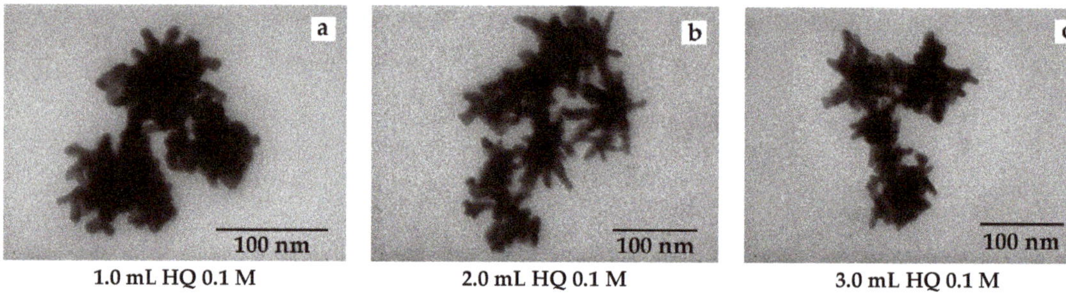

Figure 6. TEM images of GNS prepared in various HQ 0.1 M volume—(**a**) 1.0 mL, (**b**) 2.0 mL, and (**c**) 3.0 mL.

Table 6. The influence of HQ on morphology, average core and branches of GNS.

Volume HQ 0.1 M (mL)	Morphology	Core (nm)	Branches (nm)
1.0	Short multi-branches	51.22 ± 6.67	19.58 ± 5.19
2.0	Long multi-branches	64.85 ± 6.79	76.11 ± 14.23
3.0	Long multi-branches	49.85 ± 7.42	55.15 ± 10.68

3.2. *Ultrasound Combined Surfactant-Free Preparation of GNS*

3.2.1. Effect of Sonication Amplitude

UV–Vis results (Figure 7) and TEM analysis (Figure 8) were used to study the effect of amplitude sonication on the morphology and size of GNS synthesized in permanent time sonication, for 5 min. It was noticeable that the intensity of SPR of sonicated GNS was more intent than without the sonication handled sample. The intensity of SPR increased from 0.26 to 0.53, while the SPR absorbance rose from 831 nm to 877 nm, due to the changing amplitude from 20 to 60 μm (Table 7). Adjusting the amplitude to 80 μm, the branches of multi-branched particles shortened dramatically to 27.88 ± 5.83 nm, while the core diameter increased to 73.10 ± 24.66 nm. In addition, the aggregation of multi-branched particles was revealed due to the interaction of the spherical core with each other. Additionally, both SPR absorbance and intensity of SPR still had a high value, 874 nm and 1.14, respectively. This could be assigned to the greater contribution of the UV–Vis spectrum because of aggregation rather than the formation of surface plasmon resonance branches. The intensity of SPR significantly declined to 0.80, when amplitude was adjusted to d 100 μm, and the SPR absorbance fell to 690 nm. Synthesized GNS in 60 μm amplitude sonication have long and sharp multi-branches, whereas in prepared GNS in higher amplitude, the branches were short and un-harp (Table 8). Sonication power is the electrical energy supplied to the probe sonicator and transformed into mechanical energy. This is executed by exciting the piezoelectric crystals moving in the longitudinal direction, where the mechanical energy result in the probe vibrating up and down [39]. The stronger amplitude applied, the higher is the acoustic energy produced. When suitable amplitude sonication was applied, the reduction of Au^{+} to Au^{0} were accelerated, which resulted in higher reaction yields than without sonication-assisted sample, in the same time reaction. Nevertheless, an overly strong amplitude generated a high acoustic energy that led to uncontrolled rapid reduction, so the overall isotropic small particles reaction dominated the over multi-branched anisotropic.

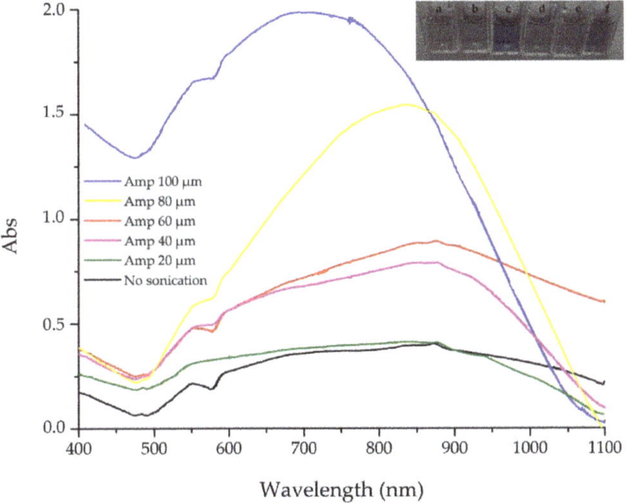

Figure 7. Absorption spectra of GNS prepared in different amplitudes—(**a**) 0 μm, (**b**) 20 μm, (**c**) 40 μm, (**d**) 60 μm, (**e**) 80 μm and (**f**) 100 μm.

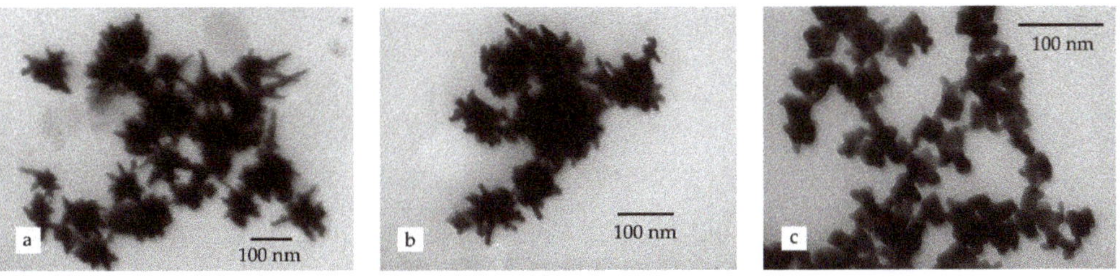

Figure 8. TEM images of GNS prepared in different amplitudes—(**a**) 60 μm, (**b**) 80 μm and (**c**) 100 μm.

Table 7. The influence of amplitude to surface plasmon resonance and intensity absorbance.

Amplitude (μm)	SPR (nm)	Absorbance
0	831	0.20
20	832	0.26
40	874	0.53
60	877	0.37
80	874	1.24
100	690	0.80

Table 8. The influence of amplitude sonication to morphology, average core and branches of GNS.

Amplitude (μm)	Morphology	Core (nm)	Branches (nm)
60	Long multi-branches	62.08 ± 7.32	67.73 ± 22.73
80	Short multi-branches	73.10 ± 24.66	27.88 ± 5.83
100	Few branches	28.26 ± 3.66	8.99 ± 2.58

3.2.2. Effect of Sonication Time

Figure 9 (the UV–Vis results) and Figure 10 (TEM images) show the influence of time sonication on the morphology and size of GNS prepared with a constant amplitude 50 μm. Both intensity and SPR absorbance increased when a longer sonication time was applied, and it reached the highest value at 6 min. Nevertheless, both decreased rapidly, despite a prolonged time sonication. Table 9 indicates that the intensity from 0.32 rose to 1.15 and SPR absorbance from 831 nm shifted to 833 nm, with an elongated time sonication of 6 min. The SPR absorbance shifted to the NIR region because both the size of core and the branched length increased. Synthesized GNS at 4 min exhibited an average branched length of 31.32 ± 7.62 nm and a core diameter of 52.02 ± 9.95 nm, that from 6 min had branches of 65.01 ± 11.39 nm and an average core of 75.34 ± 18.37. Intensity of SPR dropped to 0.44, while the SPR shifted significantly down to 630 nm. The time sonication of 8 min resulted in GNS with short and unsharp branches, besides, there was a rise of core diameter of multi-branched particles due to the interaction of each particle. Meanwhile, longer time sonication obtained gold nanoparticles that were short rod particles (Table 10). It is possible that the influence of time sonication might be based on cavitation bubbles. Sonication for a prolonged period that generated more cavitation bubbles led to an uncontrolled and rapid reduction, due to higher acoustic energy. As a result, the gold nanoparticles had a smaller core, and fewer and shorter branches were formed [40].

Table 9. The influence of time sonication to surface plasmon resonance and intensity absorbance.

US Time (min)	SPR (nm)	Absorbance
0	831	0.32
2	833	0.39
4	832	1.13
6	833	1.15
8	746	0.96
10	630	0.44

3.3. *Investigation of Interaction between CS and GNS*

The FT–IR of pure CS and CS-coated GNS spectra are shown in Figure 11. In the spectrum of pure CS, there was a band at around 2883 cm^{-1} that corresponded to the stretching vibration of the C-H groups. Two peaks located at 1649 cm^{-1} and 1324 cm^{-1} related to the C=O stretching vibration of amide I and C-N stretching of amide III, respectively [41]. These bands confirmed the N-acetyl groups of CS [42]. The band at 1589 cm^{-1} corresponded to the bending vibration of the N-H groups (amide II) [42].

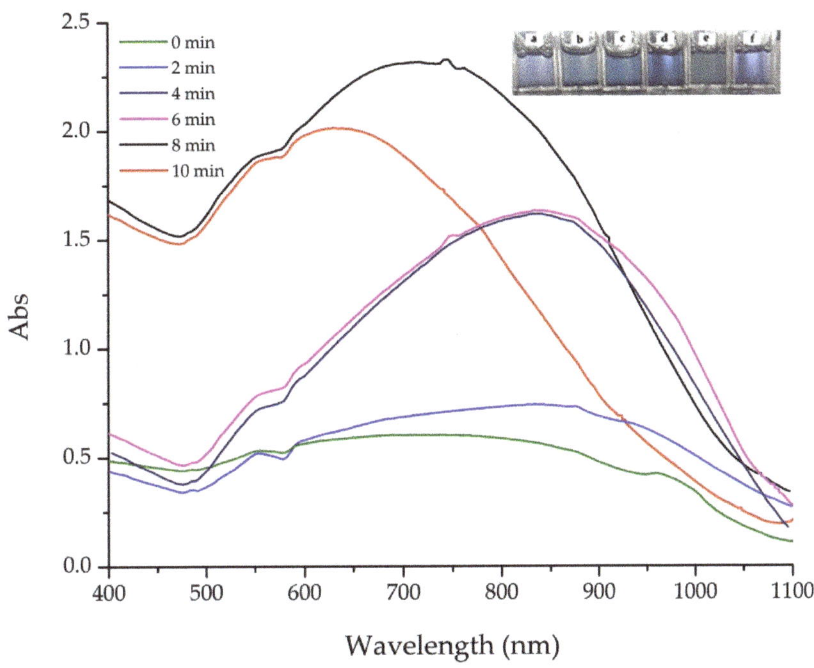

Figure 9. Absorption spectra of GNS prepared in different sonication time—(**a**) 0 min, (**b**) 2 min, (**c**) 4 min, (**d**) 6 min, (**e**) 8 min and (**f**) 10 min.

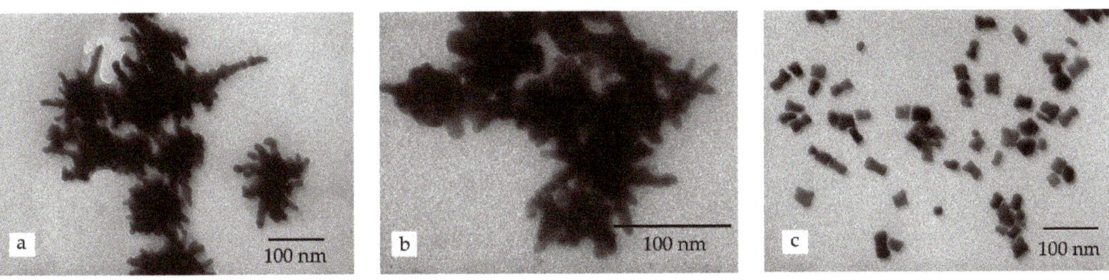

Figure 10. TEM images of GNS prepared in different sonication time—(**a**) 6 min, (**b**) 8 min and (**c**) 10 min.

Table 10. The influence time sonication to morphology, average core and branches of GNS.

US Time (min)	Morphology	Core (nm)	Branches (nm)
4	Long multi-branches	52.02 ± 9.95	31.32 ± 7.62
6	Long multi-branches	75.34 ± 18.37	65.01 ± 11.39
8	Short multi-branches	87.44 ± 11.08	18.65 ± 5.01
10	Short rod-shape	-	-

The FT–IR spectrum of CS-capped GNS was the shift of bands observed in pure CS. The C-H stretching shifted to 2880 cm^{-1}, while the bending of N-H bonds moved to 1557 cm^{-1}. Additionally, stretching of C=O (amide I) and C-N (amide III) was located at 1642 cm^{-1} and 1310 cm^{-1}. These shifts indicated the interaction between the functional groups of CS and GNS [42].

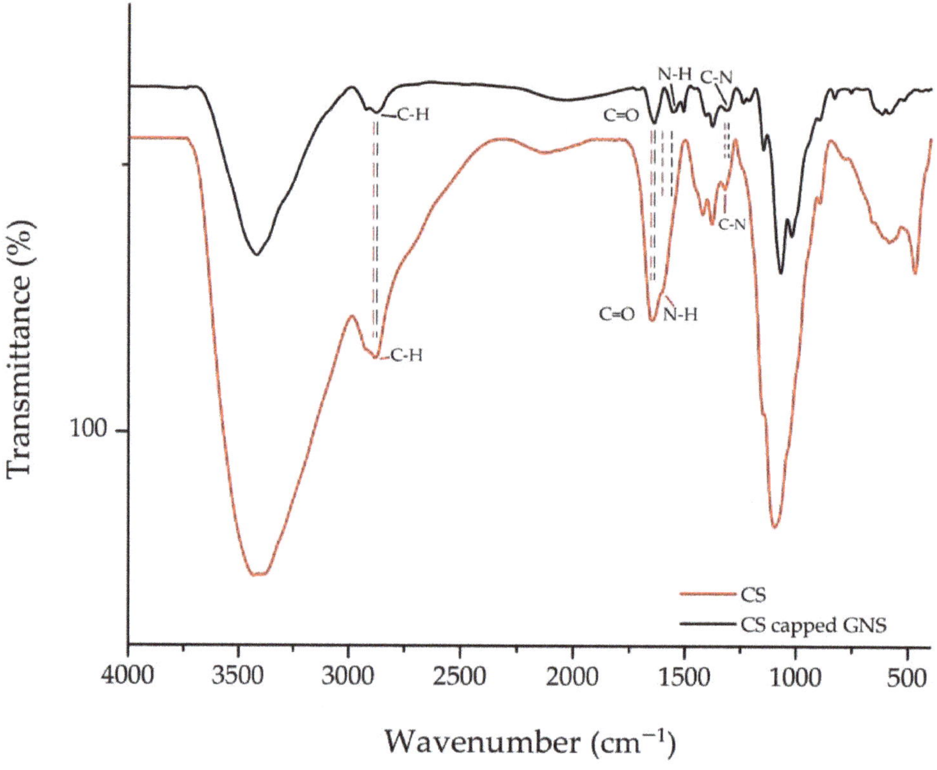

Figure 11. FT–IR spectrum of CS and CS-coated GNS.

3.4. The XRD Diagram of GNS

Figure 12 is the XRD diagram of CS-coated GNS. The recorded pattern exhibited peaks located at 38.0°, 44.9°, 65.2° and 77.4°, which correspond to the (111), (200), (220), (311) and (222) planes of the gold face-centered-cubic (fcc) crystalline structure, respectively [43]. It was clear that there was an intense peak located at 38.0°, which was indexed to the (111) plane. Additionally, a weaker peak for the (200) plane at 65.2° and another at 77.3° for the (311) plane was observed. Finally, a very weak peak located at 77.3° related to the (311) plane.

3.5. Cytotoxicity of the CS-Capped GNS

Biocompatibility is an important property for biomedical applications. MTT assay was used to study the cell compatibility of CS-coated GNS. The cell viability of normal rat fibroblast (NIH/3T3) and normal human fibroblast (BJ-5ta), exposed to various concentrations of multi-branched nanoparticles are shown in Figure 13. The proliferation of both NIH/3T3 and BJ-5ta cell lines treated in various GNS concentration were still around 90%, even at a high concentration of 200 μg/mL, except for BJ-5ta, which had a concentration of over 82%. The results indicate that CS-coated GNS was a good biocompatible agent and is a prospective material for use in biomedical applications. GNS have wide biomedical applications in Raman scattering sensing [44], stem cell tracking [45], bioimaging [46], photothermal treatment [47] and immunotherapy [48].

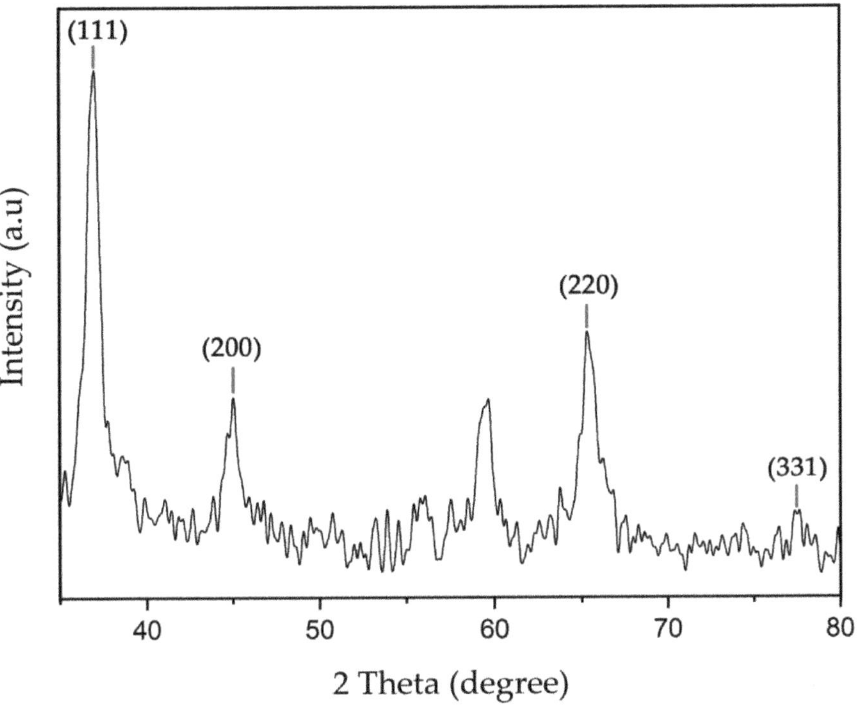

Figure 12. XRD pattern of the CS-coated GNS.

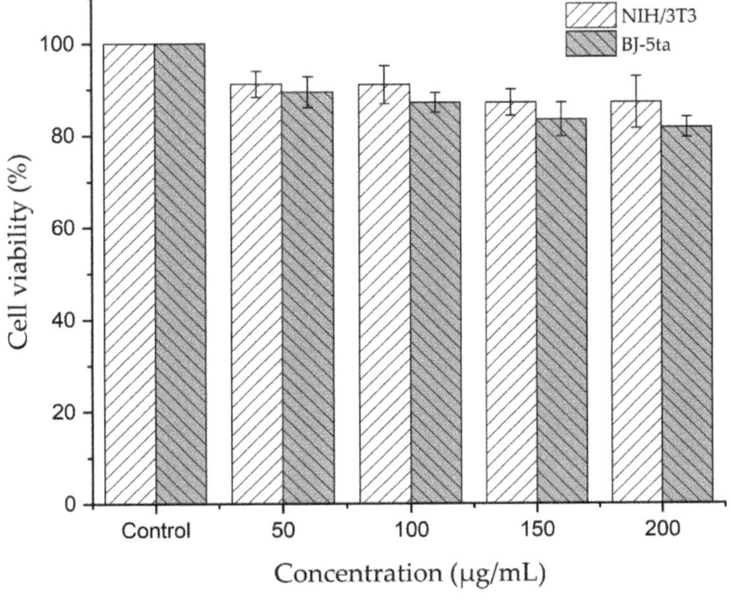

Figure 13. Cell viability of NIH/3T3 and BJ-5ta treated with GNS.

4. Conclusions

This study presents a rapid and green preparation of GNS using the seedless, free surfactant, and ultrasound-assisted method. The GNS particles obtained had long and sharp multi-branches with an average core of 67.85 ± 6.79 nm and a branched length of 76.11 ± 14.23 nm, through an adjusted conditional reaction, such as pH, mass concentration of CS, HQ concentration, as well as amplitude and time sonication. The influences of the conditions above and properties of the prepared GNS were characterized using UV–Vis absorption, FTIR, TEM, XRD, to determine the standard procedure for synthesizing GNS. Furthermore, cytotoxicity of the CS-coated GNS were investigated by the MTT assay on two cell lines, including NIH/3T3 and BJ-5ta. The results indicated that CS-coated GNS was a biocompatible agent, due to its high cell viability. Multi-branched gold nanoparticles are prospective materials not only for biomedical applications but also in cosmetics.

Author Contributions: Investigation, G.D.N., K.T.L.T. and T.M.H.; Investigation, writing—original draft preparation, P.T.H.; review and editing, V.Q.L.; writing—review and editing, project administration and funding acquisition, T.V.K.N. All authors have read and agreed to the published version of the manuscript.

Funding: This research was funded by Saigon Hi-tech Park under grant number 01/2020/HĐNVTX-KCNC-TTRD and the Project of Department of Science and Technology of Ho Chi Minh (118/2019/HĐ-QPTKHCN).

Institutional Review Board Statement: Not applicable.

Informed Consent Statement: Not applicable.

Data Availability Statement: All data is contained within the article.

Conflicts of Interest: The authors declare no conflict of interest.

References

1. Ortiz-Castillo, J.E.; Gallo-Villanueva, R.C.; Madou, M.; Perez-Gonzalez, V.H. Anisotropic gold nanoparticles: A survey of recent synthetic methodologies. *Coord. Chem. Rev.* **2020**, *425*, 213489. [CrossRef]
2. Marişca, T.; Leopold, N. Anisotropic Gold Nanoparticle-Cell Interactions Mediated by Collagen. *Materials* **2019**, *12*, 1131. [CrossRef]
3. Piotto, V.; Litti, L.; Meneghetti, M. Synthesis and Shape Manipulation of Anisotropic Gold Nanoparticles by Laser Ablation in Solution. *J. Phys. Chem. C* **2020**, *124*, 4820–4826. [CrossRef]
4. Burrows, N.D.; Vartanian, A.M.; Abadeer, N.S.; Grzincic, E.M.; Jacob, L.M.; Lin, W.; Li, J.; Dennison, J.M.; Hinman, J.G.; Murphy, C.J. Anisotropic Nanoparticles and Anisotropic Surface Chemistry. *J. Phys. Chem. Lett.* **2016**, *7*, 632–641. [CrossRef]
5. Tsoulos, T.V.; Han, L.; Weir, J.; Xin, H.L.; Fabris, L. A closer look at the physical and optical properties of gold nanostars: An experimental and computational study. *Nanoscale* **2017**, *9*, 3766–3773. [CrossRef]
6. Kohout, C.; Santi, C.; Polito, L. Anisotropic Gold Nanoparticles in Biomedical Applications. *Int. J. Mol. Sci.* **2018**, *19*, 3385. [CrossRef] [PubMed]
7. Kashani, A.S.; Badilescu, S.; Piekny, A.; Packirisamy, M. Differing Affinities of Gold Nanostars and Nanospheres toward HeLa and HepG2 Cells: Implications for Cancer Therapy. *ACS Appl. Nano Mater.* **2020**, *3*, 4114–4126. [CrossRef]
8. Wang, H.-N.; Crawford, B.M.; Fales, A.M.; Bowie, M.L.; Seewaldt, V.L.; Vo-Dinh, T. Multiplexed Detection of MicroRNA Biomarkers Using SERS-Based Inverse Molecular Sentinel (iMS) Nanoprobes. *J. Phys. Chem. C* **2016**, *120*, 21047–21055. [CrossRef] [PubMed]
9. Singh, A.; Khatun, S.; Gupta, A.N. Anisotropy versus fluctuations in the fractal self-assembly of gold nanoparticles. *Soft Matter* **2020**, *16*, 7778–7788. [CrossRef]
10. Chirico, G.; Borzenkov, M.; Pallavicini, P. *Gold Nanostars: Synthesis, Properties and Biomedical Application*; Springer International Publishing: Cham, Switzerland, 2015.
11. Chatterjee, S.; Ricciardi, L.; Deitz, J.I.; Williams, R.E.A.; McComb, D.W.; Strangi, G. Manipulating acoustic and plasmonic modes in gold nanostars. *Nanoscale Adv.* **2019**, *1*, 2690–2698. [CrossRef]
12. Shao, L.; Susha, A.S.; Cheung, L.S.; Sau, T.K.; Rogach, A.L.; Wang, J. Plasmonic Properties of Single Multispiked Gold Nanostars: Correlating Modeling with Experiments. *Langmuir* **2012**, *28*, 8979–8984. [CrossRef] [PubMed]
13. Mir-Simon, B.; Reche-Perez, I.; Guerrini, L.; Pazos-Perez, N.; Alvarez-Puebla, R.A. Universal One-Pot and Scalable Synthesis of SERS Encoded Nanoparticles. *Chem. Mater.* **2015**, *27*, 950–958. [CrossRef]
14. Liebig, F.; Henning, R.; Sarhan, R.M.; Prietzel, C.; Schmitt, C.N.Z.; Bargheer, M.; Koetz, J. A simple one-step procedure to synthesise gold nanostars in concentrated aqueous surfactant solutions. *RSC Adv.* **2019**, *9*, 23633–23641. [CrossRef]

15. Jia, Y.P.; Shi, K.; Liao, J.F.; Peng, J.R.; Hao, Y.; Qu, Y.; Chen, L.J.; Liu, L.; Yuan, X.; Qian, Z.Y.; et al. Effects of Cetyltrimethylammonium Bromide on the Toxicity of Gold Nanorods Both In Vitro and In Vivo: Molecular Origin of Cytotoxicity and Inflammation. *Small Methods* **2020**, *4*, 1900799. [CrossRef]
16. Rusanov, A.L.; Luzgina, N.G.; Lisitsa, A.V. Sodium Dodecyl Sulfate Cytotoxicity towards HaCaT Keratinocytes: Comparative Analysis of Methods for Evaluation of Cell Viability. *Bull. Exp. Biol. Med.* **2017**, *163*, 284–288. [CrossRef]
17. De Silva Indrasekara, A.S.; Johnson, S.F.; Odion, R.A.; Vo-Dinh, T. Manipulation of the Geometry and Modulation of the Optical Response of Surfactant-Free Gold Nanostars: A Systematic Bottom-Up Synthesis. *ACS Omega* **2018**, *3*, 2202–2210. [CrossRef]
18. Yuan, H.; Khoury, C.G.; Hwang, H.; Wilson, C.M.; Grant, G.A.; Vo-Dinh, T. Gold nanostars: Surfactant-free synthesis, 3D modelling, and two-photon photoluminescence imaging. *Nanotechnology* **2012**, *23*, 075102. [CrossRef]
19. Zheng, B.; Liu, X.; Wu, Y.; Yan, L.; Du, J.; Dai, J.; Xiong, Q.; Guo, Y.; Xiao, D. Surfactant-free gold nanoparticles: Rapid and green synthesis and their greatly improved catalytic activities for 4-nitrophenol reduction. *Inorg. Chem. Front.* **2017**, *4*, 1268–1272. [CrossRef]
20. Wall, M.; Harmsen, S.; Pal, S.; Zhang, L.; Arianna, G.; Lombardi, J.; Drain, C.; Kircher, M. Gold Nanoparticles: Surfactant-Free Shape Control of Gold Nanoparticles Enabled by Unified Theoretical Framework of Nanocrystal Synthesis (Adv. Mater. 21/2017). *Adv. Mater.* **2017**, *29*. [CrossRef]
21. Duan, H.; Wang, D.; Li, Y. Green chemistry for nanoparticle synthesis. *Chem. Soc. Rev.* **2015**, *44*, 5778–5792. [CrossRef]
22. Kadziński, M.; Cinelli, M.; Ciomek, K.; Coles, S.R.; Nadagouda, M.N.; Varma, R.S.; Kirwan, K. Co-constructive development of a green chemistry-based model for the assessment of nanoparticles synthesis. *Eur. J. Oper. Res.* **2018**, *264*, 472–490. [CrossRef] [PubMed]
23. Ratan, Z.A.; Haidere, M.F.; Nurunnabi, M.; Shahriar, S.M.; Ahammad, A.J.S.; Shim, Y.Y.; Reaney, M.J.T.; Cho, J.Y. Green Chemistry Synthesis of Silver Nanoparticles and Their Potential Anticancer Effects. *Cancers* **2020**, *12*, 855. [CrossRef] [PubMed]
24. Kumar, A.; Kuang, Y.; Liang, Z.; Sun, X. Microwave chemistry, recent advancements, and eco-friendly microwave-assisted synthesis of nanoarchitectures and their applications: A review. *Mater. Today Nano* **2020**, *11*, 100076. [CrossRef]
25. Lagashetty, A.; Ganiger, S.K.; Preeti, R.K.; Reddy, S.; Pari, M. Microwave-assisted green synthesis, characterization and adsorption studies on metal oxide nanoparticles synthesized using Ficus Benghalensis plant leaf extracts. *New J. Chem.* **2020**, *44*, 14095–14102. [CrossRef]
26. Babu, S.G.; Neppolian, B.; Ashokkumar, M. Ultrasound-Assisted Synthesis of Nanoparticles for Energy and Environmental Applications. In *Handbook of Ultrasonics and Sonochemistry*; Springer: Singapore, 2016; pp. 423–456.
27. Lv, W.; Gu, C.; Zeng, S.; Han, J.; Jiang, T.; Zhou, J. One-Pot Synthesis of Multi-Branch Gold Nanoparticles and Investigation of Their SERS Performance. *Biosensors* **2018**, *8*, 113. [CrossRef] [PubMed]
28. Zhang, X.-L.; Zheng, C.; Zhang, Y.; Yang, H.-H.; Liu, X.; Liu, J. One-pot synthesis of gold nanostars using plant polyphenols for cancer photoacoustic imaging and photothermal therapy. *J. Nanoparticle Res.* **2016**, *18*, 174. [CrossRef]
29. Mulder, D.W.; Phiri, M.M.; Jordaan, A.; Vorster, B.C. Modified HEPES one-pot synthetic strategy for gold nanostars. *R. Soc. Open Sci.* **2019**, *6*, 190160. [CrossRef]
30. Sáez, V.; Mason, T.J. Sonoelectrochemical synthesis of nanoparticles. *Molecules* **2009**, *14*, 4284–4299. [CrossRef]
31. Kumar, R.; Badilescu, S.; Packirisamy, M. Sonochemical Stabilization of Gold Nanostars: Effect of Sonication Frequency on the Size, Shape and Stability of Gold Nanostars. *Adv. Nano Bio Mater. Devices* **2017**, *1*, 172–181.
32. Vigderman, L.; Zubarev, E.R. High-Yield Synthesis of Gold Nanorods with Longitudinal SPR Peak Greater than 1200 nm Using Hydroquinone as a Reducing Agent. *Chem. Mater.* **2013**, *25*, 1450–1457. [CrossRef]
33. Unser, S.; Bruzas, I.; He, J.; Sagle, L. Localized Surface Plasmon Resonance Biosensing: Current Challenges and Approaches. *Sensors* **2015**, *15*, 5684. [CrossRef] [PubMed]
34. Casu, A.; Cabrini, E.; Donà, A.; Falqui, A.; Diaz-Fernandez, Y.; Milanese, C.; Taglietti, A.; Pallavicini, P. Controlled Synthesis of Gold Nanostars by Using a Zwitterionic Surfactant. *Chem. A Eur. J.* **2012**, *18*, 9381–9390. [CrossRef] [PubMed]
35. Tezcan, T.; Hsu, C.-H. High-sensitivity SERS based sensing on the labeling side of glass slides using low branched gold nanoparticles prepared with surfactant-free synthesis. *RSC Adv.* **2020**, *10*, 34290–34298. [CrossRef]
36. Jana, D.; Matti, C.; He, J.; Sagle, L. Capping Agent-Free Gold Nanostars Show Greatly Increased Versatility and Sensitivity for Biosensing. *Anal. Chem.* **2015**, *87*, 3964–3972. [CrossRef]
37. Sirajuddin; Mechler, A.; Torriero, A.A.J.; Nafady, A.; Lee, C.-Y.; Bond, A.M.; O'Mullane, A.P.; Bhargava, S.K. The formation of gold nanoparticles using hydroquinone as a reducing agent through a localized pH change upon addition of NaOH to a solution of HAuCl4. *Colloids Surf. A Physicochem. Eng. Asp.* **2010**, *370*, 35–41.
38. Liu, K.; Bu, Y.; Zheng, Y.; Jiang, X.; Yu, A.; Wang, H. Seedless Synthesis of Monodispersed Gold Nanorods with Remarkably High Yield: Synergistic Effect of Template Modification and Growth Kinetics Regulation. *Chem. A Eur. J.* **2017**, *23*, 3291–3299. [CrossRef]
39. Shojaeiarani, J.; Bajwa, D.; Holt, G. Sonication amplitude and processing time influence the cellulose nanocrystals morphology and dispersion. *Nanocomposites* **2020**, *6*, 41–46. [CrossRef]
40. Pradhan, S.; Hedberg, J.; Blomberg, E.; Wold, S.; Odnevall Wallinder, I. Effect of sonication on particle dispersion, administered dose and metal release of non-functionalized, non-inert metal nanoparticles. *J. Nanopart. Res.* **2016**, *18*, 285. [CrossRef]
41. Phan, T.T.V.; Nguyen, V.T.; Ahn, S.-H.; Oh, J. Chitosan-mediated facile green synthesis of size-controllable gold nanostars for effective photothermal therapy and photoacoustic imaging. *Eur. Polym. J.* **2019**, *118*, 492–501. [CrossRef]

42. Fernandes Queiroz, M.; Melo, K.R.; Sabry, D.A.; Sassaki, G.L.; Rocha, H.A. Does the Use of Chitosan Contribute to Oxalate Kidney Stone Formation? *Mar. Drugs* **2015**, *13*, 141. [CrossRef]
43. Oliveira, M.J.; De Almeida, M.P.; Nunes, D.; Fortunato, E.; Martins, R.; Pereira, E.; Byrne, H.J.; Águas, H.; Franco, R. Design and Simple Assembly of Gold Nanostar Bioconjugates for Surface-Enhanced Raman Spectroscopy Immunoassays. *Nanomaterials* **2019**, *9*, 1561. [CrossRef] [PubMed]
44. Evcimen, N.I.; Coskun, S.; Kozanoglu, D.; Ertas, G.; Unalan, H.E.; Nalbant Esenturk, E. Growth of branched gold nanoparticles on solid surfaces and their use as surface-enhanced Raman scattering substrates. *RSC Adv.* **2015**, *5*, 101656–101663. [CrossRef]
45. Shammas, R.L.; Fales, A.M.; Crawford, B.M.; Wisdom, A.J.; Devi, G.R.; Brown, D.A.; Vo-Dinh, T.; Hollenbeck, S.T. Human Adipose-Derived Stem Cells Labeled with Plasmonic Gold Nanostars for Cellular Tracking and Photothermal Cancer Cell Ablation. *Plast Reconstr Surg* **2017**, *139*, 900e–910e. [CrossRef] [PubMed]
46. Mousavi, S.M.; Zarei, M.; Hashemi, S.A.; Ramakrishna, S.; Chiang, W.-H.; Lai, C.W.; Gholami, A. Gold nanostars-diagnosis, bioimaging and biomedical applications. *Drug Metab. Rev.* **2020**, *52*, 299–318. [CrossRef] [PubMed]
47. Song, C.; Li, F.; Guo, X.; Chen, W.; Dong, C.; Zhang, J.; Zhang, J.; Wang, L. Gold nanostars for cancer cell-targeted SERS-imaging and NIR light-triggered plasmonic photothermal therapy (PPTT) in the first and second biological windows. *J. Mater. Chem. B* **2019**, *7*, 2001–2008. [CrossRef]
48. Vo-Dinh, T. Shining Gold Nanostars: From Cancer Diagnostics to Photothermal Treatment and Immunotherapy. *J. Immunol. Sci.* **2018**, *2*, 1–8. [CrossRef]

Article

Fluid–Solid Coupling Model and Simulation of Gas-Bearing Coal for Energy Security and Sustainability

Shixiong Hu, Xiao Liu and Xianzhong Li

[1] School of Energy Science and Engineering, Henan Polytechnic University, Jiaozuo 454003, China; 211702020003@home.hpu.edu.cn (S.H.); lixianzhong@hpu.edu.cn (X.L.)

[2] Collaborative Innovation Center of Methane (Shale Gas) in Central Plains Economic Zone, Jiaozuo 454003, China

* Correspondence: liuxiao@hpu.edu.cn;

Received: 10 January 2020; Accepted: 19 February 2020; Published: 24 February 2020

Abstract: The optimum design of gas drainage boreholes is crucial for energy security and sustainability in coal mining. Therefore, the construction of fluid–solid coupling models and numerical simulation analyses are key problems for gas drainage boreholes. This work is based on the basic theory of fluid–solid coupling, the correlation definition between coal porosity and permeability, and previous studies on the influence of adsorption expansion, change in pore free gas pressure, and the Klinkenberg effect on gas flow in coal. A mathematical model of the dynamic evolution of coal permeability and porosity is derived. A fluid–solid coupling model of gas-bearing coal and the related partial differential equation for gas migration in coal are established. Combined with an example of the measurement of the drilling radius of the bedding layer in a coal mine, a coupled numerical solution under negative pressure extraction conditions is derived by using COMSOL Multiphysics simulation software. Numerical simulation results show that the solution can effectively guide gas extraction and discharge during mining. This study provides theoretical and methodological guidance for energy security and coal mining sustainability.

Keywords: fluid–solid coupling; coal containing gas; permeability; energy safety; sustainability

---

1. Introduction

Since the 21st century, global resource shortage and environmental pollution have become difficult problems in human sustainability development [1,2]. Policy makers in various countries have focused on sustainable energy and low-carbon development by proposing sustainable strategies and methods [3]. With the advent of low-carbon economy, coalbed methane, as a clean, efficient, and safe energy source, has been eliciting considerable research attention. This unconventional natural gas is mainly present in coal seams in free and adsorption states [4]. Coal is a typical dual-porosity/permeability system containing a porous matrix surrounded by fractures. The coal matrix is separated by a natural fracture network composed of butt cleats and face cleats. The cleat system provides an effective flow channel for gas. The developed pore structure is the main space for coalbed methane. During gas percolation of porous media, the effective stress of the porous medium skeleton changes because of the change in pore pressure; the porosity and permeability of porous media also change to a certain extent [5]. These changes can affect the gas flow in pores and the redistribution of gas pressure within a certain range. Therefore, in studying the migration rule and deformation characteristics of gas in a porous medium, such as coal, the interaction between the gas flow in a porous medium and the deformation of the porous medium body should be considered. The mutual coupling between the gas seepage flow field and the stress field in the porous medium should also be considered.

Through relevant laboratory tests and field practice, people have gradually realized that the seepage property of gas in coal is related to the mechanical properties of coal, gas pressure, in situ stress, temperature, and other factors. Sommerton et al. studied the effect of stress on the gas seepage property of coal [6]. Brace investigated the law of permeability change of rock mass under stress [7]. McKee et al. analyzed the relationship between stress and coal porosity and permeability and studied the phenomenon in which the depth of coal seams and effective stress increase, the width of cleats in coal seam decreases, and permeability decreases exponentially [8]. Enever et al. obtained the influence rule between the effective stress and permeability of coal [9]. In accordance with the generalized form of the power law, Sun established a mathematical model for the flow of compressible gas in coal seams. This model, which is called the nonlinear gas flow model, was based on the measured gas parameters in the Zhongmacun Coal Mine of Jiaozuo Mining Bureau. A numerical simulation of the pressure distribution of the homogeneous gas seepage flow field was carried out using various models [10–12]. Li et al. studied the relationship among coal adsorption expansion, deformation, porosity, and permeability in consideration of the adsorption deformation characteristics of the coal skeleton and obtained the relationship among porosity, permeability, and swelling deformation [13,14]. Tao et al. analyzed the problems existing in the theory of nonlinear gas flow and theory of fluid–solid coupling of coal-bed gas. The research achievements in the fields of nonlinear gas flow theory, gas flow theory of the geothermal field effect, and coal-bed gas fluid–solid coupling theory have been scrutinized extensively [15,16]. Zhu et al. considered the Klinkenberg effect and proposed a coupled mathematical model of solid deformation and gas flow [17–20].

However, the fluid–solid coupling model that scholars have established still has certain limitations. For example, the Klinkenberg and adsorption expansion effects on gas migration in coal were not considered simultaneously. Existing research shows that strain on adsorption expansion occurs after the coal body adsorbs gas. When the strain is limited by certain factors, adsorption expansion stress follows, which causes a certain degree of primary deformation of the coal skeleton and affects the development of coal pores. Thus, the effect of adsorption expansion on gas migration should not be ignored [21]. On the basis of previous studies, the author comprehensively considers the influence of the two factors of migration of coalbed methane in coal and establishes a mathematical model of fluid–solid coupling in a low-permeability coal seam. Thereafter, relevant partial differential equations are derived. The establishment of the model expands the theory of fluid–solid coupling of gas-bearing coals under multi-field conditions and clarifies the law of gas occurrence and seepage in coal. The establishment of the fluid–solid coupling equation can characterize gas flow in coal seams from the perspective of time and improve the research method of gas dynamic flow law. The rationality of the established mathematical model is verified by using a specific example of the effective extraction radius of coal mine gas. This model provides a theoretical basis for the design and layout of gas drainage boreholes in coal mining and a reasonable reference for decision makers to control coal mine gas effectively [22,23].

2. Mathematical Model

2.1. Basic Assumptions

(1) Coal containing gas can be regarded as an isotropic elastic medium.
(2) The coal seam is considered homogeneous, that is, the physical properties of each part of the coal seam are similar everywhere and do not change with a change in position.
(3) The coal seam temperature is constant.
(4) The gas contained in the coal seam is regarded as ideal gas and obeys the ideal gas state equation. The gas desorption obeys the Langmuir equation.
(5) The seepage characteristic of coalbed methane in coal meets the Klinkenberg effect.
(6) The deformation of coal is small.
(7) The overall deformation of coal rock consists of pore and fracture deformation.

(8) Single-phase saturated gas fluid exists in the coal seam, and only free and adsorbed states are available.
(9) The model is isolated from the outside world and does not undergo any form of energy and material exchange.

2.2. Physical Property Parameter Model of Coal

The compression and adsorption expansion of deformation result in different degrees of deformation of the coal skeleton because of the changes in conditions of the coal seam, such as crustal stress and gas pressure. Owing to the different depths of the coal seam, the gas pressure and crustal stress also change to varying degrees, resulting in changes in coal seam porosity and permeability.

2.2.1. Deformation Mechanism of Coal Containing Gas

According to previous research results [24], two kinds of deformation mechanisms exist under the joint action of internal and external stresses of coal containing gas.

(1) Structural deformation: Relative dislocation occurs between coal particles because of external stress. The particle arrangement becomes increasingly compact, which causes the compression of the coal particle skeleton. The deformation caused by the two is called structural deformation, as shown in Figure 1.
(2) Body deformation: The deformation of coal particles is mainly caused by the adsorption expansion and desorption shrinkage of coalbed methane, the compression of the coal skeleton by gas pressure, and the thermal expansion and contraction effect of temperature.

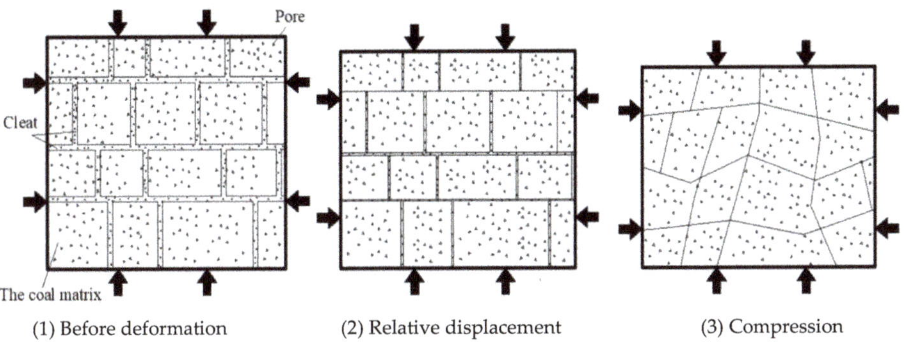

Figure 1. Structural deformation.

This study assumes that the coal seam is thermostatic. Thus, the deformation of the particles under the influence of temperature changes is ignored.

2.2.2. Porosity Mathematical Model

According to the relevant definition of porosity, the porosity change of coal can be expressed as follows:

$$\varphi = \frac{V_p}{V_b} = \frac{V_{p0} + \Delta V_p}{V_{b0} + \Delta V_b} = 1 - \frac{V_{s0} + \Delta V_s}{V_{b0} + \Delta V_b} = 1 - \frac{1 - \varphi_0}{1 + e}\left(1 + \frac{\Delta V_s}{V_{s0}}\right), \tag{1}$$

where φ is coal porosity, φ_0 is the initial porosity of coal, V_p is the pore volume of coal, V_{p0} is the initial pore volume of coal, V_b is the total apparent volume of coal, V_{b0} is the initial total apparent volume of coal, ΔV_p is the variation in the pore volume of coal, ΔV_b is the total apparent volume change in coal, V_s is the volume of the coal skeleton, ΔV_s is the volume variation of the coal skeleton, and e is the volumetric strain of coal.

In the actual state, the volume strain increment of coal particles caused by the bulk deformation of coal particles is mainly composed of three parts. The first part is the strain increment caused by pore gas pressure compressing the coal particles. The second part is the increment in expansion strain caused by coal particles adsorbing the coalbed methane. The third part is the strain increment caused by thermoelastic expansion. The strain increment caused by thermoelastic expansion is zero because we assume that the temperature of the coal seam is constant. In reference to Figure 2, the relationship among the three parts can be expressed as follows:

$$\frac{\Delta V_s}{V_{s0}} = \frac{\Delta V_{sp}}{V_{s0}} + \frac{\Delta V_{sf}}{V_{s0}} + \frac{\Delta V_{st}}{V_{s0}}, \tag{2}$$

where $\frac{\Delta V_{sp}}{V_{s0}}$ is the strain increment caused by pore gas pressure compressing the coal particles, $\frac{\Delta V_{sf}}{V_{s0}}$ is the swelling strain increment caused by coal particles adsorbing the coal bed gas, and $\frac{\Delta V_{st}}{V_{s0}}$ is the strain increment caused by thermal elastic expansion. $\frac{\Delta V_{st}}{V_{s0}}$ is zero because the temperature of the coal seam is assumed to be constant. The total volume strain of coal particle deformation can be expressed as follows:

$$\frac{\Delta V_s}{V_{s0}} = \frac{\Delta V_{sp}}{V_{s0}} + \frac{\Delta V_{sf}}{V_{s0}} = \frac{\varepsilon_p}{1-\varphi_0} - \frac{\Delta P}{K_s}, \tag{3}$$

where ε_p is the expansion strain generated by the adsorption of gas per unit volume.

$$\varepsilon_p = \frac{2a\rho RT}{3V_m K_s} \ln(1+bP), \tag{4}$$

where T is the thermodynamic temperature of the coal seam (K), a is the limit adsorption capacity per unit mass of combustibles under a reference pressure (m^3/Mg), b is the adsorption constant (MPa^{-1}), R is the universal gas constant (R = 8.3143 J/(mol·K)), ρ is the coal density (kg/m^3), V_m is the molar volume of gas (22.4 × 10^{-3} m^3/mol), and Ks is the volume modulus of the coal skeleton (Pa).

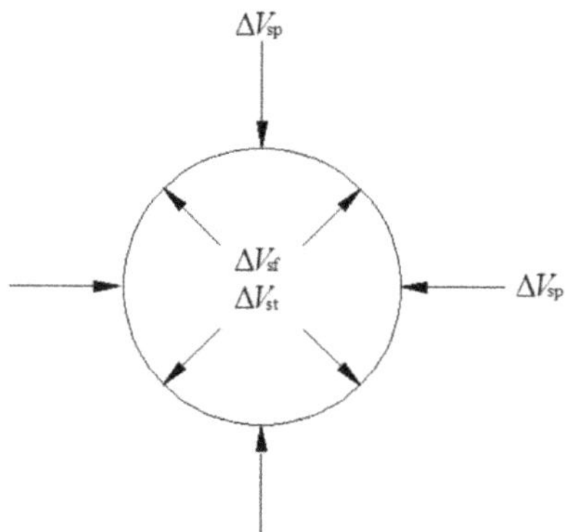

Figure 2. Deformation relationship of coal particles.

Equations (3) and (4) are substituted into Equation (1) to obtain a mathematical model of the dynamic evolution of porosity.

$$\varphi = 1 - \frac{1-\varphi_0}{1+e}\left(1 + \frac{\varepsilon_p}{1-\varphi_0} - \frac{\Delta P}{K_s}\right) = 1 - \frac{1-\varphi_0}{1+e}\left(1 + \frac{2a\rho RT}{3V_m K_s(1-\varphi_0)} \ln(1+bP) - \frac{P-P_0}{K_s}\right) \tag{5}$$

2.2.3. Permeability Evolution Mathematical Model

Permeability is an important indicator that describes the difficulty of gas migration in gas-bearing coal seams, the permeability of coal seams, and gas drainage difficulty. Therefore, a correct mathematical model of permeability evolution must be established for gas control in coal mines.

The permeability of gas-containing coal is closely related to the stress state of coal. Different stress states cause changes in coal-rock skeleton deformation and pore volume. When porosity changes, permeability also changes. To establish the relationship between coal permeability and porosity, we can refer to the Kozeny–Carman equation [25–27].

$$k = \frac{\varphi}{k_z S_p^2} = \frac{\varphi V_p^2}{k_z A_S^2}, \tag{6}$$

where k is permeability (md), k_z is a dimensionless constant (k_z = 5), S_p is the pore surface area per unit pore volume of coal (cm^2), and A_s is the total surface area of coal pore (cm^2).

The permeability in the initial state is assumed to be

$$k_0 = \frac{\varphi V_p^2}{k_z A_S^2}, \tag{7}$$

where k_0 is the initial permeability (md), A_{s0} is the total surface area of the coal pore in the initial state (cm^2).

The total volume of coal and the volume change of a single coal particle are ΔV_b and ΔV_s, respectively, when the initial state changes to a new one. Based on the definition of porosity, the new porosity is

$$\varphi = \frac{V_{p0} + (\Delta V_b - \Delta V_s)}{V_{b0} + \Delta V_b} \tag{8}$$

The new pore surface area can be expressed as follows:

$$S_p = \frac{A_{s0}(1 + \partial)}{V_{p0} + (\Delta V_b - \Delta V_s)}, \tag{9}$$

where ∂ is the increasing coefficient of the pore surface area of coal (%).

For Equations (6) and (7), the ratio of the new permeability to the original permeability is computed by the following:

$$\frac{k}{k_0} = \frac{\varphi S_{p0}^2}{\varphi_0 S_p^2} = \frac{1}{1+e} \frac{1}{(1+\partial)^2} \left(\frac{V_{p0} + \Delta V_p}{V_{P0}} \right)^3. \tag{10}$$

The total surface area of the unit volume of coal particles is almost unchanged in the stress and strain process of coal [28]. This occurrence can be ignored. Thus, ∂ is approximately zero. According to [24], Equation (10) can be simplified as follows:

$$\frac{k}{k_0} = \frac{1}{1+e} \left(\frac{V_{p0} + \Delta V_p}{V_{P0}} \right)^3 = \frac{1}{1+e} \left(1 + \frac{e}{\varphi_0} - \frac{\Delta V_s}{V_{s0}} \Delta \frac{(1-\varphi_0)}{\varphi_0} \right)^3. \tag{11}$$

The combination of Equations (3), (4) and (11) can be obtained:

$$k = \frac{k_0}{1+e} \left[1 + \frac{e}{\varphi_0} + \frac{\Delta P(1-\varphi_0)}{\varphi_0 K_s} - \frac{2a\rho RT\Delta \ln(1+bP)}{3\varphi_0 V_m K_s} \right]^3. \tag{12}$$

The above equation is a mathematical model for the evolution of the permeability of coal containing gas.

2.3. Effective Stress of Gas-Bearing Coal

Gas-containing coal is a complex deformable pore-fracture dual media. Coal has a strong adsorption capacity for gas and produces a certain adsorption expansion stress, which changes the stress distribution of coal.

Therefore, when studying the problem of fluid–solid coupling of coal containing gas and rock, the relationship between the effective stress of the coal seam and its adsorption-expansion stress should be considered simultaneously. In view of this problem, Wu et al. [29,30] derived the following formula for calculating the expansive stress of coal. The expansion stress formula of coal can be expressed as follows:

$$\sigma_p = E\varepsilon_p = \frac{2a\rho RT(1-2v)\ln(1+bP)}{3V_m}, \tag{13}$$

where σ_p is the expansion stress (Pa), v is the Poisson ratio, and E is the elastic modulus of coal (Pa).

In accordance with the effective stress law of Terzaghi and in consideration of the expansion stress absorbed by coal, the effective stress equation of gas-bearing coal can be expressed as follows:

$$\sigma'_{ij} = \sigma_{ij} - \alpha P\delta_{ij} - \sigma_p\delta_{ij}, \tag{14}$$

where σ'_{ij} is the effective stress of gas-bearing coal (MPa), σ_{ij} is the overall stress of gas-containing coal (MPa), and α is the Biot coefficient.

2.4. Establishment of a Fluid–Solid Coupling Model of Gas-Bearing Coal

2.4.1. Gas Content Equation

The existing mine gas in the coal bed occurs mostly in free, adsorption, and dissolved states. The dissolved gas content is not considered in this study because the amount is notably small. The gas content is the sum of the free-state and adsorbed-state gas contents. The adsorption gas content accounts for more than 90% of the total, and the content of adsorption gas is related to moisture, coal ash, and gas pressure. The free gas content mainly depends on coal porosity and the magnitude of gas pressure. According to previous research and the modified Langmuir adsorption equilibrium equation, the gas content of a coal seam can be obtained as follows [31]:

$$Q = \left(\frac{abcP}{1+bP} + \varphi\frac{P}{P_n}\right)\rho_n, \tag{15}$$

where P_n is the gas pressure under standard conditions (P_n = 0.10325 MPa), and ρ_n is the coalbed methane density under standard conditions (kg/m^3).

In the above equation,

$$c = \rho\frac{1-A-M}{1+0.31M},$$

where Q is the gas content (kg/m^3), c is the coal quality correction parameter (kg/m^3), A is the ash content of coal (%), and M is the coal moisture (%).

2.4.2. Stress Field Equation of Gassy Coal

Assuming that the gas-bearing coal is an isotropic linear elastic medium, the stress field changes obey the following equation.

(1) Balance equation

$$\sigma_{ij,j} + F_i = 0(i,j = 1,2,3), \tag{16}$$

where F_i is the bulk stress (N/m^3).

According to the effective stress Equation (14) of gas-bearing coal, the equilibrium differential equation expressed by effective stress is obtained by the introduction of Equation (16). Then the equilibrium differential equation can be expressed as follows:

$$\sigma'_{ij,j} + \left(\alpha P \delta_{ij}\right)_{,j} + \left(\sigma_p \delta_{ij}\right)_{,j} + F_i = 0 \tag{17}$$

(2) Geometric equation

In the spatial distribution of gas-bearing coals, let u (x, y, z), v (x, y, z), and w (x, y, z) be the displacement components in directions x, y, and z, respectively, and continuous single-valued functions of coordinates. Then, the strain and displacement components satisfy the following geometric equations, which can be expressed as tensor symbols.

$$\varepsilon_{ij} = \frac{1}{2}\left(u_{i,j} + u_{j,i}\right) (i, j = 1, 2, 3) \tag{18}$$

(3) Constitutive equation of gas-bearing coal

The constitutive equation of gas-bearing coal describes the relationship between the stress and strain of coal. The constitutive relationship in this study is based on the strains caused by the adsorption expansion of gas-bearing coal, compression of the coal particle body, and rock stress.

The linear strain caused by gas adsorption by coal particles is as follows:

$$\varepsilon_{PX} = \frac{2a\rho RT}{9V_m K_s} \ln(1 + bP) \tag{19}$$

The linear compression strain of coal particles caused by the change in pore gas pressure is as follows:

$$\varepsilon_{PY} = -\frac{\Delta P}{3K_s}. \tag{20}$$

According to Hooke's law, the strain due to crustal stress is computed using the following:

$$\varepsilon_D = \frac{1}{2G}\left(\sigma' - \frac{v}{1+v}\Theta'\right). \tag{21}$$

According to the above expressions, the total strain of gas-bearing coal is as follows

$$\varepsilon = \varepsilon_{PX} + \varepsilon_{PY} + \varepsilon_D = \frac{2a\rho RT}{9V_m K_s} \ln(1 + bP) - \frac{\Delta P}{3K_s} + \frac{1}{2G}\left(\sigma' - \frac{v}{1+v}\Theta'\right). \tag{22}$$

By using the above formula as a reference, the following equation can be derived.

$$\sigma' = 2G\varepsilon + \frac{v}{1+v}\Theta' - 2G\left(\frac{2a\rho RT \ln(1 + bP)}{9V_m K_s} - \frac{\Delta P}{3K_s}\right) \tag{23}$$

where G is the shear modulus (MPa), and Θ' is the effective volume stress.

The following equation can be obtained by arranging Equation (23) after introducing the Lame constant.

$$\sigma' = 2G\varepsilon + \lambda e + \frac{2G\Delta P}{3K_s} - \frac{4Ga\rho RT \ln(1 + bP)}{9V_m K_s}, \tag{24}$$

where λ is the Lame constant.

Assuming that the coal is a linear elastic medium, the constitutive equation of gas-bearing coal-rock deformation conforms to Hooke's law, as follows:

$$\sigma'_{ij} = \lambda e \delta_{ij} + 2G\varepsilon_{ij}. \tag{25}$$

According to the stress–strain relationship of coal in Equation (32) and combined with the above formula, the effective stress constitutive equation of gas-bearing coal expressed in tensor form is derived as follows:

$$\sigma'_{ij} = \lambda e \delta_{ij} + 2G\varepsilon_{ij} + \frac{2G\Delta P}{3K_s}\delta_{ij} - \frac{4Ga\rho RT\ln(1+bP)}{9V_m K_s}\delta_{ij}. \tag{26}$$

(4) Stress field equation of gas-bearing coal

Substituting Equation (26) into Equation (17) yields the following:

$$G\sum_{j=1}^{3}\frac{\partial^2 u_i}{\partial x_i^2} + \frac{G}{1-2v}\sum_{j=1}^{3}\frac{\partial^2 u_i}{\partial x_i x_j} + \left[\alpha + \frac{2G}{3K_s} + \left(\frac{1-2v}{V_m} - \frac{2G}{3V_m K_s}\right)\frac{2ab\rho RT}{3(1+bP)}\right]\frac{\partial P}{\partial x_i} + F_i = 0. \tag{27}$$

This formula is the fluid–solid coupling stress field equation of gas-containing coal, and the deformation field of gas-bearing coal is represented by displacement. The variation of the strain field of gas-containing coal caused by crustal stress, gas adsorption by coal particles, and gas pressure change are considered in the equation.

2.4.3. Fluid–Solid Coupling Gas Seepage Equation of Gas-Bearing Coal

(1) Gas flow equation

The previous experimental results reveal that when gas migrates in low-permeability gas-bearing coal seams, the gas molecules near the surface of the coal wall show the phenomenon of non-zero velocity, which does not conform to Darcy's law [32]. This occurrence is called the slippage effect or Klinkenberg effect in seepage mechanics. Its permeability can be expressed as follows:

$$k = k_g\left(1 + \frac{4\omega\lambda_1}{r}\right) = k_g\left(1 + \frac{m}{P}\right), \tag{28}$$

where k_g is the Klinkenberg permeability, ω is the scale factor, λ_1 is mean free path of gas molecules, r is the average pore radius, m is the Klinkenberg coefficient (MPa), and ∇P is the gas pressure gradient in the coal seam (Pa/m).

Therefore, when the Klinkenberg effect is considered, the equation of gas flow in the coal seam can be expressed by the following:

$$q = -\frac{k_g}{\mu}\left(1 + \frac{m}{P}\right)\nabla \mathrm{P}, \tag{29}$$

where q is the velocity vector of gas flow (m/s), and μ is the gas dynamic viscosity (1.08×10^{-5} Pa·s).

(2) Continuity equation

According to the hypothesis, if the model is isolated from the outside and no exchange of substance and energy in any form occurs, the gas flow in the coal seam will conform to the law of conservation of mass, expressed in the form of a differential equation:

$$\frac{\partial \mathrm{Q}}{\partial t} + \nabla\cdot\left(\rho_g q\right) = I, \tag{30}$$

where Q is the gas content in coal, ρ_g is the gas density when the gas pressure is P (kg/m^3), and I is the source sink term.

(3) Gas seepage field equation of gas-bearing coal

According to the state equation of coalbed methane in [33], the gas content equation (15), and the gas flow equation (29), the equation of gas seepage flow field can be obtained by combining them with Equation (30). The result is as follows:

$$\frac{M_g}{RT}\left[\varphi+\frac{abcP_n}{(1+bP)^2}+\frac{(1-\varphi_0)P}{(1+e)K_s}-\frac{2ab\rho RTP}{3V_mK_s(1+bP)(1+e)}\right]\frac{\partial P}{\partial t}-\frac{M_gP}{RTZ}\left[\frac{k_g}{\mu}\left(1+\frac{m}{P}\right)\nabla P\right]=-\frac{M_gP}{RT}\cdot\varphi\frac{\partial e}{\partial t}, \quad (31)$$

where M_g is the molar mass of gas (kg/mol). Z is the gas compressibility factor, and its value is approximately 1 in the case of a small temperature difference.

In summary, Equations (5), (12), (27), and (30) constitute a fluid–solid coupling model of gas-bearing coal.

3. Numerical Simulation of the Model and Analysis of Its Results

Deduction and analysis show that the fluid–solid coupling mathematical model of gas-bearing coal is a complex nonlinear equation group. The coupled numerical solution requires the use of a numerical method and COMSOL Multiphysics simulation software.

3.1. Geometric Model Establishment

The geometric model starts from the 370 m position of the transportation lane on the 14221 working face of Xin'an coal mine. The coal mine is an experimental site for determining the effective influence radius of the drilling borehole down the seam. The main mining coal seam is No. 21 coal, and the average coal seam thickness is 4.22 m. The coal seam has a simple structure and extremely low mechanical strength. The soft coal structure is developed and generally classified as classes III–V. The coal is powdery and easily polluted.

The size of the geometric model is 14 m × 14.22 m. Figure 3 shows the geometric diagram of the model. The model is divided into three layers, in which the coal seam is located in the two rock layers, the thickness is 4.22 m, and the upper part is loaded with a stable load of 11.7 MPa. The borehole is located in the middle of the model and has a diameter of 0.089 m and a negative pressure of 13 kPa. The initial gas pressure in the coal seam is 0.9 MPa, and the gas only migrates in the coal seam. The initial time is t = 0 day, and the simulation time is 100 days.

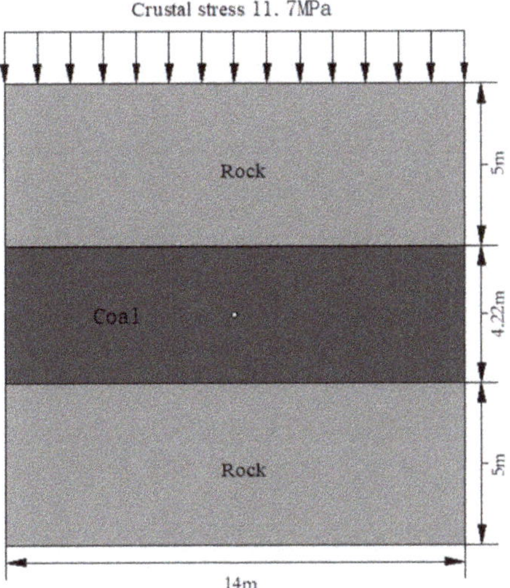

Figure 3. Coal and gas coupling geometric model.

According to the field measurement, the drainage radius of the drilling borehole down the seam is the effective extraction radius of 1.5–1.8 m when the gas is extracted for 30 days. The measured drainage radius of the mine is the effective extraction radius of 3 m when the gas is extracted for 80 days. These data are used to verify the practical application value of the model.

3.2. *Model Parameters*

Table 1 shows the relevant parameters of the fluid–solid coupling model of gas-bearing coal.

Table 1. Material parameters of gas-bearing coal.

Parameter Name	Symbol	Unit	Numerical Value
Coal density	ρ	(kg·m^{-3})	1400
Poisson's ratio of coal	v	/	0.32
Elastic modulus of coal	E	GPa	2.74
Initial porosity of coal	φ_0	/	0.0485
Initial permeability of coal	k_0	m^2	4.59×10^{-17}
Adsorption constant a	a	m^3·Mg^{-3}	35
Adsorption constant b	b	MPa^{-1}	0.762
Gas dynamic viscosity coefficient	μ	Pa·s	1.08×10^{-5}
Moisture	M	%	0.64
Ash	A	%	13.81
Gas density in standard state	ρ_g	kg·m^{-3}	0.716
Initial gas pressure	P_0	MPa	0.9
Drainage negative pressure	P_1	Pa	13,000

3.3. *Input of the Mathematical Model*

The mathematical model in Chapter 2 is inputted into the COMSOL Multiphysics numerical simulation software (COMSOL Multiphysics 5.4.0.388, COMSOL Inc., Stockholm, Sweden, 1986) to verify the rationality of the mathematical model. Figure 4 shows the specific input.

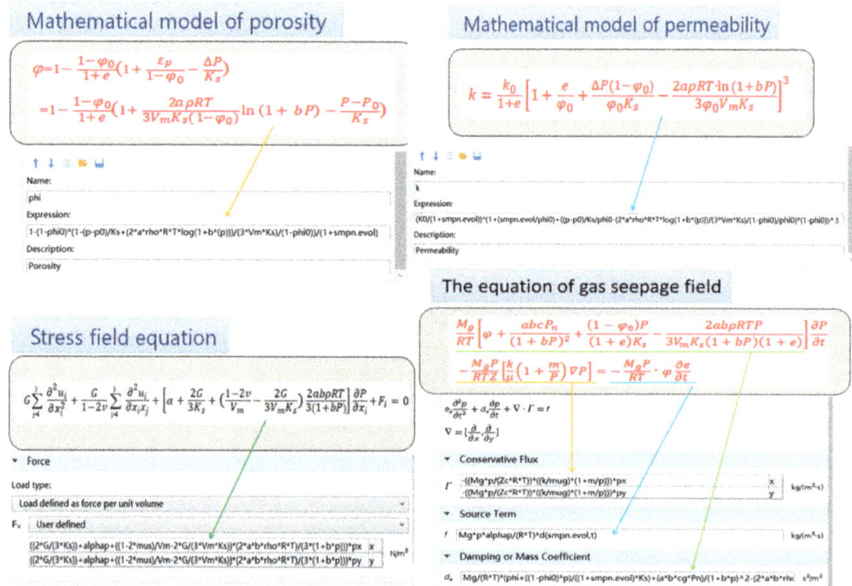

Figure 4. Mathematical model input.

3.4. Initial Conditions and Boundary Conditions

(1) Initial conditions of gas seepage flow field

$$P|_{t=0} = P_0, \tag{32}$$

where P_0 is the initial gas pressure in the coal seam.

(2) Boundary conditions

When $t = 0$, the displacement boundary conditions of the model are:

$$u|_L = u_0, \tag{33}$$

where u_0 is the initial displacement on the boundary.

(3) Stress boundary conditions

$$\rho_{ij}n_j = f_i, \tag{34}$$

where f_i is the surface force on the boundary.

$$q|_L = q_m, \tag{35}$$

where q_m is the gas flow on the boundary.

The roof and floor of the coal seam in the model are rock layers with poor gas permeability. Thus, both are assumed to be flow boundaries, and the flux is zero.

(4) Pressure boundary conditions

$$P|_L = P_m, \tag{36}$$

where P_m is the gas pressure on the boundary.

3.5. Analysis of the Numerical Simulation Results of the Model

Figure 5 shows the distribution of instantaneous gas pressure during the fluid–solid coupling deformation of coal. Figure 6 shows the contour figure of gas pressure for 30 and 80 days of extraction. The graph indicates that the maximum gas pressure in the coal seam is distributed on both sides of the geometric model. When t = 1 day, the gas pressure is 0.6 MPa at 0.32 m away from the drilling hole. With the increase in extraction time, the gas pressure around the borehole decreases gradually. When t = 30 days, the gas pressure is 0.6 MPa at 1.73 m from the drilling hole. When t = 80 days, the gas pressure is 0.6 MPa at 2.76 m away from the drilling hole.

According to Henan's regulations on the prevention and control of coal and gas, the critical value of gas pressure should not be more than 0.6 MPa. This value can be used as a reference for the effective extraction radius of extraction boreholes. The numerical simulation results show that when t = 30 days, the effective extraction radius of the extraction borehole is 1.73 m. When t = 80 days, the effective extraction radius of the extraction borehole is 2.76 m. The drilling drainage radius measured by Henan University of Science and Technology is 1.5–1.8 m for 30 days and 3 m for 80 days [34,35]. The numerical simulation results are basically consistent with the measured results, indicating their strong practical application value.

Figure 7 shows the instantaneous gas pressure evolution curve of the fluid–solid coupling of coal during deformation. The graph shows that the closer the distance from the drainage hole, the faster the gas pressure drops and the more noticeable the pressure relief effect is. The pressure relief of coalbed methane is a nonlinear process. Within a certain limit, the change in gas pressure gradient decreases with the increase in time.

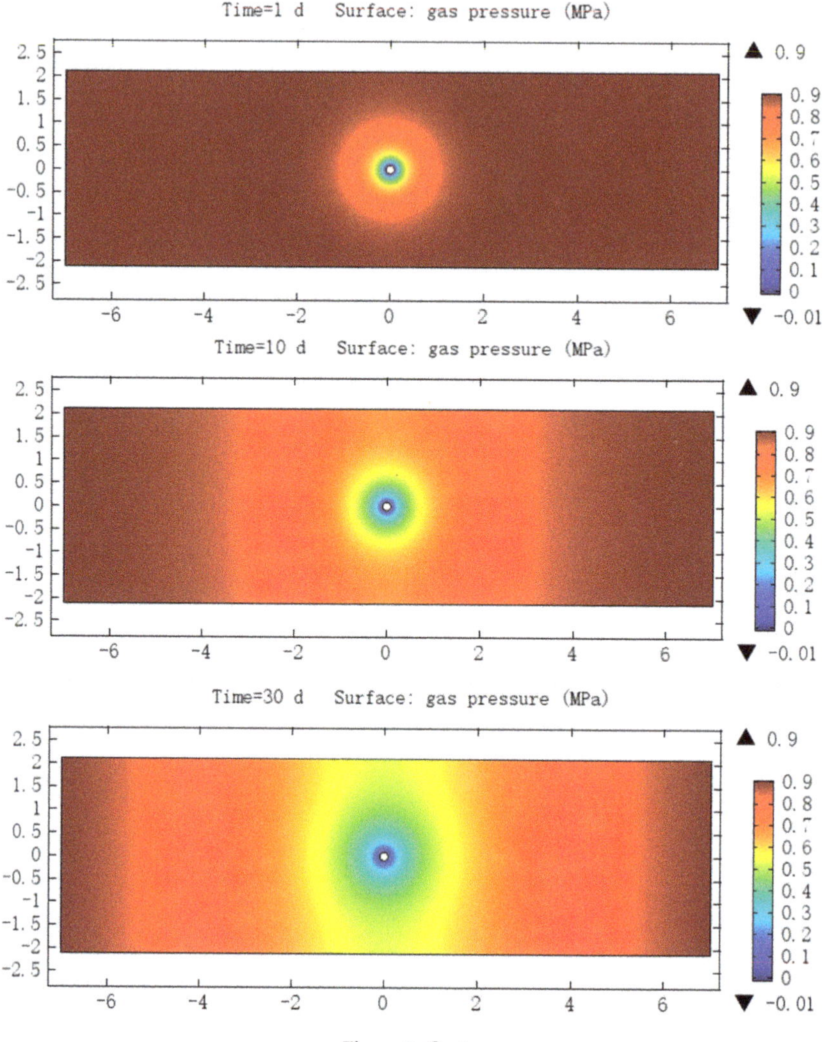

Figure 5. *Cont.*

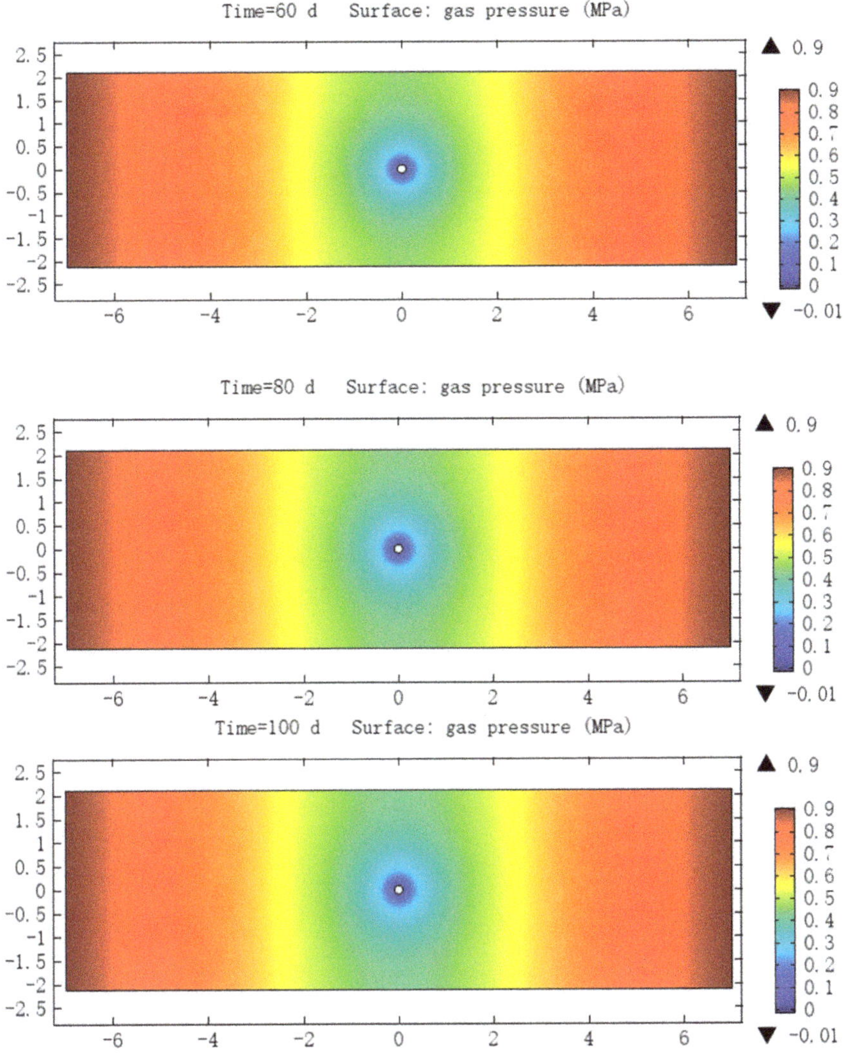

Figure 5. Gas pressure profile at different times.

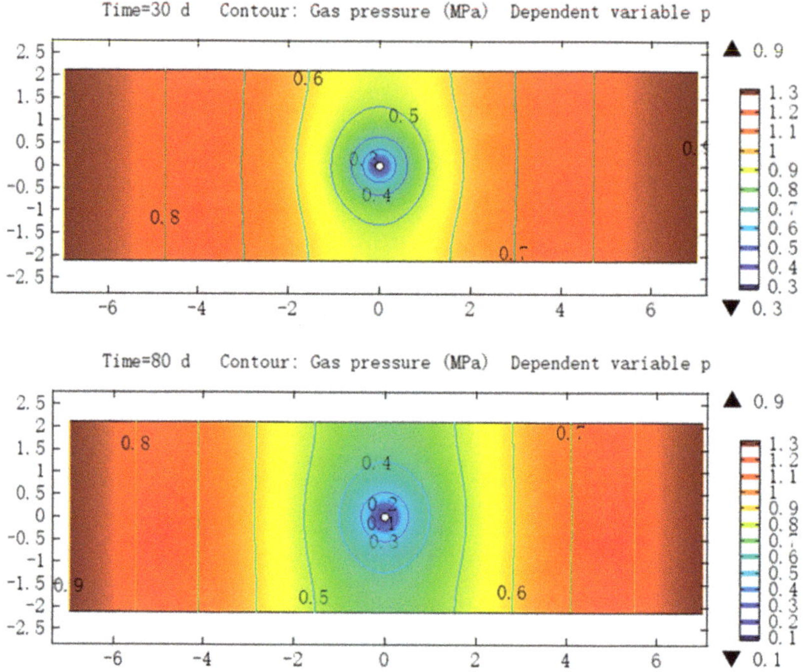

Figure 6. Contour maps of gas pressure at different times.

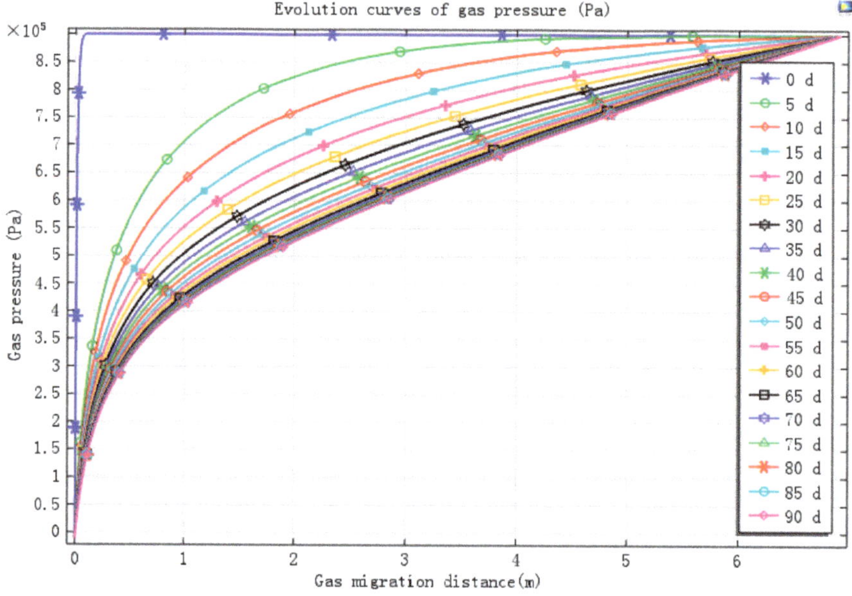

Figure 7. Evolution curves of gas pressure.

Figures 8 and 9 show the evolution curves of the porosity and permeability of coal around the borehole after drilling along the seam, respectively. The graph shows that the minimum values of porosity and permeability is lower than the initial ones at the initial state on day 0 because of the stress concentration area around the drill hole. In this area, the pores are compressed, the pore channel of gas migration and production is small, and the permeability is reduced. With the increase in extraction time, the closer the distance from the drilling hole, and the larger the values of porosity and permeability of coal are. Coal porosity and permeability increase with the increase in time, but the rate of increase declines gradually. The changing trend of permeability is almost similar to that of porosity because the closer the distance from the borehole, the more evident the coal disruption by artificial disturbance is. The coal rock breaks and forms a new pore, and its permeability increases. After gas drainage, the coal gas pressure near the extraction borehole, the gas content, and the gas adsorbed by coal particles are reduced. The coal body shrinks, coal pores and fissures are developed, and the permeability increases. The shrinkage and deformation of the coal–rock matrix are also the main factors determining coal adsorption characteristics. Although the regional pressure of the coal seam cannot be reduced, the gas dissolution in coal induces the shrinkage of the coal matrix, which enlarges the fissures and creates internal ones in the coal seam. In this way, coal porosity, the channel of gas migration and output, and permeability increase.

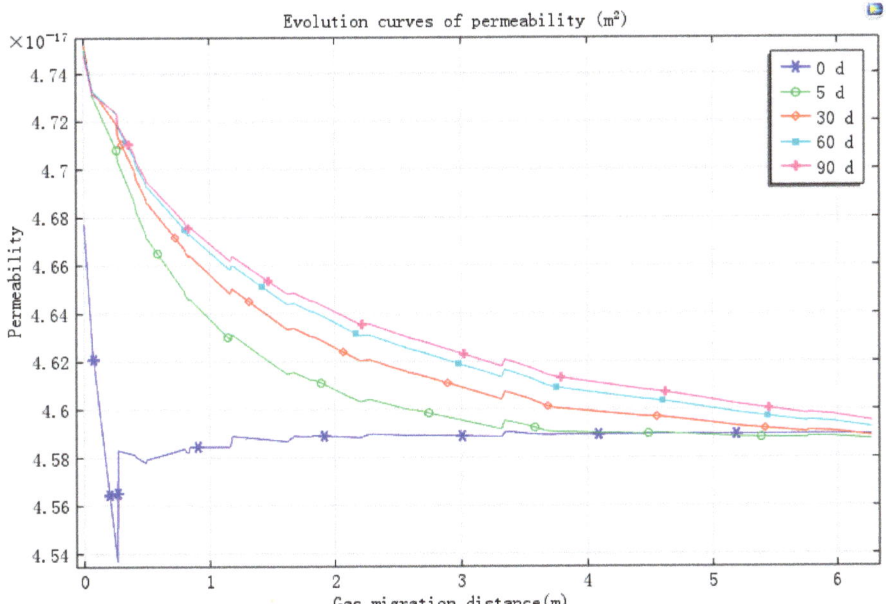

Figure 8. Evolution curves of porosity.

Figures 8 and 9 show that the porosity and permeability of coal also increase from both sides of the coal seam to the direction of the extraction boreholes, respectively. The closer to the extraction boreholes, the greater the increase in range. With the passage of time, the increase rate of permeability and porosity decreases significantly. This situation is mainly due to the artificial disturbance around the drilling hole. The coal becomes unstable and pressure-relieved, and pores and cracks increase. With the decrease in gas content in the coal seam pore, the effective stress of coal increases, the coal compresses, and porosity and permeability decrease. At the same time, the absorbed gas is constantly used to supply pore gas, and the volume shrinkage of coal particles increases the porosity, which affects the increase in coal permeability. With the continuation of extraction time, the increase in porosity and permeability of coal decreases under the two effects.

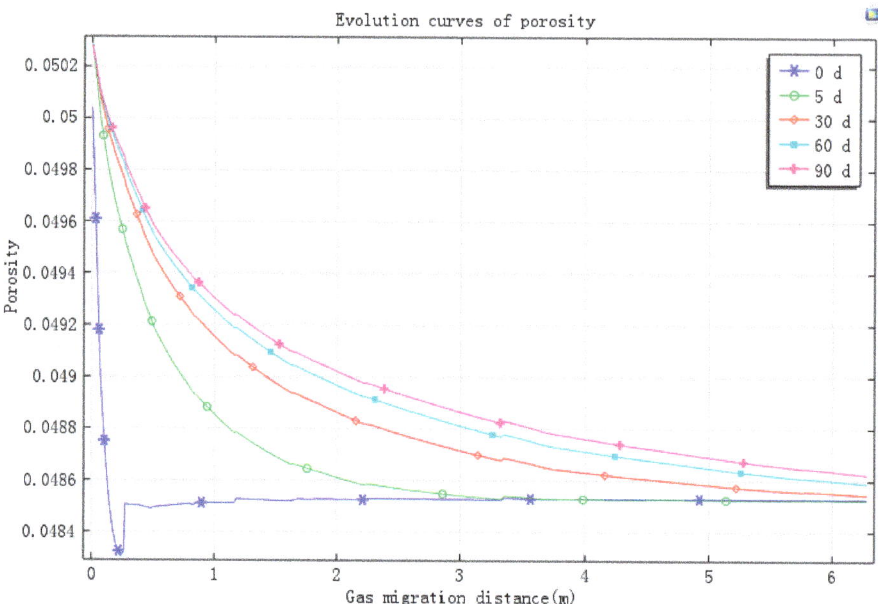

Figure 9. Evolution curves of permeability.

3.6. Discussion

After establishing the fluid–solid coupling model of gas-bearing coal, the model's field practicability needs to be verified. A good model should be applied to practical applications to reflect its importance and guiding significance for practical design and production work. The SF6 gas tracing method is mainly used to measure the effective extraction radius of the bedding borehole in this coal mine. This method overcomes many of the limitations of traditional measuring methods, such as numerous drilling holes, complicated working procedures, high requirement for sealing quality, long time, and large deviation, leading to a highly accurate, reliable, and simple measurement [36]. The measured effective extraction radius is 1.5–1.8 m at 30 days of extraction. The relevant parameters of the measured coal seams are inputted into the mathematical model, and the coupled numerical solution is calculated with COMSOL Multiphysics simulation software. The calculation results show that the effective extraction radius is 1.73 m. Afterward, the effective extraction radius is measured independently in the mine. The measured effective extraction radius is 3 m at 80 days, and the effective extraction radius is 2.76 m by numerical simulation.

A comparison with the results of field measurement shows that the numerical simulation findings of this model are close to actual conditions. The results provide reference values for the rational layout of extraction boreholes in coal mines. Based on the fluid–solid coupling equation, the effective extraction radius simulation of drilling boreholes along the seam reveals the gas pressure distribution state at different extraction times. This condition eliminates the blank belt of gas drainage with the time effect, changes the current disadvantage of the restricted effective radius test of gas extraction by many factors, and ensures the safe production of mines to a great extent. Moreover, the process of establishing the model reveals the mechanism of gas migration in the coal seam and describes factors one by one from the perspectives of coal pore, coal particle adsorption expansion deformation, compression deformation, stress field, and gas seepage flow field. Then, a relatively complete practical mathematical model is obtained. This model considers the effects of adsorption expansion and the Klinkenberg effect on gas migration in coal seams. The model improves the application of fluid–solid coupling theory in coal. The numerical simulation results show the dynamic changes in gas pressure in the coal seam and the dynamic evolution of porosity and permeability. The model can simulate the dynamic evolution law of coalbed methane in the corresponding coal seam of the coal mine based on different coal seam parameters, such as the influence of different types of water and ash on gas migration and various extraction negative pressures on the drainage effect. The numerical simulation shows the extraction effect and scope of gas drainage boreholes and provides important theoretical support and basis for the rational optimization and layout of gas drainage boreholes in mines.

4. Conclusions

The mathematical model combines the findings of previous studies on the effects of adsorption expansion and the Klinkenberg effect on gas migration in coal seams. On the basis of the basic definitions of porosity and permeability, a mathematical model of dynamic evolution of porosity and permeability is derived. On the basis of fluid–solid coupling theory, a mathematical model of fluid–solid coupling with gas-bearing coal is established.

According to the results of the numerical simulation, when t = 30 days and t = 80 days, the effective extraction radius of the bedding borehole reaches 1.73 m and 2.76 m, respectively. The simulation results are consistent with the actual measured extraction radius values. Figure 5 shows the effect of gas pressure reduction around the borehole with the continuous change in simulation time. The gas pressure decreases, and the porosity and permeability of coal increase with the increase in gas extraction time. In addition, the growth rate of permeability and porosity decreases with the increase in gas extraction time. These results are consistent with the field permeability test law and can be used as reference to further understand the mechanism of gas extraction and mine gas control. The research results also have theoretical significance and practical application value.

The dynamic evolution mathematical model of fluid–solid coupling for gas bearing coal can reflect the coalbed methane migration in a mining area of the mine through the coal seam parameters measured by coal miners. According to the simulation results of the model, the effective extraction radius of the borehole can be predicted. Thus, the extraction borehole can be optimized and reasonably arranged for safety reasons and scientific purposes, to effectively control mine gas, and to provide strong support for decision-makers as they formulate efficient coal mining schemes.

Although the model is an extension of the theory of fluid–solid coupling, its simulation results can be suitable for field applications. However, several problems should be considered in the multi-field coupling model, thus indicating the need for further research and improvement. For example, with the increase in mining depth, the influence of temperature on gas adsorption and migration in the coal seam considerably influences gas extraction. Another example is the influence of movable and residual water in the coal seam on expansion stress in the gas-bearing coal seam [37]. Studying these problems is crucial for the further improvement of the model.

Author Contributions: S.H. developed the analytical models, analyzed the data, and wrote the paper; X.L. (Xianzhong Li) gave valuable advice and contributed to the manuscript editing. X.L. (Xiao Liu) provided financial support for this paper. All authors have read and agreed to the published version of the manuscript.

Funding: This research was funded by Project of science and technology of Henan Province in 2015 (152102310095) and Project of science and technology of Jiaozuo science and Technology Bureau (Applied Basic Research) (2014400018).

Acknowledgments: This work was financially supported by the National Natural Science Foundation of Henan Science and Technology Project in 2015 (152102310095) and Science and Technology Research Project of Jiaozuo Science and Technology Bureau (Applied Basic Research) (2014400018). The support provided by these entities is gratefully acknowledged. I would also like to thank Liu and Li for their earnest guidance.

Conflicts of Interest: The authors declare no conflict of interest.

References

1. Liu, C.; Zhu, Q.; Wei, F.; Rao, W.; Liu, J.; Hu, J.; Cai, W. An integrated optimization control method for remanufacturing assembly system. *J. Clean. Prod.* **2019**, *248*, 119261. [CrossRef]
2. Liu, C.; Zhu, Q.; Wei, F.; Rao, W.; Liu, J.; Hu, J.; Cai, W. A review on remanufacturing assembly management and technology. *Int. J. Adv. Manuf. Technol.* **2019**, *105*, 4797–4808. [CrossRef]
3. Cai, W.; Liu, C.; Jia, S.; Chan, F.T.; Ma, M.; Ma, X. An emergy-based sustainability evaluation method for outsourcing machining resources. *J. Clean. Prod.* **2019**, *245*, 118849. [CrossRef]
4. Hou, Z.; Xie, H.; Zhou, H.; Were, P.; Kolditz, O. Unconventional gas resources in China. *Environ. Earth Sci.* **2015**, *73*, 5785–5789. [CrossRef]
5. Zhu, W.C.; Wei, C.H.; Liu, J.; Xu, T.; Elsworth, D. Impact of gas adsorption induced coal matrix damage on the evolution of coal permeability. *Rock Mech. Rock Eng.* **2013**, *46*, 1353–1366. [CrossRef]
6. Sommerton, W.J.; Soylemezoglu, I.M.; Dudley, R.C. Effect of stress on permeability of coal. *Int. J. Rock Mech. Min. Sci. Geomech. Abstr.* **1975**, *12*, 129–145. [CrossRef]
7. Brace, W.F. A note on permeability changes in geologic material due to stress. *Pure Appl. Geophys.* **1978**, *116*, 627–632. [CrossRef]
8. Mckee, C.R.; Bumb, A.C.; Koenig, R.A. Stress-dependent permeability and porosity of coal. In *Rocky Mountain Association of Geologists Guidebook*; Fassett, J.E., Ed.; Rocky Mountain Association of Geologists: Denver, CO, USA, 1988; pp. 143–153.
9. Enever, J.R.E.; Henning, A. The relationship between permeability and effective stress for Australian coal and its implications with respect to coalbed methane exploration and reservoir model. In *Proceedings of the 1997 International Coalbed Methane Symposium*; University of Alabama: Tuscaloosa, AL, USA, 1997; pp. 13–22.
10. Sun, P.D. Study on the Flow Law of coal seam gas flow field. *J. Coal Mine* **1987**, *12*, 74–82.
11. Sun, P.D. Research on gas dynamics model. *Coalf. Geol. Explor.* **1993**, *21*, 32–40.
12. Sun, P.D. Coal gas dynamics and its applications. *Sci. Geol. Sin.* **1994**, *3*, 66–72.
13. Li, X.C.; Guo, Y.Y.; Wu, S.Y. The relationship among coal adsorption swelling deformation, porosity and permeability. *J. Taiyuan Univ. Technol.* **2005**, *36*, 264–265.
14. Harpalani, S.; Chen, G. Estimation of Changes in Fracture Porosity of Coal with Gas Emission. *Fuel* **1995**, *75*, 1491–1498.
15. Yunqi, T.A.O.; Jiang, X.U.; Shuchun, L.I.; Shoujian, P.E.N.G. Advances in study on seepage flow property of coalbed methane. *Coalf. Geol. Surv.* **2009**, *37*, 1–5.
16. Yin, G.Z.; Wang, D.K.; Zhang, D.M.; Hang, G. Solid-gas coupling dynamic model and numerical simulation of coal containing gas. *Rock Soil Mech.* **2008**, *30*, 1431–1435.
17. Zhu, W.C.; Liu, J.; Sheng, J.C.; Elsworth, D. Analysis of coupled gas flow and deformation process with desorption and Klinkenberg effects in coal seams. *Int. J. Rock Mech. Min. Sci.* **2007**, *44*, 971–980. [CrossRef]
18. Xia, T.Q.; Gao, F.; Kang, J.H.; Wang, X.X. A fully coupling coal–gas model associated with inertia and slip effects for CBM migration. *Environ. Earth Sci.* **2016**, *75*, 582. [CrossRef]
19. Lu, S.Q.; Cheng, Y.P.; Li, W. Model development and analysis of the evolution of coal permeability under different boundary conditions. *J. Nat. Gas Sci. Eng.* **2016**, *31*, 129–138. [CrossRef]
20. Chen, Y.; Xu, J.; Peng, S.; Yan, F.; Fan, C. A Gas-Solid-Liquid Coupling Model of Coal Seams and the Optimization of Gas Drainage Boreholes. *Energies* **2018**, *11*, 560. [CrossRef]
21. Liu, Y.B.; Cao, S.G.; Li, Y.; Wang, J.; Guo, P.; Xu, J.; Bai, Y.J. Experimental study on the deformation effect of coal adsorptive expansion. *J. Rock Mech. Eng.* **2010**, *29*, 2484–2491.
22. Yang, C.; Li, X.; Ren, Y.; Zhao, Y.; Zhu, F. Statistical analysis and countermeasures of gas explosion accident in coal mines. *Procedia Eng.* **2014**, *84*, 166–171.
23. Xie, Z.C.; Zhang, D.M.; Song, Z.L.; Li, M.H.; Liu, C.; Sun, D.L. Optimization of Drilling Layouts Based on Controlled Presplitting Blasting through Strata for Gas Drainage in Coal Roadway Strips. *Energies* **2017**, *10*, 1228. [CrossRef]
24. Tao, Y.Q. THM coupling model of gas-bearing coal and Simulation of coal and gas outburst. Ph.D. Thesis, Chongqing University, Chongqing, China, 2009.
25. Guo, P.; Cao, S.G.; Zhang, Z.G.; Li, Y.; Liu, Y.B.; Li, Y. Solid-gas coupling mathematical model and numerical simulation of gas-bearing coal. *J. Coal Sci.* **2012**, *37*, 330–336.

26. Ettinger, I.L. Swelling stress in the gas-coal system as an energy source in the development of gas bursts. *Sov. Min. Sci.* **1979**, *15*, 494–501. [CrossRef]
27. Lu, P.; Shen, Z.; Zhu, G.W. Effective stress and mechanical deformation failure characteristics of coal containing gas. *J. Univ. Sci. Technol. China* **2001**, *3*, 687–693.
28. Lu, P.; Shen, Z.W.; Zhu, G.W.; Fang, E.C. Percolation Characterization and Experimental Study of Rock Sample in the Whole Process of Stress and Strain. *J. China Univ. Sci. Technol.* **2002**, *32*, 678–684.
29. Wu, S.Y.; Zhao, W. Effective stress analysis of coal containing adsorbed coalbed methane. *J. Rock Mech. Eng.* **2005**, *24*, 1674–1678.
30. Wu, S.Y.; Zhao, W. Analysis of effectives in adsorbed methane-coal system. *Chin. J. Rock Mech. Eng.* **2005**, *24*, 1674–1678.
31. Sun, P. *Introduction to Multiphysical Field Coupling Model. and Numerical Simulation*; China Science and Technology Press: Beijing, China, 2007.
32. Zhou, S.; Lin, B.Q. *Theory of Coal Seam Gas Occurrence and Flow*; Coal Industry Press: Beijing, China, 1990.
33. Yue, G.W.; Liu, J.; Wang, Z.F. Study on Coal Seam Gas Content Based on Real Gas State Equation. *J. Saf. Environ.* **2014**, *14*, 73–77.
34. Liu, S.J.; Ma, G.; Lu, J.; Lin, B.Q. Relative pressure determination technology for effective radius found on gas content. *J. China Coal Soc.* **2011**, *36*, 1715–1719.
35. Liu, Q.Q.; Cheng, Y.P.; Wang, H.F.; Liu, H.Y.; Liu, J.J. Numerical Resolving of Effective Methane Drainage Radius in Drill Hole along Seam. *Coal Min. Technol.* **2012**, *17*, 5–7.
36. Chen, J.Y.; Ma, P.L.; Kong, Y.H.; Ma, C. SF6 gas tracer method for determining effective radius of gas drainage in boreholes. *Saf. Coal Mines*. [CrossRef]
37. Hu, Y.; Shao, Y.; Lu, Y.; Zhang, Y.F. Experimental study on occurrence models of water in pores and the influencing to the development of tight gas reservoir. *Nat. Gas Geosci.* **2011**, *22*, 176–181.

Article

An Optimization Approach Considering User Utility for the PV-Storage Charging Station Planning Process

Yingxin Liu [1,*], Houqi Dong [1], Shengyan Wang [1], Mengxin Lan [1], Ming Zeng [1], Shuo Zhang [1], Meng Yang [2] and Shuo Yin [2]

1 School of Economics and Management, North China Electric Power University, Beijing 102206, China; dong_ncepu@163.com (H.D.); wangshengyan222@163.com (S.W.); lanmengxin163@163.com (M.L.); zengmingbj@vip.sina.com (M.Z.); zhangshuo@ncepu.edu.cn (S.Z.)

2 Economic Research Institute of State Grid Henan Electric Power Company, Zhengzhou 450052, China; yangmeng0372@163.com (M.Y.); yinshuo1111@163.com (S.Y.)

* Correspondence: ttkllyx1993@163.com; Tel.: +86-010-6177-3141

Received: 28 November 2019; Accepted: 6 January 2020; Published: 8 January 2020

Abstract: Based on the comprehensive utilization of energy storage, photovoltaic power generation, and intelligent charging piles, photovoltaic (PV)-storage charging stations can provide green energy for electric vehicles (EVs), which can significantly improve the green level of the transportation industry. However, there are many challenges in the PV-storage charging station planning process, making it theoretically and practically significant to study approaches to planning. This paper promotes a bi-level optimization planning approach for PV-storage charging stations. First, taking PV-storage charging stations and EV users as the upper- and lower-level problems, respectively, during the planning process, a bi-level optimization model for PV-storage charging stations considering user utility is established for capacity allocation and user behavior-based electricity pricing. Second, the model is converted into a single-level mixed-integer linear programming model using the piecewise linear utility function, Karush–Kuhn–Tucker (KKT) conditions, and linearization methods. Finally, to verify the validity of the proposed model and the solution algorithm, a commercial solver is used to solve the optimization model and obtain the planning scheme. The results show that the proposed bi-level optimization model can provide a more economical and reasonable planning scheme than the single-level model, and can reduce the investment cost by 8.84%, operation and maintenance cost by 13.23%, and increase net revenue by 5.11%.

Keywords: PV-storage charging stations; green energy; planning process; bi-level optimization; user utility

---

1. Introduction

Advances in energy storage technology and grid intelligence and increased electric vehicle (EV) ownership have greatly promoted the development of EV charging infrastructure. However, the existing charging stations are neither low-carbon nor friendly to the distribution system because they have no energy storage facilities and must obtain electric power from the distribution network [1]. In order to take advantage of the bi-directional flow between renewable energy and energy storage systems, green photovoltaic (PV)-storage charging stations, installed with both a photovoltaic power generation system and an energy storage system, were developed based on the existing traditional charging stations. A PV-storage charging station is a microgrid that integrates the technologies of photovoltaic power generation, energy storage (ES), and smart charging station (SCS). The associated operation between photovoltaic power generation and EV charging and discharging can help promote the efficient consumption of renewable energy on-site and fulfill the EV load demand. Also, the introduction of an energy storage system can effectively alleviate the impact of EV charging on the regional distribution network. This microgrid is an organically integrated source–storage–load system and meets the requirements for new-generation power systems: clean and efficient, green and low carbon, safe and controllable. At present, many countries, such as the United States, the United Kingdom, the Netherlands, and Malaysia, have built large-scale solar-powered EV charging stations. China has also launched some PV-storage charging EV station demonstration projects in cities such as Dongguan, Shanghai, and Qingdao. PV-storage charging stations will develop rapidly with advanced PV-storage technology and decreased economic costs in the coming days.

There are currently many studies on EV charging station planning. One study [2] summarized the main theoretical methods and research directions. Traditional charging station planning mainly aims to meet the increasing EV load demand and focuses on the siting and sizing of charging stations in the distribution network. However, with the integration of renewable energy and energy storage systems, capacity allocation and operational optimization for PV-storage charging stations have become hot research topics.

The literature on PV-storage charging station planning is mainly divided into two categories. The first category focuses on exploring the location and capacity allocation of charging stations from the perspective of distribution networks [3–5] or transportation networks [6–8], in the case of already known network structure. Charging station planning in a distribution network considers factors such as the environment, power quality [3], distribution feeder layout and availability [4], and operation safety and cost optimization [5]. When it comes to the transportation network, charging station planning considers the temporal and spatial dynamics of EV movement [6] and the spatial distribution of EVs [7], and a planning model was established using queuing theory and graph theory [8]. Considering the dual factors of distribution and traffic networks, some studies [9–12] carried out charging station planning with the objective of maximizing benefits and minimizing energy loss, while other studies [13,14] evaluated the planning results.

The second category focuses on the optimization of the internal design and energy capacity of PV-storage charging stations. This kind of study focuses mainly on internal charging stations, and rarely consider the constraints of the external network, which is usually regarded as an infinite source. Internal optimization is aimed at determining the compactness of internal facilities at charging stations, including the number of facilities and photovoltaic units and the ES capacity [15]. Commonly targeted at minimizing operation cost, the optimization model was established based on constraints such as operation, cost, and equipment utilization [16] and then was solved using corresponding particle swarm optimization algorithms [17] and NSGA-II (Non-dominated Sorting Genetic II Algorithm) [18]. In addition to cost, another study [19] also considered queuing time. It can be easily seen from the above studies that current PV-storage charging station planning rarely considers uncertain factors such as distributed generation, user behavior, and electricity price.

This paper, therefore, aims to study the internal optimization of PV-storage charging stations under uncertain conditions. Taking the uncertain factors—the charging station operator (CSO) and EV users—as the upper- and lower-level problems, a user behavior-based bi-level optimization model for the PV-storage model was established to determine the capacity allocation and electricity pricing. In the upper-level model, EV charging capacity constraints are considered, and there is an assumption that each EV user charges at a charging station only during a single time period t, meaning the EV can be charged to the maximum storage capacity during that time period, in order to simplify the planning problem. In the lower-level model, the real-time electricity price is considered to calculate the expected revenue of PV-storage charging stations. The model was then converted into a single-level mixed-integer linear programming using the piecewise linear utility function, Karush–Kuhn–Tucker (KKT) conditions, and linearization methods. The obtained linear programming problem was solved to compare and analyze the quantitative influence of uncertain variables on charging station planning.

2. Bi-Level Optimization Model for PV-Storage Charging Station Planning

2.1. PV-Storage Charging Station System

The PV-storage charging station system, the upper-level problem studied in this paper, is shown in Figure 1. A PV-storage charging station is usually composed of multiple EV charging facilities (CFs), energy storage (ES) units, and photovoltaic units. A charging station that is connected to an external distribution network via a power electronic converter can obtain its energy supply from the main power grid in case of insufficient power generation. ES units are installed in the system in order to make full use of renewable energy (RES). When the PV output power is higher than the charging demand of the connected EV, the ES can start power charging to store excess energy. When the PV output is insufficient, the ES can perform power discharging to supply energy. Therefore, the PV-storage charging station can achieve minimal energy consumption by coordinately controlling the ES and PV output. All of the above components are assumed to be integrated into the charging station based on the AC interface and controlled centrally by the system operator.

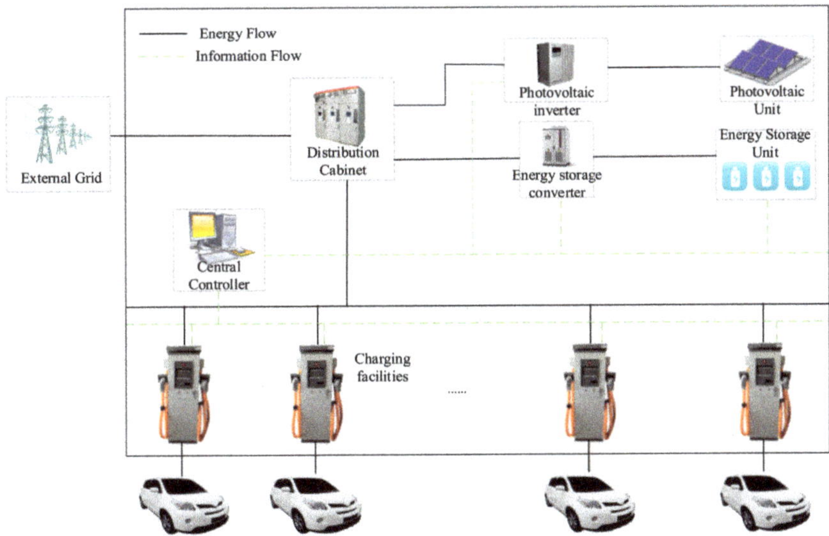

Figure 1. Internal structure of photovoltaic (PV)-storage charging station and energy and information flows.

Both the charging station operator (CSO) and EV users are considered in this paper. The CSO is responsible for the construction and operation of the charging station and will consider the EV users' selection strategy in charging storage planning. The CSO considers the construction cost, operation, and maintenance cost, and power sales revenue to determine capacity allocation and electricity pricing, based on the premise of satisfying the internal balance and security constraints. EV users can choose from the electricity pricing schemes offered by the CSO to obtain better economic benefits.

This paper assumes that the investment and operation of the PV-storage charging station are performed by the CSO. The PV-storage charging station earns revenue by providing EV users with energy from PV units or the external power grid. In addition, the CSO can also profit from trading with the power grid via PV power generation in a deregulated environment. The electricity exchange price is determined through bilateral negotiation between the CSO and the main power grid and remains unchanged during the contract period. The operational performance of the charging station varies with floating EV charge demand. The CSO must fully consider the potential response of EV users, participate in the design of electricity pricing, and coordinately determine the optimal configuration and control strategies in order to maximize profits.

On the other hand, EV users have to decide on their charging load at the charging station each time they charge (provided that the minimum charging demand is large enough to power the next driving distance), and their charging demand is affected by the electricity price. Therefore, the optimal decision for EV users depends not only on their own preference but also on the CSO's electricity pricing, which may conflict with their goals.

The optimal planning and decision problem for the CSO can be expressed as a bi-level planning model, as shown in Figure 2. The upper-level problem represents the CSO's decision in relation to the PV, ES, CS configuration, and electricity pricing to maximize the expected revenue. The lower-level problem considers the CSO's pricing scheme and predicts the EV users' response by calculating the optimal charging benefits for each user. The price implemented by the CSO is in real-time and can be adjusted to influence the users' behavior so that they can be encouraged to actively respond to the RES output curve. The construct of this model will help to obtain an optimal capacity configuration and price scheme.

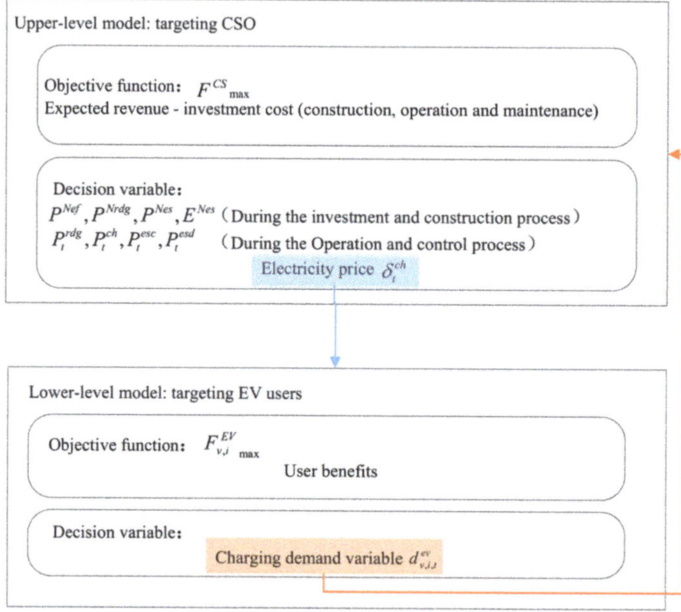

Figure 2. Bi-level model framework for optimizing the design and decision-making of PV-storage charging stations. CSO, charging station operator. EV, electric vehicle.

2.2. Upper-Level Model of PV-Storage Charging Station Planning

2.2.1. Objective Function

According to the description of the model framework in the previous section, the upper-level optimization problem involves optimal configuration and pricing of PV-storage charging stations, i.e., determining PV, ES, and CS capacity and pricing. The objective function for the upper-level model is shown in Equation (1):

$$\underset{\substack{P^{Nrdg}, P^{Ncf}, P^{Nes}, E^{Nes}, \delta_t^{ch} \\ P_t^{rdg}, P_t^{esc}, P_t^{esd}, P_t^{ch}, P_t^{\text{int}}}}{\text{Max}} \quad F^{CS} = B^{Ope} - C^{Inv} \tag{1}$$

$$C^{Inv} = k^{cf}c^{cf}P^{Ncf} + k^{rdg}c^{rdg}P^{Nrdg} + k^{es}(c^{esp}P^{Nes} + c^{ese}E^{Nes}) \tag{2}$$

$$B^{Ope} = \theta \cdot \sum_{t=1}^{T}\left(\delta_t^{ch}P_t^{ch} - \delta_t^{int}P_t^{int}\right)\Delta t - \left(c^{cfm}P^{Ncf} + c^{rdgm}P^{Nrdg} + c^{esm}E^{Nes}\right) \tag{3}$$

$$k = \zeta(1+\zeta)^d / \left[(1+\zeta)^d - 1\right] \tag{4}$$

In the bi-level model, the objective function for the upper-level model is the maximum revenue of the CSO, and expected revenue is the difference between the expected revenue (B^{Ope}) and the investment cost (C^{Inv}). In Equation (2), C^{Inv} represents the investment cost, which is correlated with the construction cost of the CF, PV, ES facilities, land lease, and other relevant expenses. P^{Ncf}, P^{Nrdg}, P^{Nes}, and E^{Nes} represent the rated installed power of CF and PV, ES, and the installed storage capacity of ES, respectively. c^{cf}, c^{rdg}, c^{esp}, and c^{ese} represent the unit investment cost of CF, PV, and ES facilities. k^{cf}, k^{rdg} and k^{es} represent the capital recovery factor of CF, PV, and ES, which can be calculated by Equation (4). In Equation (3), B^{Ope} represents expected revenue, including the charging service revenue from EV users and revenue from interactive transactions with the grid. θ represents the number of days in a year. δ_t^{ch} and δ_t^{int} represent the electricity price per kWh sold by the CSO and traded with the main grid, respectively. P_t^{ch} and P_t^{int} represent the charging power of EV users at time t and power during the interactive transaction with the main power grid. Δt represents the time of each period, which is defined 0.5 h in this paper. When P_t^{int} is positive, electricity is purchased from the main power grid. When P_t^{int} is negative, electricity is sold to the main power grid. c^{cfm}, c^{rdgm} and c^{esm} represent the annual unit operation cost of the CF, PV, and ES facilities, respectively. In Equation (4), k represents the capital recovery factor, and d and ζ represent the facility lifetime and discount rate, respectively.

2.2.2. Constraints

The CSO needs to consider constraints during facility construction and operation, including constraints of investment, balance, and security. The specific constraints are shown in Equations (5)–(19).

Equations (5)–(9) are investment and price constraints:

$$0 \le P^{Ncf} \le P_{\max}^{Ncf} \tag{5}$$

$$0 \le P^{Nrdg} \le P_{\max}^{Nrdg} \tag{6}$$

$$0 \le P^{Nes} \le P_{\max}^{Nes} \tag{7}$$

$$0 \le E^{Nes} \le E_{\max}^{Nes} \tag{8}$$

$$0 \le \delta_t^{ch} \le \delta_{\max}^{ch} \ \forall t = 1, \cdots, T \tag{9}$$

Equations (5)–(7) represent the maximum installable rated power constraints for CF, PV, and ES in the charging station, and their investment capacity has an upper limit due to capital or space limitations. Equation (8) represents the maximum installable storage capacity of ES. Equation (9) represents the price constraints by the CSO that can help ensure the stability of the EV charging market.

Equations (10)–(19) are operation constraints:

$$-P_{\max}^{tr} \le P_t^{int} \le P_{\max}^{tr} \ \forall t = 1, \cdots, T \tag{10}$$

$$0 \le P_t^{rdg} \le P^{Nrdg}\gamma_t^{rdg} \ \forall t = 1, \cdots, T \tag{11}$$

$$0 \le P_t^{ch} \le P^{Ncf} \ \forall t = 1, \cdots, T \tag{12}$$

$$0 \le P_t^{esc} \le P^{Nes} \ \forall t = 1, \cdots, T \tag{13}$$

$$0 \le P_t^{esd} \le P^{Nes} \ \forall t = 1, \cdots, T \tag{14}$$

Equation (10) specifies that the exchange power between the charging station and the main power grid should not exceed the substation transformer capacity. Equations (11) and (12) specify the maximum PV scheduling power and CF capacity constraint, where γ_t^{rdg} represents the photovoltaic output factor. Equations (13) and (14) represent the charging and discharging constraints of the ES, where P_t^{esc} and P_t^{esd} represent charging and discharging power, respectively, at time t.

$$E_t^{es} = E_{t-1}^{es} + P_t^{esc}\eta^{esc}\Delta t - P_t^{esd}\Delta t/\eta^{esd} \ \forall t = 1, \cdots, T \tag{15}$$

$$E^{Nes}SOC_{\min}^{es} \le E_t^{es} \le E^{Nes}SOC_{\max}^{es} \ \forall t = 1, \cdots, T \tag{16}$$

$$E_0^{es} = E_T^{es} \tag{17}$$

$$P_t^{rdg} + P_t^{int} = P_t^{ch} + P_t^{esc} - P_t^{esd} \ \forall t = 1, \cdots, T \tag{18}$$

$$P_t^{ch}\eta^{cf}\Delta t = \sum_{v\in\Omega_V} d_{v,t}^{ev} f_{v,t} \ \forall t = 1, \cdots, T \tag{19}$$

Equations (15) and (16) show the energy change characteristics and their charge and discharge state (SOC) constraints during operation of the ES unit, where $SOC_{\min}^{es}$ and $SOC_{\max}^{es}$ represent the minimum and maximum charge and discharge states, and η^{esc} and η^{esd} represent charge and discharge efficiency. Equation (17) specifies that the available capacity (E_T^{es}) of the ES at the end of the dispatching operation must be consistent with the start capacity (E_0^{es}) [20] to ensure the sustainable performance of the charging station. Equation (18) is the power balance constraint of the PV-storage charging station. Equation (19) shows that the power load provided by the CSO must always satisfy the charging demand of the EV user, where $d_{v,t}^{ev}$ represents the power demand of a v-type electric vehicle at time t, and $f_{v,t}$ represents the number of v-type vehicles in use. Equation (19) is a constraint that connects the upper and lower levels, including the upper variable P_t^{ch} and lower decision variable, where Δt represents the time scale, set as 0.5 h in this paper. For the sake of simplicity, it is assumed that each EV user charges in the charging station only during a single time period t, regardless of the charging behavior across multiple time periods.

2.3. Lower-Level Model for PV-Storage Charging Station Planning

2.3.1. Objective Function

The lower-level optimization problem in Figure 2 is the behavior selection problem of EV users based on real-time pricing. The charging demand at each period is determined. The objective function of the lower-level problem is shown in Equation (20):

$$\underset{d_{v,t}^{ev}}{\mathrm{Max}} F_v^{EV} = [U_v(d_{v,t}^{ev}) - \delta_t^{ch} d_{v,t}^{ev}] \tag{20}$$

The objective function for the lower-level problem is the maximum EV user benefits, and the benefits are expressed as the difference between user utility $U_v(d_{v,t}^{ev})$ and the electricity charging cost. U_v in the equation represents the utility function of the electric vehicle user, and its specific form will be described in detail in the next section, and the other symbols have the same meanings as in the previous section.

2.3.2. Constraints

Equation (21) is the EV power demand constraint, which is used to ensure that the user's charging demand meets the minimum charging load and maximum charging limit, i.e., the EV capacity is kept between the minimum mileage power and the rated battery capacity.

$$d_{v,\min}^{ev} \le d_{v,t}^{ev} \le d_{v,\max}^{ev} \tag{21}$$

The essence of the lower-level optimization is that EV users make their choices according to the electricity price, i.e., users just need to consider their own benefits and costs and do not need to consider the constraints of the charging station and the power grid, since the capacity configuration and price-based optimal charging load in each period are determined by the CSO in the upper-level model.

3. Algorithms for Bi-Level PV-Storage Charging Station Planning

3.1. KKT Algorithm Analysis

In the previous section, a bi-level planning model was established to solve user behavior-based PV-storage charging station planning, but the specific form of the benefits function for the lower-level problem was not described. This section elaborates EV user benefits using the piecewise linear function, where the lower-level problem is linearized and, based on this, the KKT conditions are used to convert the bi-level optimization problem into a single-level optimization problem. Then linear approximation is used to transform the single-level problem into a mixed-integer linear programming problem that is finally solved using the classical solutions.

As shown in Figure 3, the bi-level optimization problem is converted into a single-level mixed-integer linear programming problem through the following three steps: first, the lower-level function is piecewise linearized to represent user utility. Second, the KKT conditions, which are used to replace the lower-level problem, are included in the upper-level problem, thus obtaining an equivalent single-level problem. Finally, the nonlinear terms in the obtained single-level problem are linearized to obtain a single-level mixed-integer linear program. In this way, the original problem is transformed and can be efficiently solved.

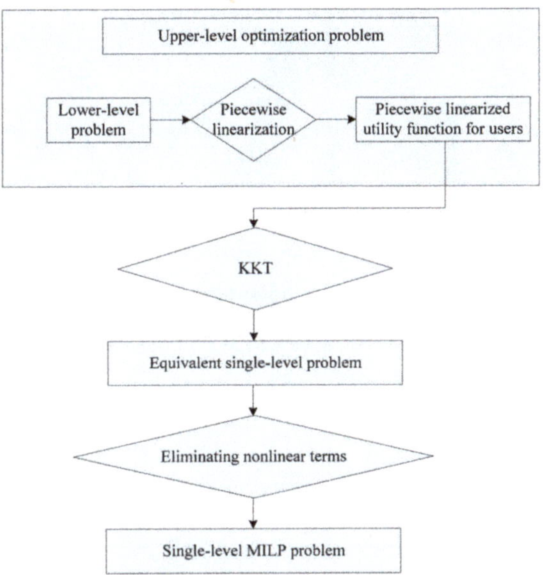

Figure 3. Bi-level planning algorithm flowchart. KKT, Karush–Kuhn–Tucker. MILP, mixed-integer linear programming.

3.2. Bi-Level Optimization Model Solution

3.2.1. Linear Description of Lower-Level Problem

The utility function is used to measure consumer satisfaction as a function of consumption of a vested commodity combination. The consumers in this paper are EV users, and the commodity is charging power. According to the utility function, the utility increases as a function of increasing purchased commodities, while the marginal utility of unit commodity decreases.

As shown in Figure 4, this paper assumes that EV user utility is a piecewise linear function of charge capacity, given that the piecewise function is a typical function representing the relationship between welfare utility and energy consumption [21,22], and linearization of the EV user benefits utility can be applied to the proposed bi-level optimization model and solution method. Without loss of generality, this paper assumes that the segmentation function is divided into five segments (M = 5), and each segment corresponds to a predetermined marginal utility value. $d^{ev}_{v,t,m}$ represents the charging demand of EV users in each segment, and the total utility from the CSO can be expressed as piecewise linearized using Equations (22) and (23):

$$\sum_{m=1}^{M} d^{ev}_{v,t,m} = d^{ev}_{v,t} \tag{22}$$

$$U_v(d^{ev}_{v,t}) = \sum_{m=1}^{M} u^{ev}_{v,m} d^{ev}_{v,t,m} \tag{23}$$

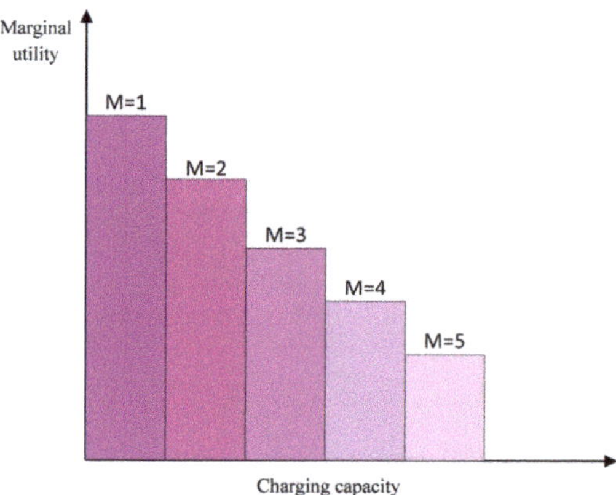

Figure 4. EV user utility function graph.

The original lower-level model can be expressed as Equation (24):

$$\underset{d^{ev}_{v,t,m}, d^{ev}_{v,t}}{\text{Maximize}} \; F^{EV}_v = \delta^{ch}_t d^{ev}_{v,t} - \sum_{m=1}^{M} \left(u^{ev}_{v,m} d^{ev}_{v,t,m} \right) \tag{24}$$

Equations (25)–(29) are charging constraints for EV users. Equation (25) is a logical constraint that associates the decision variable $d^{ev}_{v,t}$ with its piecewise form $d^{ev}_{v,t,m}$. The total capacity of EV users is assumed to meet the daily driving distance requirements without exceeding the battery capacity.

$$\sum_{m=1}^{M} d^{ev}_{v,t,m} = d^{ev}_{v,t} : \lambda_{v,t} \tag{25}$$

$$d^{ev}_{v,t} - d^{ev}_{v,max} \leq 0 : \mu^1_{v,t} \tag{26}$$

$$-d^{ev}_{v,t} + d^{ev}_{v,\min} \leq 0 : \mu^2_{v,t} \tag{27}$$

$$d^{ev}_{v,t,m} - d^{ub}_{v,t,m} \leq 0 \; \forall m \in M : \mu^3_{v,t,m} \tag{28}$$

$$d_{v,t,m}^{ev} \geq 0 \; \forall m \in M : \mu_{v,t,m}^{4} \tag{29}$$

$$d_{v,max}^{ev} = E_{rated}^{ev}\left(SOC_{max}^{ev} - SOC_{v,in}^{ev}\right) \tag{30}$$

$$d_{v,min}^{ev} = l_{v}^{tot}\varepsilon^{ev} + E_{rated}^{ev}\left(SOC_{min}^{ev} - SOC_{v,in}^{ev}\right) \tag{31}$$

where E_{rated}^{ev} and ε^{ev} represent EV battery capacity and capacity consumption per kilometer, l_{v}^{tot} represents the expected the daily travel distance of v-type EVs, and SOC_{min}^{ev}, SOC_{max}^{ev} and $SOC_{v,in}^{ev}$ represent the minimum and maximum SOCs of EV batteries and the initial SOC at the charging station. Equation (28) indicates that the EVs' charge capacity in each segment is smaller than the upper limit of this segment. Equation (29) indicates that the charge capacity of each segment is positive. It should be noted that the subsequent variable of each equation is its dual variable.

Based on this, the linear expression of the bi-level problem can be obtained.

3.2.2. Reducing a Bi-Level Problem to a Simple Level Problem

The lower-level problem in the previous section was linearized. In the proposed bi-level problem, according to its structural and form transformation, the upper-level variable can be considered as a parameter in the lower-level problem, and for a given variable δ_t, each lower-level problem is a linear optimization problem that is continuous and convex in structure. Therefore, the KKT conditions can be used to replace the lower-level problems in Equations (24)–(29) with their corresponding KKT optimality conditions and then be included in the upper-level problem. By doing this, the original bi-level model is transformed into an equivalent single-level problem [23].

The equations for the KKT conditions are as follows:

Equation (32) represents the Lagrangian function of the lower-level problem, and $\lambda_{v,t}$, μ_t^1, μ_t^2, $\mu_{t,m}^3$, and $\mu_{t,m}^4$ represent the Lagrange multipliers of the corresponding lower-level constraints. Equations (35)–(38) are relaxed complementary constraints, where $0 \leq \mu \perp d \geq 0$ represents $0 \leq \mu, d \geq 0, \mu d = 0$. It should be noted that Equations (35)–(38) have a bilinear term by multiplying the dual variable and the original variable.

$$\begin{gathered} L(d_{v,t}^{ev}, d_{v,i,t,m}^{ev}, \lambda, \mu) = d_{v,t}^{ev}\delta_t^{ch} - \sum_{m=1}^{M} d_{v,t,m}^{ev}u_{v,t,m}^{ev} + \lambda_{v,t}(d_{v,t}^{ev} - \sum_{m=1}^{M} d_{v,t,m}^{ev}) \\ +\mu_{v,t}^{ev}(d_{v,t}^{ev} - d_{v,t,\min}^{ev}) + \mu_{v,t}^{2}(-d_{v,t}^{ev} + d_{v,t,\max}^{ev}) + \mu_{v,t,m}^{3}(d_{v,t,m}^{ev} - d_{v,t,m}^{\max}) - \mu_{v,t,m}^{4}d_{v,t,m}^{ev} \end{gathered} \tag{32}$$

$$\frac{\delta L}{\delta d_{v,t}^{ev}} = \delta_t^{ch} + \lambda_{v,t} + \mu_{v,t}^{1} - \mu_{v,t}^{2} = 0 \; \forall t \in T \tag{33}$$

$$\frac{\delta L}{\delta d_{v,t,m}^{ev}} = u_{v,t,m}^{ev} + \lambda_{v,t} + \mu_{v,t,m}^{3} - \mu_{v,t,m}^{4} = 0 \; \forall t \in T, \; m \in M \tag{34}$$

$$0 \leq \mu_{v,t}^{1} \perp (d_{v,t}^{ev} - d_{v,t,\min}^{ev}) \geq 0 \; \forall t \in T \tag{35}$$

$$0 \leq \mu_{v,t}^{2} \perp (-d_{v,t}^{ev} + d_{v,t,\max}^{ev}) \geq 0 \; \forall t \in T \tag{36}$$

$$0 \leq \mu_{v,t,m}^{3} \perp (d_{v,t,m}^{ev} - d_{v,t,m}^{ub}) \geq 0 \; \forall t \in T, \; m \in M \tag{37}$$

$$0 \leq \mu_{v,t,m}^{4} \perp d_{v,t,m}^{ev} \geq 0 \; \forall t \in T, \; m \in M \tag{38}$$

By substituting Equations (24)–(29) for the lower-level problem with Equations (33)–(38), a single-level model equivalent to the bi-level optimization can be obtained.

3.2.3. Linear Description of Lower-Level Problem

We can transform the bi-level model into a single-level problem, but this single-level problem is bilinear and has nonlinear terms, and cannot obtain an exact solution using the classical algorithm. Therefore, the nonlinear terms in the single-level problem described in the previous section should be linearized. The nonlinear terms in the single-level constraints are generated when the KKT conditions are introduced to achieve single-level transformation. The relaxing complementarity conditions introduced in the constraint have nonlinear terms by multiplying two variables, and these nonlinear terms can be transformed into a mixed-integer linear program [24]. The specific transformation for $0 \le \mu \perp d \ge 0$ is as follows:

$$\begin{gathered} \mu \ge 0, d \ge 0 \\ \mu \le (1-w)M \\ d \le wM \\ w \in \{0,\ 1\} \end{gathered} \tag{39}$$

where M is a sufficiently large constant, and w is a variable between 0 and 1. It should be pointed out that since the value of M will affect the accuracy of the problem and the efficiency of the calculation, decision-makers need to make appropriate choices. Some studies have given alternatives.

By using the above method, the nonlinear complementarity constraints in Equations (35)–(38) can be rewritten as follows:

$$\mu^1_{v,t} \ge 0, d^{ev}_{v,t,\min} - d^{ev}_{v,t} \ge 0 \tag{40}$$

$$\mu^1_{v,t} \le \left(1 - w^1_{v,t}\right) M_1, d^{ev}_{v,t,\min} - d^{ev}_{v,t} \le w^1_{v,t} M_1 \tag{41}$$

$$\mu^2_{v,t} \ge 0, d^{ev}_{v,t} - d^{ev}_{v,t,\max} \ge 0 \tag{42}$$

$$\mu^2_{v,t} \le \left(1 - w^2_{v,t}\right) M_2, d^{ev}_{v,t} - d^{ev}_{v,t,\max} \le w^2_{v,t} M_2 \tag{43}$$

$$\mu^3_{v,t,m} \ge 0, d^{ub}_{v,t,m} - d^{ev}_{v,t,m} \ge 0 \tag{44}$$

$$\mu^3_{v,t} \le \left(1 - w^3_{v,t}\right) M_3, d^{ub}_{v,t,m} - d^{ev}_{v,t,m} \le w^3_{v,t,m} M_3 \tag{45}$$

$$\mu^4_{v,t,m} \ge 0, d^{ev}_{v,t,m} \ge 0 \tag{46}$$

$$\mu^4_{v,t} \le \left(1 - w^4_{v,t}\right) M_4, d^{ev}_{v,t,m} \le w^4_{v,t,m} M_4 \tag{47}$$

$$w^1_{v,t}, w^2_{v,t}, w^3_{v,t,m}, w^4_{v,t,m} \in \{0,1\} \tag{48}$$

In addition to the upper-level objective function, the above process linearizes the nonlinear constraints using an equivalent mixed-integer linear programming model. Here we linearized the nonlinear terms in the upper-level objective function using a classical linear approximation method. Owing to its simplicity and easy access to a solution, this method has been fully applied in some studies [25,26]. The specific method is as follows.

According to the linearized model and assuming $z = a \times b$ represents a bilinear term, $a \in [a_{\min}, a_{\max}]$, $b \in [b_{\min}, b_{\max}]$, the bilinear term $z = a \times b$ can be linearized by the following form:

$$\begin{gathered} z \ge a_{\min} b + b_{\min} a - a_{\min} b_{\min} \\ z \ge a_{\max} b + b_{\max} a - a_{\max} b_{\max} \\ z \le a_{\min} b + b_{\max} a - a_{\min} b_{\max} \\ z \le a_{\max} b + b_{\min} a - a_{\max} b_{\min} \end{gathered} \tag{49}$$

Applying this method to the upper-level objective function, the converted objective function can be expressed as:

$$B^{Ope} = \theta \cdot \sum_{t=1}^{T} \left(z^{ch}_t - \delta^{int}_t P^{int}_t\right) \Delta t - \left(c^{cfm} P^{Ncf} + c^{rdgm} P^{Nrdg} + c^{esm} E^{Nes}\right) \tag{50}$$

where constraints are added correspondingly:

$$z^{ch}_t \ge \delta^{ch}_{\max} P^{ch}_t + z^{cf}_t - \delta^{ch}_{\max} P^{Ncf} \ \ \forall t = 1, \cdots, T \tag{51}$$

$$z_t^{ch} \leq z_t^{cf} \ \forall t = 1, \cdots, T \tag{52}$$

$$z_t^{ch} \leq \delta_{\max}^{ch} P_t^{ch} \ \forall t = 1, \cdots, T \tag{53}$$

$$z_t^{cf} \geq \delta_{\max}^{ch} P^{Ncf} + P_{\max}^{Ncf} \delta_t^{ch} - \delta_{\max}^{ch} P_{\max}^{Ncf} \ \forall t = 1, \cdots, T \tag{54}$$

$$z_t^{cf} \leq P_{\max}^{Ncf} \delta_t^{ch} \ \forall t = 1, \cdots, T \tag{55}$$

$$z_t^{cf} \leq \delta_{\max}^{ch} P^{Ncf} \ \forall t = 1, \cdots, T \tag{56}$$

$$z_t^{ch} \geq 0 \ \forall t = 1, \cdots, T \tag{57}$$

$$z_t^{cf} \geq 0 \ \forall t = 1, \cdots, T \tag{58}$$

The nonlinear terms in the upper-level objective function are effectively linearized using the above method. As a result, the nonlinear bi-level optimization programming model is transformed into a single-level mixed-integer linear program, which can be effectively solved through commercial solvers such as CPLEX (12.7.1, IBM, Armonk, NY, USA, 2018).

4. Case Analysis

Numerical simulations of the model mentioned in the first two sections were conducted to obtain planning results of power pricing and capacity allocation. The planning results of the existing bi-level optimization model were compared with those of the common single-level optimization model and their differences were summarized. The superiority of the bi-level model was verified.

4.1. Basic Data

In this paper, solar photovoltaic panels and lithium-ion batteries were used as photovoltaic power generation and energy storage equipment. Relevant data shown in Table 1 were obtained from the website and merchant research. Assuming that the total area of the PV-storage charging station is 3000 m^2, the maximum installation power and capacity of CF, PV, and ES is 5000 kW, 500 kW, and 900 kW (2500 kWh), respectively, and the basic discount rate is 6%, based on compound interest.

Table 1. Basic parameters for facilities. SOC, charge and discharge state.

Equipment Type	Technical Parameters	Cost Parameters
Photovoltaic	d = 20 years	c^{rdg} = \$870/kW c^{rdgm} = \$12/kW/years
Lithium battery	d = 20 years $\eta^{esc} = \eta^{esd}$ = 93% $SOC_{\min}^{es}$ = 30% $SOC_{\max}^{es}$ = 90%	c^{esp} = \$200/kW c^{ese} = \$143/kWh c^{esm} = \$0.8/kWh/years
Charging pile	d = 20 years η^{cf} = 95%	c^{cf} = \$100/kW c^{cfm} = \$6/kW/years

The transformer–capacitor connecting the PV-storage charging station to the external power grid is set as 4000 kVA, the real-time price of a certain power market is set as the real-time electricity price, and the output parameter curve of a certain PV plant is used the PV output factor. The specific parameters are shown in Figure 5.

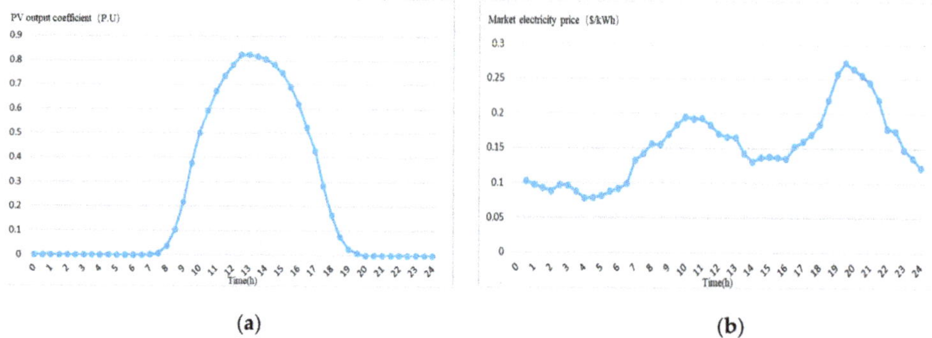

Figure 5. (**a**) Photovoltaic output coefficient and (**b**) market electricity price parameter.

For simplicity, the simulation example uses Nissan's electric vehicles to represent the EV user community. Each EV has a battery capacity of 40 kWh and a power consumption of 0.18 kWh/km. The lower and upper limits of SOC are set as 30% and 90%, respectively. To be consistent with the proposed model framework, EV users in the charging station system are assumed to be classified into three categories based on daily driving distance: remote range (LR), medium-range (MR), and short-range (SR). In addition, the CSO can predict EV traffic and the proportion of EV users in the three driving distance groups at the charging station at different times of the day. The numbers of LR, MR, and SR EV users are shown in Figure 6, and their parameters are shown in Table 2.

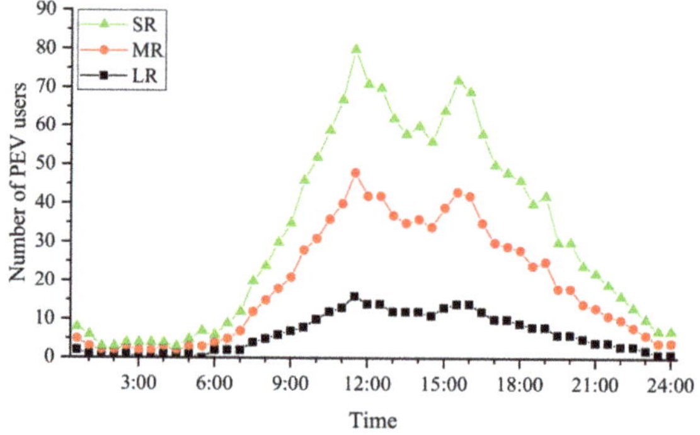

Figure 6. Numbers of different types of electric vehicle users.

Table 2. Basic parameters of different types of electric vehicle (EV) users. SR, short-range. MR, medium-range. LR, long-range.

Parameter	User Type		
	SR	**MR**	**LR**
SOC_{in}^{ev}	30%	40%	50%
l^{tot}	20 km	50 km	100 km

4.2. Simulation Example Results

Based on the above basic data, the commercial solver CPLEX (12.7.1, IBM, Armonk, New York, USA, 2018) was used to perform the example simulation in the MATLAB environment (2018a, MathWorks, Natick, MA, USA, 2018), and the obtained PV-storage charging planning results and economic cost are shown in Table 3. The real-time electricity pricing profile is shown in Figure 7.

Table 3. PV-storage charging station planning results. CF, charging facility. ES, energy storage.

	PV Installed Power (kW)	CF Installed Power (kW)	ES Installed Power and Storage Capacity (kW/kWh)	Daily Electricity Sales (kWh)
Planning Results	500	862	616/900	8764
	Annual Investment Cost ($)	**Annual Operation and Maintenance Cost ($)**	**Annual Revenue ($)**	**Annual Net Revenue ($)**
	79,650	11,839	844,029	752,485

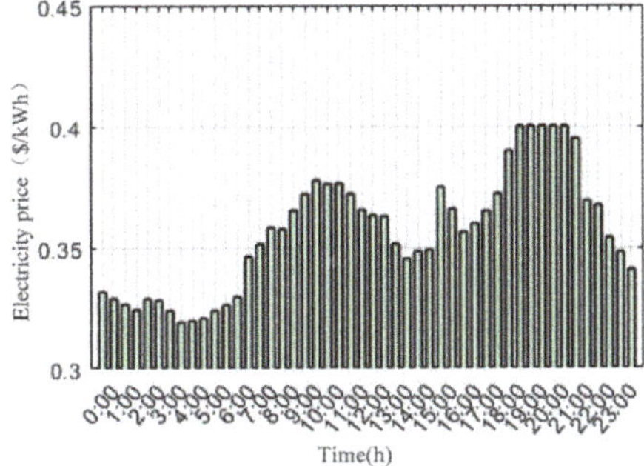

Figure 7. Real-time electricity price profile.

It can be seen from Table 3 that the daily sales of the PV-storage charging station is 8.764 MWh, with an annual net revenue of $752,485. The photovoltaic installation is at the boundary of the maximum installed capacity, indicating that during operation of the charging station, photovoltaic power generation can achieve higher revenue compared with purchasing electricity from the main power grid. Therefore, the planning results show that under sufficient load, the PV-storage charging station should use more available space to install photovoltaics. In this operation, the photovoltaic output accounts for nearly 30% and can be fully consumed owing to more energy storage facilities, indicating that overall advances in energy storage technology can help cost reduction.

It can be seen from Figure 7 (the real-time electricity price profile) that the highest electricity price in a day is $0.40/kWh, which occurs between 19:00 and 21:00, while the lowest electricity price is $0.32/kWh, which occurs between 03:00 and 04:00. From the perspective of the price changing trend, the electricity price peaks from 09:00 to 11:00 and from 19:00 to 21:00, during which it reaches the highest limit peak. There is a small fluctuation in the afternoon, which is consistent with the charging behavior of EV users. EV users charge their batteries mainly in the morning and evening, so the charging load is greater during those periods. The photovoltaic output is low and even reaches zero at night, and the electricity from the main power grid is also priced higher during this period. Therefore, the PV-storage charging station has to turn to the main power grid or energy storage system to meet the users' load demand. At this time, the CSO should set a higher electricity price to ensure the profitability of the system.

The planning results of the simulation of the PV-storage charging station system are shown in Figure 8. The red bars represent the power demand of EV users, the blue and orange bars represent the charge and discharge states of the ES and the charge and discharge amounts, respectively, and the purple bars represent PV output. If a green bar is positive, it means the charging station purchased electricity from the main grid. If the green bar is negative, it means the charging station sold electricity to the main grid. The figure is based on the internal power balance of PV-storage charging stations. The area above the time axis represents the power source of the charging station, while the area below represents the power output. Total power generation is equal to total power consumption, so the figure has vertical symmetry.

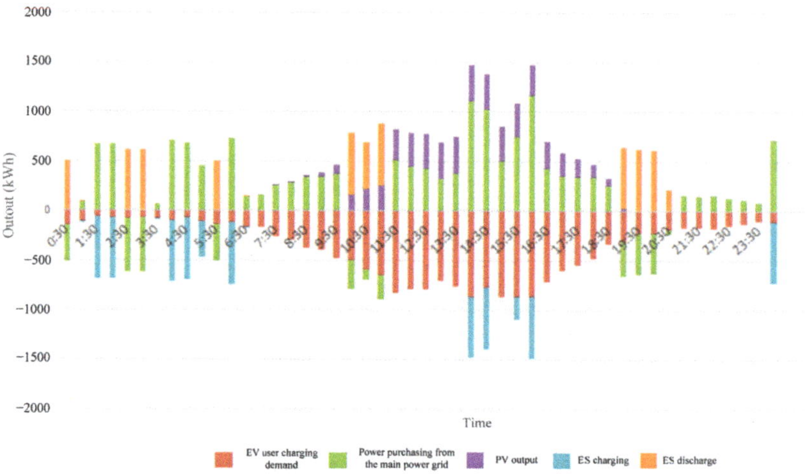

Figure 8. Operation strategy of PV-storage charging station.

It can be seen from Figure 8 that the EV load is greater in the morning and the evening, and smaller at night. From the perspective of time, the energy storage system works more between 22:00 and 05:00 the next day, because the electricity price of the main power grid is lower at this time, and the CSO purchases electricity from the main grid and continuously charges electricity in the ES system to maintain storage power. In addition, three ES discharges occur between 00:00 and 02:00 at night. Since the night electricity price is floating, the CSO is profitable by using low-price charging and selling high-priced electricity at the next moment. In general, the charging station performs ES at night, and between 05:00 and 09:00, the EV load is relatively small and the photovoltaic power generation is also very small. At this time, the load is mainly satisfied by the power supply purchased from the main power grid without using the stored energy in the ES system. However, during the peak charging period in the morning, the CSO calls up a large amount of nighttime energy storage and uses PV power to meet the EV charging demand, while the charging station can transfer electricity to the external grid until the SOC of the ES reaches the lower limit. This is the main period when the charging station can generate a profit. From 11:00 to 13:00, the CSO directly purchases electricity from the main power grid and photovoltaic power to meet the EV user load without activating the ES system. The ES is charged to store energy until 13:00–15:00 when the PV output is large and the electricity purchased from the main power grid is low-priced. The situation from 16:00 to 18:00 is similar to that from 11:00 to 13:00. From 18:00 to 21:00, the CSO meets the needs of EV users by discharging the energy stored in the afternoon and sending electric power to the main grid to obtain greater benefits.

Looking at Figures 7 and 8, we discuss the impact of electricity price on PV-storage charging station operation strategy. When the electricity price and load are low at night, charging stations carry out arbitrage by providing ES charging at a low price and discharging at a high price. From 19:00 to 21:00, the charging demand increases and the PV output is low, and the charging station raises the price, making a profit from the price difference between purchase and sale. Electricity prices are also appropriately raised as demand increases further when PV power generation increases from 21:00 to 00:00. The CSO benefits mainly from low-cost PV power generation and ES. From 12:00 to 18:00, the electricity price fluctuates steadily, controlling the stable fluctuation of the charging load. At this time, the revenue mainly comes from PV power generation and the price difference between purchase and sale. At night, since the main grid has a high electricity purchase price and no PV power generation, PV charging stations raise their electricity prices and reduce some of the load. Charging profits mainly come from the discharge of ES.

Figure 9 shows the daily SOC of the ES system that can reflect its operating state in the PV-storage charging station. SOC can be roughly divided into the following parts. The ES system has rapidly changing SOC at night. As seen by the operating state in Figure 9, it can be concluded that the PV-storage charging station utilizes the nighttime price spread by performing fast charging and discharging to maximize profits. From 03:00 to 05:00, the ES system is in the fast charging state and the energy storage reaches the upper limit of SOC, then is maintained at 56% after proper discharge. From 09:00 to 11:00, the discharge energy storage is in a lower state and is recharged at 11:00 and 13:00 to the upper limit of SOC, then is discharged in the afternoon to maintain at 56%. At night, it is discharged again to remain at the lower limit of SOC.

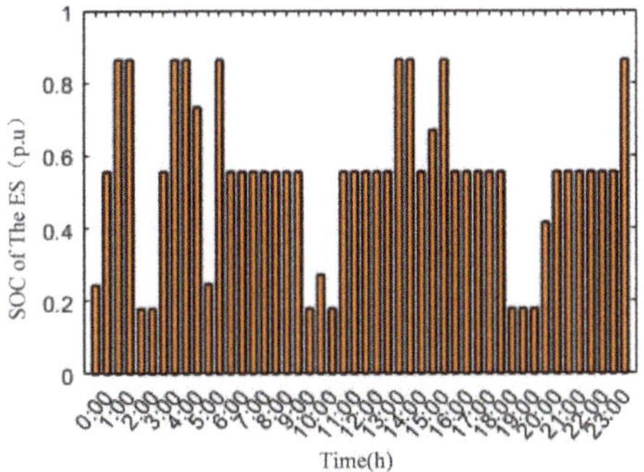

Figure 9. SOC of the ES.

Figure 10 shows the PV/load ratio of the PV-storage charging station. From this figure, it can be seen that PV has a contribution rate of over 40% from 10:00 to 17:00, which indicates that the charging station can make full use of PV power generation. The reasonable combination of PV and ES not only improves the charging station's economy but also better consumes and utilizes renewable energy.

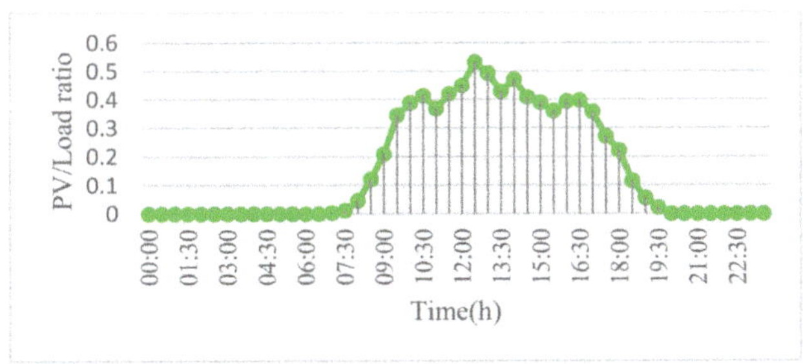

Figure 10. PV/load ratio of PV-storage charging station.

The following conclusions can be drawn by analyzing the above results. First, in the case of mature technology and reduced cost, PV-storage charging stations should make full use of space to invest in and construct photovoltaic and energy storage facilities. Secondly, CSOs should set reasonable electricity prices and adjust the charging demand of EV users via electricity pricing means to achieve balanced benefits. Finally, CSOs need to make full use of the characteristics of photovoltaic power generation and especially energy storage, and properly adjust charging and discharging based on the electricity price to promote the economic efficiency of system operation.

4.3. Models Comparison

This section compares and analyzes the effects of the bi-level model and the common single-level planning model. The single-level model considers that the charging demand of EV users $d_{v,t}^{ev}$ is fixed and is equal to the calculated demand for the bi-level optimization model, and assumes that the CSO can accurately predict EV user types and the number of electric vehicles at different time points in advance, i.e., every EV user has a fixed charging demand, regardless of choice.

Based on the above assumptions and design, this section constructs two schemes, shown in Table 4. Scheme I builds a single-level model where the CSO's electricity sales price is set at 0.35 \$/kWh. Scheme II is a bi-level model that takes account of capacity configuration and electricity price optimization under uncertainty conditions.

Table 4. Scheme settings.

	Scheme I	Scheme II
Planning type	Single-level planning	Bi-level planning
Electricity price	0.35 \$/kWh	Optimized price

The results and economic characteristics of schemes using CPLEX are shown in Table 5.

Table 5. Planning results of different schemes.

	Scheme I	Scheme I Without ES	Scheme II
CSO electricity price (\$/kWh)	0.35	0.35	Figure 7
CF installed power (kW)	1154	1154	862
PV installed power (kW)	500	500	500
ES installed power (kW)	846	-	545
ES installed storage capacity (kWh)	900	-	900
Investment cost (\$)	87,376	56,691	79,650
Operation and maintenance cost (\$)	13,644	12,924	11,839
Total revenue (\$)	802,991	590,572	844,029
Net revenue (\$)	701,971	520,975	752,485
PV output (kWh)	2625	2625	2625
PV output (%)	22.6%	22.6%	29.9%

As can be seen from Table 5 and Figure 11, different planning schemes result in different configurations of PV-storage charging stations. More specifically, compared with the bi-level model of Scheme II, the single-level model of Scheme I has a larger CF installation capacity and the electricity price is fixed, and the larger installation capacity leads to increased investment operation and maintenance costs and decreased net revenue. The reason for these results is that the charging demand of EV users in the single-level model is set as a fixed value, regardless of electricity price. The CSO needs to invest in and build more charging facilities to meet the needs of EV users, so compared with the EV user choice-based model, this scheme will lead to increased cost.

Compared with Scheme I, in Scheme I without ES, the investment cost is reduced, but the operating return and total revenue are also reduced, which indicates that the installation of ES is beneficial to the charging station. The reason is that the charging station can use ES derived from the power grid or PV and discharge electricity for EV users, obtaining revenue through price differences.

In addition, the PV investment for both schemes is the set upper limit, indicating that the extensive installation of PVs within a limited capacity is beneficial to obtain greater revenue.

The above comparison reveals the importance of EV user behavior in the optimal design of PV-storage charging stations. Since the charging demand of EV users usually varies with the electricity price, it is important to consider user choice when evaluating the profitability of the PV-storage charging station planning scheme.

This section may be divided by subheadings. It should provide a concise and precise description of the experimental results, their interpretation as well as the experimental conclusions that can be drawn.

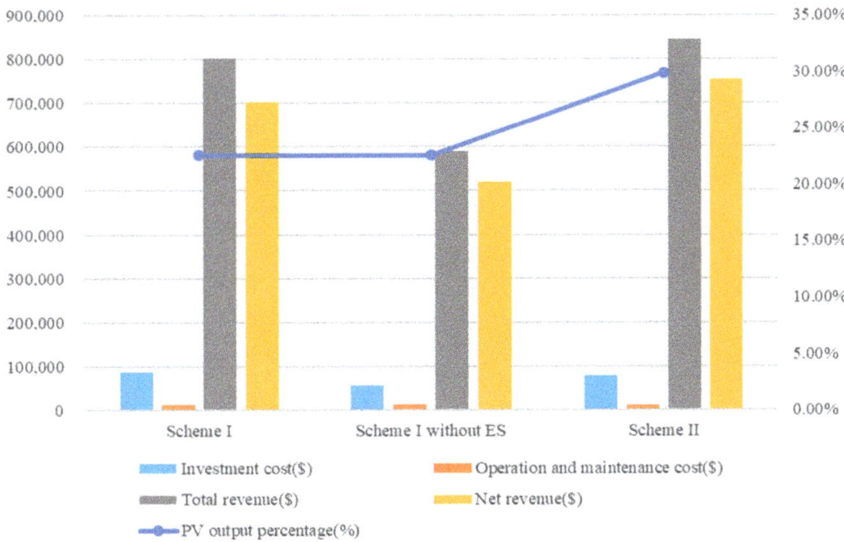

Figure 11. Economic effects of the two schemes.

5. Conclusions

In order to propose an effective PV-storage charging station planning scheme, this paper established a bi-level planning model for charging stations that considers the needs of EV users. Using dual theory, KKT conditions, and linearization tools, the problem was transformed into mixed-integer linear programming. The model was solved using a commercial solver and a case study was performed. The main research results of this paper are as follows:

(1) We describe and analyze the systematic structure and model framework of a PV-storage charging station. Taking the PV-storage charging station and EV users as the main upper- and lower-level problems, respectively, a bi-level optimization model for capacity allocation and electricity pricing in the PV-storage charging station is established based on constraints such as systematic planning and operation and user demand, with the objective of maximizing revenue and EV user benefits. The bi-level problem is transformed into a single-level mixed-integer linear programming model using the piecewise linear utility function, KKT conditions, and a linearization method.

(2) Using CPLEX (12.7.1, IBM, Armonk, NY, USA, 2018) software for simulation, the validity and practicability of the model are verified by analyzing the planning results, electricity price, operation states, and ES system state.

(3) Through case analysis, it can be concluded that PV-storage charging stations can reduce cost by investing in green PV and carrying out price arbitrage by allocating ES to improve revenue, which is shown in Table 5. By using these methods, the benefits of PV-storage charging stations can be maximized.

This research provides an optimal operational scheme for PV-storage charging stations, including initial investment and construction cost, the output installment capacity of each facility, and a real-time electricity pricing scheme, and has a high economy and practicality.

Author Contributions: In this research activity, all authors were involved in the research work, Conceptualization, Y.L. and M.Z.; Methodology, H.D.; Software, H.D.; Validation, Y.L. and S.Z.; Formal Analysis, S.Z.; Investigation, S.W.; Resources, M.L.; Data Curation, M.L.; Writing—Original Draft Preparation, H.D. and S.W.; Writing—Review

& Editing, Y.L.; Visualization, H.D.; Supervision, Y.L.; Project Administration, M.Y.; Funding Acquisition, S.Y. All authors have read and agreed to the published version of the manuscript.

Funding: The work described in this paper was supported by National Natural Science Foundation of China (71601078) and the Science and Technology Project of SGCC, Research on the competition situation and operation mode of distribution companies in the park (52170018000S).

Acknowledgments: This paper was completed with the help of many teachers and classmates. We would like to express our gratitude to them for their help and guidance.

Conflicts of Interest: The authors declare no conflict of interest.

Nomenclature

Parameters	
$c^{cf}/c^{rdg}/c^{esp}$	Unit power investment cost of CF, PV, and ES facilities ($/kW)
c^{cfm}/c^{rdgm}	Annual unit operation cost of CF, PV facilities ($/kW/years)
c^{ese}	Annual unit operation cost of ES facilities ($/kWh/years)
c^{esm}	Annual unit operation cost of ES facilities ($/kWh/years)
$d^{ev}_{min}/d^{ev}_{max}$	Lower/upper limit of EV charge level (kWh)
$E^{Nes}_{max}/P^{Nes}_{max}$	Maximum installed storage capacity/power of ES (kWh; kW)
$k^{cf}/k^{rdg}/k^{es}$	Annualization operators
$P^{Ncf}_{max}/P^{Nrdg}_{max}$	Maximum capacity of installed CF/RDG (kW)
P^{tr}_{max}	Rated capacity of distribution transformer (kW)
$SOC^{es}_{min}/SOC^{es}_{max}$	Lower/upper SOC limit for ES (%)
u^{ev}	Marginal utility of EV users ($/kWh)
δ^{ch}_{max}	Upper limit of charging tariff ($)
Δt	Duration of time period (0.5 h)
η^{cf}	Charging efficiency of CF (%)
η^{esc}/η^{esd}	ES operation efficiency (%)
θ	Number of days in a year
f^{ev}	EV uptake
γ^{rdg}	Ratio of RDG output to installed capacity
δ^{int}	market price ($/kWh)
D	Facility lifetime (years)
ζ	Discount rate (%)
Variables	
d^{ev}	Required EV charge level (kWh)
E^{es}	Stored energy in ES units (kWh)
E^{Nes}/p^{Nes}	ES storage capacity/power (kWh; kW)
p^{ch}	Total EV charging power (kW)
p^{esc}/p^{esd}	ES charging/discharging power (kW)
p^{int}	Exchanged power with grid (kW)
p^{Ncf}/p^{Nrdg}	Installed capacity of CF/RDG (kW)
p^{rdg}	Total power output of RDG (kW)
δ^{ch}	Offered charging tariff ($/kWh)

References

1. Cai, W.; Lai, K.-H.; Liu, C.; Wei, F.; Ma, M.; Jia, S.; Jiang, Z.; Lv, L. Promoting sustainability of manufacturing industry through the lean energy-saving and emission-reduction strategy. *Sci. Total Environ.* **2019**, *665*, 23–32. [CrossRef] [PubMed]
2. Rahman, I.; Vasant, P.M.; Singh, B.S.M.; Abdullah-Al-Wadud, M.; Adnan, N. Review of recent trends in optimization techniques for plug-in hybrid, and electric vehicle charging infrastructures. *Renew. Sustain. Energy Rev.* **2016**, *58*, 1039–1047. [CrossRef]
3. Huang, F.T.; Weng, G.Q.; Nan, Y.R.; Yang, X.D.; Chen, D. Optimization of Electric Vehicle Charging Stations Based on Improve Cloud Adaptive Particle Swarm in Distribution Network with Multiple DG. *Chin. Soc. Electr. Eng.* **2018**, *38*, 514–525.

4. Liu, C.; Liu, H.; Li, X.L.; Zhang, J.; Li, K.; Zhang, J.; Zhang, X. Multi-objective EV charging station planning with consideration of road network reliability and distribution network reliability. *Electr. Power Autom. Equip.* **2017**, *37*, 28–34.
5. Zheng, Y.; Dong, Z.Y.; Xu, Y.; Meng, K.; Zhao, J.H.; Qiu, J. Electric Vehicle Battery Charging/Swap Stations in Distribution Systems: Comparison Study and Optimal Planning. *IEEE Trans. Power Syst.* **2013**, *29*, 221–229. [CrossRef]
6. Ge, S.Y.; Zhu, L.W.; Liu, H.; Li, T.; Liu, C. Optimal Deployment of Electric Vehicle Charging Stations on the Highway Based on Dynamic Traffic Simulation. *Trans. China Electrotech. Soc.* **2018**, *33*, 2991–3001.
7. Feng, Y.; Zhao, X.; Ren, G.; Zhao, J. Planning Method for Urban Centralized Charging Stations. *Proc. CSU EPSA* **2018**, *30*, 58–61.
8. Yang, Z.Z.; Gao, Z.Y. Location Method of Electric Vehicle Charging Station Based on Data Driven. *J. Transp. Syst. Eng. Inf. Technol.* **2018**, *18*, 143–150.
9. Sadeghi-Barzani, P.; Rajabi-Ghahnavieh, A.; Kazemi-Karegar, H. Optimal fast charging station placing and sizing. *Appl. Energy* **2014**, *125*, 289–299. [CrossRef]
10. Xiang, Y.; Liu, J.; Li, R.; Li, F.; Gu, C.; Tang, S. Economic planning of electric vehicle charging stations considering traffic constraints and load profile templates. *Appl. Energy* **2016**, *178*, 647–659. [CrossRef]
11. Zhang, X.W.; Qiang, W.; Wen, X.L.; Li, H.Z. Planning of Electric Vehicle Charging Station Considering V2G Mode. *Mod. Electr. Power* **2019**, *36*, 71–78. [CrossRef]
12. Liu, Z.P.; Wen, F.Z.; Xue, Y.S.; Xin, J.B. Optimal Siting and Sizing of Electric Vehicle Charging Stations. *Electr. Power Autom.* **2012**, *36*, 54–59.
13. Guo, S.; Zhao, H. Optimal site selection of electric vehicle charging station by using fuzzy TOPSIS based on sustainability perspective. *Appl. Energy* **2015**, *158*, 390–402. [CrossRef]
14. Sun, Y.; Ding, M.S.; Liu, J.S.; Yang, Y.B.; Xu, X.H.; Xu, Q.S.; Shi, S.S. Optimal Planning of the Capacity and Site of EV Charing Facilities by the Analytical Hierarchy Process. *Electr. Meas. Instrum.* **2014**, *51*, 114–119.
15. Cheng, J. Optimal Configuration and Protection of Optical Storage Capacity of Photovoltaic Electric Vehicle Charging Station. Master's Thesis, Nanjing Normal University, Nanjing, China, May 2018.
16. Lu, J.L.; Yang, Y.; Wang, Y.; He, T.X. Copula-based Capacity Configuration of Energy Storage System for A PV-Assisted Electric Vehicles Charging Station. *Acta Energ. Sol. Sin.* **2016**, *37*, 780–786.
17. Li, W.C.; Tong, Y.B.; Zhang, W.G. Energy Storage Capacity Allocation Method of Electric Vehicle Charging Station Considering Battery Life. *Adv. Technol. Electr. Eng. Energy* **2019**, *39*, 55–63.
18. Chen, Z.; Xiao, X.N.; Lu, X.Y.; Liu, N.; Zhang, J.H. Multi-Objective Optimization for Capacity Configuration of PV-Based Electric Vehicle Charging Stations. *Trans. China Electrotech. Soc.* **2013**, *28*, 238–248.
19. Meng, X.Y.; Zhang, W.G.; Bao, Y.; Huang, M.; Yuan, R.M.; Chen, Z. Optimal Configuration of Charging Facility for Electric Vehicle Fast Charging Station Considering Charging Power. *Electr. Power Autom. Equip.* **2018**, *38*, 28–34.
20. Gunter, S.J.; Afridi, K.K.; Perreault, D.J. Optimal Design of Grid-Connected PEV Charging Systems with Integrated Distributed Resources. *IEEE Trans. Smart Grid* **2013**, *4*, 956–967. [CrossRef]
21. Nguyen, D.T.; Nguyen, H.T.; Le, L.B. Dynamic Pricing Design for Demand Response Integration in Power Distribution Networks. *IEEE Trans. Power Syst.* **2016**, *31*, 3457–3472. [CrossRef]
22. Kazempour, S.J.; Conejo, A.J.; Ruiz, C. Strategic Bidding for a Large Consumer. *IEEE Trans. Power Syst.* **2015**, *30*, 848–856. [CrossRef]
23. Sinha, A.; Malo, P.; Deb, K. A Review on Bilevel Optimization: From Classical to Evolutionary Approaches and Applications. *IEEE Trans. Evol. Comput.* **2018**, *22*, 276–295. [CrossRef]
24. Fortuny-Amat, J.; Mccarl, B. A Representation and Economic Interpretation of a Two-Level Programming Problem. *J. Oper. Res. Soc.* **1981**, *32*, 783–792. [CrossRef]
25. Sherali, H.D.; Tuncbilek, C.H. New reformulation linearization/convexification relaxations for univariate and multivariate polynomial programming problems. *Oper. Res. Lett.* **1997**, *21*, 1–9. [CrossRef]
26. Sherali, H.D.; Dalkiran, E.; Liberti, L. Reduced RLT representations for nonconvex polynomial programming problems. *J. Glob. Optim.* **2012**, *52*, 447–469. [CrossRef]

Article

Experimental and Numerical Simulation Study on Co-Incineration of Solid and Liquid Wastes for Green Production of Pesticides

Bin Zhang *, Jinjie He, Chengming Hu and Wei Chen

College of Electromechanical Engineering, Qingdao University of Science and Technology, Qingdao 266061, China; hjjhjjhe@126.com (J.H.); hcm426@126.com (C.H.); cset2019@qust.edu.cn (W.C.)
* Correspondence: zb-sh@qust.edu.cn; Tel.: +86-131-5628-9096

Received: 16 August 2019; Accepted: 17 September 2019; Published: 23 September 2019

Abstract: A large amount of solid and liquid waste is produced in pesticide production. It is necessary to adopt appropriate disposal processes to reduce pollutant emissions. A co-incineration scheme for mixing multi-component wastes in a rotary kiln was proposed for waste disposal from pesticide production. According to the daily output of solid and liquid wastes, the proportion of mixing was determined. An experiment of the co-incineration of solid and liquid wastes was established. Experimental results showed that the mixed waste could be completely disposed at 850 °C, and the residence time in the kiln exceeded 1 h. A model method for mixture and diesel oil-assisted combustion was proposed. Numerical simulation was performed to predict the granular motion and reveal the combustion interactions of the co-incineration of mixed wastes in the rotary kiln. Simulation results reproduced movements, such as rolling and cascading, and obtained the optimum rotational speed and diesel oil flow for the rotary kiln incineration operation. The simulation showed that the temperature in the kiln was maintained at 850 °C, and the mass fraction of CO and O_2 at the outlet reached the standard for the complete combustion of the waste. Finally, the rotary kiln incineration and flue gas treatment processes were successfully applied in engineering for green production of pesticides.

Keywords: emission reduction; green production; waste disposal; co-incineration; rotary kiln

---

1. Introduction

China is not only a major country in the use of pesticides, it is also a producer [1], with a total production of 2.491 million tons of pesticide in 2017. However, an unintended consequence is that this pesticide production also produces considerable multi-component wastes such as liquid and solid wastes. Liquid and solid wastes generated during pesticide production are hazardous wastes [2]. Solid wastes produced by pesticides are the residual sludge of biochemical systems, semi-solid pentylamine and aniline, viscous kettle residue, waste packaging materials and activated carbon. Meanwhile, pesticide liquid wastes have a high chemical oxygen demand, high salt content and high toxicity [3]. Their characteristics of complex sources, wide variety and difficult degradability of waste lead to great difficulty in hazardous waste disposal. If disposed improperly, these hazardous wastes are harmful to the environment. Therefore, the disposal of liquid and solid wastes in pesticide production is still a difficulty and hot spot of waste disposal at present.

Common disposed methods include landfilling, physical disposed processes, biological conversion technologies, and incineration. Landfills contain pollutants that can also cause groundwater pollution [4]. Physical processing can only simplify the separation of different components of hazardous wastes [5]. Bioconversion technology can only convert biodegradable waste into high quality products, a process which has limitations [6]. It has been reported that incineration is an advanced choice in comparison with other disposed methods. The main purpose of incineration is to disposal of hazardous wastes while minimizing environmental impact and recovering energy. Incineration is an effective means of realizing harmlessness, reduction and resource utilization [7–9].

On the marketisation level, the high-temperature incineration method for solid wastes is the main disposal method [10,11]. However, the incineration of liquid wastes in combustion plants is not common, and only a few cases have been reported. In liquid waste combustion, inorganic salts in waste liquor will melt at a high temperature, which will aggravate the denudation of refractories and the slagging of ash on the wall. The acid gas produced by incineration not only pollutes the atmosphere but also reduces the dew point of flue gas, which causes corrosion and ash accumulation in the furnace. When the viscosity of liquid wastes is high or contains some impurities, the liquid wastes need to be filtered to remove such impurities. Hence, waste liquid incineration has the problems of high cost and low profit [12,13].

In recent years, the disposal method of co-incineration of hazardous waste has been paid increasing attention. The co-incineration of solid wastes and sewage has been intensively applied in many countries. Wang [14] assessed the environmental impacts of the sewage treatment scheme. Their results showed that the co-incineration of sewage and municipal solid waste is beneficial to reduce greenhouse gas emissions. Hu [15] analyzed the combustion characteristics of municipal solid waste, paper mill sludge and their mixture. When the mixing ratio of sludge was less than 30%, there was a significant synergistic effect in co-incineration. Lin [16] simulated the conditions of the co-incineration of municipal solid waste and sewage in an incinerator. The simulation results showed that the hybrid fuel could be completely burned to meet pollutant emission standards. The co-incineration of solid wastes and sewage has been widely used; however, that of solid and liquid wastes has rarely been reported, although it has a good application prospect in the waste disposal of pesticide production.

In comparison with other waste incineration equipment, the advantages of hazardous waste processing in rotary kilns include long residence time, high burning temperature, stable burning state, neutralization of acid waste gas and low cost [17–20]. Rotary kiln incinerators are widely used in many large factories as a safe method to disposal of hazardous waste. They are a general furnace for the disposal of solid, liquid, gaseous and complex combustible wastes [20,21]. Bujak [22] treated medical waste through rotary kiln incineration and analyzed environmental, economic and energy aspects. The results showed that the rotary kiln disposal produces significant environmental and economic benefits. Bujak [23] also introduced the preliminary results of the use of rotary kilns in plastic waste incineration. The actual atmospheric emission from the heat treatment of plastic waste is lower than the current emission standards. Mellmann [24] predicted the different forms of transverse bed motion in a rotary cylinder to understand particle behavior for efficient industrial production. The theory and technology for treating waste in rotary kilns has matured; however, the co-incineration of mixed multi-component wastes using rotary kilns has rarely been reported.

In the present work, a co-incineration scheme for mixed multi-component wastes in a rotary kiln was proposed for waste disposal from pesticide production. Firstly, the proportion of mixing of solid and liquid wastes was determined, and the elemental composition, calorific value and ash melting point of the mixture were measured. The experiment of co-incineration of solid wastes and liquid wastes was established. The slag was collected at different incineration times, and the slag clinker ignition loss was measured. This study proposes a model method for mixture and diesel oil-assisted combustion, explores these particle motion characteristics, and reveals the combustion interactions of the co-incineration of mixed wastes. The results obtained could help to further understand waste co-incineration. Finally, the rotary kiln and the entire treatment system were reconstructed and commissioned. The co-incineration of multi-component wastes disposal technology in a rotary kiln could realize the harmless treatment of hazardous waste, a process which has better social, environmental, and economic benefits.

2. Materials and Methods

2.1. Material Characteristics

As shown in Figure 1, materials produced in pesticide production involve a complex composition (i.e., aniline, amyl amine, kettle residue, kettle substrate, floccus, pendimethalin, activated carbon, and mother liquor of methomyl). In material pretreatment, liquid and solid wastes were thoroughly mixed in a mass ratio of 4:1 and were fed into the kiln inlet through a screw feeder. A constant temperature oven, muffle furnace, scanning electron microscope, calorimeter, crucible holder and electronic balance were used to measure material characteristics. The experimental measurement equipment is shown in Figure 2.

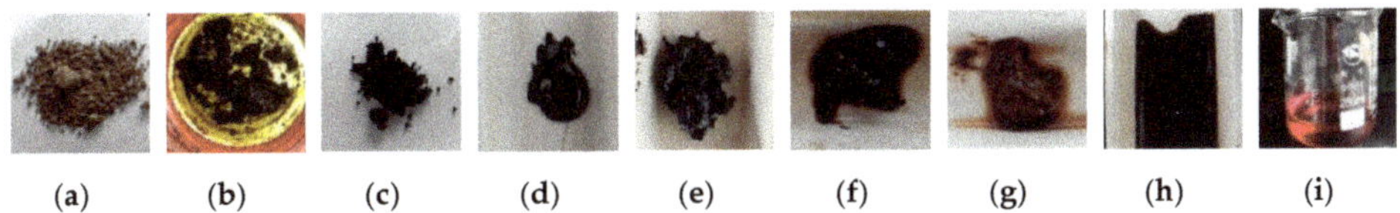

(a) (b) (c) (d) (e) (f) (g) (h) (i)

Figure 1. Solid wastes and liquid wastes in the pesticide production. (**a**) Sludge; (**b**) activated carbon; (**c**) pendimethalin; (**d**) amyl amine; (**e**) kettle residue; (**f**) aniline; (**g**) floccus; (**h**) kettle substrate; (**i**) and mother liquor of methomyl.

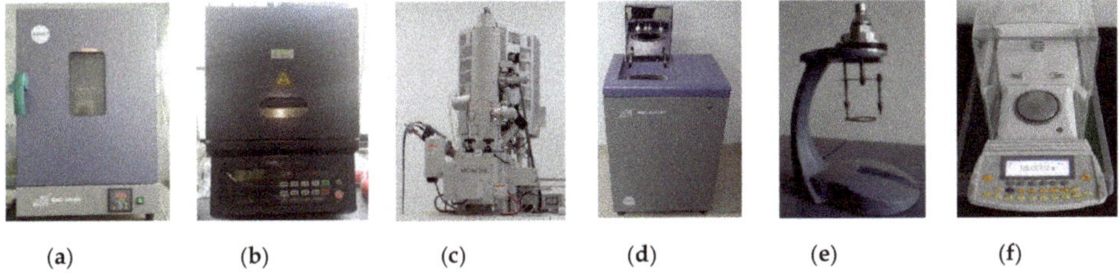

(a) (b) (c) (d) (e) (f)

Figure 2. Experimental measurement equipment. (**a**) Constant temperature oven; (**b**) muffle furnace; (**c**) scanning electron microscope; (**d**) calorimeter; (**e**) crucible holder; and (**f**) electronic balance.

According to Chinese Standards GB/T 212-2008, the moisture content of the mixed waste is 23.6%, and the ash content of the mixed waste is 20.4%. Moisture content was measured by heating the waste in a constant temperature oven at 105 °C, and the ash content was measured by igniting the mixed waste in a muffle furnace at 850 °C to a constant weight. The elements in the mixed waste were measured by a scanning electron microscope. The element composition is shown in Table 1.

Table 1. Proximate and ultimate analyses of wastes.

Proximate Analysis (wt%)				Ultimate Analysis (wt%)				$Q_{gr,ad}$ (MJ/kg)
M_{ad}	A_{ad}	FC_{ad}	V_{ad}	C_{ad}	H_{ad}	O_{ad}	N_{ad}	
23.6	20.4	20	36	56.74	2.20	36.20	4.86	4.05

The calorific value of waste is an important physical parameter for incineration. The calorific value of the mixed waste was determined using the oxygen bomb method. The standard benzoic acid is a combustion additive with a calorific value of 26,479 J/g. The waste was placed on a crucible and ignited with an ignition wire. The calorific value of the mixed waste was measured by a calorimeter. Solid wastes in total and wastes in total were calculated by weighted average. The daily production, mass fraction and calorific value of the material components are summarized in Table 2. Table 2 indicates that the daily production of liquid wastes was 4000 kg, which was the main part of waste disposal. The daily production of solid wastes in total was 1145 kg, which was only 22% of the total waste. The caloric value of total liquid wastes was low, whereas that of the total solid wastes was high. The calorific value of mixed wastes was 4050 kJ/kg, as determined by the weighted average method (Equations (1) and (2)). The calorific value was low, and diesel oil needs be added in auxiliary combustion.

$$Q_t = Q_s f_s + Q_l f_l \tag{1}$$

$$Q_s = \sum_{i=1}^{n} Q_i f_i \tag{2}$$

The ash fusion temperature, which was measured by a pyramid method (GB/T 219-2008), maintained its original shape before 1400 °C and melted directly to the molten state at 1450 °C. Consequently, the material of the ash fusion temperature was higher than 1400 °C, and the temperature difference between deformation and softening temperatures was less than 100 °C. The normal working temperature range of the solid-state rotary kiln was 800–1000 °C. Thus, the temperature in the kiln was lower than the ash fusion temperature, and the ash was not easily melted to avoid slagging on the inner wall of the furnace.

Table 2. Daily production, mass fraction and calorific value of the material components.

Material Name	Daily Production/kg	Mass Fraction/%	Calorific Value/(10^3 kJ/kg)
Aniline	20	0.38	25.8
Amyl amine	10	0.19	36.6
Kettle residue	100	1.94	23.7
Kettle substrate	15	0.29	38.4
Floccus	70	1.36	42.0
Pendimethalin	100	1.94	25.2
Activated carbon	230	4.47	18.5
Sludge	600	11.66	4.5
Solid wastes in total [1]	1145	22.26	14.1
Mother liquor of methomyl	4000	77.74	1.2
Wastes in total [2]	5145	100	4.05

[1] Total solid wastes indicate the sum of all materials without the mother liquor of methomyl. [2] Total waste indicates the sum of all materials.

2.2. Co-Incineration Experiment in the Rotary Kiln

A co-incineration experimental system (Figure 3) was established to explore whether mixed wastes could be completely burned in a rotary kiln and to analyze the incineration process. The internal diameter and length of the rotary kiln were 150 and 1630 mm, respectively. The entrance of the screw feeding machine was also the air inlet. Cement pouring was used for the thermal insulation layer, and the kiln inclination angle range was 0–5°. The cylinder of the rotary kiln was driven by a motor, and the speed range was 0.5–5 r/min. A thermocouple that could monitor the central temperature of the furnace in real time was installed in the middle of the kiln. Moreover, the control box could automatically adjust the power of the electric heater by setting the furnace temperature to control the temperature in the kiln.

The constant temperature of the kiln was kept at 850 °C, and the materials from the total wastes were continuously fed by imports. Moreover, Table 2 indicates that the state of the wastes was semi-fluid. The feeding speed and time were 0.05 kg/min and 10 min, respectively, and the total feeding was 0.5 kg. In the experiment, the slag appeared after the mixed waste was sent to the kiln for 60 min. Thus, the slag was collected after 60, 70, 80, 90, 100 and 110 min from when the material was sent. The slag was labelled #1, #2, #3, #4, #5 and #6. The clinker ignition loss was determined by a close roaster. The result is shown in Figure 4.

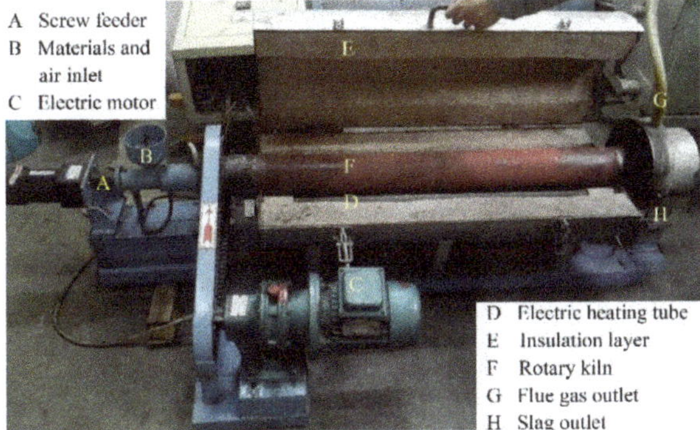

Figure 3. Integral structure of the experimental rotary kiln.

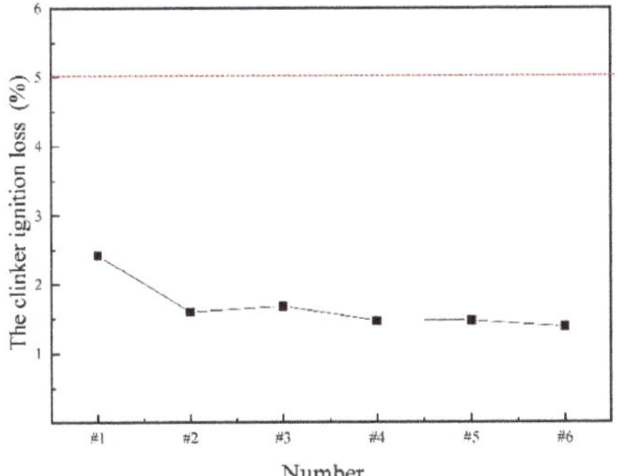

Figure 4. Clinker ignition loss.

The residence time in the kiln directly affects the oxidative decomposition of harmful substances. In order to ensure that the mixed waste can be completely burned, the rotary kiln incineration time should be more than 1 h. As shown in Figure 4, the clinker ignition loss of the slag was less than 5%. When burned for one hour, the clinker ignition loss was about 2.5%. The clinker ignition loss of the slag decreased slightly with the increase in the material residence time in the kiln. After 90 min of discharge time, the clinker ignition loss remained basically unchanged. The particle size varied significantly at the different times, and slag with various particle sizes was observed in the different batches, as shown in Figure 5. The results confirmed that disposing solid and liquid wastes together at 850 °C in a rotary kiln was feasible.

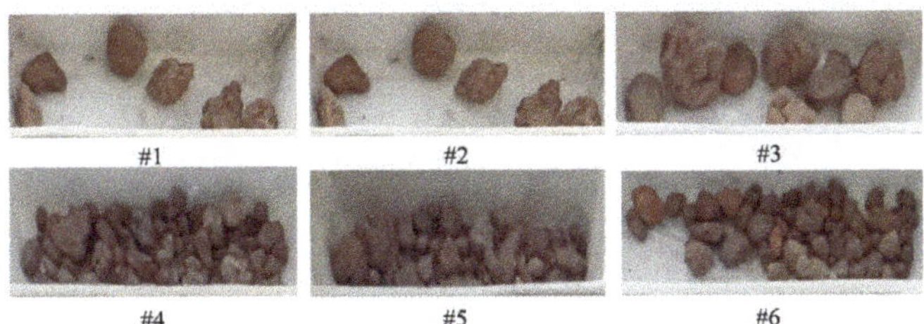

Figure 5. Clinker with various particle sizes in the different batches.

3. Numerical Method

It is necessary to consider the multiphase nature of fuel combustion, which involves the gases and particles and their interactions.

3.1. Gas Phase

The continuity equation and momentum equation are represented by Equations (3) and (4), respectively.

$$\frac{\partial \rho}{\partial t}+\frac{\partial(\rho v_j)}{\partial x_j}=S_m \quad (3)$$

$$\frac{\partial(\rho v_i)}{\partial t}+\frac{\partial(\rho v_j v_i)}{\partial x_j}=-\frac{\partial P}{\partial x_i}+\frac{\partial}{\partial x_j}[\mu(\frac{\partial v_j}{\partial x_i}+\frac{\partial v_i}{\partial x_j}-\frac{2}{3}\frac{\partial v_k}{\partial x_k}\delta_{ij})]+\rho g_i+S_m \quad (4)$$

where δ_{ij} is the Kronecker symbol, for $i=j$, $\delta_{ij}=1$, S_m is the mass source.

The energy model used in this study is as follows.

$$\frac{\partial(\rho E)}{\partial t}+\frac{\partial(\rho v_j E)}{\partial x_j}=-P\frac{\partial v_j}{\partial x_j}+\frac{\partial}{\partial x_j}(\lambda\frac{\partial T}{\partial x_j})+\tau_{ij}\frac{\partial v_i}{\partial x_j}+S_m \quad (5)$$

where λ is the thermal conductivity, T is the temperature, E is the internal energy, and τ_{ij} is the stress tensor.

3.2. Particle Phase

The trajectory of solid particles is described by the Lagrange model, and the particles are tracked by random discrete particle model.

$$\frac{du_w}{d_t}=F_D(u_g-u_w)+\frac{g_x(\rho_w-\rho)}{\rho_w}+F_x \quad (6)$$

where $F_D(u_g-u_w)$ is the drag force per unit particle mass.

3.3. Combustion Model

The complex composition of the mixed waste itself makes the combustion process very complicated. The waste particles are heated by diesel oil combustion in a rotary kiln and go through various reaction processes [25].

The devolatilization process is described by the single rate model. This model supposes that the devolatilization rate is first order and depends on the residual volatiles in particles [26].

$$-\frac{dm_p}{dt}=k[m_p-(1-f_{vo,0})(1-f_{eo,0})m_{p,0}] \quad (7)$$

$$k=A_0e^{-(E_a/RT)} \quad (8)$$

The kinetic/diffusion-limited rate model was used to simulate char reaction. The char combustion rate is written by Equation (9) [27].

$$\frac{dm_p}{dt}=-A_p p_{ox}\frac{D_0\Re}{D_0+\Re} \quad (9)$$

$$\Re=C_2e^{-(E_a/RT_p)} \quad (10)$$

The non-premixed combustion model is allowed to contain three streams, namely one oxide flow and two different fuel flows. In this paper, the simulation assumed that mixed waste was the fuel stream and diesel oil was the secondary stream. In the process of solving the combustion, the combustion is simplified to a mixed problem, and the mixture fraction/ Probability Density Function (PDF) model was considered. The concentration of the individual components was solved on the basis of the predicted distribution of the mixed fraction.

The mixing fraction, f, is the local mass fraction of th' aze combusted and unburned fuel stream elements in all components. A combustion process consists of a fuel and an oxidant, in which the mixing fraction can be expressed as [28].

$$f = \frac{Z_x - Z_{x,ox}}{Z_{x,fuel} - Z_{x,ox}} \tag{11}$$

Diesel oil is assumed to be a secondary stream, and the mixture fraction of the secondary stream needs to be considered. The sum of these quality scores in the system is always equal to 1.

$$f_{waste} + f_{diesel} + f_{ox} = 1 \tag{12}$$

4. Numerical Simulation

4.1. Numerical Simulation of Material Movement in Kiln

The movements of the wastes were tested on the side of the kiln to further determine the operating conditions of the kiln. Different initial conditions (e.g., rotating speed and particle diameter) were used to test the presented numerical model. The simulation parameters are shown in Table 3. The simulations were performed in a two-dimensional rectangular space using the Computational Fluid Dynamics (CFD) code. The computational mesh is shown in Figure 6a. The Euler multiphase method assumes that both the gas phase and the particle phase are incompressible continuous fluid. The interfacial force between particles and the kiln wall is described by particle dynamics theory. The two-phase distribution of the initial state is shown in Figure 6b.

Table 3. Parameters of numerical simulation.

Description	Value				
Rotating speed (rpm)		2	3	4	
Particle diameter (cm)	1	2	3	4	5
Rotary kiln diameter (m)			1.2		
Solid density (kg/m^3)			1150		
Air density (kg/m^3)			1.225		
Air pressure (10^5 N/m^2)			1.01		
Particle–wall restitution coefficient			0.8		
Particle–particle restitution coefficient			0.9		

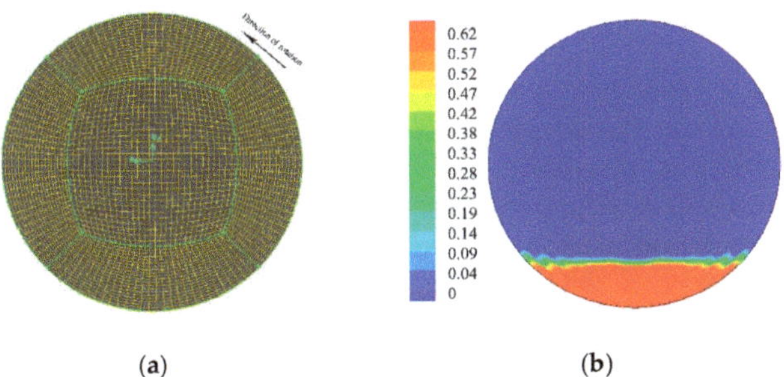

Figure 6. Profile of the rotary kiln. (**a**) Rotary kiln mesh; (**b**) volume distribution of materials at the initial time.

At the initial moment, the particle surface tends to be flat and evenly distributed at the bottom of the rotary kiln. The volume distribution under different particle diameters in stable state is shown in Figures 7–9. The results showed the surface of the material presented an inclined angle to the horizontal plane. The material surface gradually tilted and became slightly S-shaped. With the increase of the rotational speed and the diameter of the particle, the particle movement accelerated and experienced sliding, slumping, rolling, cascading and cataracting movements. Centrifugal motion did not occur because of the rotational speed limit. When the particle diameter was 1 cm, the particle movement was only slumping, and when the particle diameter was 5 cm, the particle movement could achieve a cataracting motion.

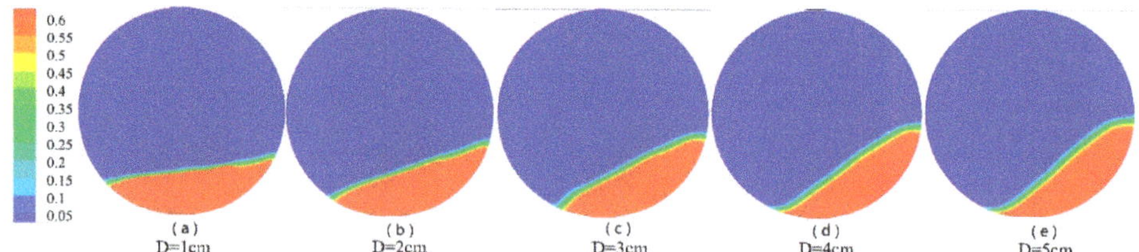

Figure 7. Volume distribution under different particle diameters (rotation speed = 2 rpm).

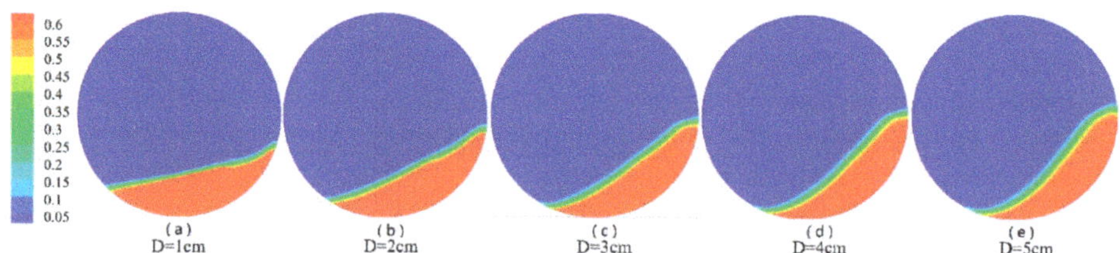

Figure 8. Volume distribution under different particle diameters (rotation speed = 3 rpm).

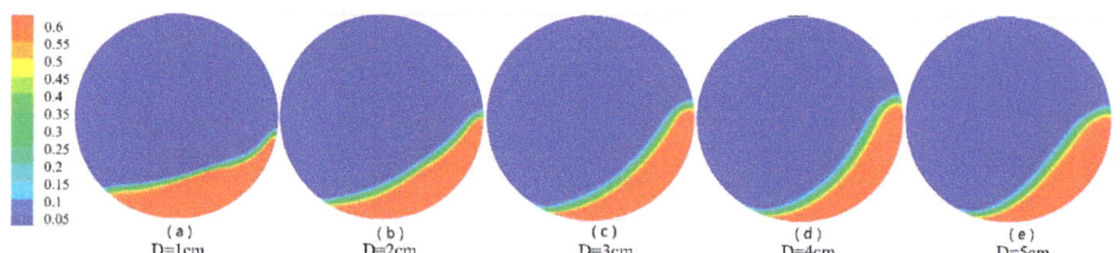

Figure 9. Volume distribution under different particle diameters (rotation speed = 4 rpm).

The rolling bed is preferred in most cases; it is beneficial to improve the heat transfer between the material beds and ensure that the waste is fully mixed and incinerated [24]. In addition, the operation speed of the rotary kiln itself should not be excessively large; otherwise, the transmission structure could have large power consumption. Given that the particle state was rolling and cascading, which is suitable for the rotary kiln working condition, the diameter of the wastes should not be extremely large, and the rotational speed of the rotary kiln should not be exceedingly fast. Therefore, the rotary kiln rotation speed was determined to be 3 rpm.

The velocity vector of particles can directly reflect the motion characteristics of particles in the kiln. The particle area is evidently layered, and the surface speed of the material is considerably larger than the internal stacking speed (Figure 10). This layer is the active layer of the material particles, which is the main area of material mixing and heat exchange. The internal relative speed is small; it is the passive layer, which is the area where the mixing and heat exchange are weak. As shown in Figure 10, the direction of the particle velocity in the active layer is opposite to the direction of the kiln wall movement. When the particles reach the bottom, they begin to change direction in accordance with the movement direction of the kiln wall. The velocity direction of particles in the passive layer is the same as that of the kiln wall, and the particles re-enter the active layer of the surface when they reach the top of the material bed on the right side. The velocity at the free surface of the particle is large, and the velocity vector at the top of the active layer is at an angle to the bed surface. Thus, particle collisions may occur at the upper right of the rotary kiln. The particles fly out under the action of rotational inertia and rejoin the material layer under the influence of gravity, which affects the surface of the active layer. The simulation results are consistent with the material movement theory of the rotary kiln.

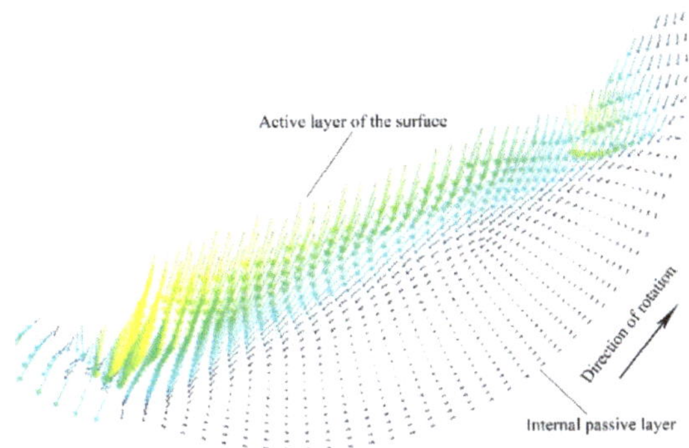

Figure 10. Velocity vector field of particles.

4.2. Numerical Simulation of Waste Rotary Incineration

Computational fluid dynamics was used to formulate a 3D steady combustion problem. Figure 11 shows the geometric model and boundary conditions of the simulated rotary kiln. The simulated rotary kiln was assumed to be cylindrical. The meshes of the rotary kiln were generated by Gambit with 1,100,000 cells. The actual diameter and length of the rotary kiln were 1200 and 8000 mm, respectively. For simplicity, only the diesel oil and the mixed waste inlet were set in front of the kiln. Both inlets were 325 mm in diameter. The inclination angle of the rotary kiln was 1.5°, and the rotational speed was 3 rad/min.

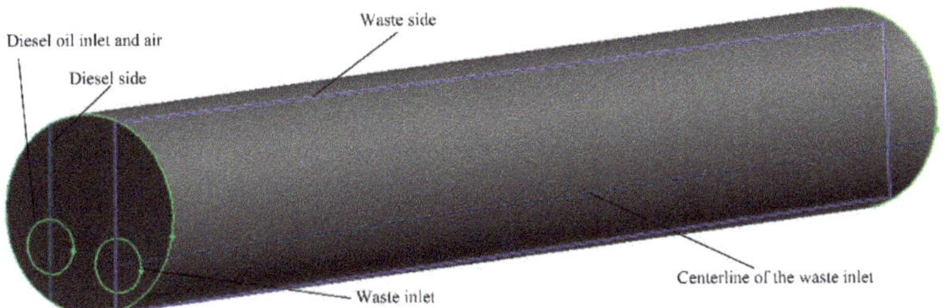

Figure 11. Geometry model of the rotary kiln.

Fuel combustion is a complex phenomenon that includes physical and chemical exchanges, multi-phase flow, heat transfer, momentum transfer, and mass transfer. The Eulerian–Lagrangian model was used to describe the flow of the gas and particle phases, and RNG $k - epsilon$ model took into account the rotational effect of material in the kiln. P-1 radiation model was used to describe the radiation heat transfer in the rotary kiln. The discrete model was used to solve the coupling of heat, mass and momentum between the phases. The particle size obeyed the Rosin–Rammler distribution from 1 to 5 cm. The initial data of solid wastes and diesel fuel were added to create a PDF file for non-premixed combustion. The SIMPLEC algorithm was chosen for the pressure–velocity coupling. The inlet parameters used for the four simulation cases are shown in Table 4.

Table 4. Inlet parameters.

Temperature (k)		Case (1)	Case (2)	Case (3)	Case (4)
		Mass Flow Rates (kg/s)			
Air	300	1.5	1.5	1.5	1.5
Diesel inlet	500	0.02	0.03	0.04	0.03
Waste inlet	300	0	0	0	0.06

Figure 12 shows the temperature curves in the kiln under three calculated cases of burning diesel oil only. The results showed that the three cases had almost the same trend of temperature change. As the mass flow of diesel oil increased, the temperature gradually rose in the kiln. The maximum combustion temperature of diesel oil with a mass flow rate of 0.04 kg/s reached 1400 °C. This temperature may have exceeded the ash melting point; thus, the ash became molten, and slag formed on the kiln wall. The combustion temperature of diesel oil with a mass flow rate of 0.02 kg/s was relatively low, and the temperature in the kiln was only 700 °C, which may have resulted in the insufficient combustion of the wastes. The temperature field with a mass flow rate of 0.03 kg/s was moderately long, and the temperature in the middle of the flame reached 1200 °C. A large central combustion zone was formed at a distance of 2 m from the kiln inlet, which was conducive to the full mixing combustion of the wastes. The outlet temperature of the kiln was less than 1000 °C. Diesel fuel with a mass flow rate of 0.03 kg/s can not only maintain a high combustion temperature but could also save fuel.

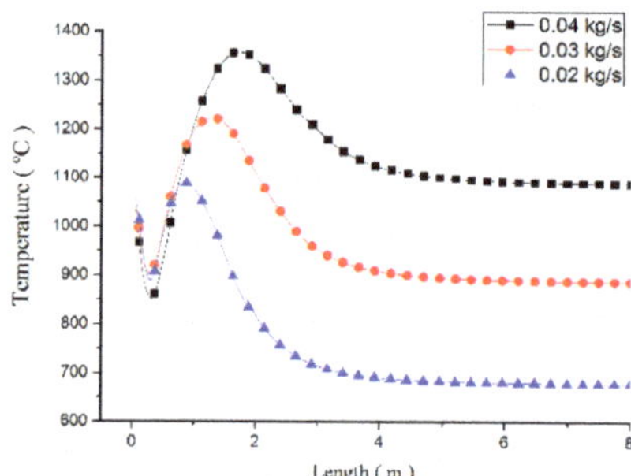

Figure 12. Temperature curve of diesel oil combustion under different mass flow rates.

The temperature field of the rotary kiln with waste combustion was simulated, as shown in Figures 13 and 14. The diesel and waste side are shown in Figure 11. The rotary kiln had two high-temperature zones. One zone was obtained by diesel oil injection and combustion, and the other one, which was obtained by mixed wastes that roll over and forward in the axial direction, was burned by gas from the diesel oil combustion of the rotary kiln. The temperature of the material gradually decreased from top to bottom. The lower temperature at the bottom of the kiln was due to the cascading movement of the waste, which burned on the surface of the bed. The heat released by waste combustion was relatively low. The diesel oil combustion temperature was high, and the wastes were rapidly heated in the kiln. The temperature of the wastes gradually rose, reaching the ignition temperature at a distance of 2 m from the kiln inlet, and the wastes began to burn. Then, the fuel bed went into the main combustion stage. The temperature of the flame area was up to 900 °C, and the mixed wastes could be sufficiently burned before leaving the kiln.

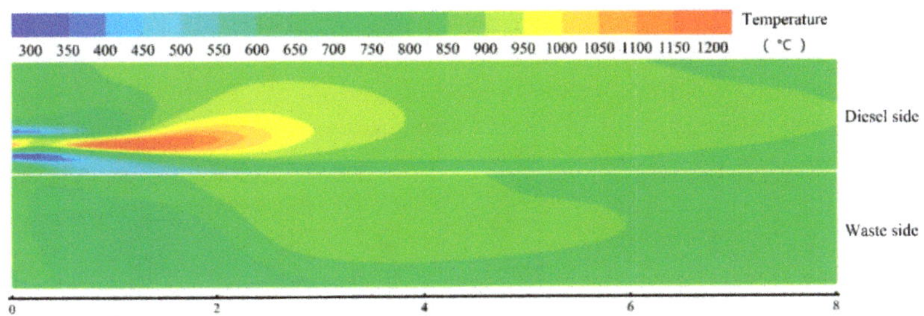

Figure 13. Temperature field of waste side and diesel side in the rotary kiln.

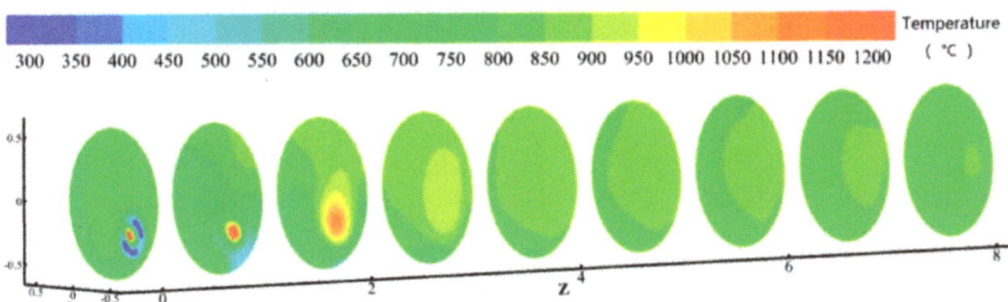

Figure 14. Temperature field of different cross sections in the rotary kiln (each section interval is 1 m).

Figures 13 and 14 show that the material moved forward along the axial direction in the kiln, and the diesel oil and wastes reached the highest temperature in the range of 1–3 m, where combustion was sufficient. The diesel oil reached the highest temperature at 2 m from the front of the kiln, and the solid wastes reached the highest temperature at 3 m.

Figure 15 shows the gas mass distribution curve on the centerline of the waste inlet (Figure 10). The study of the distribution of CO concentration in the rotary kiln could obtain the combustion status inside the kiln. The CO in the kiln was mainly derived from the pyrolysis of the wastes. The CO concentration was mainly concentrated at 1.5 m from the waste inlet, whereas the content in other regions was small. Given the sufficient oxygen in the kiln to mix well with the CO, the CO reaction gradually became complete, and the CO content at the final outlet tended to zero. CO_2 mainly came from the combustion of volatile matter and fixed carbon and reached a high concentration at 2 m in front of the kiln. From the O_2 concentration distribution, the concentration of oxygen from the inlet was reduced due to the large amount of combustion consumption of the volatile matter. In this situation, the incomplete combustion of the waste produced the most CO. In the latter half of the rotary kiln, only a few combustible components existed due to the burning of waste; thus, the oxygen content was high.

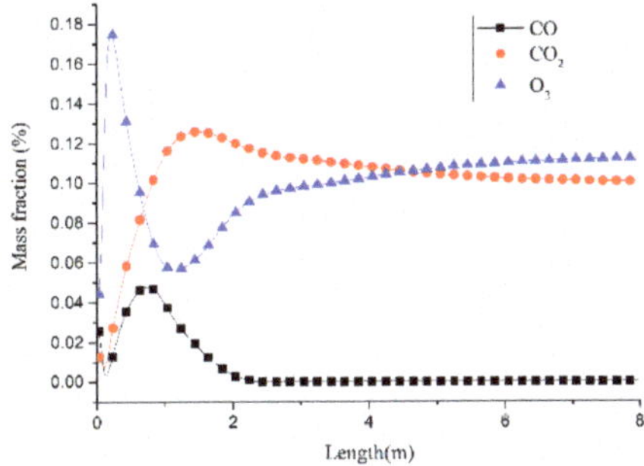

Figure 15. Gas mass fraction in the rotary kiln.

5. Engineering Application of Rotary Kiln

The rotary kiln and the entire system have been operated in a pesticide plant in Shandong, China. The incineration system is mainly composed of pretreatment, waste incineration disposal, and a flue gas treatment system. Figure 16 shows the practical engineering application of our rotary kiln.

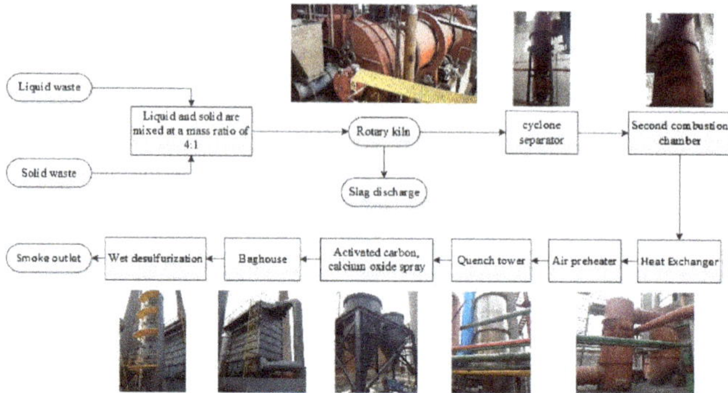

Figure 16. Incineration process and flue gas treatment process.

The incineration disposal disposes the pesticide waste liquid and the solid wastes after pretreatment. Under the continuous rotation of the rotary kiln, the mixed wastes are continuously turned, heated, dried, vaporized and burned in the kiln. The combustion temperature in the rotary kiln is kept at about 850 °C, and the residence time of the wastes in the kiln is more than 1 h. Primary dust removal is performed in high-temperature cyclone separator to reduce the particle content in flue gas produced by combustion. After dust removal, the gas enters the secondary combustion chamber. The gas products of the rotary kiln are mixed with the combustion-supporting air of the secondary combustion chamber, and the unburned combustible gas in the rotary kiln is completely burned to remove the toxic organic component. The combustion temperature of the second combustion chamber reaches 1100 °C, and the residence time of the flue gas in the second combustion chamber is no more than 2 s. The discharged gas enters the flue gas treatment system.

The flue gas treatment system is composed of heat exchanger, air preheater, quench tower, activated carbon and a calcium oxide injection system, baghouse, and wet desulphurization. The flue gas discharged from the second combustion chamber initially enters the heat exchanger and air preheater, which can fully utilize the waste heat of high-temperature flue gas, save energy and reduce consumption. The outlet gas temperature is 500–550 °C. Thereafter, the flue gas enters the quench tower. The high-temperature flue gas contacts directly with the atomized cooling water in the quench tower, and the flue gas temperature drops rapidly to 200 °C within 1 s, thereby avoiding dioxin regeneration. Before the flue gas from the quenching tower enters the baghouse, calcium oxide and activated carbon are sprayed successively into the flue. The calcium oxide is used to neutralize the acidic substances in the flue gas. The activated carbon is used to absorb the heavy metal and dioxins in the flue gas. In the baghouse, the suspended particles in the flue gas are intercepted by the filter bag and discharged in the form of fly ash. The flue gas enters the wet absorption tower of the desulphurization unit and contacts with sprayed limestone slurry droplets. The acid substances in the flue gas are absorbed.

An Atmos FIR Fourier infrared multicomponent gas analyzer was used to measure the concentration of SO_X, NO_X and HCl at the chimney outlet according to GB/T16157-1996. In order to ensure the measurement accuracy, the gas analyzer was calibrated with the target standard gas before measuring. Similarly, automatic smoke sampler with sampling gun was used to measure dust. According to the USA EPA23a method, flue gas sampling was isokinetically collected, and dioxins were preprocessed and measured [29]. The final flue gas concentration (Table 5) meets the emission standard of waste incineration.

Table 5. Measured value of pollutant concentration in flue gas.

Pollutants	Daily Average Concentration	Chinese Standard 18485–2014
Dust/(mg/Nm^3)	15	20
SO_x/(mg/Nm^3)	40	80
NO_x/(mg/Nm^3)	120	250
HCl/(mg/Nm^3)	25	50
Dioxins/(ngTEQ/Nm^3)	0.015	0.1

The main power consumption equipment in the incineration process of the rotary kiln re the screw conveyor, the motor driving device of the rotary kiln, the draft fan, the ignition burner, the secondary burner, the blower, the slag discharging motor, the quench pump, the lye pump and so on. The total power of the power consumption equipment is 98.29 kW. The hourly waste disposal capacity of the rotary kiln is 1000 kg/h. It takes 353 MJ energy to dispose of a ton of waste. The low calorific value of diesel oil is 42 MJ/L. The diesel oil consumption is 200 L/h. The economic cost of the rotary kiln incineration system mainly includes equipment operation cost, diesel fuel cost, and maintenance cost. The rotary kiln works continuously for 10 h. The power consumption of a rotary kiln is 982.9 kWh per day. The diesel oil consumption per day is 2000 L. There are three workers in total. It costs 691 RMB to treat a ton of waste.

6. Conclusions

A large amount of liquid and solid wastes generated during pesticide production are urgently needed for disposal. A co-incineration scheme for mixing multi-component wastes in a rotary kiln was proposed for waste disposal from pesticide production. The co-incineration experimental results showed that the residence time of solid and liquid mixtures in the kiln is more than 1 h, and the clinker ignition loss of the slag is less than 5%. Therefore, the co-incineration of solid and liquid wastes is feasible. A numerical study of different particle diameters and different wall rotational speeds was performed to reproduce the configurations of solid flows (i.e., sliding, slumping, rolling, cascading, cataracting, and centrifuging). The rolling or cascading mode was selected as the best operating condition. The numerical simulation revealed the combustion interactions of the co-incineration of mixed wastes, and the results showed that the co-incineration of solid and liquid wastes is feasible by adding auxiliary fuel. The rotary kiln incineration can keep the kiln temperature, mass residue and pollutants within the allowable range. Finally, the rotary kiln incineration process and flue gas treatment process were operated and commissioned. The liquid and solid wastes can be completely burned, and the flue gas concentration after treatment can meet the emission standard of waste incineration.

Author Contributions: Investigation, B.Z. and J.H.; methodology, B.Z.; software, C.H.; data curation, W.C.; writing—original draft, J.H.; writing—review and editing, B.Z.

Funding: This research was funded by Key Research and Development Program of Shandong Province of China (No. 2019GHY112002).

Conflicts of Interest: The authors declare no conflict of interest.

Nomenclature

A_{ad}	Ash content of the mixed waste, %	P	Static pressure, Pa
A_0	Arrhenius type pre-exponential factor	p_{ox}	Partial pressure of oxidant species in the gas, Pa
A_p	Surface area of the particle, m^2	Q_l	Calorific value of mother liquor of methomyl
C_2	Arrhenius factor	Q_i	Calorific value of solid composition, MJkg^{-1}
C_{ad}	Carbon content in mixed waste, %	Q_s	Calorific value of total solid wastes, MJkg^{-1}

D_0	Diffusion rate coefficient	Q_t	Calorific value of total mixed waste, $MJkg^{-1}$
E	Internal energy, Jkg^{-1}	u_w	Particle velocities, ms^{-1}
E_a	Activation energy, $Jmol^{-1}$	u_g	Gas velocities, ms^{-1}
FC_{ad}	Fixed carbon of the mixed waste, %	u_w	Particle velocities, ms^{-1}
F_x	Additional forces, N	S_m	Mass source, kgs^{-1}
f	Mixing fraction	T	Temperature, K
f_{diesel}	Mixture fraction of the secondary stream diesel oil	T_p	Particle temperature, K
$f_{eo,0}$	Mass fraction of evaporating material	V_{ad}	Volatiles content of the wastes, %
f_i	Mass fraction of solid composition in the total solid wastes	v_i	Velocity vector in the x direction, ms^{-1}
f_l	Mass fraction of mother liquor in the total waste	v_j	Velocity vector in the y direction, ms^{-1}
f_{ox}	Mixture fraction of oxidant	v_k	Velocity vector in the z direction, ms^{-1}
f_{waste}	Mixture fraction of the waste fuel	Z_x	Mass fraction for element
f_s	Mass fraction of total solid wastes in the total waste	$Z_{x,fuel}$	Elemental mass fraction of fuel inlet
$f_{vo,0}$	Mass fraction of the volatiles in the initial particle	$Z_{x,ox}$	Elemental mass fraction of oxidant inlet
H_{ad}	Hydrogen content in wastes, %	τ_{ij}	Stress tensor, Pa
k	Kinetic rate, s^{-1}	$\mathfrak{R}$	Kinetic rate, s^{-1}
M_{ad}	Moisture content of the wastes, %	μ	Dynamic viscosity, Pas
m_p	Particle mass, kg	δ_{ij}	Kronecker symbol
$m_{p,0}$	Initial particle mass, kg	ρ	Gas phase density, kgm^{-3}
N_{ad}	Nitrogen content in mixed waste, %	λ	Thermal conductivity, Wm^{-1} K^{-1}
O_{ad}	Oxygen content in mixed waste, %	ρ_w	Density of particle phase, kgm^{-3}
g_i	Gravitational body force, ms^{-2}, Greek letters		

References

1. Jin, F.; Wang, J.; Shao, H. Pesticide use and residue control in china. *J. Pestic. Sci.* **2010**, *35*, 138–142. [CrossRef]
2. Chen, Z.L.; Dong, F.S.; Jun, X.U. Management of pesticide residues in China. *J. Integr. Agric.* **2015**, *14*, 2319–2327. [CrossRef]
3. Xu, D.; Wang, S.; Zhang, J. Supercritical water oxidation of a pesticide wastewater. *Chem. Eng. Res. Des.* **2015**, *94*, 396–406. [CrossRef]
4. Renou, S.; Givaudan, J.G.; Poulain, S. Landfill leachate treatment: review and opportunity. *J. Hazard. Mater.* **2008**, *150*, 468–493. [CrossRef] [PubMed]
5. Gao, Q.; Wang, L.; Li, Z. Adsorptive Removal of Pyridine in Simulation Wastewater Using Coke Powder. *Processes* **2019**, *7*, 459. [CrossRef]
6. Migliori, M.; Catizzone, E.; Giordano, G. Pilot Plant Data Assessment in Anaerobic Digestion of Organic Fraction of Municipal Waste Solids. *Processes* **2019**, *7*, 54. [CrossRef]
7. Nie, Y. Development and prospects of municipal solid waste (MSW) incineration in China. *Front. Environ. Sci. Eng. China* **2008**, *2*, 1–7. [CrossRef]
8. Jabłońska, B.; Kiełbasa, P.; Korenko, M.; Dróżdż, T. Physical and Chemical Properties of Waste from PET Bottles Washing as A Component of Solid Fuels. *Energies* **2019**, *12*, 2197. [CrossRef]
9. Tabasová, A.; Kropáč, J.; Kermes, V. Waste-to-energy technologies: Impact on environment. *Energy* **2012**, *44*, 146–155. [CrossRef]
10. Ghouleh, Z.; Shao, Y. Turning municipal solid waste incineration into a cleaner cement production. *J. Clean. Prod.* **2018**, *195*, 268–279. [CrossRef]
11. Guo, Y.; Glad, T.; Zhong, Z. Environmental life-cycle assessment of municipal solid waste incineration stocks in Chinese industrial parks. *Resour. Conserv. Recycl.* **2018**, *139*, 387–395. [CrossRef]
12. Chen, H.C.; Zhao, C.S.; Li, Y.W.; Lu, D.F. NO_x emission from incineration of organic liquid waste in a circulating fluidized bed. *Korean J. Chem. Eng.* **2007**, *24*, 906–910. [CrossRef]

13. Ma, J.; Liu, D.; Chen, Z.; Chen, X. Agglomeration characteristics during fluidized bed combustion of salty wastewater. *Powder Technol.* **2014**, *253*, 537–547. [CrossRef]
14. Wang, N.Y.; Chun, H.S.; Peite, C. Environmental effects of sewage sludge carbonization and other treatment alternatives. *Energies* **2013**, *6*, 871–883. [CrossRef]
15. Hu, S.; Ma, X.; Lin, Y. Thermogravimetric analysis of the co-combustion of paper mill sludge and municipal solid waste. *Energy Convers. Manag.* **2015**, *99*, 112–118. [CrossRef]
16. Lin, H.; Ma, X. Simulation of co-incineration of sewage sludge with municipal solid waste in a grate furnace incinerator. *Waste Manag.* **2012**, *32*, 561–567. [CrossRef]
17. Zhong, Q.; Zhang, J.; Yang, Y. Thermal Behavior of Coal Used in Rotary Kiln and Its Combustion Intensification. *Energies* **2018**, *11*, 1055. [CrossRef]
18. Shi, H.; Si, W.; Li, X. The Concept, Design and Performance of a Novel Rotary Kiln Type Air-Staged Biomass Gasifier. *Energies* **2016**, *9*, 67. [CrossRef]
19. Weinberg, A.V.; Varona, C.; Chaucherie, X. Extending refractory lifetime in rotary kilns for hazardous waste incineration. *Ceram. Int.* **2016**, *42*, 17626–17634. [CrossRef]
20. Bai, Y.; Bao, Y.B.; Cai, X.L. Feasibility of disposing waste glyphosate neutralization liquor with cement rotary kiln. *J. Hazard. Mater.* **2014**, *278*, 500–505. [CrossRef]
21. Huber, F.; Blasenbauer, D.; Mallow, O. Thermal co-treatment of combustible hazardous waste and waste incineration fly ash in a rotary kiln. *Waste Manag.* **2016**, *58*, 181–190. [CrossRef] [PubMed]
22. Bujak, J. Thermal treatment of medical waste in a rotary kiln. *J. Environ. Manag.* **2015**, *162*, 139–147. [CrossRef] [PubMed]
23. Bujak, J. Thermal utilization (treatment) of plastic waste. *Energy* **2015**, *90*, 1468–1477. [CrossRef]
24. Mellmann, J. The transverse motion of solids in rotating cylinders—forms of motion and transition behavior. *Powder Technol.* **2001**, *118*, 251–270. [CrossRef]
25. Wang, M.; Liao, B.; Liu, Y. Numerical simulation of oxy-coal combustion in a rotary cement kiln. *Appl. Therm. Eng.* **2016**, *103*, 491–500. [CrossRef]
26. Badzioch, S.; Hawksley, P.G.W. Kinetics of Thermal Decomposition of Pulverized Coal Particles. *Ind. Eng. Chem. Process Des. Dev.* **1970**, *9*, 521–530. [CrossRef]
27. Baum, M.M.; Street, P.J. Predicting the Combustion Behaviour of Coal Particles. *Combust. Sci. Technol.* **1971**, *3*, 231–243. [CrossRef]
28. Ghenai, C.; Lin, C.X.; Ebadian, M.A. Numerical Investigation of Oxygen-Enriched Pulverized Coal Combustion. *Heat Transf.* **2003**, *2*. [CrossRef]
29. Pham, M.T.N.; Anh, H.Q.; Nghiem, X.T. Characterization of PCDD/Fs and dioxin-like PCBs in flue gas from thermal industrial processes in Vietnam: A comprehensive investigation on emission profiles and levels. *Chemosphere* **2019**, *225*, 238–246. [CrossRef]

Article

Flow and Diffusion Characteristics of Typical Halon Extinguishing Agent Substitute under Different Release Pressures

Jiaming Jin, Renming Pan *, Ruiyu Chen, Xiaokang Xu and Quanwei Li *

School of Chemical Engineering, Nanjing University of Science and Technology, Nanjing 210094, China; jmjin_njust@163.com (J.J.); crynjust@njust.edu.cn (R.C.); xiaokangxu@njust.edu.cn (X.X.)
* Correspondence: panrenming@njust.edu.cn (R.P.); liqw83@163.com (Q.L.); Tel.: +86-025-8431-5511 (R.P.)

Received: 3 December 2020; Accepted: 16 December 2020; Published: 21 December 2020

Abstract: To provide guidance towards reducing the weight of the HFC-125 storage vessel by reducing the release pressure and to reveal the effects of release pressure on the extinguishing efficiency of HFC-125, we investigated the flow and diffusion characteristics of HFC-125 under six release pressures in the present study. The influence of release pressure on the degree of superheat, injection duration, pressure loss, jet angle, and concentration distribution were analyzed. Results show that the degree of superheat and the injection duration both decreased with the release pressure. The bubble expansion in the HFC-125 could slow down the pressure decrease in the storage vessel. The flow process in the pipeline can be divided into three phases: pipeline filling, stable flow, and mixed gases release. Both of the maximum and mean values of the pipeline pressure loss increased with the release pressure. The maximum concentration value decreased with the increase of the distance from the nozzle. The maximum concentration value in the near field from the nozzle increased with the release pressure. The concentration and holding time (duration above 17.6% volume concentration) of HFC-125 in the near field from the nozzle met the requirements of minimum performance standards (MPS) for HFC-125.

Keywords: Halon candidate substitute; aircraft weight reduction; HFC-125; release pressure; flow; diffusion

1. Introduction

Halon fire extinguishing agents, especially for Halon 1301, negatively impact the ozone layer and have a high global warming potential (GWP) index, which goes against the environment protection and the sustainable development of the world [1]. The search for appropriate alternatives for the halon fire extinguishing agents has received increasing attention in the past decades [2,3]. Pentafluoroethane (HFC-125) does not destroy the ozone layer and its GWP index is only half of that of Halon 1301. It is regarded as a candidate substitute for Halon 1301 in the engine nacelle and auxiliary power unit of commercial aircrafts at low temperatures [4,5]. However, when HFC-125 is applied in aircrafts, owing to the lower extinguishing efficiency of HFC-125 compared with Halon 1301, more HFC-125 agents (an increase of approximately 80% in weight) and larger fire extinguishing agent storage vessels (approximately 2.3 to 4.3 times in volume larger than that of Halon 1301) are needed [6]. Such an increase in weight and volume will pose a great challenge to the aircraft design in the aspect of cost control, fuel consumption, and safety [7], which goes against the sustainable development of the aviation industry. It is noted that if the release pressure is appropriately reduced on the condition of meeting the requirements of the airworthiness provisions for the volume concentration and holding time of the fire extinguishing agent, the weight of the agent storage vessel can be reduced, which is considerably beneficial to the weight reduction of the aircraft. The weight reduction can greatly reduce the greenhouse gas emissions generated by the fuel consumption. In order to evaluate whether the requirements of the minimum performance standards (MPS) [8] for HFC-125 can be satisfied or not on the condition of reducing the release pressure, it is of great necessity to study the flow characteristics in the pipeline and the diffusion behaviors in the power nacelle of HFC-125 under different pressures.

Some attention was devoted to the flow behaviors of fire extinguishing agents in pipelines. However, few studies have been reported on HFC-125. Williamson studied the flow behaviors of Halon 1301 in pipelines under 2.48 MPa [9]. He found that the pressure decreased in a nonlinear way when the Halon 1301 agent flowed in the pipeline, and the agent boiling would slow down the pressure decrease. Moreover, the release rate of the Halon 1301 agent increased with the bottle volume. In the case of 5.2 MPa, Elliott et al. [10] proposed a homogeneous and equilibrium model of two-phase (gas and liquid phase) flow to estimate the flow behaviors of Halon 1301 in the pipeline. It was found that the predicted data based on the model were in accordance with the experimental data. Yang et al. [11] presented a two-phase (gas and liquid phase) equilibrium model to calculate the thermodynamic properties and filling conditions of five selected agents: HFC-227ea, CF3I, FC-218, HFC-125, and CF3Br. The accuracy of the presented model was verified under the release pressures from 2.8 to 4.1 MPa. The predicted values based on the two-phase quilibrium model were found to be in good agreement with the measured values. Tuzla et al. [12] developed a computer code to predict the single-phase and two-phase flow behaviors of the fire extinguishing agents in the pipeline under 5.71 MPa on the basis of the multi-phase flow algorithms generally used in the nuclear power plant. Kim et al. [13] employed FLUENT software to simulate the flow behaviors of Halon 1301 in the fire extinguishing system under 4.1 MPa. The volume percentage of the Halon 1301 agent in the pipeline and the outlet was obtained. In addition, the release behaviors of the Halon 1301 agent in the case of different surface areas of the rupture disk were analyzed. The results indicated that the release rate of the Halon 1301 agent increased with the surface area of the rupture disk. Moreover, it was found that little influence of the pipeline diameter was exerted on the release process of the Halon 1301 agent in the storage vessel. However, the release rate of the Halon 1301 agent at the pipeline outlet increased with the pipeline diameter. Some studies were reported on the diffusion behaviors of fire extinguishing agents in enclosure spaces, with the majority being numerical simulation studies. Among them, the study concerning the diffusion of HFC-125 in enclosure spaces was rare. Sarkos [14] studied the diffusion behaviors of Halon 1301 in a full-scale aircraft cabin under 2.48 MPa. The profiles of the agent concentration, visibility, pressure, temperature, noise, etc. were measured. The results indicated that the influence of the ultralow-pressure, over-temperature, and the agent concentration overshoot on passengers can be alleviated by the air disturbance. In addition, the air disturbance was helpful in the diffusion of the agent. The fire extinguishing agent released from the ceil can finally enter the lavatory and other complex areas through diffusion. Niu et al. [15] employed fire dynamics simulator (FDS) to study the diffusion behaviors of the Halon 1301 agent in the helicopter engine nacelle in the case of no-ventilation. The concentration distribution of the Halon 1301 agent in 6–10 s in the case of different mass flow rates and injection time were measured. It was found that reducing the mass flow rate of the fire extinguishing agent was beneficial to improve the system reliability and reduce the amount of the fire extinguishing agent. Using the lumped parameter approach, Kurokawa et al. [16,17] proposed a one-dimensional model to predict the volume concentration of the Halon 1301 fire extinguishing agent with the assumption that the profile of the flow rate was as a ladder shape. The proposed model was found to acceptably predict the volume concentration of the fire extinguishing agent. Adopting Fluent software, Zaparoli [18] investigated the diffusion behaviors of the Halon 1301 agent in the cases of three air flow rates in the cargo hold of the aircraft. It was indicated that the agent concentration decreased continuously with time when the air flow rate was 0.08 kg/s. Using Hflowx and Fluent, Lee [19] simulated the diffusion of the fire extinguishing agent (HFC-125, CF3I, and Halon 1301) in the engine nacelle and auxiliary power unit (APU) nacelle on the basis of a one-dimensional two-phase flow algorithm. It was found that small differences occurred between the simulation results and the experimental results. In summary, the flow and diffusion behaviors of the Halon fire extinguishing agent have been analyzed under single release pressure in the previous studies. However, scarce attention was focused on the flow and diffusion behaviors of HFC-125. Moreover, it is indicated in the previous studies that the flow and diffusion characteristics of the fire extinguishing agent are in close relationship with the volume of the storage vessel, the pipe diameter, the nozzle location and configuration, the outlet flow rate, the degree of phase transition, etc., whose design are greatly dependent on the initial release pressure. Therefore, considering the urgent need of the alternatives for the halon fire extinguishing agents and the deduction of the aircraft weight, it is of considerable necessity and importance to study the flow and diffusion behaviors of HFC-125 under different release pressures.

In the present study, the flow behaviors in the pipeline and the diffusion behaviors in the enclosure spaces of HFC-125 were studied under different release pressures using a full-scale airborne fire extinguishing system in the engine nacelle. Many parameters including the degree of superheat, the injection duration, the jet structure, and the concentration distribution were measured and discussed. The effects of release pressure on the above-mentioned parameters were then analyzed.

2. Experiment Apparatus

Figure 1 shows the schematic diagram of the full-scale airborne fire extinguishing system. It was mainly composed of 3 parts: agents release system, enclosure space, and data acquisition system. The agents release system consisted of a high-pressure storage vessel with the working capacity of 1.4 L, a vessel head valve, a pipe, and a nozzle. The storage vessel was pressurized by nitrogen to drive the agents in the storage vessel. In the present study, the storage vessel was filled with HFC-125 of 0.95 kg. In order to saturate the nitrogen dissolved in the fire extinguishing agent, we increased the vessel pressure slowly by nitrogen to 2.41, 2.76, 3.1, 3.45, 3.79, and 4.14 MPa under 294.25 K. The vessel head valve was installed at the outlet of the vessel to control the opening and close of the fire extinguishing system. The downstream pipe was 2400 mm long and 15.6 mm in diameter and employed the same straight-through nozzle. The diffusion characteristics of the fire extinguishing agent jet was proceeded in an enclosure space (2200 mm (long) × 2300 mm (wide) × 2000 mm (high)) with a pressure relief port.

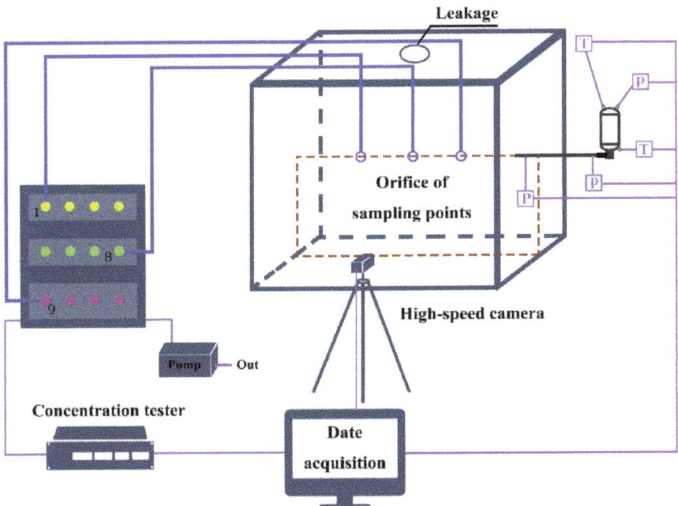

Figure 1. The full-scale airborne fire extinguishing system.

The data acquisition system consisted of two K-type thermocouples, three pressure transmitters, a fire extinguishing agent concentration tester, and a high-speed camera. The variations of pressure in the vessel and pipeline and the equilibrium temperature of the agents throughout the release process were monitored by the pressure transmitter and the thermocouple, respectively. The thermocouples were installed at the top and bottom of the vessel. Three pressure transmitters with the range of 0–5 MPa were installed to record the vessel pressure at the top of the vessel (P) and the pressure loss of the HFC-125 agent at the inlet of the pipe (P_{in}) and the inlet of the nozzle (P_{out}) in the pipeline. The length between P_{in} and P_{out} in the downstream pipeline was 2 m. There were 12 channels in the fire extinguishing agent concentration tester, and the measurement range was 0–80%. It was noted that the concentration mentioned here refers to the volume concentration of the gaseous fire extinguishing agent. The concentration distribution of the HFC-125 at 3 imaginary isometric sections (shown in Figure 2) was measured and analyzed in the present study. Therein, Section II was the central section of the enclosure space, and Section I and III were 550 mm away from the left and right of Section II, as shown in Figure 2. Point 2, point 4, and point 10 were located on the circle with a diameter of 109.7 mm, while point 3, point 5, and point 11 were located on the circle with a diameter of 219.4 mm. Point 12 was located on the circle with a diameter of 329.1 mm. The orifices of the sampling pipe were fixed at the 12 points located in the enclosure space to measure the concentration of the fire extinguishing agent. A Photron FASTCAM UX50 high-speed camera with the frame rate of 1000 fps was used to study the diffusion behaviors of the fire extinguishing agent in the ejection process in the enclosure space.

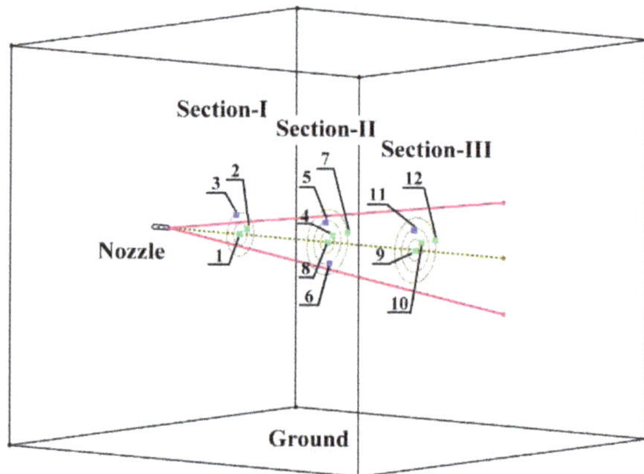

Figure 2. The location of sampling points.

3. Fire Extinguishing Agent Properties and Experimental Conditions

HFC-125 is named pentafluoroethane, whose molecular formula is C_2HF_5. Its molecular weight is 120.02 and its boiling point at standard atmospheric pressure is 224.7 K. It is in a gas state under normal temperature and pressure, while it can be liquefied when it is pressured. The value of ozone depletion potential (ODP) of HFC-125 is far below 0.001, and it has been recognized as the candidate substitute for the halon fire extinguishing agents by the EPA (United States Environmental Protection Agency)'s Significant New Alternatives Policy (SNAP).

In the present study, the structure of the storage vessel and filling conditions were consistent with those of the airborne APU fire extinguishing system. The flow and diffusion characteristics of HFC-125 were studied under six release pressures of 2.41, 2.76, 3.1, 3.45, 3.79, and 4.14 MPa.

4. Results and Discussion

4.1. Degree of Superheat under Different Release Pressures

It is reported that the flow and diffusion characteristics will be greatly affected in a complex unsteady flow field yielded by the different release pressures [20]. Different from the conventional liquid jet, which is greatly dependent on the surface evaporation of the broken liquid droplets, the thermodynamic parameters are the main factors that affect the jet characteristics of the gas fire extinguishing agent. Figure 3 shows the relationship between the saturation vapor pressure and the saturation temperature of HFC-125; the boiling point of HFC-125 is 224.7 K (−48.45 °C). Even during the release process, the liquidus temperature of HFC-125 still greatly exceeds its boiling point. That is, the HFC-125 agent is always in the state of superheat and the phenomenon of jet expansion will occur in the release process. Since a mass of bubbles are produced due to the nucleation of the HFC-125 agent in the superheated state, the HFC-125 agent expands in the vessel. The expansion of the HFC-125 agent in the flow process is influenced by the release pressure. Meanwhile, the bubble expansion can always retard the pressure drop during the release process.

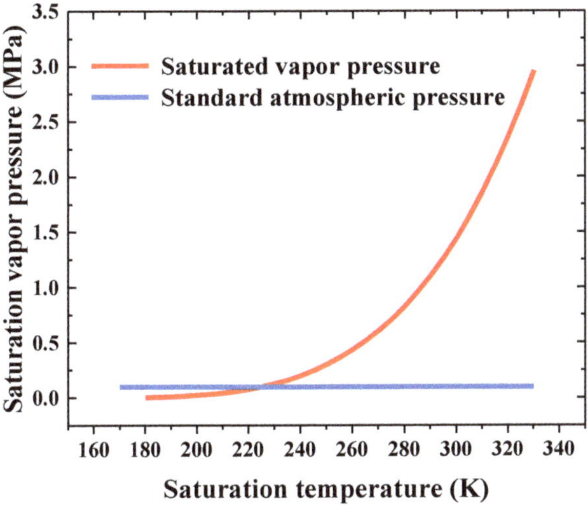

Figure 3. The saturated vapor pressure versus saturation temperature of HFC-125.

In general, the phenomenon of superheat can be quantitatively characterized in two forms, namely, the degree of superheat (D_{sup}) and the dimensionless degree of superheat (d_{sup}). They are generally adopted to determine the generation and growth rate of bubbles in the liquid. There is a logarithmic relationship between them, by which they can be transformed into each other [21]. The influence of the release pressure on the expansion degree of the HFC-125 agent can be revealed by either of them. The expressions for D_{sup} and d_{sup} are shown in Equations (1) and (2), respectively.

$$D_{\text{sup}} = T_{agent} - T_{sat}, \tag{1}$$

$$d_{\text{sup}} = P_{air}/P_{sat}, \tag{2}$$

where D_{sup} denotes the degree of superheat (K), d_{sup} represents a dimensionless number, T_{agent} denotes the temperature of the liquid in the vessel (K), and T_{sat} represents the saturation temperature at the ambient pressure (K). D_{sup} was selected for characterizing the influence of release pressure on the degree of superheat in the current study. Figure 4 and Equation (3) exhibit the correlation of the degree of superheat of the HFC-125 agent with the release pressure.

$$D_{\text{sup}} = -17.95P + 3.7, \quad (3)$$

where P denotes the release pressure (MPa).

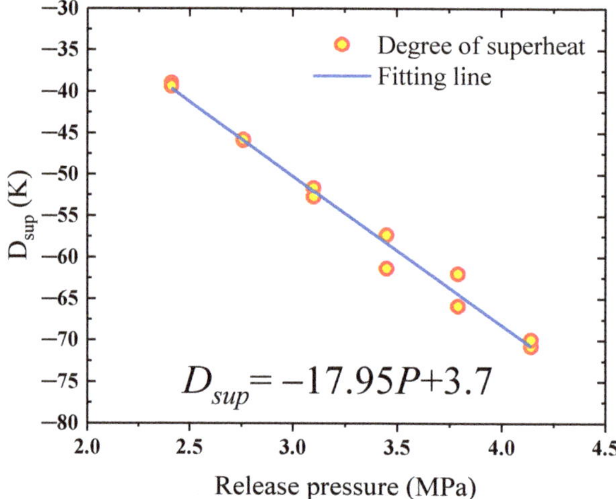

Figure 4. D_{sup} of the HFC-125 agent as a function of release pressure.

Figure 4 shows that there was a linear relationship between D_{sup} of and the release pressure, and D_{sup} decreased with the release pressure. Moreover, the value of D_{sup} in the case of 4.14 MPa was two times larger than that in the case of 2.41 MPa. It can be deduced that in the case of high release pressures, the nucleation and growth of the generated bubbles were relatively mitigatory, and the expansion of HFC-125 in the vessel was feebler. As a consequence, the injection of the HFC-125 agent can be completed in a short time frame in the case of high release pressures, which may indicate that the fire extinguishing efficiency of HFC-125 is higher at high release pressures compared with that at low release pressures.

4.2. Injection Duration under Different Release Pressures

The injection duration is one of the important factors that must be considered to evaluate the performance of the fire extinguishing system. It is worth noting that the injection duration obtained from the pressure change curve in the vessel and the time required to reach the extinguishing concentration differ in the description. The injection duration of the fire extinguishing system refers to the time difference between the moment of sudden pressure drop in the vessel at the dawning of injection and the moment when the pressure starts to level off. Figure 5 illustrates the injection duration of the HFC-125 agent at different release pressures. The results indicate that the injection duration of the HFC-125 agent jet in the superheated state decreased with the release pressure as an exponential decay function, and the equation is presented in Equation (4).

$$T_{in} = 3.35 \exp(-P/1.026) + 1.154, \quad (4)$$

where T_{in} represents the injection duration (s).

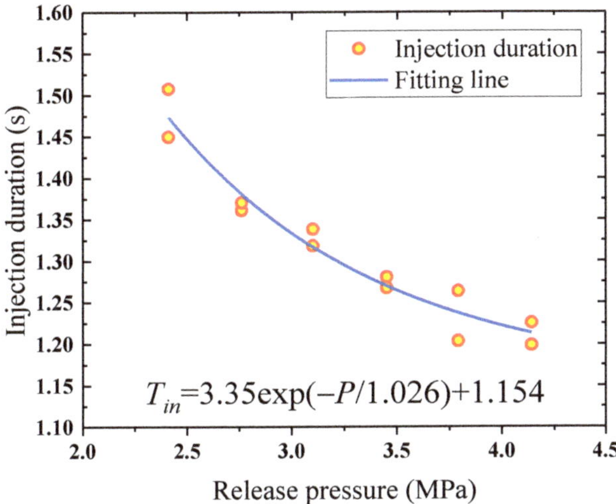

Figure 5. The injection duration of HFC-125 under different release pressures.

The reasons for the negative correlation between the injection duration of the HFC-125 agent and the release pressure are presented as follows. First and foremost, low pressures occurred in the pipeline and in the vessel during the early stage of the release process in the case of low release pressures. The ambient pressure of the HFC-125 agent was much lower than its corresponding saturated vapor pressure at 294.25 K when the fire extinguishing system was pressured to a lower level. As a consequence, the dimensionless degree of superheat ($d_{\mathrm{sup}} = P_{air}/P_{sat}$) declined and a more intense phase transition and more bubbles were generated in the liquidus HFC-125 agent. The generation of bubbles supplemented the pressure in the vessel and the pipeline, as illustrated in Figure 6. That is, the decline of the pressure drop rate in the vessel was caused by the generation of a mass of bubbles. When the pressure in the vessel no longer declined, the HFC-125 agent in the pipeline changed from the gas-liquid phase to the gas phase. Nevertheless, when the fire extinguishing system was pressured to a higher level, which was still lower than the saturated vapor pressure at 294.25 K, the dimensionless degree of superheat ($d_{\mathrm{sup}} = P_{air}/P_{sat}$) was larger than that in the case of the lower level. The expansion of the HFC-125 agent at the high release pressures was comparatively tender and would not evaporate as intensively as the case of the low release pressures. The supplement of the pressure in the vessel and pipeline was not obvious. Therefore, when the fire extinguishing system was pressured to the high level, the release pressure had little effect on the injection duration of the HFC-125 agent and the jet process was completed in less time compared with that at the low release pressures. Furthermore, the fire extinguishing system pressured in the low level meant that there was relatively little difference in the pressure between the fire extinguishing agent in the storage vessel and the environment, which led to the energy of the HFC-125 agent, which was converted into the kinetic energy, decreasing. The speed of the HFC-125 agent flow from the storage vessel to the pipeline declined consequently, and the injection duration was correspondingly prolonged.

Moreover, the bubble expansion occurred earlier when the release pressure was higher. It was close to the pressure dropping process in the storage vessel, which was mainly based on the emptying and was supplemented by the bubble expansion. In the case of the high release pressures, the pressure supplement was obviously weaker than the emptying, and a significant pressure drop occurred. On the contrary, in the case of the low release pressures, the pressure supplement was obviously strengthened, which neutralized part of the pressure drop attributed to the emptying process. The balance between the supplement and the emptying would be maintained for a longer time with the decline of the release pressure, leading to the phenomenon of the bubble expansion supplement being delayed.

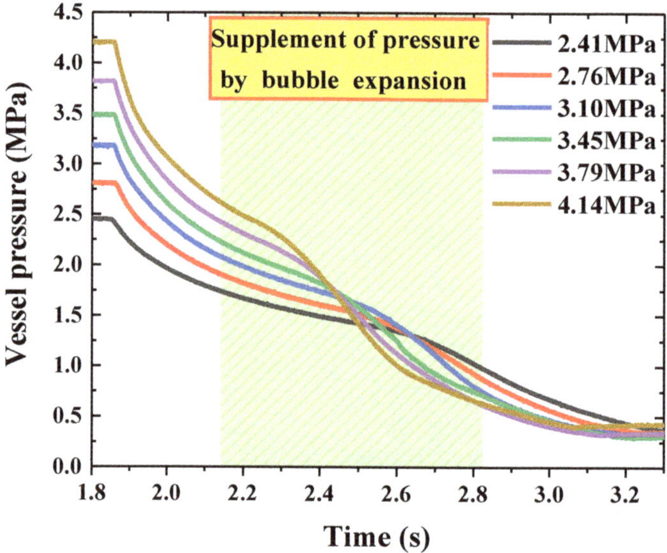

Figure 6. The vessel pressure of HFC-125 agent under various release pressures.

4.3. Pressure Loss in Pipeline under Different Release Pressures

In the fire extinguishing system, the fire extinguishing agent cannot be transported without pipeline. However, the flow of the fire extinguishing agent in the pipeline is bound to cause energy loss because of the interaction between the agent molecules and the contact between the agent with the rough wall of the pipe and the local components. In order to study the influence of release pressure on the pressure loss in the pipeline, we compared the monitoring data of P_{in} pressure transmitter at the beginning of the pipeline with P_{out} pressure transmitter at the end of the pipeline in this section, as shown in Figure 7. At a phenomenological level, the differential pressure evolution during the two-phase (liquid and gas) release process under different release pressures exhibited the same trend.

As mentioned in Sections 4.1 and 4.2, the phase transition definitely affected the release process of the fire extinguishing agent in the superheated state. This effect can be observed in the variations of the pipeline differential pressure as a function of time. It was found in Figure 7 that the release process of the HFC-125 agent can be divided into three phases. In the current study, in order to illustrate the three phases, we have displayed the differential pressure as a function of time in the case of 4.14 MPa in Figure 8.

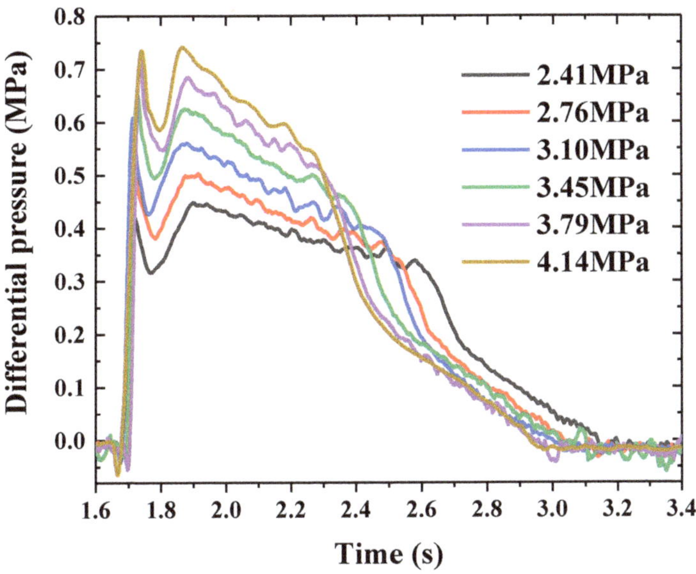

Figure 7. The differential pressure of HFC-125 under different release pressures.

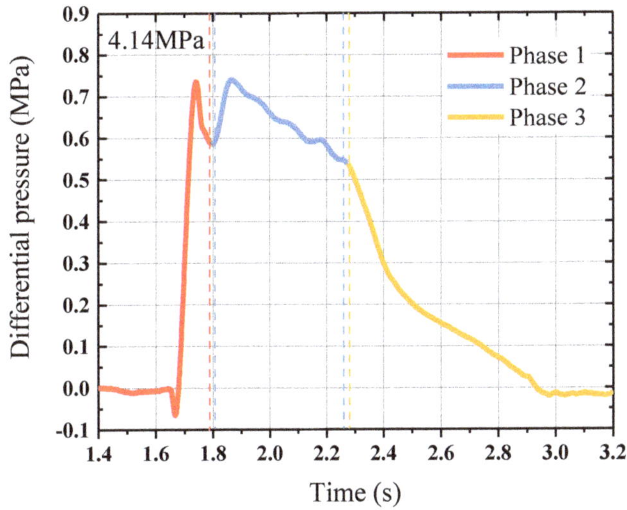

Figure 8. Phase division determined by the differential pressure curve in the pipeline at 4.14 MPa.

(1) Phase 1: Pipeline filling

Phase 1 denotes the filling process of the HFC-125 agent into the pipeline. In this phase, the valve in the storage vessel was opened and the HFC-125 agent was filled into the pipeline. As the storage vessel was highly pressured, the HFC-125 agent was released quickly and the P_{in} increased sharply. The rarefaction waves formed at the front of the jet passed through the position of the P_{in} and P_{out} pressure sensors in sequence along the pipeline. Because the P_{out} was far away from the storage vessel, the spread of the rarefaction wave showed a time difference between the inlet and the outlet, which caused the differential pressure to rise instantly and to reach the first peak at 2.119 s. Subsequently, the value of ΔP decreased due to the continual supplement of the HFC-125 agent to the position of the P_{out} pressure sensor. When the differential pressure reached a local minimum value at 2.179 s, the pipe was totally filled with the HFC-125 agent. It can be observed in Figure 8 that phase 1 merely occupied a short region of the differential pressure curve, indicating that the agent was injected into the pipeline at a quite fast speed.

(2) Phase 2: Stable Flow

Phase 2 represents the stable jet process. Intense phase transition of the superheated HFC-125 agent happened in the pipeline ahead of the rarefaction waves of the jet that left the nozzle. Because the formed gas supplemented the pressure in the pipeline gradually, P_{in} was supplemented and P_{out} was emptied. Thus, the differential pressure between the P_{in} and P_{out} increased. Meanwhile, the release process also contained the phenomenon of the nitrogen precipitation. The peak of the differential pressure in Phase 2 was attributed to the combination of the above-mentioned two reasons. Because of the differential pressure between the fire extinguishing system and the enclosure space, the pressure in the storage vessel decreased and the fire extinguishing agent sprayed continuously when the rarefaction waves left the nozzle. In this process, the differential pressure smoothly declined with time. A dynamic balance between the release rate of the fire extinguishing agent and the rate of the gas produced by the phase transition occurred in this phase.

(3) Phase 3: Mixed Gases Release

Phase 3 represents the release of the gas phase agent and the residual nitrogen. The liquid agent in the storage vessel was nearly emptied at the end of Phase 2. The production rate of gas was gradually deficient. In this phase, due to the emptied HFC-125 agent, the supplemented amount of the liquid HFC-125 agent at the P_{in} position decreased sharply, causing the differential pressure to decline rapidly. Then, it entered the stable decline process. Unlike Phase 2, this process corresponds to the smooth emptying of the mixture of gaseous fire extinguishing agent and residual nitrogen. Finally, only a little gas remained in the fire extinguishing system.

The linear increase of the mean and maximum values of the differential pressure in the pipeline with the release pressure is demonstrated in Figure 9a, and the specific functions are shown in Equations (5) and (6). In addition, Figure 9b shows that the largest value of differential pressure that occurred in Phase 2. Under all release pressures, it was found that the pipeline pressure loss mainly occurred in the phases of stable flow and pipeline filling. This was owing to the fact that a mass of bubbles was generated in these two phases due to the superheat of the HFC-125 agent, which caused the agent volume to expand and the density of the mixture of the liquid and vapor to decrease. The above-mentioned reasons caused the decline of the pressure at P_{out}. The differential pressure was consequently high. Moreover, the mean value of differential pressure also increased with the release pressure at each phase. The occurrence of the above-mentioned phenomena is closely related to the bubbles generated by the superheat of the HFC-125 agent, and it will be illustrated as follows.

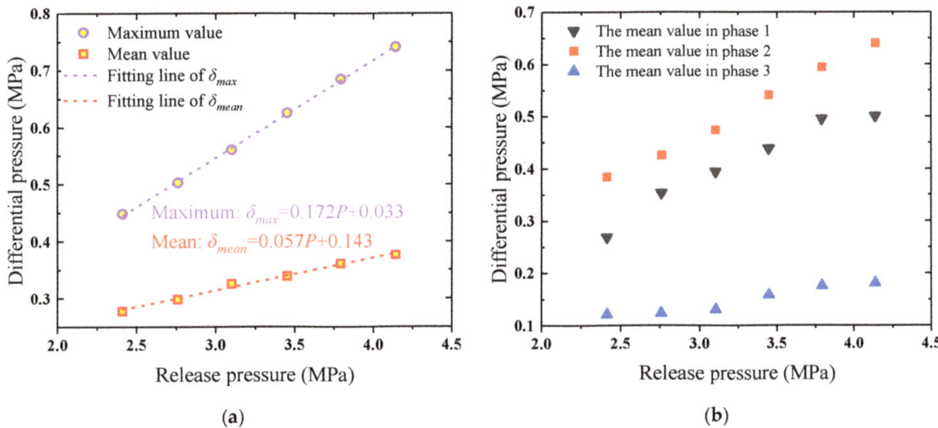

Figure 9. (**a**) The maximum and mean values of differential pressure in the whole process. (**b**) The mean value of differential pressure in each phase.

At low release pressures, as shown in Figure 3, the saturated vapor pressure of HFC-125 at 294.25 K was much higher than its pressure in the vessel. As a result, more bubbles were generated due to the intense phase transition resulting from the smaller dimensionless degree of superheat ($d_{\text{sup}} = P_{air}/P_{sat}$) in the liquid agent. This caused the gas–liquid ratio in the pipeline to increase and the viscosity of the fluid to decrease. As a consequence, the pressure loss was maintained at a low level at low release pressures. On the contrary, in the case of the higher release pressures, although the saturated vapor pressure of HFC-125 at 294.25 K was still much higher than its pressure in the vessel, there were relatively less bubbles generated by the intense phase transition resulting from the relatively larger dimensionless degree of superheat ($d_{\text{sup}} = P_{air}/P_{sat}$) in the liquid agent compared to the case of lower release pressures. The bubble expansion occurring in the HFC-125 agent was relatively gentle. Therefore, the gas–liquid ratio in the pipeline was relatively small, which made the viscosity of the HFC-125 agent increase. The value of the pressure loss at high release pressures was higher than that at low release pressures.

$$\delta_{\text{max}} = 0.172P + 0.033, \tag{5}$$

$$\delta_{mean} = 0.057P + 0.143, \tag{6}$$

where δ_{max} denotes the maximum value of differential pressure (MPa), and δ_{mean} represents the mean value of differential pressure (MPa).

4.4. The Variation Characteristics of the Jet Structure in Near Field

The jet angle, which is defined as the angle between the edges of the jet formed in the enclosure space, is generally used to characterize the jet structure of the fire extinguishing agent. It can be acquired by the images obtained by the high-speed camera. It is labeled as θ_s in the present study, as shown in Figure 10.

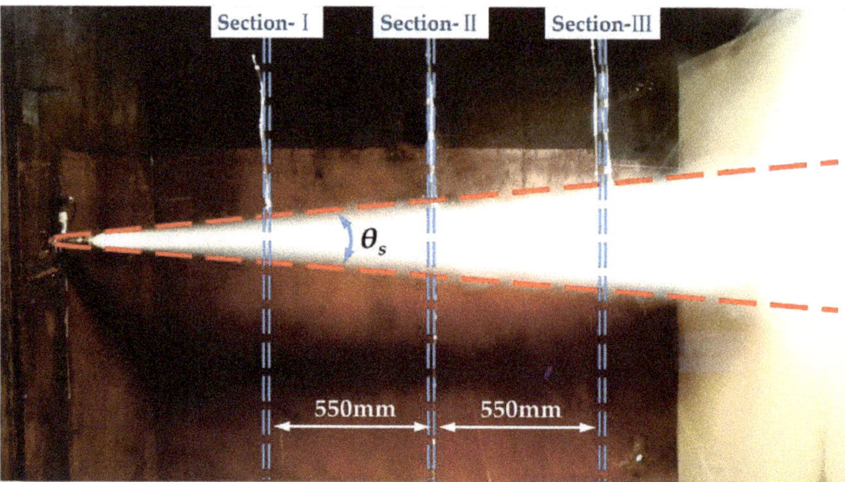

Figure 10. Jet angle.

Figure 11 shows the variation characteristics of the jet structure with time during the ejection process of HFC-125 at 4.14 MPa. The starting moment of t + 0 ms is defined as the moment when the fire extinguishing agent jetted from the nozzle. The jet process can be roughly divided into three stages: the development stage, the stable release stage, and the rapid decay stage. The development stage is from the beginning of the release process to about 46 ms, which corresponds to the rising region of Phase 2 shown in Figure 8. The differential pressure between the inside and the outside of the nozzle outlet led to the violent phase transition of the HFC-125 agent in the enclosure space. At t + 0 ms, the "blade"-shaped jet core area could be observed in the near field near the nozzle outlet, which was due to the small amount of the HFC-125 agent and its rapid evaporation. Only the transparent "blade"-shaped jet could be observed in the high-speed images. With the supplement of the HFC-125 agent from the storage vessel, the jet of the HFC-125 agent continuously penetrated into the enclosure space, and the "blade"-shaped jet core area could be clearly observed in the near field. The liquid HFC-125 agent in the jet core area was in the state of superheat, and the intense phase transition occurred when the liquid HFC-125 agent was sprayed along the axial direction of the jet, which made the core area break up gradually with white plumes forming around the core area. With the continuous supplement of the HFC-125 agent, the jet angle increased. This was due to the fact that the jet angle increased with the degree of phase transition [22,23]. The pressure at the nozzle decreased gradually with the progress of the release, which led to the dimensionless degree of superheat ($d_{\text{sup}} = P_{air}/P_{sat}$) to decrease, and more severe phase transition occurred in the core area of the jet. As a result, the jet angle gradually increased and developed into a stable conical shape. After that, the ejection process entered the stable release stage (t + 46 ms~t + 612 ms), which corresponds to the falling region of Phase 2 shown in Figure 8. The jet angle reached the maximum value and remained relatively stable for about 0.5 s. The time range above t + 612 ms was the rapid decay stage of the ejection process, which corresponded to Phase 3 shown in Figure 8. In the storage vessel, the pressure decreased gradually with the advance of release, and the residual liquid HFC-125 agents entrained with the gases were gradually emptied. The jet angle decreased rapidly, and the white gas–liquid mixed jet column gradually disappeared. Finally, the ejection process ended.

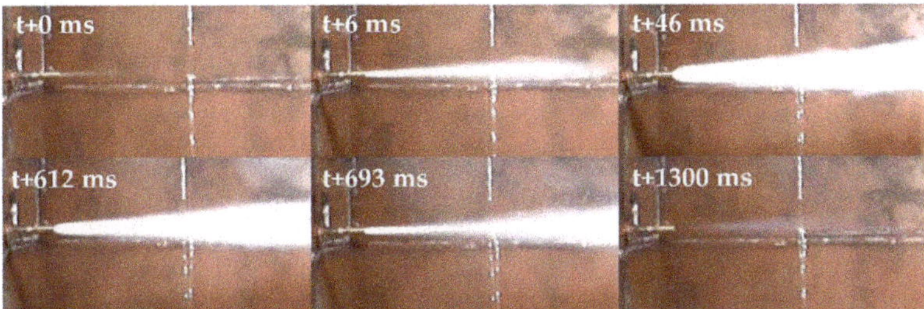

Figure 11. The injection process of HFC-125 agent obtained by the high-speed camera.

Figure 12 indicates that the maximum jet angle of the HFC-125 agent decreased linearly with the release pressure, and Equation (7) was obtained. The main reason for this was that when the release pressure was low, as discussed in Section 4.1, the jet velocity of the HFC-125 agent was low, resulting in a severe phase transition. Compared with the case of liquid state, HFC-125 showed better dispersibility and was easier to move outside the jet in the form of small droplets or gas. Therefore, the jet angle was larger in the case of low release pressures compared with that in the case of high release pressures.

$$\theta_{smax} = -0.843P + 9.53, \tag{7}$$

where θs_{max} denotes the maximum jet angle (°).

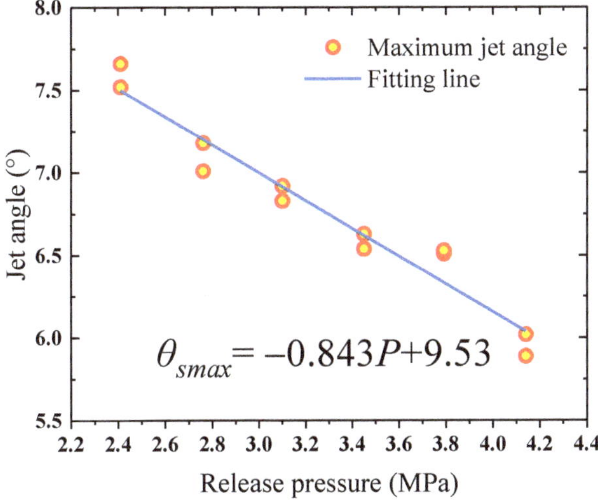

Figure 12. The maximum jet angle versus the release pressure.

4.5. Concentration Distribution of HFC-125 Agent under Different Release Pressures

4.5.1. Concentration Distribution in Radial Section

Figures 13 and 14 show the concentration distribution of the HFC-125 agent in the radical in Sections II and III in the case of 3.45 MPa, respectively. Section II is the radical annular cross section at the center of the enclosure space, and points 4, 5, 6, 7, and center point 8 are the concentration sampling positions, as shown in Figure 2. Two peaks in the variations of concentration with time were presented at points 4 and 8. The first peak at about 6 s formed because of the large differential velocity close to the nozzle caused by the front of the agent jet.

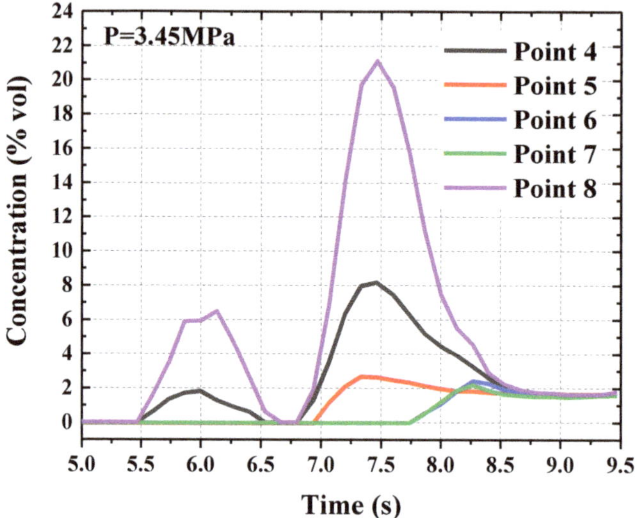

Figure 13. Concentration as a function of time at points 4, 5, 6, 7, and 8 in Section II in the case of 3.45 MPa.

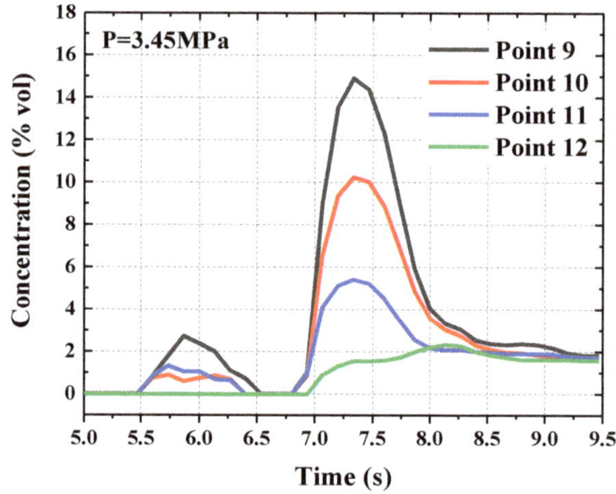

Figure 14. Concentration versus time at points 9, 10, 11, and 12 in Section III in the case of 3.45 MPa.

As shown in Figure 13, the variations of concentration at point 8 were similar with those at point 4. However, the maximum value at point 8 was almost three times larger than that at point 4, and only the maximum concentration of point 8 exceeded 17.6% in all points in Section II in the entire injection process. In addition, the concentrations at points 5, 6, and 7 were much lower, because they were far away from the center line. At last, the equilibrium concentration of 1.8% was achieved at each point after injection.

It is indicated in Figure 13 that, although point 5 and point 6 were symmetrically located above and below the central point, their concentration variations with time showed large differences. The value of concentration at point 5 increased earlier than point 6, and the peak value of point 5 occurred almost at the same time as point 8 and point 4. This phenomenon indicates that the buoyancy of the jet was more effective than the gravity during the injection process. For the points that were symmetrically located above and below the center line, the presentation of suppression effect in the area above the center line was earlier than that below the center line, being of great significance to the fire suppression design in the enclosure space. The trend of slight upward deflection of the jet during the development process in Figure 11 can further verify the inference of buoyancy.

In addition, as shown in Figure 2, Section III was the radical annular cross-section in the enclosure space, and points 10, 11, 12, and center point 9 were the concentration sampling positions. Figure 14 indicates that the maximum value of the concentration at point 9 was the largest, followed by that of point 10, whereas the concentration value at point 11 was the minimum. The concentration with time at points 9, 10, and 11 was quite similar. According to Fick's law, the evaporation rate of the liquid HFC-125 agent outside the jet core area was faster than that in the core area, and the corresponding white plumes shown in the high-speed images disappeared faster. This is related to the distance from the jet core. Therefore, the concentration far away from the center line was lower than that at the center line. However, the variations of the concentration at point 12 were quite different from those at points 9, 10, and 11. It can be illustrated by the high-speed images that when the HFC-125 agent jet reached the wall, some agents turned around due to the restriction of the wall, and the backflow formed consequently, as shown in Figure 15. The backflow entrained air in the jet and mixed with the HFC-125 agent, resulting in the lower concentration of the HFC-125 agent in the area affected by the backflow.

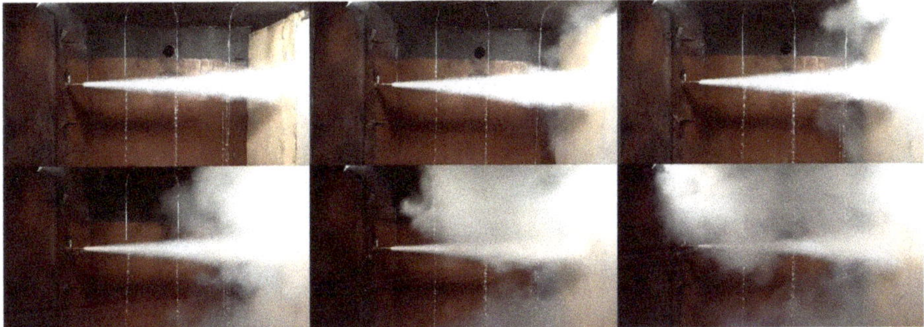

Figure 15. The backflow recorded by the high-speed camera.

4.5.2. Concentration Distribution along the Jet under Different Release Pressures

Figure 16a,c,d exhibits the concentration variations of the HFC-125 agent with time at points 1, 8, and 9 under different release pressures. It is indicated that the maximum concentration value decreased with the distance from the nozzle. This was due to the fact that the front end of the jet constantly entrained the air and the air mixed with the HFC-125 agent in the jet, which diluted the concentration of the HFC-125 agent and made the HFC-125 agent concentration lower in the area far away from the nozzle.

In addition, Figure 16b shows that the maximum concentration value increased with the release pressure at point 1. This may have been due to the fact that in the case of low release pressures (2.41–3.1 MPa), the jet angle decreased with the increase of release pressure, which made the distribution of HFC-125 more concentrated and the maximum concentration value increase rapidly with the increase of release pressure. However, it can be seen from Figure 16b that when the release pressure was 3.1 MPa, the concentration of the HFC-125 agent reached more than 90%, which indicates that the jet was mainly liquid. Although the jet angle was still smaller with the increase of release pressure, the phase transition of the agent was much lower than that in the case of low release pressures. Therefore, in the case of higher release pressures above 3.1 MPa, the jet mostly existed in liquid phase at point 1, which was less mixed with air compared with that at low release pressures. Therefore, the maximum concentration value at point 1 remained almost unchanged in the case of higher release pressures (3.1–4.14 MPa).

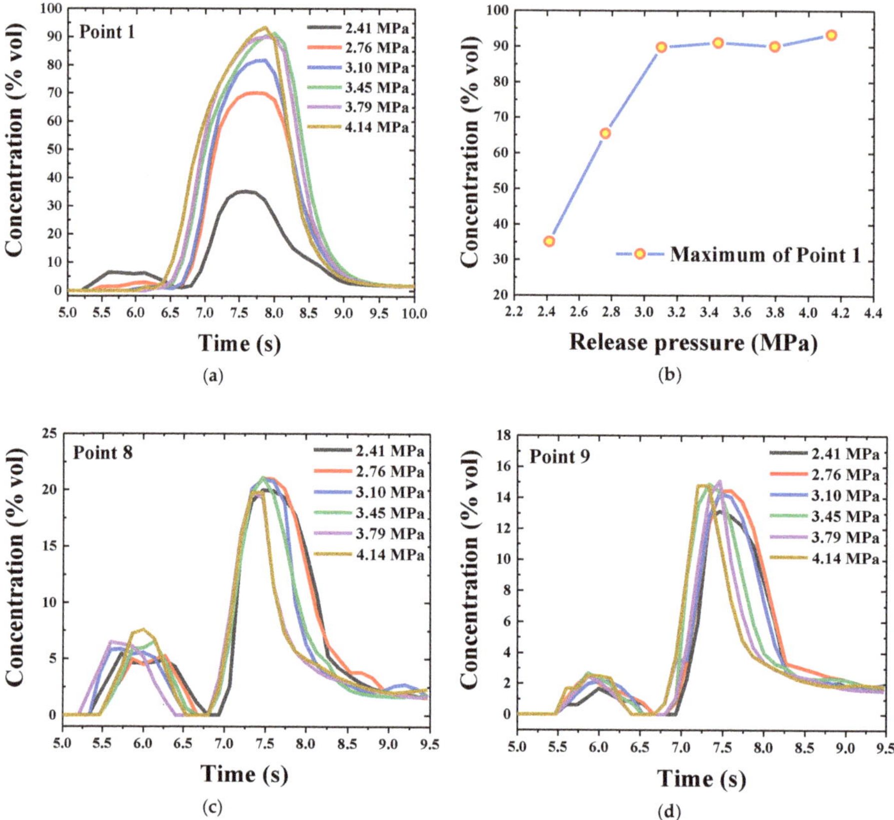

Figure 16. (**a**) Concentration of HFC-125 as a function of time at point 1; (**b**) the maximum value of concentration value with different release pressure at point 1; (**c**) concentration of HFC-125 as a function of time at point 8; (**d**) concentration of HFC-125 as a function of time at point 9.

However, as Figure 16c,d shows, little effect of release pressure on the maximum concentration values at points 8 and 9 were demonstrated. A possible explanation for this is that the HFC-125 agents were nearly all transformed into the gaseous state at points 8 and 9 and mixed with air. In addition, as mentioned above, the backflow disturbance forming at point 9 may have also been responsible for this phenomenon.

In the present study, holding time was defined as the time span in which the volume concentration of the HFC-125 agent was above 17.6%. Since the volume concentration at point 9 could not reach 17.6%, Figure 17 only shows the holding time at points 1 and 8 at different release pressures. It is indicated that the holding times at point 1 all exceeded 0.5 s at all employed release pressures, and the holding time at point 8 could exceeds 0.5 s only in the case of low release pressure (2.41–3.1 MPa). It can be inferred that the holding time of the HFC-125 agent can achieve 0.5 s only in some areas with low release pressure in the jet cone, even in the center line of the jet with the most concentrated fire extinguishing agent. In the case of the release pressure of 2.41 MPa, the value of the holding time at points 1 and 8 was the closest.

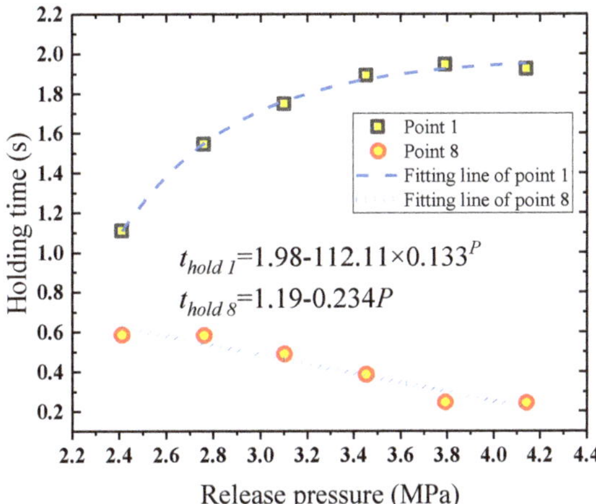

Figure 17. The holding time at points 1 and 8 at different release pressures.

In the case of point 1, which was located close to the nozzle, the holding time increased with the release pressure from 2.41 to 3.1 MPa and then tended to be flat from 3.1 to 4.14 MPa. It was similar to the variations of the maximum concentration value with the release pressure at point 1. This may have been due to the fact that in the range from 2.41 to 3.1 MPa, the holding time may have been mainly determined by the degree of phase transition. In the case of low release pressures, the dimensionless degree of superheat ($d_{\mathrm{sup}} = P_{air}/P_{sat}$) was relatively small and the degree of phase transition was severe, resulting in a larger jet angle, as described in Section 4.4. As a result, the larger jet angle made the HFC-125 agent mix with more air and reduce the concentration. Therefore, the holding time was short. In addition, the jet angle decreased with the increase of release pressure, which made the distribution of HFC-125 more concentrated. Thus, the holding time increased with the increase of release pressure below 3.1 MPa. In the range from 3.1 to 4.14 MPa, the concentration of the HFC-125 agent reached more than 90%, as shown in Figure 16b, which indicates that the jet was mainly liquid. Although the jet angle was still smaller with the increase of release pressure, the phase transition of the agent was much lower than that in the case of low release pressures. Therefore, in the case of higher release pressures above 3.1 MPa, the jet mostly existed in liquid phase at point 1, which was less mixed with air compared with that at low release pressures. Therefore, the holding time in the point 1 remained almost unchanged in the case of higher release pressures (3.1–4.14 MPa).

However, as Figure 17 shows, the variations of the holding time with time at point 8 was opposite to those at point 1. The holding time decreased with the release pressure. A possible explanation for this was that point 8 was slightly far away from the nozzle. The HFC-125 agent reaching point 8 was mainly in the gaseous state. In the range from 2.41 to 3.1 MPa, the holding time was only a little longer than 0.5 s. In the range from 3.1 to 4.14 MPa, the holding time could not achieve 0.5 s. It was due to this fact that there was not much liquid HFC-125 agent left at point 8, and the effect of phase transition on holding time was weak. The holding time was only related to the injection duration. The velocity of the jet was quite high in the case of high release pressures. The HFC-125 agent in the vessel was emptied in a short period of time. The existence time of the jet in the enclosure space decreased with the release pressure. Then, the holding time decreased with the release pressure. The empirical equations of the holding time with the release pressure at points 1 and 8 can be obtained, as shown in Equations (8) and (9).

Point 1:

$$t_{hold1} = 1.98 - 112.11 \times 0.133^{P}, \tag{8}$$

Point 8:

$$t_{hold8} = -0.234P + 1.19, \tag{9}$$

where $t_{hold\ 1}$ is the holding time of point 1 and $t_{hold\ 8}$ is the holding time of point 8 (s).

5. Conclusions

The flow and diffusion characteristics of a typical candidate substitute for the halon fire extinguishing agents, namely, HFC-125 under different release pressures (2.41, 2.76, 3.1, 3.45, 3.79, and 4.14 MPa) were investigated in the present study. A series of parameters including the degree of superheat, the injection duration, the differential pressure in the pipeline, the jet structure, and the concentration distribution in the enclosure space were measured and analyzed. The major results and conclusions are summarized as follows.

1. The degree of superheat of the HFC-125 in the storage vessel decreased linearly with the release pressure, indicating that high release pressure can assist in improving the fire suppression efficiency.
2. The injection duration approximately decreased with the release pressure as a decay exponential function. In addition, the bubble expansion can slow down the pressure drop in the storage vessel, and the bubble expansion occurred earlier with the increase of the release pressure. Moreover, the release process of the HFC-125 agent was a two-phase (liquid and gas) flow.
3. The flow process of the HFC-125 agent in the pipeline can be divided into three phases: pipeline filling, stable flow, and mixed gases release. The maximum and mean values of the differential pressure in the pipeline increased linearly with the release pressure. The pipeline pressure loss mainly occurred in the first two phases (stable flow and pipeline filling). The mean value of the differential pressure at each phase also increased with the release pressure.
4. A typical jet plume structure occurred outside the nozzle. The jet process of the HFC-125 agent in the enclosure space can be divided into three stages: the development stage, the stable release stage, and the rapid decay stage.
5. The maximum concentration value decreased with the increase of the distance from the nozzle. Moreover, the jet would deflect upward due to the effects of buoyancy. With the increase of release pressure, the maximum concentration value in the near field from the nozzle increased, and the maximum concentration value in the far field from the nozzle varied little. Furthermore, the backflow resulted by the restriction of the wall could decrease the concentration of the HFC-125 agent.
6. The concentration and holding time of the HFC-125 agent could meet the requirement of MPS in some areas. The holding time of HFC-125 could achieve 0.5 s in some areas at low release pressures in the near field of the jet cone. With the increase of the release pressure, the holding time in the near field from the nozzle increased first, and then almost remained constant, which was opposite to that in the far field from the nozzle.

It can be concluded from the present study that one nozzle is not enough to meet the requirement of MPS for HFC-125. Several nozzles and appropriate nozzle distribution design in the target enclosure space are needed.

Author Contributions: Conceptualization, J.J. and R.P.; methodology, J.J. and Q.L.; validation, R.P., Q.L., and R.C.; resources, R.P.; data curation, J.J. and X.X.; writing—original draft preparation, J.J.; writing—review and editing, R.P., Q.L., and R.C.; visualization, J.J.; supervision, R.P.; project administration, R.P.; funding acquisition, R.C. All authors have read and agreed to the published version of the manuscript.

Funding: This research was funded by the National Natural Science Foundation of China (no. 51806106) and the Natural Science Foundation of Jiangsu Province, China (no. BK20170838).

Conflicts of Interest: The authors declare no conflict of interest. The funders had no role in the design of the study; in the collection, analyses, or interpretation of data; in the writing of the manuscript; or in the decision to publish the results.

References

1. Li, H.; Min, X.; Dai, M.; Dong, X. The biomass potential and GHG (Greenhouse Gas) emissions mitigation of straw-based biomass power plant: A case study in Anhui Province, China. *Processes* **2019**, *7*, 608. [CrossRef]
2. Jung, K.; Ro, D.; Park, Y.K. Estimation, and Framework Proposal of Greenhouse Gas Emissions of Fluorinated Substitutes for Ozone-Depleting Substances by Application Area in the Republic of Korea. *Sustainability* **2020**, *12*, 6355. [CrossRef]
3. Jou, R.-C.; Chen, T.-Y. Willingness to Pay of Air Passengers for Carbon-Offset. *Sustainability* **2015**, *7*, 3071. [CrossRef]
4. Ingerson, D.A. FAA Engine Compartment Halon Replacement Project Background, Overview, and Current Status. In *International Aircraft Fire and Cabin Safety Research Conference*; Federal Aviation Administration: Washington, DC, USA, 1998.
5. Ingerson, D.A. *Simulating the Distribution of Halon 1301 in an Aircraft Engine Nacelle With HFC-125*; Federal Aviation Administration: Washington, DC, USA, 1999.
6. Bennett, M.V.; Bennett, J.M. *Aircraft Engine/APU Fire Extinguishing System Design Model (HFC-125)*; Defense Technical Information Center: Fort Belvoir, VA, USA, 1997.
7. Lu, Z.; Zhuang, L.; Dong, L.; Liang, X. Model-Based Safety Analysis for the Fly-by-Wire System by Using Monte Carlo Simulation. *Processes* **2020**, *8*, 90. [CrossRef]
8. FAA. *Minimum Performance Standards for Halon 1301 Replacement in the Fire Extinguishing Agents/Systems of Civil Aircraft Engine and Auxiliary Power Unit Compartments(rev04)*; Federal Aviation Administration: Washington, DC, USA, 2010.
9. Williamson, H. V Halon 1301 flow in pipelines. *Fire Technol.* **1976**, *12*, 18–32. [CrossRef]
10. Elliott, D.G.; Garrison, P.W.; Klein, G.A.; Moran, K.M.; Zydowicz, M.P. Flow of nitrogen-pressurized Halon 1301 in fire extinguishing system. *JPL Publ.* **1984**, *84–62*, 1–124.
11. Yang, J.C.; Huber, M.L.; Boyer, C.I. A model for calculating alternative agent/nitrogen thermodynamic properties. In Proceedings of the Halon Options Technical Working Conference, Albuquerque, NM, USA, 9–11 May 1995; Volume 11.
12. Tuzla, K.; Palmer, T.; Chen, J.C.; Sundaram, R.K.; Yeung, W.S. *Development of Computer Program for Fire Suppressant Fluid Flow*; Lehigh University: Bethlehem, PA, USA, 1998.
13. Kim, J.; Baek, B.; Lee, J. Numerical analysis of flow characteristics of fire extinguishing agents in aircraft fire extinguishing systems. *J. Mech. Sci. Technol.* **2009**, *23*, 1877–1884. [CrossRef]
14. Sarkos, C.P. *Characteristics of Halon 1301 Dispensing Systems for Aircraft Cabin Fire Protection*; National Aviation Facilities Experimental Center: Atlantic City, NJ, USA, 1975.
15. Niu, X.; Xie, Y.; Hasemi, Y. Analysis of Fire Spread and Fire Extinguishing Agent Distribution in Nacelle of Helicopter under No-ventilation Condition. *Procedia Eng.* **2013**, *62*, 1073–1080. [CrossRef]
16. Kurokawa, F.Y.; De Andrade, C.R.; Zaparoli, E.L. Numerical simulation of the concentration of fire extinguishing agent in aircraft cargo compartment. In Proceedings of the 10th Brazilian Congress of Thermal Sciences and Engineering, Rio de Janeiro, Brazil, 29 November–4 December 2004.
17. Kurokawa, F.Y.; De Andrade, C.R.; Zaparoli, E.L. Modeling of aircraft fire suppression system by the lumped parameter approach. *Aircr. Eng. Aerosp. Technol.* **2016**, *88*, 535–539. [CrossRef]
18. Zaparoli, E.L. Numerical simulation of an aircraft cargo compartment fire suppression system using cfd tool. In Proceedings of the 19th International Congress of Mechanical Engineering, Gramado, Brazil, 15–20 November 2009.
19. Lee, J. Simulation Method for the Fire Suppression Process Inside the Engine Core and APU Compartments. In Proceedings of the Fourth Triennial International Aircraft Fire and Cabin Safety Research Conference, Lisabon, Portugal, 15–18 November 2004.
20. Xue, R.; Ruan, Y.; Liu, X.; Chen, L.; Liu, L.; Hou, Y. Numerical Study of the Effects of Injection Fluctuations on Liquid Nitrogen Spray Cooling. *Processes* **2019**, *7*, 564. [CrossRef]
21. Zeng, W. *Characteristics and Mechanisms of Direct-Injection Liquid Jet and Flash-Boiling Spray*; Shanghai Jiao Tong University: Shanghai, China, 2012.

22. Geanette, P.; Arne, E.H.; George, M. General review of flashing jet studies. *J. Hazard. Mater.* **2010**, *173*, 2–18. [CrossRef]
23. Lamanna, G.; Kamoun, H.; Weigand, B.; Steelant, J. Towards a unified treatment of fully flashing sprays. *Int. J. Multiph. Flow* **2014**, *58*, 168–184. [CrossRef]

Publisher's Note: MDPI stays neutral with regard to jurisdictional claims in published maps and institutional affiliations.

MDPI

Article

On the Way to Integrate Increasing Shares of Variable Renewables in China: Activating Nearby Accommodation Potential under New Provincial Renewable Portfolio Standard

Yinhe Bu [1] **and Xingping Zhang** [1,2,3,*]

[1] School of Economics and Management, North China Electric Power University, Beijing 102206, China; byh@ncepu.edu.cn
[2] Beijing Key Laboratory of New Energy and Low-Carbon Development, North China Electric Power University, Beijing 102206, China
[3] Research Center for Beijing Energy Development, Beijing 102206, China
* Correspondence: zxp@ncepu.edu.cn; Tel.: +86-10-61773096

Abstract: More than 1.2 billion kW wind and solar power generation will be integrated in China by 2030. The new provincial renewable portfolio standard, officially implemented in 2020, establishes an efficient bridge between rapid capacity growth and limited accommodation capability. A data-driven prospect analysis framework was proposed to evaluate the activated potential under two kinds of nearby accommodation approaches and to explore the completion prospect of this new obligated quota from provincial levels. Empirical results illustrate diverse prospects across regions. Particularly, it is hard for two kinds of provinces to complete their obligated quotas merely via the single nearby accommodation approach: The first one is close to renewable energy resources but lacks flexible peak regulation capability in Northeast and Northwest China, and the other is close to the nationwide load center but lacks nearby integration from renewables in Southeast, North, and Middle China. Therefore, the pathway for the former is to activate more provincial accommodation potential either via releasing system flexibility or by substituting generation right, and the pathway for the latter is to introduce trans-regional or trans-provincial accommodation and import more renewable energy power.

Keywords: renewable energy; renewable portfolio standard; nearby accommodation; peak regulation; prospect analysis

Citation: Bu, Y.; Zhang, X. On the Way to Integrate Increasing Shares of Variable Renewables in China: Activating Nearby Accommodation Potential under New Provincial Renewable Portfolio Standard. *Processes* **2021**, 9, 361. https://doi.org/10.3390/pr9020361

Academic Editors: Yan Wang, Conghu Liu, Zhigang Jiang and Wei Cai

Received: 21 January 2021
Accepted: 13 February 2021
Published: 16 February 2021

1. Introduction

Explosive capacity booms have occurred in wind and solar power in China over the past decade. It will maintain high annual capacity growth and play an important role in speeding up transformation of China's energy structure and mitigating global climate change, corresponding to the carbon emission reduction targets declared by China for 2030 and 2060. However, the rapid capacity growth of integrated variable renewable energy (VRE) such as wind and solar has brought apparent power curtailment in Northwest and Northeast China since 2015, and will inevitably bring more under future high VRE penetration.

Much research in recent years has focused on large-scale VRE accommodation in China. It is hard for China to decrease curtailment level. The inverse distribution of energy supply and demand, concentrated development mode of energy resources, inflexible power supply structure, limited balancing capability of power grid, and weak growth of electricity consumption has caused this situation, typically from 2015 to 2016 [1]. Though China has taken a set of actions to accommodating increasing share of variable wind and solar power, and there has been a significant decline in curtailment level from 2017 to 2019, and these obstacles have not been completely removed [2]. For further analysis, numerous researchers put their scenario back to 2015, and have established that not only the physical limit in power system but also the lack of a more powerful plan-based policy together with coordinating a market-based mechanism has led to this problem [3,4].

The new provincial renewable portfolio standard (RPS) issued by China in May 2019 was officially implemented in 2020, and is a landmark policy towards China's accommodation of renewable energy. This scheme is quite different with the worldwide adoption not only in assessing subjects but also in tracking completion progress of obligated quota. It is a demand-based policy with provincially distributed quota, and is implemented by directly assessing electricity selling companies or consumers to sell or buy an increasing proportion of their electricity from renewable sources. The RPS adopted outside China is usually a supply-based policy, which is the promotion scheme of a quota obligation on electricity suppliers to supply an increasing proportion of their electricity from renewables [5]. It is accompanied with voluntary tradable green certificates (TGC) and tradable surplus accommodation to complement the fulfillment of quotas, while in most cases outside China, RPS policy is accompanied with compulsory TGC created by government to track the fulfillment of the quotas [5].

A major focus in China's localized RPS design progress from 2007 to 2019 has been how to evaluate the effects of RPS adoption. It is generally accepted that great changes will happen to China's power supply structure, carbon emissions reduction, electricity prices, renewable investment decisions, social welfare, and governmental expenditure on subsidies when RPS, together with TGC, is practically adopted [6,7]. Particularly, Chinese researchers pay more attention to the connection between RPS and existing energy policies, such as the substitution effect of RPS and TGC on feed-in tariff (FIT), and the inside cooperation among RPS policy, electricity market, and carbon trading [8–10]. The special effect towards cutting curtailment and promoting accommodation of VRE in China is also separately proved [11].

After the announcement of this new RPS scheme, several alternative effect evaluations were developed. In particular, the inevitable effect of RPS on retail electricity market was studied [12]. One creative study analyzed the feasibility of this new RPS from the perspectives of VRE supply and demand, and found that the curtailment of renewables reduced by RPS [13]. Based on this research, further prospect analysis of the new RPS in China from provincial level is necessary. Besides, previous work has focused more on policy evaluation but less on the dual effect between RPS completion and VRE accommodation.

This paper aims to fill this gap and seek out a pathway for activating accommodation potential and fulfilling the new provincial RPS scheme. The data-driven prospect analysis framework highlighted in this study contains a simplified calculation method, which is applied to quantitatively emphasize the specific relationship between accommodation and demand-side quota obligation issued by RPS policy. Two prominent nearby approaches from both technical and marketable perspectives are simulated in scenario analysis combined with two case studies of Northeast and Northwest China. The ideal potential for activating accommodation and achieving obligated quota is evaluated based on the typical data from 2015 to 2018. Results illustrated in the case and extended for nationwide application reveal a clear pathway for activating accommodation potential and accelerating RPS completion in different or similar provinces. The application evidence of these nearby accommodation approaches and prospect analysis for completing provincial quota will provide a valuable reference and useful mode for China to integrate increasing shares of VRE in the next ten years.

This paper focused on the RPS completion in 30 provincial administrative regions (Tibet is not contained in the assessment) across mainland China and is organized as follows. Section 2 firstly introduces the evolution of RPS policy in China, then summarizes two basic characteristics of the newest scheme and illustrates its specific assessment mechanism. Section 3 introduces a calculation and simplification method of accommodation potential. Section 4 conducts scenario analysis within two cases based on several nearby accommodation approaches. Section 5 discusses the results of nationwide application of the above approaches and reveals several facts. Section 6 concludes the paper and proposes related policy recommendations.

2. Framework and RPS Scheme

2.1. Framework of This Paper

The framework of this paper is specifically illustrated as Figure 1.

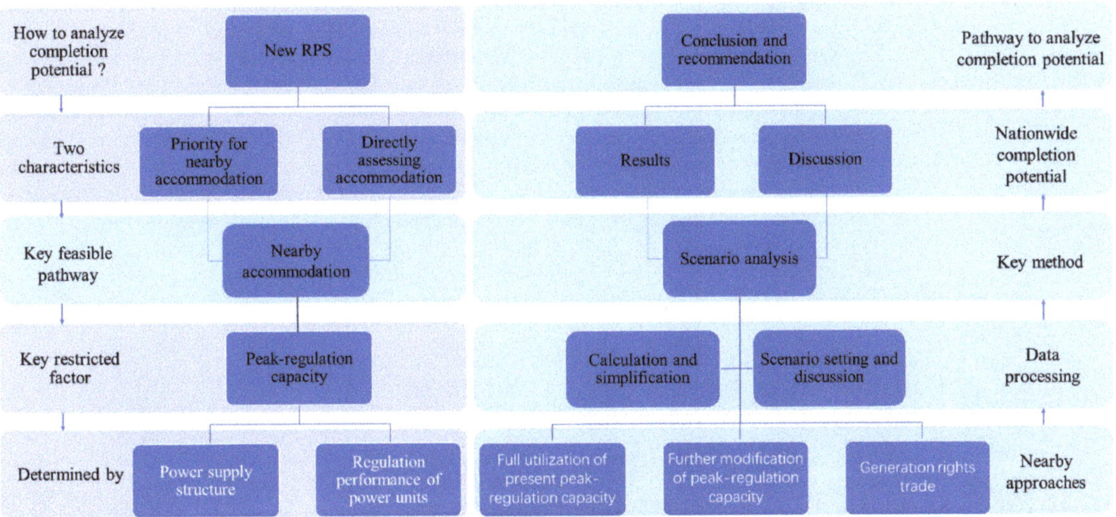

Figure 1. Framework of this paper.

This paper starts by analyzing the two basic characteristics of the new RPS policy in China, which leads to the fact that the key feasible pathway for implementing RPS is to preferentially conduct nearby accommodation approaches. Peak regulation capacity, determined by power supply structure and regulation performance of power units, is emphasized as the key restricted factor for releasing nearby accommodation potential. The nearby accommodation approaches proposed in this paper are summarized as follows: Full utilization or further modification of present peak regulation capacity, and generation rights trading between captive power plants and renewable power units. They will either technically optimize the peak regulation capability, or economically motivate all responsible market players to alternatively buy or sell more electricity from VRE based on physical restrictions of China's power system. Both background and implementation situation of these representative approaches are respectively analyzed in corresponding case studies on Northeast and Northwest China. These two case studies include several scenario analyses where the provincial accommodation potential is calculated by reasonable simplification according to the proposed method. All representative approaches are simply simulated and most of the results are discussed regionally or provincially. It is also emphasized which provinces or regions are more suitable for applying these approaches. Part of the results is nationwide extended and further discussed to obtain a relatively clear completion prospect of the new RPS in China.

2.2. Evolution of RPS Policy in China

Evolution for introduction and localization of RPS in China could be divided into two stages. It was firstly introduced in 2007, beginning with assessing electricity suppliers on the quota of renewable energy's installed capacity and generating electricity. It was then updated in 2012 and 2014, stimulating that provincial government had the administrative responsibility while power grid corporation and generation enterprise were respectively obligated for implementation and cooperation. These policy schemes concentrated on the capacity or electricity proportion on the generating side, and delivered large construction booms of wind and photovoltaics (PV) power station. However, their effects towards accommodating renewable energy power were undesirable. Then the assessment subject and index were switched from the generating side to the demand side, when the first amendment of RPS policy was issued on 23 March 2018 [14]. The assessment was solely conducted by compulsory TGC. The second amendment in 2018 was soon proposed and was more specific in obligation distribution and trading rules of TGC than the first one [15]. On 13 November 2018, the third amendment issued as *Notice on trial implementation of renewable portfolio standard* was published [16]. Instead of compulsory TGC, the obligated quota was mainly assessed by actual accommodation, and complementarily assessed by both accommodation trading and voluntary TGC. Then in May 2019, the newest amendment *Determination of obligation quota and assessment method of accommodation for renewable energy power* was finally published [17]. It further emphasized the direct assessment of accommodation. The evolution of RPS policy in China is illustrated in Figure 2.

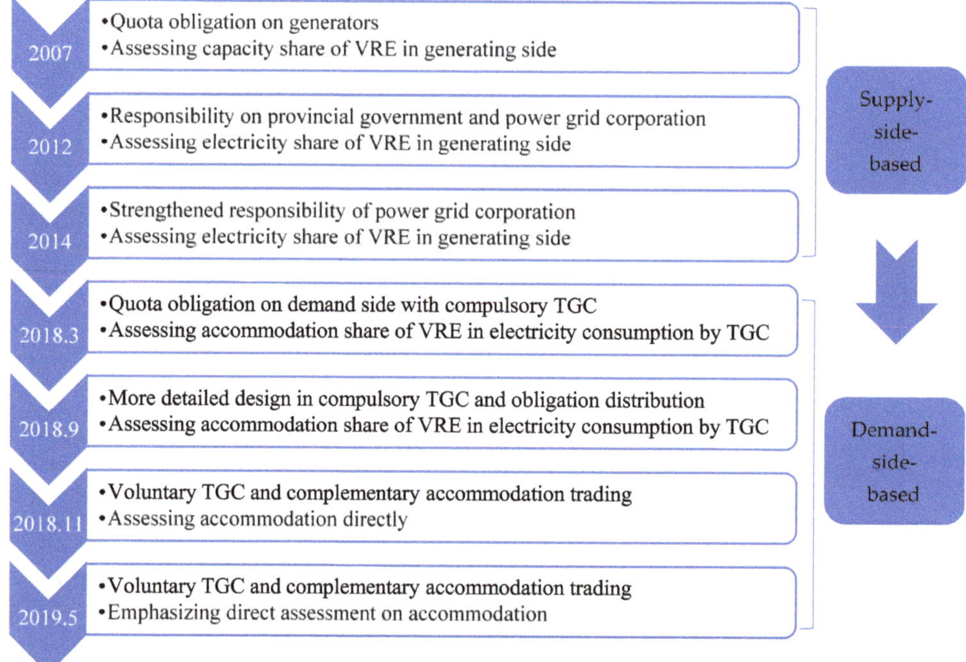

Figure 2. Evolution of renewable portfolio standard (RPS) policy: Characteristics on quota obligation and assessment, 2007–2019.

The newest RPS policy in China stipulated that the RPS target will be annually distributed, and provincially implemented and assessed. The distributed provincial RPS target in the assessment year is annually allocated one year before. It is calculated by estimating the provincial renewable energy electricity which is generated and absorbed locally, predictable net import of renewable energy electricity which is generated outside but absorbed locally, and predictable provincial electricity consumption. Responsible subjects involved in RPS assessment include independent power supply companies that possesses no operation right of distribution network, dependent ones which belong to power grid corporation and directly supplies power to end users, power consumers who purchase electricity through the wholesale electricity market, and enterprises that possess their captive power plants. All involved subjects are encouraged by the provincial energy bureau to correspondingly buy or sell the same proportion of renewable energy power in their total power demand or supply as the provincial RPS target of their located province. This new RPS scheme announced the official launch of demand side-based provincial RPS in China and was simpler and more feasible in assessment. At present, both the forcible target and motivated one from 2018 to 2020 have been specifically given.

2.3. *Two Main Characteristics of New RPS in China*

2.3.1. Nearby Accommodation Is Preferred

Various challenges for accommodation always exist in China. China's energy resources and demand are distributed inversely. VRE resources are large in scale and concentrated in provinces throughout North China (NC), Northeast China (NEC), and Northwest China (NWC). These three regions, collectively called the Three-North region, are far away from the nationwide load center in East and South China. In contrast, VRE is mainly developed in a decentralized way and accommodated nearby in Europe and America. The basic pattern of energy development for China seems to be sending electricity from the west to the east, from the north to the south, and allocating energy throughout the country.

The truth is that China's electricity has been balanced by provinces for a long time and the power generation plan is formulated under the leadership of the local government. Under the plan-based mode, administrative intervention driven by provincial economic interests hinders the trans-provincial exchange of electricity. Although China is actively promoting the orderly liberalization of electricity generation and utilization plans, the proportion of planned electricity in 2017 is still as high as 80% of the benchmark hours. The rigid implementation of the thermal power generation plan and market transaction plan reduces the adjustment flexibility of real-time dynamic balance and restricts the power output of VRE. In addition, the physical power transmission capacity for VRE in the Three-North region is insufficient. The power transmission capacity only accounts for 22% of the installed VRE capacity, and is further occupied by the transmission task from the coal-fired power bases. Though China is speeding up the mechanism reform of the electricity industry, the national unified electricity market has not yet been fully established. The market mechanism, which is conducive to breaking the barriers between provinces and promoting the trans-regional and trans-provincial consumption of electricity, has not yet been formed. There is still a long way to go to give full play to the decisive role of the market in the optimal allocation of resources. Therefore, in this market transition stage, a wide range of VRE allocation and accommodation is restricted. Nearby accommodation merely remains to be a more feasible way.

The newest RPS policy in China tends to encourage all responsible subjects in RPS assessment to deeply develop the potential of nearby accommodation. It stimulates that priority should be given to maximize nearby accommodation in those provinces which finally export electricity. Other provinces, which finally import electricity outside, should preferentially fully accommodate nearby VRE and then maximize trans-provincial or trans-regional accommodation.

Where are the potentials for activating more physical nearby accommodation? Total consumption market is insufficient enough to support the rapid capacity growth of all power sources including VRE, especially in the Three-North region. Peak regulation capability of provincial power system is the major constrained factor related to support enormous absorption of random PV or wind power. The maximum limit of this capability is determined by both power supply structure and peak regulation performance of a single power unit. However, not only the lack of flexible regulated power supplies and limited flexibility of the conventional power units, especially those coal-fired power units, but also the negative participation captive power plants involved in peak regulation have made the power system in China inflexible to accommodate more VRE. Much potential in flexible operation remains to be fully developed for China's power system in contrast with Spain,

Denmark, Germany, and the United States. The capacity share of flexible and coal-fired power in power supply structure is illustrated in Table 1. Therefore, the basic principles for diverse nearby accommodation approaches are either flexibility-based optimization of power supply structure or improvement of units' peak regulation performance. Specific approaches are discussed in the fourth section of this paper.

Table 1. Flexible and coal-fired power in power supply structure, 2016.

Regions or Countries	Three-North	NC	NEC	NWC	Spain	Germany	America	Portugal
Flexible power (%)	3.9	7.6	1.5	0.8	34.3	17.5	48.7	34.0
Coal-fired power (%)	69.9	80.1	70.1	56.8	11.7	31.1	30.1	9.5

Source: State Grid Corporation of China.

2.3.2. Using Direct Assessment on Accommodation Instead of Compulsory TGC

TGC and RPS usually complement each other especially in those countries with mature electricity market. The operation mechanism of RPS scheme in California, Japan, and Britain is to replace physical measurement with green certificates, and to reflect the performance of quota-bearing entities with the number of certificates. The cost of realizing the quota obligation taken by power companies is channeled out through the terminal sales electricity price. If the power company fails to meet the quota target, it shall pay a fine which is higher than the cost of purchasing renewable energy or certificates.

However, the complementary relationship between RPS and TGC seems to be weakened according to the latest scheme issued in China. China creatively designs RPS policy by directly assessing provincial physical accommodation together with market-based accommodation. Provinces are preferentially encouraged to complete their targeted accommodation by physical electricity balance, under the organization and technological support of provincial power grid corporation, and the unified management of the provincial energy bureau. Both nearby accommodation and active trans-provincial or trans-regiona1 accommodation contribute to the physical assessment part. As for the market-based part, China designs two auxiliary market-based measures for responsible subjects to realize their proportion targets which are more independent than nationwide compulsory TGC. One alternative measure is to directly purchase the surplus accommodation from market entities who have exceeded their obligation. The other one is to conduct voluntary trading of TGC with green power producers, and the equivalent amount of accommodation corresponding to purchased certificates is also recorded as supplementary RPS completion. All the identified supplementary completion of single responsible subjects contributes to the total market-based accommodation of a province.

3. Methodology

3.1. Fundamental Conditions

The overall completing progress of RPS for a province can be assessed by Equations (1)–(3):

$$Com_i^T = A_i^T + RPS_i^T + TGC_i^T - w_i^T, \tag{1}$$

$$Ind_i^T = \frac{Com_i^T}{\left(C_i^T - w_i^T\right)}, \tag{2}$$

$$A_i^T = L_i^T + E_i^T, \tag{3}$$

where Com_i^t is the amount index of RPS completion for province i in year T. Ind_i^T is the proportion index of RPS completion for province i in year T. Both are defined according to the latest RPS scheme issued in May 2019. A_i^T is the actual provincial accommodation of renewable energy electricity for province i in year T, which physically participates in provincial electricity balance. A_i^T is calculated by two parts. L_i^T is the part of accommodation that is locally generated and consumed. E_i^T is the other part of accommodation that is generated outside but consumed locally, and it is set to be negative if the province finally exports non-hydro renewable energy power. As for the market-based accommodation which is complementarily included in RPS completion, RPS_i^T is the total assigned amount of directly-purchased accommodation from all responsible subjects located in province i. TGC_i^T is the total equivalent amount of accommodation corresponding to the voluntarily green certificates purchased by all subjects located in province i. w_i^T is the part of electricity consumption which is closely related to public welfare and declared to be out of assessment for province i in year T. C_i^T is the annual electricity consumption in the whole province i for year T.

Once the new RPS is strictly implemented, provinces will either raise physical accommodation capability, or encourage responsible subjects to purchase a high enough amount of RPS completion by direct assigned transaction of accommodation and voluntary subscription of TGC. The targeted physical accommodation amount can be simply illustrated by GA_i^T, as shown in Equation (4). It can be calculated by the distributed provincial RPS target typed as Tar_i^T and electricity consumption. All nearby accommodation approaches proposed in this paper can be adopted by provincial manager and organizer of RPS implementation to make A_i^T close to or exceed GA_i^T.

$$GA_i^T = Tar_i^T \cdot C_i^T, \tag{4}$$

To calculate the completion potential of new provincial renewable portfolio standards in China based on scenario analysis of nearby accommodation approaches, the following section introduces several basic formulas.

The power output of VRE in province i at the time t, typed as PN_i^t, should meet the requirement of instantaneous power balance in the provincial power system. The power balance is illustrated by Equations (5)–(7):

$$PN_i^t = PC_i^t - PA_i^t - PFX_i^t, \tag{5}$$

$$PA_i^t = PAC_i^t + PAF_i^t, \tag{6}$$

$$PFX_i^t = PNU_i^t + PTR_i^t + PCA_i^t + POT_i^t, \tag{7}$$

where PC_i^t is defined as the provincial consumption load, PA_i^t is defined as the power output of adjustable generating units which is the sum of adjustable conventional units' power output typed as PAC_i^t and flexible units' power output typed as PAF_i^t, and PFX_i^t is the power output of generating units that cannot be adjusted, which is the sum of nuclear power units' power output typed as PNU_i^t, imported power through trans-provincial transmission lines typed as PTR_i^t, off-managed captive power plants' power output typed as PCA_i^t, and other uncontrollable power outputs typed as POT_i^t.

The adjustable conventional units in China are mainly coal-fired power units and hydropower with reservoir storage. The ratio of flexible power units such as natural gas generation and pumped storage in China's power supply structure is too low to apparently influence peak regulation capability of provincial power system. For simplification, the power output of adjustable generating units mentioned above can be changed into Equation (8).

$$PA_i^t = PCO_i^t + PHY_i^t, \tag{8}$$

where PCO_i^t and PHY_i^t are the power output of coal-fired units and hydropower with reservoir storage, respectively.

The power output of hydropower with reservoir storage is usually stable. In addition, curtailment for hydropower is forbidden. The uncontrollable power output showed in Equation (7) is fixed according to plan-based management. Specially, the power of the external transmission line is set according to the daily planned value and could not be adjusted, which means the actual import power must be accommodated by receiving provinces. The basic provincial electricity balance is illustrated in Table A1. Therefore, the more generating space coal-fired power units assign, the more generating space VRE acquires, which can be illustrated by Equation (9).

$$|\Delta PN_i^t| = |\Delta PCO_i^t|, \tag{9}$$

This paper assumes that the traditional power system reserve is merely undertaken by coal-fired power units. After deducting the system reserve, the remaining part of coal-fired power units is used for tracking and balancing the fluctuation of VRE's power output. Other forms of thermal power together with runoff hydropower were not considered unless specially mentioned.

The power output of coal-fired units should be limited between minimum technical output and operated installed capacity after deducting system reserve. This is shown in Equation (10):

$$CAP_i^t \cdot (1 - \sigma_i^t) \cdot (1 - \beta_i^t) \leq PCO_i^t \leq CAP_i^t \cdot (1 - \sigma_i^t), \tag{10}$$

where CAP_i^t is defined as the installed capacity of coal-fired units, β_i^t represents the average peak regulation depth of all kinds of operated coal-fired power units in province i at time t, and σ_i^t is defined as the system reserve rate undertaken by coal-fired units.

3.2. Calculation of Completion Potential

The maximum generating space for VRE which is assigned from coal-fired power units can be shown in Equation (11), if other conditions remain the same and peak regulation performance is considered only.

$$Max|\Delta PN_i^t| = |PCO_i^t - CAP_i^t \cdot (1 - \sigma_i^t) \cdot (1 - \beta_i^t)|, \tag{11}$$

This paper merely took nearby accommodation approaches into consideration, while the change of trans-provincial a trans-regional transmission plan could be ignored. Respectively, the imported part of accommodation typed as E_i^T remained unchanged. Therefore, the maximum accommodation amount of VRE which was assigned from coal-fired power units could be calculated by integrating both sides, typed as Equation (12).

$$Max\left|\Delta A_i^T\right| = \int |PCO_i^t - CAP_i^t \cdot (1 - \sigma_i^t) \cdot (1 - \beta_i^t)|dt, \tag{12}$$

The peak regulation depth of a single power unit is closely related to its capacity and peak regulation mode. This paper merely considered the low-load peak regulation mode and typed five categories of coal-fired power by the capacity of a single unit. Each category was respectively configured into a unified peak regulation depth. The peak regulation depth of different categories is illustrated in Table 2. The weighted average peak regulation depth of all kinds of operated coal-fired power units can be calculated. The specific proportion of various coal-fired power units in different provinces is illustrated in Table A2. The setting for peak regulation depth in Table 2 is based on the investigation results illustrated in Table A3.

Table 2. Calculation setting of peak regulation depth for coal-fired power units typed by capacity.

Capacity of a Single Unit/GW	Minimum Technical Output/%	Peak Regulation Depth/%
≥1	45	55
0.6–1	52	48
0.3–0.6	56	44
0.2–0.3	70	30
<0.2	85	15

The calculation setting of the system reserve rate undertaken by coal-fired units is based on the *Notice on issuing early warning of coal and electricity planning and construction risks in 2020* [18].

In this notice, red warning provinces have obviously-installed redundancy of coal-fired units, and the system reserve rate is set by its boundary value of interval; orange ones have relatively abundant installed capacity, and the system reserve rate is set by average value of interval; green ones have basic balance or a slight gap between power supply and demand, and the system reserve rate is also taken by the boundary value of the interval; those undefined provinces' system reserve rates are consulted by the reasonable reserve rate. The reserve capacity is calculated with the product of the maximum power generation load and actual system reserve rate. These setting data can be also seen in Table A2.

For province i in year T, the maximum physically-rising potential of the proportion index for RPS completion can be further simplified by Equation (13).

$$Max\left|\Delta Ind_i^T\right| = \left|\frac{ECO_i^T - \tau_i^T \cdot CAP_i^T \cdot (1-\sigma_i^T) \cdot \sum_j^5 \delta_{ij}^T \cdot \left(1-\beta_{ij}^T\right)}{C_i^T}\right|, \quad (13)$$

where ECO_i^T is defined as the annual generating electricity of coal-fired power units, τ_i^T is the annual utilization hour of coal-fired power units, CAP_i^T is the total annual installed capacity of coal-fired power units, σ_i^T is provincial power system reserve rate, δ_{ij}^T is the capacity proportion of five different coal-fired units marked by j, and β_{ij}^T is its corresponding peak regulation depth.

For further consideration of heating units, also known as cogeneration units or combined heat and power (CHP), the peak regulation of them was uniformly set to be 75%. The provincial capacity proportion of heating units is illustrated in Table A2. The heating period was set to be 3624 h from November to March of the following year, and the non-heating period was set to be 5136 h from April to October. The actual utilization hours of heating units during these two periods was respectively estimated by their corresponding generating electricity and operated capacity. The capacity proportion of five types of units was assumed to be the same among heating units and non-heating units. Then the formula could vary into Equation (14).

$$Max\left|\Delta Ind_i^T\right| = \left|\frac{ECO_i^T - CAP_i^T \cdot (1-\sigma_i^T) \cdot \sum_j^5 \delta_{ij}^T \cdot \left(1-\beta_{ij}^T\right) \cdot \left[\tau_{iw}^T + (1-\alpha_i^T) \cdot \tau_{im}^T\right] - \tau_{in}^T \cdot \alpha_i^T \cdot CAP_i^T \cdot (1-\beta_{in}^T)}{C_i^T}\right|, \quad (14)$$

where τ_{iw}^T is the average utilization hours of all coal-fired units during a non-heating period, τ_{im}^T is the average utilization hours of non-heating units during a heating period and τ_{in}^T is the average utilization hours of heating units during a heating period, α_i^T is the provincial capacity proportion of heating units, and β_{in}^T is the peak regulation depth of heating units during a heating period.

The basic principle of this calculation was to make one variable change as the scenario analysis needed and to keep the other ones to be the same. The above formulas regard peak regulation performance of coal-fired power units as the basic variable, because it is the core factor that restricts nearby accommodation. With these calculations, the prospect for completing a provincial obligated quota in China could be estimated.

3.3. Feasibility and Rationality of the Method

Accurate calculation of VRE absorption capacity is an effective way to improve the level of accommodation, which could theoretically predict the maximum potential for provinces achieving their obligated quotas. At present, the calculation method studied in many literatures can be mainly summarized into two types: The typical daily analysis method and time series production simulation analysis method.

The typical daily analysis method usually selects the extreme situation where the power output of VRE is the largest while the load is the smallest, to determine the conservative capacity of integrated VRE [19]. The time series production simulation analysis method usually takes the month or year as the calculation time length, and simulates the power and electricity balance of the power grid time by time [20].

The calculation method proposed in this paper absorbs the advantage in simplification of the first method and avoids the disadvantage in complex data requirements of the second one. It concentrates on the macroscopic average level represented by two kinds of a typical day in a heating period and non-heating period. The basic principle of this method is to control other variables to remain unchanged and to study how the variety of the major variable influences the accommodation. It is especially suitable and more rational to be included in a nearby accommodation framework, due to relatively stable provincial power supply structure and inter-provincial power balance. Though it is idealistic both in assumptions and simplification, the calculation results are valuable enough for analyzing the RPS completion potential from both technical and marketable ways.

In particular, the method proposed in this paper is quite different from those traditional predicting methods. Traditional ones usually predict values of several related variables tied with the targeted variable and give out an estimated value for an exact future date. In contrast, this method keeps these variable values unchangeable except for the studied one prepared for simulation and scenario analysis, and gives out the most ideal evolution for the targeted variable based on present structural data from the actual power system. The differential value for targeted variables between the present actual value and the most ideal one represents the potential remains to be activated. It is not a prediction for an exact date or time series but reflects a present prospect outlook for future development.

Thus, the validity of simulation results on this basis could be proved in another way. As for the traditional methods, the structure validity of their simulation results could be proved by structural validity procedures including boundary adequacy, structure verification, dimensional consistency, parameter verification, and extreme conditions [21,22], while the behavioral validity could be demonstrated by comparison between data output from the model and data from the actual power system, and the test procedures include but are not limited to trend comparison and removal [23–25]. However, there were no actual reference data for this estimated ideal evolution calculated in this paper; thus, the results could be not easily proved by their comparison. In addition, results were not for an exact future date or time series and could not be validated by fitting historical data with trend extension. Therefore, to guarantee the simulation results calculated from the proposed method complied with structural validity rules, all balance formulas applied in this paper were referred from the Renewable Energy Production Simulation Model developed by the China Electric Power Research Institute (CEPRI) which has been successfully applied in VRE integration optimization and off-grid multi-good micro-grid optimization for nearly ten years [26]. These formulas illustrate the basic principle that the more generation space released or transferred from thermal power units, the more VRE accommodation and RPS completion become. Under the most ideal circumstance, thermal power units could operate at a lower minimum technical power output and more generation rights to trade between captive thermal power plants and VRE units could be fulfilled, and an explosive boom could happen to VRE accommodation. In addition to these formulas with valid logical structure and consistent dimensions, the basic data for scenario-based case study were collectively sorted out either from official published data books or from investigation of the State Grid Corporation of China [27–34]. The adequacy for boundary conditions and parameters could be also guaranteed [21]. In contrast, to demonstrate the behavioral validity, this paper compared the overlapping variables values acquired from proposed method and similar researches developed by CEPRI and State Grid Energy Research Institute (SGERI) [26,35].

4. Approaches for Activating Nearby Accommodation and Accelerating RPS Completion

Nearby accommodation is the key feasible pathway for absorbing more VRE in the electricity market transition period. Peak regulation capacity, determined by power supply structure and regulation performance of power units, is the key restricted factor for nearby accommodation. However, it is difficult to change the power supply structure in a short time. Inflexible coal-fired power still occupies a core position in the generation structure. Therefore, utilizing and improving peak regulation performance of power units, especially coal-fired power units, becomes a practicable way. Technically, full use and further modification of present peak regulation capacity both contribute to removing the physical limit of nearby accommodation. As for a marketable approach, generation rights trade is the most representative one carried out in China, which motivates thermal power units to make room for VRE power generation based on the peak regulation capacity of present power system. These two series of approaches contribute to completing the obligated quota issued by RPS policy, and the accommodation potential activated by them is separately evaluated in two case studies.

4.1. Full Utilization and Further Modification of Peak Regulation Capacity of Coal-Fired Units in Northeast China

4.1.1. Background and Implementation

Since 2010, electricity consumption growth has been slow, while the power supply has continued to grow at a relatively fast rate, and the contradiction between supply and demand has become more prominent. From 2010 to 2016, the installed power supply in Northeast China increased by 47%, 26 percentage points higher than the demand load growth in the same period, and the installed power supply in 2016 was 2.2 times the maximum load. The utilization rate of power generation equipment in the northeast power grid continued to decline. In 2016, the overall utilization hours fell to 3432 h, which was lower than the national average of 353 h. Thermal power occupied 4068 utilization hours, which was 692 h lower than that in 2010, and wind power possessed 1689 h, which was 386 h lower.

In addition, heating units accounted for an enormous share of thermal power in Northeast China. The peak regulation performance of heating units has been extremely restricted by operation mode because they must meet the heating load demand while producing electric power. For instance, when the heating load gradually increased, the 300 MW CHP unit had to enlarge the capacity of suction; consequently, the minimum technical output rose while the maximum technical output dropped, and its adjustable load range varied from 51.1% to 7.1%, lower than that of non-heating units. When the strong wind period overlapped with the heating period during winter and spring, both limited peak regulation capability and declining electricity consumption together with inadequate transmission infrastructure brought a huge curtailment of wind power.

In response, the dispatching department of the northeast power grid strictly controlled the startup mode and power output of heating units in accordance with the minimum operation mode approved by the Energy Supervision Bureau, dynamically monitored the heating information online in real time, calculated the peak regulation capacity, and arranged the units to participate in deep peak regulation to the maximum extent. In this way, the thermal power plant operated with minimum technical output in 2016, giving additional accommodation room for wind power of 6.372 billion kWh.

In addition, the *Northeast China Power Peak Regulation Auxiliary Service Market* was established in 2014 and updated in 2016. It stimulated that peak regulation capacity of all generators, except the capacity with obligatory regulation task, ought to participate in the market. Whenever a thermal power unit was operating below the minimum technical output it could it obtain compensation, otherwise it would be regarded as a beneficiary of the peak regulating auxiliary service and therefore share the service cost [36]. This market-based approach has successfully motivated generators to reduce the minimum power output of their CHP units. The auxiliary service electricity in Northeast China became 8.571 billion kWh in 2017, and wind power additionally generated 2.499 billion kWh due to the peak regulating auxiliary service market.

The above two approaches, especially adopted by Northeast China, concentrated on the full utilization of present peak regulation capacity from two pathways. One is to technologically optimize operation dispatching, the other one is to economically motivate coal-fired units to actively reduce their power output.

Further optimization of present peak regulation capacity is also urgent for developing more peak regulation capacity from coal-fired power units. The thermal storage transformation of CHP units and flexibility modification of pure condensing units has been sped up nationwide. Since the end of 2017, total modified capacity has become 9.18 million, and 930 million kWh additional VRE has been accommodated. More than half of the modification has been done by Liaoning Province. By the end of 2020, 133 million kW of CHP units and 82 million kW of pure condensing units are planned to be modified, and 45 million kW of peak regulation capacity should be increased in the Three-North region. Another 185 million kW of deep peak regulation modification is planned to be completed in East and Middle China.

4.1.2. Scenario Analysis

The first two scenario settings in this part only considered the operation mode of coal-fired units based on the present maximum peak regulation depth. It was firstly calculated regardless of the characteristics of heating units, under the condition that all coal-fired units are dispatched to operate on their present minimum technical power output. The results for three provinces in Northeast China are illustrated in Table 3, typed as S_1. If the heating units were also taken into consideration, and all heating units were operating at their present minimum technical power output in heating period while other units still operated according to their initial minimum technical output, the results are illustrated as S_2 in Table 3.

Table 3. Flexibility modification scenarios of Northeast China [1].

Province	Tar_i^{2018}	Tar_i^{2019}	Tar_i^{2020}	I_i^{2015}	I_i^{2016}	I_i^{2017}	I_i^{2018}	S_1	S_2	S_3	S_4
Jilin	15	15	16.5	12.1	13.7	16.4	17	17.0	16.1	18.1	17.1
Liaoning	10	10	10.5	7.7	8.6	9.2	11.7	11.2	10.4	11.7	10.9
Heilongjiang	15	17.5	20.5	11.2	12.4	15.8	16.2	15.5	14.8	16.5	15.8

Source: Calculated by the data from National Energy Administration (NEA) [17,30–34]. [1] I_i^T (%) is the actual accommodation proportion from 2015 to 2018.

The second two scenario settings considered the rising potential of peak regulation depth by flexibility modification. It was firstly calculated regardless of the heating units under condition that all coal-fired units were dispatched to operate on 10% lower than the present minimum technical power output. The results are illustrated in Table 3, typed as S_3. Similarly, heating units were then considered, and all heating units together with non-heating units were operating on 10% lower than the initial setting, regardless of whether they were in or out of the heating period, and the results are illustrated as S_4 in Table 3.

4.1.3. Discussion

In this case, both full utilization and further modification of peak regulation capacity of coal-fired units raise the RPS completion potential in Northeast China. The maximum physically-rising potential of the proportion index for RPS completion in all scenarios is close to or more than the obligated quotas in 2018, 2019, and 2020. The actual accommodation proportion in 2018 and 2019 is close to or more than the potential one calculated in this scenario analysis. This can be explained by two possibilities. One is that the potential from flexibility modification of coal-fired units has been completely developed in Northeast China. The other is that all the nearby accommodation approaches adopted in Northeast China work in coordination with each other and bring a higher rising potential than a single approach. The results from the consideration of heating units are always lower than that of the other. This can be explained because the existence of a huge amount of heating units in Northeast China dramatically reduces the completion potential of RPS gained from the flexibility modification of coal-fired units.

It has been proven to be a necessary approach for activating nearby accommodation, and is especially suitable for those provinces with a huge amount of heating units. However, some obstacles still exist during the adoption of these approaches. On one hand, the dispatching capability of power grid corporations needs further improvement, so that present peak regulation capacity can be fully used for VRE accommodation. On the other hand, the auxiliary service market still needs to be further improved, even in Northeast China during the reform of the power industry, let alone some provinces that have not yet established a peak regulation auxiliary service market. Both generators and power grid corporations in these provinces are not motivated enough to actively conduct flexibility modification.

4.2. Generation Rights Trading between Captive Power Plants and Renewable Power Units in Northwest China

4.2.1. Background and Implementation

Northwest China once faced more serious curtailment of VRE than Northeast China. It not only similarly possesses a huge amount of CHP units, but also has a huge number of captive power plants. A captive power plant is a power plant owned, used, and managed by an industrial or commercial energy user for its own energy consumption, and is widely used by high-electricity consuming industries in Northwest China [36]. It can be integrated to the power grid for exchanging excess power, or be used off-grid for selfishly meeting one's own electricity demand but escaping from peak regulation obligation. Whether the captive power plants participate in peak regulation makes a difference to system flexibility as well. The installed capacity and power generation of captive power plants has grown rapidly. They have forced public power plants and non-hydro renewable energy power suppliers to further press their power output. For instance, the captive power plants in Xinjiang Province have generated more electricity than public ones since 2014, and the usage hour of them is always 1000 h more than public power plants from 2012 to 2016. In response, Northwest China has been exploring a substitution trade mode to fully release the peak regulation capacity of captive power plants since 2016. It is usually called the generation rights trading between captive power plants and renewable power units. By the end of 2017, the additional accommodation benefitted from this trade in Shaanxi, Gansu, Qinghai, Ningxia, and Xinjiang respectively achieved 0.39, 1.89, 0.54, 1.22, and 7.9 billion kWh. This total trade achieved 14.2 billion kWh in 2019.

4.2.2. Scenario Analysis

In this scenario setting, the maximum physically-rising potential of the proportion index for RPS completion was directly calculated in three represent provinces in Northwest China, by assigning 15%, 25%, and 35% of the generation rights for captive power plants to renewable energy generators with the actual accommodation data in 2016. The results are respectively illustrated as S_5, S_6, and S_7 in Table 4.

Table 4. Generation rights trading scenarios of Northwest China [1].

Province	Tar_i^{2018}	Tar_i^{2019}	Tar_i^{2020}	I_i^{2015}	I_i^{2016}	I_i^{2017}	I_i^{2018}	TCA_i^{2016}	ECA_i^{2016}	ICA_i^{2016}	S_5	S_6	S_7
Gansu	14.5	17	19	11.4	12.5	13.8	13.4	2475	24790	9.9	16.0	18.3	20.7
Ningxia	18	18	20	13.4	19.1	21	22.3	815	21390	3.8	22.7	25.1	27.6
Xinjiang	11.5	12	13	7.8	11.1	13.1	14.7	6120	96500	6.3	19.2	24.6	29.9

Source: Calculated from the data from NEA [17,30–34]. [1] TCA_i^{2016} (GWh) is the trading electricity of generation rights trading in 2016, ECA_i^{2016} (GWh) is the total generation of captive power plants in 2016, ICA_i^{2016} (%) is the actual assigning share in 2016.

4.2.3. Discussion

In this case, the completion potential, developed from generation rights trading between captive power plants and renewable power units in Northwest China, is far lower than the installed capacity of captive power plants. The trading shares in these three provinces are all below 10% which illustrates that there remains much more potential to be further developed. The maximum proportion index for RPS completion in all scenarios is far higher than actual accommodation shares in electricity consumption from 2015 to 2018 except in Ningxia Province. It has already completed the obligated quota of 2020 in advance. As for Gansu Province, it cannot complete the obligated quota of 2020 before singly assigning the generation rights of captive power plants for 27.9% or more. In contrast, the same index for Xinjiang is 3.5%, which means the maximum proportion index for RPS completion in Xinjiang province is more sensitive to this approach.

This approach is also suitable to be applied in Shandong Province and Inner Mongolia Province which also possess huge installed capacity of captive power plants. Once the trading is adopted, the supplementary peak regulation capacity of captive power plants would further raise the completion potential of RPS in these provinces. However, some argue that this approach is a helpless choice for renewable power units because they will have to face extraordinary power rationing if they do not participate in the transaction. In addition, a vast majority of VRE generators must clinch a deal with zero electricity price, with the average electricity price reduced by 0.294 yuan/kWh, if they intend to participate in the transaction. This benefits high-energy-consuming enterprises a lot. But if the government subsidy is considered, it is fair enough for both sides of the trade. This will be a relatively feasible way for accommodating VRE as much as possible in the transition period. After that, the unified management and sharing mechanism of peak regulation capacity need to be completely established to maximize the peak regulation capability.

5. Results and Discussions

5.1. Result Discussions and Prospect Analysis

The two representative nearby accommodation approaches for activating completion potential of new RPS in China have distinct concerns. The first one focused on either full utilization of present peak regulation capacity through minimum operation mode control and motivation from auxiliary service market, or further modification of peak regulation capacity through flexibility modification of pure condensing units and thermal storage transformation of CHP units. The other one concentrated on market-based substitution of generation rights between captive power plants and renewable power units with no change in present peak regulation capacity. Both of them certainly contributed to activating completion potential. The generation rights trade seemed to be a little more effective due to the direct substitution of the electricity amount.

In the related case separately discussed above, Northeast China adopted all approaches of the first type and each element of the single approach contributed to releasing the completion potential. It continuously would contribute more, but the deep digging space for Northeast China has been almost run out. In contrast, its nationwide application would benefit more provinces, especially in North China and Northwest China, as illustrated in Table 5.

The nationwide simulation results from scenarios 1 to 4 illustrate that in most cases, the growth of the actual accommodation proportion from 2015 to 2018 has dramatically exceeded the maximum limit of rising potential which is calculated by the proposed method based on the data in 2015. This can be explained from two possibilities. One is that the motivation of one single approach is far weaker than that of approach portfolios. There possibly remains positive synergy between different approaches, especially inside the first category. While provinces use diverse approaches during the same period, the progress of index improvement might be out of imagination. The other one is that all the results calculated in the first four scenarios are the extreme evolution developed from the basic year 2015. Neither new growth of installed capacity nor new-added electricity consumption from 2015 to 2018 completely involves the fundamental assumptions and calculations. Therefore, not only do the approach portfolios give more actual growth of the completion index, but also the accommodation priority in the new-added electricity market deeply releases more potential for completing RPS.

Table 5. Simulation results for the index of RPS completion by single approach of flexibility modification [1,2].

Region	Province	I_i^{2015}	I_i^{2018}	T_i^{2020}	S_1	S_2	S_3	S_4
North China	Beijing	7.6	11.7	15	10.6	10.3	11.3	10.9
	Tianjin	7.6	11	15	11.5	10.8	12.1	11.4
	Hebei	7.6	11.3	15	11.3	10.7	11.9	11.1
	Shandong	5	9.4	10	7.5	6.9	7.9	7.2
	Shanxi	7	14.5	14.5	10	9.4	10.7	9.9
East China	Jiangsu	3.3	7	7.5	4.9	4.6	5.2	4.8
	Zhejiang	2.4	5.3	7.5	3.4	3.4	3.7	3.5
	Anhui	3.9	11	11.5	5.8	5.5	6.1	5.6
	Fujian	3.4	4.9	6	5.1	5	5.4	5.1
	Shanghai	1.6	3.3	3.0	2.4	2.4	2.6	2.5
Middle China	Hubei	3.7	7.5	10	5.6	5.4	5.9	5.5
	Hunan	2.8	10.2	13	4.3	4.2	4.5	4.3
	Henan	2.3	9.4	10.5	3.5	3.3	3.7	3.4
	Jiangxi	2.2	8.6	8	3.2	3.2	3.5	3.5
	Sichuan	1.4	4.4	3.5	2	2	2.2	2.2
	Chongqing	1.4	2.9	2.5	2	1.9	2.1	2
Northeast China	Heilongjiang	11.2	16.2	20.5	15.5	14.8	16.5	15.8
	Liaoning	7.7	11.7	10.5	11.2	10.4	11.7	10.9
	Jilin	12.1	17	16.5	17	16.1	18.1	17.1
Northwest China	Shaanxi	2.7	10.6	12	3.8	3.7	4.1	3.9
	Gansu	11.4	13.4	19	16.5	15.8	17.5	16.4
	Qinghai	13.5	18.5	25	17.6	17.6	19	19
	Inner Mongolia	12	17.3	18	17.3	16.2	18.4	17
	Ningxia	13.4	22.3	20	20.3	19.9	21.4	20.2
	Xinjiang	7.8	14.7	13	12.4	12	12.9	12.2
South China	Guangdong	1.8	3.5	4	2.9	2.6	3	2.7
	Guangxi	1	4.2	5	1.6	1.6	1.8	1.8
	Guizhou	2	4.5	5.0	3	3	3.2	3.2
	Hainan	4	5.2	5	5.6	5.6	5.9	5.9
	Yunnan	5.1	15.6	11.5	7.8	7.8	8.2	8.2

[1] Simulation results are calculated based on the structural data from 2015 to 2018 [17,27–29]. [2] Provinces not shaded have completed their obligated quota for 2020 until the end of 2018.

In addition, the nationwide completion potential of RPS activated by nearby accommodation approaches can be inferred from the results. As mentioned above, the nearby accommodation approach is currently the most feasible pathway for provincial completion of RPS target, while the peak regulation capacity is the most important restricted factor of nearby accommodation. Thus, it is inferred that it would be almost impossible for provinces to complete their physical targeted accommodation if they could not complete these four scenarios which were based on extremely idealized calculations. There remains some potential for provinces in the Three-North region for further improvement by nearby accommodation approaches, but the targets for provinces in North China would be relatively hard to complete by merely adopting nearby accommodation. In addition, most provinces outside the Three-North region seem to not be sensitive to nearby accommodation approaches. These provinces are usually far away from VRE resources but close to the nationwide load center in Southeast China. They may have run out of nearby accommodation potential and need some other pathways to raise RPS completion.

To simply illustrate the provincial completion potential of RPS brought by the single approach mentioned above, this paper chooses the maximum results of the RPS completion index among scenario 1 to scenario 4 illustrated in Table 5, and uses the maximum differential value between scenario results and the actual index in 2015 to represent the completion potential brought by a single approach of flexibility modification. Meanwhile, the gap between the same index in 2015 and the targeted one in 2020 was also considered. Regarding the completion potential surplus of the present gap or the actual index in the

2018 surplus the targeted one in 2020, this paper recognizes the province has completed its RPS target. It was assumed to be very hard to complete the target if the ratio of potential and gap was between 0 and 0.5. If the ratio was between 0.5 and 0.6, provinces found it hard to complete the RPS target. The same ratio was between 0.6 and 0.8 for provinces that found it easy to fulfill, while the ratio for another category of provinces recognized to be very easy to complete was between 0.8 and 1. The nationwide completion prospect is illustrated as a five-color map in Figure 3.

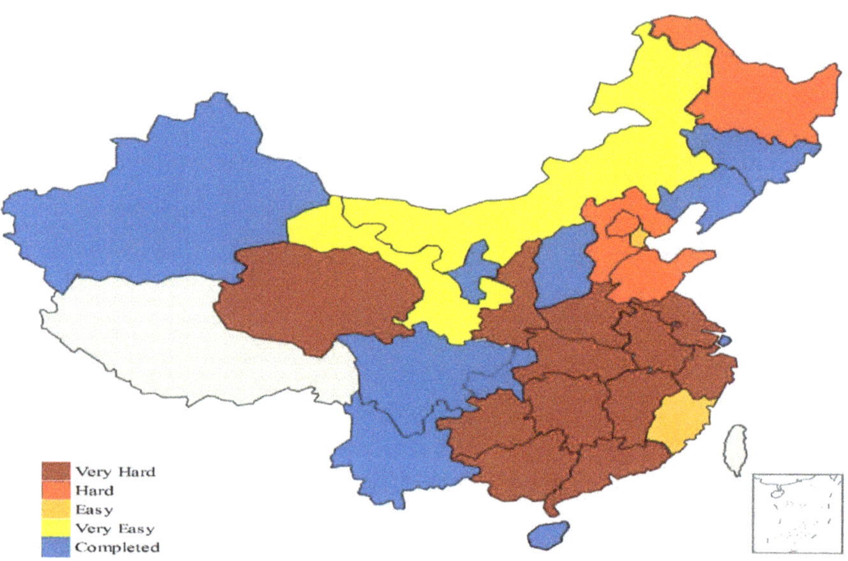

Figure 3. RPS completion potential brought by single approach of flexibility modification.

5.2. Result Comparisons and Validity

To illustrate the behavioral validity of our results, we compared them with the overlapping results for two similar studies developed by State Grid Energy Research Institute (SGERI)SGERI and China Electric Power Research Institute (CEPRI) [26,35]. These two official studies applied similar scenario analysis but simulated results based on more structural data for a complex power system (provincial online power grid). Their results could also figure out the added VRE consumption by singly adopting these two nearby accommodation approaches proposed in this paper. The first type of approach was to fully utilize and further modify peak regulation capacity of coal-fired units, and the other was to fulfill more generation rights trading between captive power plants and renewable power units. The maximum value for added VRE consumption via singly adopting the former is illustrated as A_1, while the maximum value via singly adopting the latter is illustrated as A_2. Not only were these two ideal values calculated from this paper and two similar studies compared, but also the actual added VRE consumption from 2016 to 2019.

For further demonstration, these two similar studies were simply introduced as follows. As for the first method developed by SGERI, the flexibility modification was to separately retrofit thermal power units to be so flexible that their minimum technical power output could be 30% and 40% of the rated installed capacity. The total modification plan for the whole provincial power system case covered 4 to 16 GW power units. The added VRE consumption was equal to reduced VRE curtailment and related value series could be calculated by proper trend extension based on the same modification scale as proposed in this paper. Similarly, results for A_2 could be calculated based on the relationship between substitution trading value series and VRE curtailment rate series. In contrast, the second method developed by CEPRI only illustrated the relationship between added substitution trading values and reduced VRE curtailment rate. Therefore, the results for A_2 could also be calculated by trend extension based on the same substitution plan as proposed in this paper. All these results are calculated based on similar conditions and data for the same typical years (from 2015 to 2016), compiled and organized in Table 6, so that they could be compared with each other.

Table 6. Comparison among results from different methods: Added variable renewable energy (VRE) consumption by adopting a single approach.

Provinces	Results from Different Methods: Added VRE Consumption by Adopting a Single Approach (TWh)					Actual Added VRE Consumption from 2016 to 2019 (TWh)	
	This Paper		SGERI		CEPRI		
	A_1	A_2	A_1	A_2	A_2	A_1	A_2
Heilongjiang	4.0	-	5.4	-	-		-
Liaoning	6.4	-	5.7	-	-	<12.1	-
Jilin	3.3	-	5.1	-	-		-
Gansu	5.5	9.9	5.3	7.9	5.3		6.9
Ningxia	6.0	12.6	5.4	4.3	3.1	<10.03	4.5
Xinjiang	9.5	39.6	5.0	43.0	18.8		28.9

The estimation for A_1 in this paper was assumed to be the most ideal VRE consumption a provincial power system ought to accommodate under the present conditions as long as it were to adopt the first approach immediately, and was more optimistic than that done by SGERI. In contrast, diverse results for A_2 revealed the fact that provinces with a huge amount of captive thermal power units are more sensitive to the second approach. All estimations for the Xinjiang Province are several times that of the values for Gansu and Ningxia Provinces. Meanwhile, the estimation acquired by SGERI was the most optimistic and the one done by CEPRI was the least, though estimations for two other provinces in this paper were also more than that done by others. The differential values between three methods indeed existed but all these results reflected similar behavior logic, with which these methods could estimate the evolution of VRE consumption for provincial power system. Thus, the results for scenario-based case studies in this paper were reasonable.

In addition, the actual added VRE consumption from 2016 to 2019 could indirectly reflect the rising potential for future VRE increase when the calculation was back to 2015 and 2016, considering a relatively high utilization level of VRE at present. All these actual data or upper limit for data were compiled from published data books or official reports. It is obvious that the actual added VRE consumption from adopting A_1 and A_2 was far lower than those maximum estimations acquired from three methods, which also contributed to illustrating the validity of our results.

6. Conclusions and Policy Recommendations

China's new-modified provincial RPS will play a significant role in energy revolution amidst the tension between rapid growth and limited accommodation of non-hydro renewable energy. It was officially implemented in 2020 by assessing the physically-consumed accommodation in provincial electricity balance, and the complementary market-based one via assigning redundant the accommodation or voluntary trade of green certificates. Two basic characteristics of this new policy involve priority of nearby accommodation and direct assessment on accommodation. Thus, the adoption of nearby accommodation approaches will maintain core position during the implementation of RPS in this transition period. This paper focused on the prospects of completing a new provincial renewable portfolio standard in 30 provincial administrative regions across mainland China, and proposed a data-driven analysis framework to estimate the potential activated by nearby accommodation approaches. It began by introducing the evolution of RPS policy and the specific newly-modified demand-side-based assessment mechanism towards both single responsible subjects and overall provinces; highlighted in presenting a simplified calculation method for estimating nearby accommodation potential. This was followed by several scenario analyses combined with an introduction of background and implementation for approaches applied in two case studies, and ended with results discussion. Empirical results proved that maximum use and further modification of coal-fired units' peak regulation performance and generation rights trading between captive power plants and renewable power units contribute to exploiting nearby accommodation potential and completing obligated quota. These nearby accommodation approaches are relatively suitable for those provinces in the Northeast and Northwest China which were close to renewable energy resources but less effective in provinces which were close to the nationwide load center in Southeast and North China. Other provinces in Middle China which possessing neither resource advantage nor location advantage would find it harder to complete RPS target merely via the single nearby accommodation approach.

This paper proposed policy recommendations in order to further activate the accommodation potential for VRE and fulfill the new renewable portfolio standard. First, provinces close to renewable energy resources should maintain the priority of nearby accommodation while provinces close to the load center should seek out other alternative ways such as trans-provincial or trans-regional accommodation. As the most urgent nearby accommodation approach, nationwide deep peak regulation modification of coal-fired power units should be sped up. More generation rights trade between captive power plants and renewable power units in related provinces should be encouraged. Second, further improvement of generation rights trade, peak regulation auxiliary service, and other market-based approaches such as provincial direct electricity trade and regional peak regulation aid should be conducted, in coordination with the reform of electricity market in China. Third, the approach portfolio is preferred rather over the adoption of a single measure, and further synergistic effects among various approaches should be explored. Fourth, provinces should prepare for inadequate physical accommodation and actively take part in accommodation assignment or voluntary TGC markets.

Author Contributions: Conceptualization, Y.B. and X.Z.; methodology, Y.B.; software, Y.B.; validation, Y.B.; formal analysis, Y.B.; investigation, Y.B.; resources, Y.B. and X.Z.; data curation, Y.B.; writing—original draft preparation, Y.B.; writing—review and editing, X.Z.; visualization, Y.B.; supervision, X.Z.; project administration, X.Z.; funding acquisition, X.Z. All authors have read and agreed to the published version of the manuscript.

Funding: This research was funded by the National Natural Science Fund of China, grant number 72074075.

Institutional Review Board Statement: Not applicable.

Informed Consent Statement: Not applicable.

Data Availability Statement: Data sharing not applicable. No new data were created or analyzed in this study. Data sharing is not applicable to this article.

Acknowledgments: The authors would like to thank the anonymous referees and the editor of this journal. The authors also gratefully acknowledge the administrative and technical support of the State Grid Gansu Electric Power Company and State Grid Qinghai Electric Power Company.

Conflicts of Interest: The authors declare no conflict of interest. The funders had no role in the design of the study; in the collection, analyses, or interpretation of data; in the writing of the manuscript; or in the decision to publish the results.

Appendix A

Table A1. Structural data of provincial electricity balance and accommodation, 2015 [1,2].

Province	Imp_i^t	EN_i^t	EH_i^t	ECO_i^t	ET_i^t	C_i^t	A_i^t	E_i^t	L_i^t
Inner Mongolia	−138	31.4	3.6	342.2	392.3	254.3	30.6	−0.8	31.4
Sichuan	−121.6	1.2	276.7	42.9	320.8	199.2	2.8	1.6	1.2
Yunnan	−111.4	10	217.7	27.6	255.3	143.9	7.3	−2.7	10
Guizhou	−75.7	3.3	82.7	107.1	193.1	117.4	2.3	−1	3.3
Shanxi	−72.1	10.8	3.1	231.9	245.8	173.7	12.1	1.3	10.8
Hubei	−69.1	2.3	130.3	103	235.6	166.5	6.2	3.9	2.3
Anhui	−42.2	2.4	4.9	198.9	206.2	164	6.4	4	2.4
Xinjiang	−31.8	20.8	20.3	206.7	247.8	216	16.9	−3.9	20.8
Ningxia	−28.8	12.4	1.6	102.6	116.6	87.8	11.8	−0.6	12.4
Gansu	−12.9	18.6	33.6	70.6	122.8	109.9	12.5	−6.1	18.6
Shaanxi	−9.9	2.3	8.3	121.5	132.1	122.2	3.3	1	2.3
Jilin	−5.2	6.1	5.3	59	70.4	65.2	7.9	1.8	6.1
Fujian	−3.1	4.5	43.9	110.9	188.3	185.2	6.3	1.8	4.5
Heilongjiang	−2.6	7.2	1.9	80.4	89.5	86.9	9.7	2.5	7.2
Guangxi	1.5	0.7	76.2	54.4	132	133.4	1.4	0.7	0.7
Hainan	1.7	0.8	0.9	23.4	25.5	27.2	1.1	0.3	0.8
Qinghai	8.5	8.2	37.1	12	57.3	65.8	8.9	0.7	8.2
Jiangxi	10.5	1.4	17.1	79.7	98.2	108.7	2.4	1	1.4
Chongqing	19.3	0.3	22.9	45	68.2	87.5	1.2	0.9	0.3
Hunan	19.5	2.3	52	71	125.3	144.8	4.1	1.8	2.3
Tianjin	20	0.7	0	59.4	60.1	80.1	6.1	5.4	0.7
Henan	32.1	1.5	10.9	243.5	255.9	288	6.7	5.2	1.5
Liaoning	36.6	11.3	3.2	132.9	161.9	198.5	15.2	3.9	11.3
Shandong	49.8	12.8	0.7	448.4	461.9	511.7	25.7	12.9	12.8
Beijing	53.1	0.3	0.7	41.2	42.2	95.3	7.2	6.9	0.3
Zhejiang	58.3	2.4	22.9	222.2	297.1	355.4	8.4	6	2.4
Shanghai	58.5	1.1	0	81	82.1	140.6	2.3	1.2	1.1
Jiangsu	68.9	9.6	1.2	415.2	442.6	511.5	16.9	7.3	9.6
Hebei	87.5	17.1	1.1	210.6	230.1	317.6	24.1	7	17.1
Guangdong	152.2	4.5	28.4	285.4	378.9	531.1	9.7	5.2	4.5

Source: Compiled by the authors from data yearbooks or reports issued by National Energy Administration (NEA) and China Electricity Council (CEC) [27–34]. [1] ET_i^t is total provincial electricity generated locally, Imp_i^t is the final import or export electricity outside province, and it is set to be negative if the province finally exported power. EN_i^t, EH_i^t, and ECO_i^t, respectively, refer to locally generated electricity from VRE, hydropower, and coal–fired power. The units of all variables are TWh. [2] Provinces shaded finally export electricity during provincial electricity balance.

Table A2. Structural capacity data of thermal power and maximum load of provincial power system, 2015.

Province	Type of Capacity (%) (Capacity of Single Unit, GW)					Total Capacity (GW)	Maximum Load (GW)	System Reserve Rate (%)	Heating Unit Share (%)
	≥1	0.6–1	0.3–0.6	0.2–0.3	<0.2				
Heilongjiang	0	21	37	19	23	20.4	11.9	17	64
Jilin	0	22	41	19	18	17.8	10.7	18	75
Liaoning	7	29	38	7	19	30.7	21.8	16	66
Inner Mongolia	0	40	36	10	14	72.6	27.5	24	43
Tianjin	15	16	48	9	12	12.8	11.2	22	70
Hebei	0	32	48	9	11	43.5	42.7	18.3	63
Shandong	6	16	30	3	45	87.5	27.5	21	51
Shanxi	0	35	40	10	15	59.4	29.7	20	47
Shaanxi	0	51	34	4	11	29.4	19	20	23
Gansu	0	20	63	5	12	19.3	16.3	26	50
Qinghai	0	21	39	8	32	3.2	10.9	23	0

Table A2. *Cont.*

Province	Type of Capacity (%) (Capacity of Single Unit, GW)					Total Capacity (GW)	Maximum Load (GW)	System Reserve Rate (%)	Heating Unit Share (%)
	≥1	0.6–1	0.3–0.6	0.2–0.3	<0.2				
Ningxia	10	26	46	2	16	19.8	16.2	22	17
Xinjiang	5	13	45	5	32	42	30.3	27	24
Sichuan	0	41	29	5	25	16.2	34	16	0
Chongqing	15	27	37	0	21	14.1	10.8	23	13
Shanghai	18	24	42	0	16	22.6	18.8	19	19
Zhejiang	26	29	22	3	20	62.3	40.1	18	15
Fujian	7	49	33	0	11	28.9	32.1	19	24
Guangdong	19	30	29	4	18	73.2	93.5	19	34
Guangxi	13	35	22	4	26	16.5	14.1	21	0
Yunnan	0	43	39	3	15	14	33.4	15	0
Guizhou	0	50	34	2	14	26.8	21.9	20	0
Henan	10	41	29	6	14	62.1	47.2	18.5	27
Hubei	12	30	40	2	16	25.8	37.7	19.5	23
Jiangxi	6	56	24	4	10	17.9	15	18	0
Anhui	11	55	22	0	12	46.1	27.1	14	24
Hunan	0	51	30	2	17	21.9	31.8	23	7
Jiangsu	21	30	30	1	18	83.8	73.1	16	54
Hainan	0	0	60	0	40	4.6	3.6	30	0
Beijing	0	0	36	47	17	9.7	8.6	15	72

Source: China Electricity Council (CEC) [18,27–29].

Table A3. Investigation results of peak regulation depth of different coal−fired units typed by capacity.

Capacity of Unit /MW	Minimum Technical Output/MW	Peak Regulation Depth/%
1000	450	0−55
800	470	0−41.7
600	280	0−53.3
500	290	0−42
350	180	0−48
320	180	0−43.7
300	165	0−45

Source: State grid corporation of China.

References

1. Shu, Y.B.; Zhang, Z.G.; Guo, J.B.; Zhang, Z.L. Study on key factors and solution of renewable energy accommodation. *Proc. CSEE* **2017**, *37*, 1–8. (In Chinese)
2. Li, H.; Zhang, N.; Kang, C.Q.; Xie, G.H.; Li, Q.H. Analytics of contribution degree for renewable energy accommodation factors. *Proc. CSEE* **2019**, *39*, 1009–1018. (In Chinese)
3. He, Y.X.; Xu, Y.; Pang, Y.X.; Tian, H.Y.; Wu, R. A regulatory policy to promote renewable energy consumption in China: Review and future evolutionary path. *Renew. Energ.* **2016**, *89*, 695–705. [CrossRef]
4. Bu, Y.H.; Zhang, X.P. The Prospect of New Provincial Renewable Portfolio Standard in China Based on Structural Data Analysis. *Front. Energy Res.* **2020**, *8*, 59. [CrossRef]
5. Zhang, Y.Z.; Zhao, X.G.; Ren, L.Z.; Ji, L.; Liu, P.K. The development of China's biomass power industry under feed−in tariff and renewable portfolio standard: A system dynamics analysis. *Energy* **2017**, *139*, 947–961.
6. Rouhani, O.M.; Niemeier, D.; Gao, H.O.; Bel, G. Cost−benefit analysis of various California renewable portfolio standard targets: Is a 33% RPS optimal? *Renew. Sust. Energy Rev.* **2016**, *62*, 1122–1132. [CrossRef]
7. Feng, T.T.; Yang, Y.S.; Yang, Y.H. What will happen to the power supply structure and CO2 emissions reduction when TGC meets CET in the electricity market in China? *Renew. Sust. Energy Rev.* **2018**, *92*, 121–132. [CrossRef]
8. Zhang, Q.; Wang, G.; Li, Y.; Li, H.L.; Mclellan, B.; Chen, S.Y. Substitution effect of renewable portfolio standards and renewable energy certificate trading for feed−in tariff. *Appl. Energy* **2018**, *227*, 426–435. [CrossRef]
9. Zuo, Y.; Zhao, X.G.; Zhang, Y.Z.; Zhou, Y. From feed−in tariff to renewable portfolio standards: An evolutionary game theory perspective. *J. Clean. Prod.* **2019**, *213*, 1274–1289.

10. Li, W.; Liu, L.G.; Zhang, S.; Zhang, H.Z. Will the tradable green certifications and renewable portfolio standard policy work well in China: A recursive CGE analysis. *J. Renew. Sustain. Energy* **2018**, *10*, 055904. [CrossRef]
11. Zhang, Y.Z.; Zhao, X.G.; Ren, L.Z.; Zuo, Y. The development of the renewable energy power industry under feed−in tariff and renewable portfolio standard: A case study of China's wind power industry. *J. Clean. Prod.* **2017**, *168*, 532. [CrossRef]
12. Zhu, C.P.; Fan, R.G.; Lin, J.C. The impact of renewable portfolio standard on retail electricity market: A system dynamics model of tripartite evolutionary game. *Energy Policy* **2020**, *136*, 111072. [CrossRef]
13. Fan, J.L.; Wang, J.X.; Hu, J.W.; Yang, Y.; Wang, Y. Will China achieve its renewable portfolio standard targets? An analysis from the perspective of supply and demand. *Renew. Sust. Energ. Rev.*. (in press). [CrossRef]
14. NEA. Notice on Soliciting Opinions of Renewable Portfolio Standard and Assessment Measures. Available online: http://zfxxgk.nea.gov.cn/auto87/201803/t20180323_3131.htm (accessed on 23 March 2018).
15. NEA. Notice on Soliciting Opinions of Renewable Portfolio Standard and Assessment Measures (The Second Version). Available online: http://www.sohu.com/a/256151232_776203 (accessed on 25 September 2018).
16. NEA. Notice on Trial Implementation of Renewable Portfolio Standard and Assessment Measures. Available online: http://www.nea.gov.cn/2018$-$11/15/c_137607356.htm (accessed on 15 November 2018).
17. NEA. Notice on Guarantee Mechanism of Renewable Energy Power Accommodation. Available online: http://zfxxgk.nea.gov.cn/auto87/201905/t20190515_3662.htm (accessed on 10 May 2019).
18. NEA. Notice on Issuing Early Warning of Coal and Electricity Planning and Construction Risks in 2020. Available online: http://zfxxgk.nea.gov.cn/auto84/201705/t20170510_2785.htm (accessed on 20 April 2017).
19. Matteo, Z.; Marco, G.; Marco, F.; Michele, R.; Agostino, G.; Mirko, M.; Emanuele, M. K−MILP: A novel clustering approach to select typical and extreme days for multi−energy systems design optimization. *Energy* **2019**, *181*, 1051–1063.
20. Zhang, D.; You, P.Y.; Liu, F.; Zhang, Y.H.; Zhang, Y.T.; Feng, C. Regulating cost for renewable energy integration in power grids. *Glob. Energy Interconnect.* **2018**, *1*, 544–551. (In Chinese)
21. Qudrat−Ullah, H.; Seong, B.S. How to do structural validity of a system dynamics type simulation model: The case of an energy policy model. *Energy Policy* **2010**, *38*, 2216–2224. [CrossRef]
22. Kleindorfer, G.B.; O'Neill, L.; Ganeshan, R. Validation in simulation: Various positions in the philosophy of science. *Manag. Sci.* **1998**, *44*, 1087–1099. [CrossRef]
23. Qudrat−Ullah, H. Behavior validity of a simulation model for sustainable development. *Int. J. Manag. Decis. Mak.* **2008**, *9*, 129–140.
24. Sterman, J.D. Testing Behavioral Simulation Models by Direct Experiment. *Manag. Sci.* **1987**, *33*, 1572–1592. [CrossRef]
25. Pidd, M. Why modeling and model use matter. *J. Oper. Res. Soc.* **2010**, *1*, 14–24.
26. China Electric Power Research Institute. *Production Simulation of New Energy Power System*, 1st ed.; China Electric Power Press: Beijing, China, 2019; pp. 96–183.
27. China Electricity Council. *Compilation for Statistical Data of Power Industry in 2018*, 1st ed.; China Electricity Council Press: Beijing, China, 2019; pp. 20–45.
28. China Electricity Council. *Compilation for Statistical Data of Power Industry in 2017*, 1st ed.; China Electricity Council Press: Beijing, China, 2018; pp. 20–45.
29. China Electricity Council. *Compilation for Statistical Data of Power Industry in 2015*, 1st ed.; China Electricity Council Press: Beijing, China, 2016; pp. 20–45.
30. NEA. Report on the Monitoring and Evaluation of National Renewable Energy and Electricity Development in 2015. Available online: http://zfxxgk.nea.gov.cn/auto87/201608/t20160823_2289.htm (accessed on 16 August 2016).
31. NEA. Report on the Monitoring and Evaluation of National Renewable Energy and Electricity Development in 2016. Available online: http://zfxxgk.nea.gov.cn/auto87/201704/t20170418_2773.htm (accessed on 10 April 2017).
32. NEA. Report on the Monitoring and Evaluation of National Renewable Energy and Electricity Development in 2017. Available online: http://zfxxgk.nea.gov.cn/auto87/201805/t20180522_3179.htm (accessed on 11 May 2018).
33. NEA. Report on the Monitoring and Evaluation of National Renewable Energy and Electricity Development in 2018. Available online: http://zfxxgk.nea.gov.cn/auto87/201906/t20190610_3673.htm (accessed on 4 June 2019).
34. NEA. Report on the Monitoring and Evaluation of National Renewable Energy and Electricity Development in 2019. Available online: http://zfxxgk.nea.gov.cn/2020$-$05/06/c_139059627.htm (accessed on 6 May 2020).
35. State Grid Energy Research Institute. *Analysis Report on New Energy Power Generation 2020*, 1st ed.; China Electric Power Press: Beijing, China, 2020; pp. 93–97.
36. Li, S.T.; Zhang, S.F.; Andrews−speed, P. Using diverse market−based approaches to integrate renewable energy: Experience from China. *Energy Policy* **2018**, *125*, 330–337. [CrossRef]

Article

Evaluating Production Process Efficiency of Provincial Greenhouse Vegetables in China Using Data Envelopment Analysis: A Green and Sustainable Perspective

Yuhu Liang [1], Xu Jing [2], Yanan Wang [1], Yan Shi [3] and Junhu Ruan [1,*]

[1] College of Economics and Management, Northwest A&F University, Yangling 712100, China; yuhuliang_2018@163.com (Y.L.); wyn3615@nwsuaf.edu.cn (Y.W.)

[2] College of Information Engineering, Northwest A&F University, Yangling 712100, China; jingxu@nwsuaf.edu.cn

[3] General Education Center, Tokai University, Kumamoto 862-8652, Japan; yshi@ktmail.tokai-u.jp

* Correspondence: rjh@nwsuaf.edu.cn; Tel.: +86-185-0292-4249

Received: 25 September 2019; Accepted: 17 October 2019; Published: 1 November 2019

Abstract: The evaluation of vegetable production process efficiency is of great significance for energy saving and waste reduction in production processes. However, few studies have considered the effect of greenhouse vegetable production process efficiency on energy saving and waste reduction. In this paper, data envelopment analysis (DEA) is used to analyze the production process efficiency and the effective use of input elements of greenhouse vegetables at the provincial level in China. The results reveal that many chemical fertilizers, farmyard manure, and pesticides in China are inefficient. On the other hand, the pure technical efficiency of greenhouse tomatoes and cucumbers is low in most areas of China. Meanwhile, the scale efficiency of greenhouse eggplants and greenhouse peppers is low in most areas of China. In order to save energy and develop green sustainable agriculture, we put forward some suggestions to improve the production efficiency of greenhouse vegetables in different provinces.

Keywords: production process; efficiency evaluation; greenhouse vegetables; data envelopment analysis; sustainability

---

1. Introduction

Green technology, as an emerging technology, is conducive to the transformation of the global sustainable production system. The main potential benefits of green technologies are that they can significantly reduce the cost of carbon dioxide emissions, reduce energy waste, and improve environmental performance. Some researchers have paid attention to the energy saving and waste reduction in production processes based on green technology [1,2]. However, few researchers consider that adjusting input factors can save energy and reduce waste from the perspective of green sustainability in vegetable production.

China is the largest vegetable producer and consumer in the world. The vegetable industry has been rapidly developing since 1980. Vegetable production was 190 million tons in 1980. Then, the total vegetable production was 798 million tons in 2016. Greenhouse agriculture can improve the level of automatic control and management of greenhouses to give full play to the efficiency of greenhouse agriculture [3,4]. Because of the advantages of greenhouse agriculture, vegetable greenhouses have been rapidly developed since the reform and opening up of China.China is also a vast country with vast territories and abundant resources. There are big gaps between different provinces in terms of resource input and technology level of vegetable production in greenhouses. At present, the excessive input of fertilizer, pesticide, seed, and labor in greenhouse vegetable production has caused more and more serious environmental pollution. Therefore, it is very important to estimate the input–output efficiency of greenhouse vegetables and to put forward some suggestions for dealing with the problems.

A lot of attention has been paid to the measurement of agricultural production efficiency [5,6]. Liu et al. [7] investigated the degree of efficiency and efficiency change of prefecture-level cities in north-east China from 2000 to 2012. Pang et al. [8] analyzed the agricultural eco-efficiency development level and spatial pattern in China. Baráth et al. [9] investigated relative productivity levels and examined productivity change for European agriculture between 2004 and 2013. Toma et al. [10] examined the agricultural efficiency of EU countries, through a bootstrap-data envelopment analysis approach. Raheli et al. [11] evaluated the sustainability and efficiency of tomato production and investigated the determinants of inefficiency of tomato farming in the Marand region of an east Azerbaijan province. Cardozo et al. [12] assessed the impacts of irrigation systems on sugarcane production from the perspective of the efficient use of land and water. Moreover, some researchers have focused on the efficiency of vegetable production [13,14]. These studies measured the agriculture production efficiency in specific countries or regions and found directions to increase agricultural yields. However, the sustainability of agriculture production is not well considered [15,16].

Recently, more and more researchers have taken the environment and energy into consideration. Many data envelopment analysis (DEA)-based models are adopted to study the environmental issues. These include directional distance function [17,18] , slacks-based model [19–21], DEA radial measure [22,23], DEA non-radial measure [22,24,25], etc. Fei et al. [26] employed the meta-frontier DEA method to measure agricultural energy efficiency in China's agricultural sector, and then used the Malmquist index approach to explore the energy productivity changes. Fei et al. [27] explored the integrated efficiency of inputs-outputs and unified performance in energy consumption and CO_2 emissions for the Chinese agricultural sector. Li et al. [28] calculated the relative efficiency and energy-saving potential of 30 provinces in China from 1997 to 2014. Le et al. [29] estimated the productivity change and environmental efficiency of agriculture in nine east Asian countries from 2002 to 2010. Wang et al. [30] adopted the stochastic frontiers analysis model to measure China's agricultural water use efficiency.

Through the analysis of the above literature, the following observations can be obtained. Firstly, few studies focus on the efficiency of the regional level from the perspective of greenhouse vegetables. Different from open-field agricultural production, the efficiency of greenhouse vegetable production has a weaker impact on natural resources and external environment. Thus, specific inputs should be considered for measuring the production efficiency of vegetable greenhouses. In this work, material and service costs, working days, seed costs, fertilizer costs, farmyard manure costs, and pesticide costs are used as the input indicators of efficiency evaluation.

In addition, few studies focus on the adverse effects of excessive fertilizers and pesticides in the environment of greenhouse vegetables. The development of greenhouse vegetables is beneficial to increase the income level of farmers. Although greenhouse vegetables can provide sustainable seasonal supply, excessive consumption of fertilizers and pesticides will lead to soil damage and the waste of resources in the production process of greenhouse vegetables. Therefore, we use the DEA to study the production efficiency of greenhouse vegetables from the perspective of the sustainability of agriculture production. We also put forward some suggestions on reducing the production cost of greenhouse vegetables and protecting the environment. This study is not only conducive to improving the production of greenhouse vegetables, but also conducive to the development of sustainable green agriculture.

The rest of the paper is organized as follows. Section 2 presents the methodology of the work and data processing. Empirical results and discussion are given in Section 3. The conclusions and policy implication are summarized in Section 4.

2. Methodology and Data Source

2.1. Production Function of Vegetable Industry

The vegetable production function refers to the relationship between the input and output of vegetable production. Assuming that $X_1, X_2, \ldots, X_n$ represent the input quantity of n production factors, Y represents the maximum yield of the vegetable product which is produced by a given input quantity under a given technical condition. f refers to the functional relationship between input and output, and the production function is introduced as Equation (1):

$$Y = f(X_1, X_2, \ldots, X_n). \tag{1}$$

In the input–output efficiency analysis of vegetable industry, according to the characteristics of vegetable industry production, land (S), capital (K), and labor (L) are three main factors of the production. So the production function of the vegetable industry can be written as Equation (2). The characteristics of the production function are conciseness, operability, and wide application. This model is suitable for vegetable industry. It uses the Cobb–Douglas production function and adds one factor of land (S) production.

$$Y = AK^{\alpha}L^{\beta}S^{\gamma} \tag{2}$$

Assuming that technical conditions remain unchanged, three input indices of land (S), capital (K), and labor (L) are determined. In the Equation (2), land (S) represents the area of vegetables, capital (K) represents the material production factors of input for vegetable production, labor (L) represents the number of working days of vegetable production, and Y represents the output of the main product of vegetable.

2.2. Data Envelopment Analysis Model

This paper mainly uses DEA to study the efficiency of decision-making units (DMUs). DEA is based on the concept of relative efficiency [31]. According to multiple input and multiple output indices, the relative effectiveness of the same unit is evaluated by linear programming [32]. At present, DEA is widely used in production efficiency evaluation [33–35]. The most well-known DEA models are the CCR model [36] and BCC model [37]. The CCR model is adopted when the yield of scale of production is unchanged. In the case of variable return on scale of production, the BCC model is adopted. The difference between the CCR model and BCC model is whether the return on production scale changes or not. On the other hand, when the DMUs do not run on the optimal scale, the operation results of the CCR model may be affected by scale efficiency. In order to eliminate the impact of scale efficiency on measurement results, the BCC model is selected in this paper.

According to different perspectives of the input–output researches, DEA model can be divided into the input orientated DEA model and output orientated DEA model [38]. Among them, the input orientated DEA model defines the production frontier in which input decreases proportionally under the condition of constant output, while the output orientated DEA model defines the production frontier in which output increases proportionally under the condition of constant input. Since farmers usually make the decision of maximizing output under a given input rather than the decision of minimizing input under a given output, the input orientated DEA model is adopted in this paper. Suppose there are n DMUs, there are m inputs and s outputs in each DMU, and each DMU is represented by a DMU_j. The input and output variables are expressed by X_j, Y_j respectively.

$$X_j = (x_{1j}, x_{2j}, \ldots, x_{mj})^T,\ x_{ij} > 0\ (i = 1, 2, \ldots, m;\ j = 1, 2, \ldots, n) \tag{3}$$

$$Y_j = (y_{1j}, y_{2j}, \ldots, y_{sj})^T,\ x_{rj} > 0\ (r = 1, 2, \ldots, s;\ j = 1, 2, \ldots, n) \tag{4}$$

The specific model is described as follows:

$$min\theta = V_D, \tag{5}$$

$$s.t. \sum_{j=1}^{n} \lambda_j X_j + s^- = \theta X_0, \tag{6}$$

$$\sum_{j=1}^{n} \lambda_j Y_j - s^+ = \theta Y_0, \tag{7}$$

$$\sum_{j=1}^{n} \lambda_j = 1, \tag{8}$$

$$s^- \geq 0,\ s^+ \geq 0,\ \lambda_j \geq 0\ (j = 1, 2, ..., n). \tag{9}$$

Objective (5) denotes the relative efficiency of DMU, where θ denotes the relative effective value of DMU, and V_D is a constant term. Constraint (6) is an adjustment constraint for input variables, where s^- denotes the slack variable of inputs, λ_j denotes the variable coefficient, and X_0 denotes the constant term. Constraint (7) is an adjustment constraint for output variables, where s^+ denotes the slack variable of outputs, λ_j denotes the variable coefficient, and Y_0 denotes the constant term. Constraint (8) guarantees the sum of constraints on the coefficients of variables to be 1, where λ_j denotes the variable coefficient. Constraint (9) is a constraint on the range of values of s^-, s^+ and λ_j.

There are three conditions in Objective (5) to Constraint (9):

(1) If $\theta = 1$ and $s^- \neq 0$, or if $\theta = 1$ and $s^+ \neq 0$, then decision unit j is weak DEA efficient.
(2) If $\theta = 1$, $s^- = 0$, and $s^+ = 0$, then decision unit j is DEA efficient.
(3) If $\theta < 1$, then decision unit j is inefficient.

In the BCC model, the efficiency analysis results can be divided into three parts: overall technical efficiency (OTE), pure technical efficiency (PTE), and scale efficiency (SE). OTE represents the overall level of efficiency of decision-making units. PTE represents that the efficiency brought about by the management and technical level of decision-making units. SE represents the efficiency value of the existing scale of production relative to the optimal scale of production without considering the level of technology and management. The relationship among the three efficiency values can be expressed as Equation (10):

$$OTE = PTE \times SE. \tag{10}$$

In the BCC model, the scale reward value of decision-making unit is equal to the sum of all λ which is corresponding to the given decision-making unit. The formula for calculating the return on scale of decision making units is as Equation (11):

$$k = \sum \lambda / \theta. \tag{11}$$

In this equation, k reflects that the decision-making unit is in the stage of scale reward. λ denotes the weight of DMU, and the θ denotes the relative effective value of DMU. When $k = 1$, the scale reward of the decision-making unit is constant, and the production scale of the decision-making unit is optimal. When $k < 1$, it implies that the scale reward of the decision-making unit is increasing, and the decision-making unit can get more output by properly increasing the input on the basis of the existing input. When $k > 1$, it implies that the size of the decision-making unit is diminishing, and the return of the increasing input on the existing basis is less. According to the results of BCC model analysis, we can also calculate the optimum degree of various kinds of output and the saving ratio of various input factors. In other words, we can get the rate of insufficient output and the rate of input redundancy. The formula for calculating the rate of underproduction and input redundancy is shown in Equation (12):

$$\eta_0 = minX_0/X_0. \tag{12}$$

In this equation, the η_0 denotes the relative effective of the input, the $minX_0$ represents the minimum amount of input that guarantees at least the output of the unit being evaluated, and X_0 is the original input.

2.3. Data Source and DEA Model for Efficiency Evaluation of Greenhouse Vegetables

The source of the input–output data comes from the National Collection of Agricultural Products Cost–Benefit Data, which is collected by the Price Department of the National Development and Reform Commission. The National Collection of Agricultural Products Cost–Benefit Data 2017 contains data on production cost and income of major agricultural products. This study mainly refers to the corresponding statistical indicators in the National Collection of Agricultural Product Cost–Benefit Data. In selecting data indicators, according to vegetable production function, we consider the scientificity, accuracy, and validity of vegetable production input–output data. The data of vegetable industry in the main vegetable growing provinces of China in 2016 is selected. There are 21 provinces planting greenhouse tomatoes, 21 provinces planting greenhouse cucumbers, 10 provinces planting greenhouse eggplants, and 11 provinces planting greenhouse peppers in China. Y indicates the output of greenhouse vegetables. Capital (K) is expressed in terms of material and service cost, seed cost, fertilizer cost, farmyard manure cost, and pesticide cost which are required by greenhouse vegetables. Labor (L) is expressed as the number of working days of greenhouse vegetables. Land (S) indicates the acreage of greenhouse vegetables. The specific value is derived from the National Collection of Agricultural Product Cost and Benefit Data (2017). The classification and description of variables are as follows in Table 1.

Table 1. Classification and description of variables.

Variable Classification	Variable Names	Definitions	Units
Output	OVMP	Output value of main product	Kilogram
Inputs	MSC	Material and service cost	Yuan
	SC	Seed cost	Yuan
	FC	Fertilizer cost	Yuan
	FMC	Farmyard manure cost	Yuan
	PC	Pesticide cost	Yuan
	WD	Working days	Day
	S	Acreage of greenhouse vegetables	m^2
Constraint variable	OV	Original value of an output or input variable	Kilogram or yuan or day
	RM	Radial adjustment of an input variable	Kilogram or yuan or day
	SM	Slack movement of an input variable	Kilogram or yuan or day
	PV	Target value of an output or input variable	Kilogram or yuan or day

According to the production theory and basic DEA model, we developed a DEA model to evaluate the production efficiency of greenhouse vegetables. The specific model is described as follows:

$$min\theta = V_D, \tag{13}$$

$$s.t.\ \sum_{j=1}^{n}\lambda_j(MSC+SC+FC+FMC+PC+WD)+s^- = \theta(MSC_0+SC_0+FC_0+FMC_0+PC_0+WD_0), \quad (14)$$

$$\sum_{j=1}^{n}\lambda_j OVMP - s^+ = \theta OVMP_0, \quad (15)$$

$$\sum_{j=1}^{n}\lambda_j = 1, \quad (16)$$

$$s^- = |OV_X - PV_X|, \quad (17)$$

$$s^+ = |OV_Y - PV_Y|, \quad (18)$$

$$\lambda_j \geq 0\ (j = 1, 2, ..., n). \quad (19)$$

In Constraint (14), MSC_0, SC_0, FC_0, FMC_0, PC_0, and WD_0 represent the original values of MSC, SC, FC, FMC, PC, and WD, respectively. In Constraint (15), $OVMP_0$ represents the original value of OVMP. OV_X represents the original value of the input variable, and PV_X represents the target value of the input variable in Constraint (17). OV_Y represents the original value of the output variable, and PV_Y represents the target value of the output variable in Constraint (18). According to the data in the National Collection of Agricultural Products Cost–Benefit Data, the statistical data are the output, capital input, and labor input of greenhouse vegetables per 667 m^2. In other words, the input of land is equivalent to a fixed value, so the land area is neglected in Constraint (14). According to the above equations, the efficiency evaluation results can be obtained by using DEAP2. 1 software.

3. Results and Discussion

3.1. Comparative Analysis of Vegetable Efficiency in Greenhouses

Based on DEA model under BCC assumption, the production efficiency of greenhouse vegetables is calculated. The results of the DEA model represent the relative efficiency, and the results will change with different number of variables. In order to study the input–output efficiency of greenhouse vegetable planting provinces in China, four greenhouse vegetables were selected: greenhouse tomatoes, greenhouse cucumbers, greenhouse eggplants, and greenhouse peppers. According to the input–output data, DEAP2. 1 software was run to analyze the overall technical efficiency, pure technical efficiency, and scale efficiency of the four greenhouse vegetables. The relative production efficiency of greenhouse vegetables in each province can be judged by these efficiency values. Taking the production efficiency of greenhouse tomatoes as an example, the DEA model under BCC assumption was adopted. The results are shown in Figure 1 and Table 2. OTE represents the overall technical efficiency, PTE represents the pure technical efficiency, and SE represents the scale efficiency.

Table 2. The overall technical efficiency (OTE), pure technical efficiency (PTE), and scale efficiency (SE) of greenhouse tomatoes.

Province	OTE	PTE	SE	SR
Beijing	0.959	1.000	0.959	*drs*
Tianjin	1.000	1.000	1.000	–
Hebei	0.720	0.745	0.967	*irs*
Shanxi	0.728	0.729	0.998	*drs*
Inner M	0.628	0.651	0.964	*irs*
Liaoning	0.603	0.608	0.992	*irs*
Jilin	0.725	0.782	0.927	*irs*
Heilongjiang	1.000	1.000	1.000	–
Shanghai	0.891	0.975	0.914	*irs*
Jiangsu	0.788	0.812	0.970	*irs*
Zhejiang	0.920	1.000	0.920	*irs*
Anhui	0.929	0.984	0.944	*irs*
Shandong	0.488	0.618	0.789	*irs*
Henan	0.779	0.823	0.946	*drs*
Hubei	1.000	1.000	1.000	–
Sichuan	0.906	1.000	0.906	*drs*
Shaanxi	0.756	0.848	0.890	*irs*
Gansu	0.705	0.781	0.903	*drs*
Qinghai	0.949	0.958	0.990	*irs*
Ningxia	0.980	1.000	0.980	*irs*
Xinjiang	1.000	1.000	1.000	–

Note: SR means returns to scale, *irs* means increasing returns to scale, *drs* means decreasing returns to scale, and – means no change in return on scale.

For greenhouse tomato cultivation in vegetable industry, the results are shown in Table 2 and Figure 1a. The average OTE, PTE, and SE of 21 provinces are 0.831, 0.872, and 0.950, respectively. Through the decomposition and comparison of efficiency, we can find that four provinces are effective, which are Tianjin, Heilongjiang, Hubei, and Xinjiang, respectively. However, Shandong Province has the lowest efficiency, which is only 0.488, followed by Inner Mongolia and Liaoning, with 0.603 and 0.628, respectively.

For the greenhouse cucumber cultivation in the vegetable industry, the results are shown in Figure 1b. The average OTE, PTE, and SE of 21 provinces are 0.821, 0.875, and 0.942, respectively. Through the decomposition and comparison of efficiency, we can find that seven provinces are effective, which are Beijing, Tianjin, Heilongjiang, Zhejiang, Hubei, Sichuan, and Xinjiang. However, Shanxi Province has the lowest efficiency, which is only 0.546, followed by Qinghai and Inner Mongolia, with 0.557 and 0.564, respectively. The difference between the maximum value and the minimum value is 0.454.

For greenhouse eggplant cultivation in vegetable industry, the results are shown in Figure 1c. The average OTE, PTE, and SE of 10 provinces are 0.837, 0.963, and 0.866, respectively. Through the decomposition and comparison of efficiency, we can find that four provinces are effective, which are Beijing, Tianjin, Liaoning, and Sichuan, respectively. However, Zhejiang Province has the lowest efficiency, which is only 0.458, followed by Shanxi Province and Shanghai, with 0.615 and 0.666, respectively.

For greenhouse pepper cultivation, the results are shown in Figure 1d. The average OTE, PTE, and SE of 11 provinces are 0.828, 0.928, and 0.888, respectively. Through the decomposition and comparison of efficiency, we can find that three provinces are effective, which are Beijing, Tianjin, and Sichuan. However, Gansu Province has the lowest efficiency, which is only 0.528, followed by Zhejiang Province with 0.634. The PTE is 0.645, and the SE is 0.786 in Gansu Province. The PTE is 0.807, and the SE is 0.786 in Zhejiang Province.

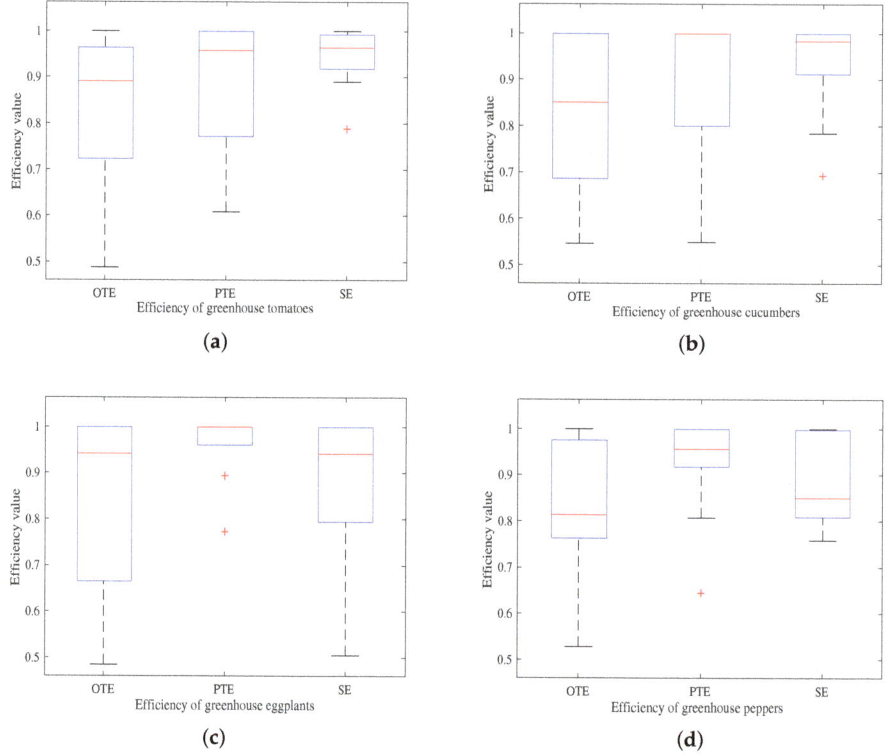

Figure 1. Boxplot of OTE, PTE, and SE of greenhouse tomatoes (**a**), greenhouse cucumbers (**b**), greenhouse eggplants (**c**) and greenhouse peppers (**d**).

From the empirical results above, we can see that the OTE of greenhouse vegetables in different provinces is quite different. Only a few provinces are effective, most of them are in the state of increasing returns to scale or decreasing returns to scale. From the perspective of the PTE, vegetable greenhouses failed to make use of current technologies to maximize output under fixed input conditions in most provinces.

3.2. Comparison and Analysis of the Efficiency of Greenhouse Vegetables at Provincial Level

Based on the overall analysis of the efficiency of the four greenhouse vegetables, we can find that greenhouse vegetable efficiencies in many provinces of China are low. In Figure 2, we analyze the PTE and the SE of the four greenhouse vegetables. Based on the analysis of the PTE and the SE, this paper explores whether the PTE or the SE leads to the inefficiency of greenhouse vegetables.

By breaking down and comparing the efficiency of greenhouse tomatoes as shown in Figure 2a, we can find that four provinces are effective, which are Tianjin, Heilongjiang, Hubei, and Xinjiang, respectively. Among the other 17 provinces without DEA efficiency, the PTE of 11 provinces is lower than the SE. In other words, the loss of PTE may lead to inefficiency in most provinces. This indicates that the main obstacle to improving the efficiency of greenhouse tomato production in most parts of China is the difficulty in improving the PTE of greenhouse tomatoes.

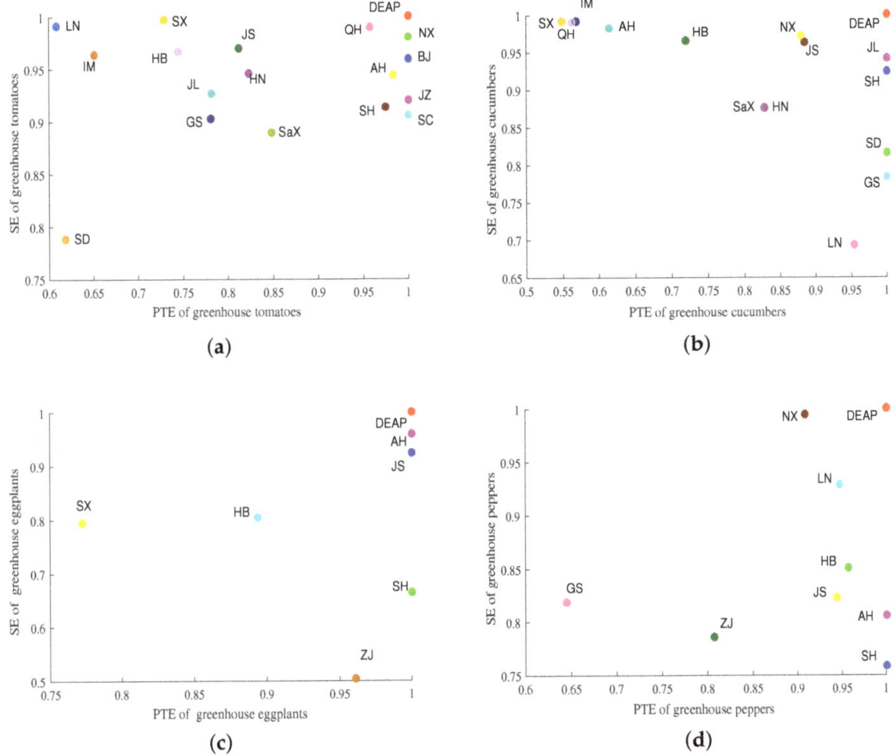

Figure 2. Efficiency distribution map of greenhouse tomatoes (**a**), greenhouse cucumbers (**b**), greenhouse eggplants (**c**) and greenhouse peppers (**d**).

By decomposing and comparing the efficiency of greenhouse cucumbers as shown in Figure 2b, it can be found that seven provinces are effective, namely Beijing, Tianjin, Heilongjiang, Zhejiang, Hubei, Sichuan, and Xinjiang, respectively. The difference between the maximum and minimum values is 0.454. Among the other 14 provinces without DEA efficiency, the PTE of nine provinces is lower than the SE. In other words, the loss of PTE may lead to inefficiency in most provinces. This shows that the difficulty of improving the PTE is the main obstacle to improve the efficiency of greenhouse cucumber production in most parts of China.

By breaking down and comparing the efficiency of greenhouse eggplants as shown in Figure 2c, we can find that four provinces are effective, namely Beijing, Tianjin, Liaoning, and Sichuan, respectively. Among the other six provinces without DEA efficiency, the SE of five provinces is lower than the PTE. In other words, the loss of SE may lead to inefficiency in most provinces. This indicates that the difficulty of increasing the SE is a major obstacle to improving the efficiency of greenhouse eggplant production in China.

By breaking down and comparing the efficiency of greenhouse peppers as shown in Figure 2d, we can find that three provinces are effective, namely Beijing, Tianjin, and Sichuan, respectively. Among the other eight provinces without DEA efficiency, the SE of six provinces is lower than the PTE. In other words, the loss of SE may lead to inefficiency in most provinces. This shows that the difficulty of improving the SE is the main obstacle to improve the production efficiency of greenhouse peppers in China.

3.3. Spatial Distribution Analysis of Efficiency

In order to examine the spatial distribution of vegetable production efficiency in greenhouse, we use GIS 10.6 software to analyze the spatial distribution of vegetable production efficiency in greenhouse. In Figure 3, the distribution of greenhouse vegetable efficiency in different provinces is studied by taking greenhouse tomatoes as an example. We divide the comprehensive technical efficiency, the fertilizer efficiency, the farm fertilizer efficiency, and the pesticide efficiency into four categories. We use different colors to represent different levels. An efficiency value of 100% is classified as DEA efficiency. An efficiency value between 80% and 100% is classified as high level. An efficiency value between 60% and 80% is classified as medium level. An efficiency value between 0% and 60% is classified as low level. The gray areas in Figure 3 are with no data, most of which are located in the south of the Yangtze river. This is because there is plenty of sunshine and rain in the south of the Yangtze River. Most regions and provinces are mainly traditional uncovered farmland, with less vegetables planted in greenhouses.

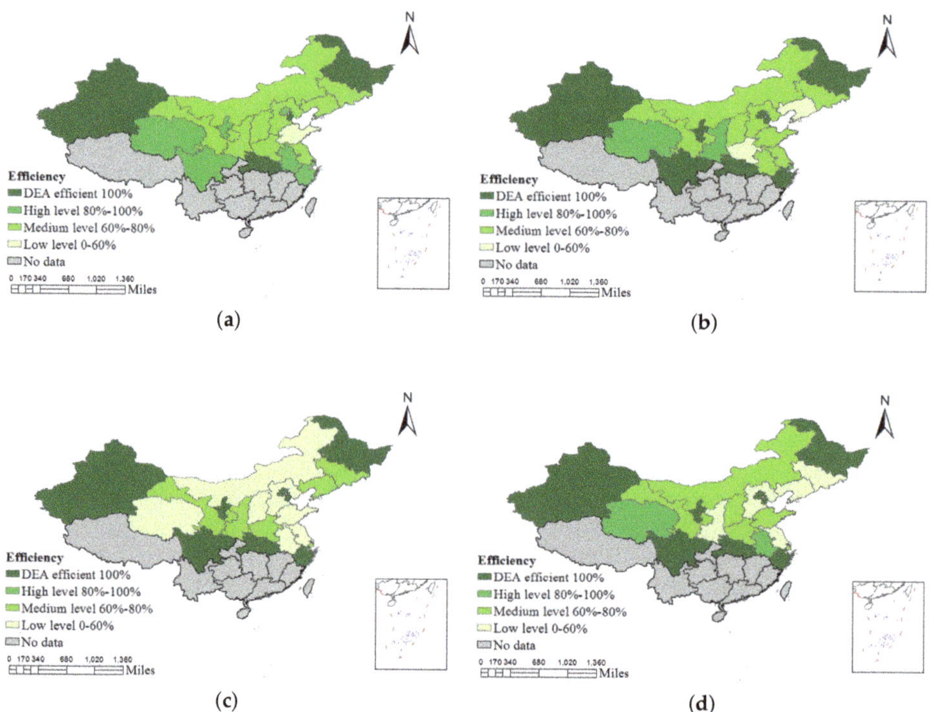

Figure 3. Spatial distribution map of OTE (**a**), fertilizer efficiency (**b**), farm manure efficiency, (**c**) and pesticide efficiency (**d**) of greenhouse tomato.

Figure 3a shows the distribution of the comprehensive efficiency of greenhouse tomatoes in various provinces of China. On the whole, the OTE in most areas of China is not high. Tianjin, Heilongjiang, Xinjiang, and Hubei are in DEA efficient. Beijing, Shanghai, Zhejiang, Anhui, Sichuan, Qinghai, and Ningxia are high level of efficiency. Hebei, Shanxi, Inner Mongolia, Liaoning, Jilin, Jiangsu, Henan, Shaanxi, and Gansu are medium-level efficiency. Shandong Province has the lowest comprehensive efficiency, which is low-level efficiency.

Figure 3b shows the distribution of the efficiency of fertilizer in various provinces of China. It can be seen that most areas are in the medium level of fertilizer use efficiency, which can be improved a lot. Beijing, Tianjin, Heilongjiang, Zhejiang, Hubei, Sichuan, Ningxia, and Xinjiang are DEA efficient. Shanghai, Shaanxi, and Qinghai are of a high level of efficiency. Hebei, Shanxi, Inner Mongolia, Jilin, Jiangsu, Anhui, Shandong, and Gansu are medium-level efficiency. Liaoning and Henan are low-level efficiency. The utilization efficiency of chemical fertilizer in Henan and Liaoning provinces is less than 60%. In Henan Province, the use efficiency of fertilizer is low mainly because of the excessive use and intensity of fertilizer and the lack of use of farmyard manure [39]. In order to pursue a high yield, most provinces may use large amounts of elemental fertilizers excessively and frequently, resulting in low fertilizer use efficiency. If farmers use chemical fertilizer unreasonably, it will not only waste resources, but also cause environmental pollution [40].

Figure 3c shows the distribution of the use efficiency of farmyard manure in various provinces of China. Beijing, Tianjin, Heilongjiang, Zhejiang, Hubei, Sichuan, Ningxia, and Xinjiang are DEA efficient. Liaoning, Jilin, Shanghai, Henan, Shaanxi, and Gansu are medium-level efficiency. Hebei, Shanxi, Inner Mongolia, Jiangsu, Anhui, Shandong, and Qinghai are low-level efficiency. The efficiency of farm manure use in 13 provinces of China is less than 80%. This shows that the efficiency of farm manure use is low in most provinces of China, and there are big problems in the use of farm manure. The effective use of farmyard manure is conducive to reducing environmental pollution and reducing the damage of fertilizer to land [41]. Therefore, it is important to improve the utilization efficiency of farmyard manure for the cultivation of greenhouse vegetables in China.

Figure 3d shows the distribution of pesticide use efficiency in various provinces of China. Beijing, Tianjin, Heilongjiang, Zhejiang, Hubei, Sichuan, Ningxia, and Xinjiang are DEA efficient. Anhui and Qinghai are a high level of efficiency. Shanxi, Inner Mongolia, Shanghai, Shandong, Henan, and Gansu are medium-level efficiency. Hebei, Liaoning, Jilin, Jiangsu, and Shaanxi are low-level efficiency. As can be seen from the Figure 3d, the utilization efficiency of pesticides is less than 60% in most areas mainly along the east coast of China, such as Liaoning, Hebei, and Jiangsu. Although these areas are economically developed, the use of pesticides is not very efficient. Due to its proximity to the ocean and their humid climate, it is easy to breed a large number of pests. Farmers often use a lot of pesticides to control pests. This can easily lead to inefficient use of pesticides [42]. In addition, the utilization efficiency of pesticide in greenhouse tomatoes in Shaanxi Province is not high. The main reason for the low utilization rate of pesticide in Shaanxi Province may be the backward application equipment [5].

3.4. Analysis and Adjustment of Inefficient Provinces

The DEA method can not only explore the reasons why the decision-making unit is ineffective, but also give corresponding improvement methods. DEAP2.1 software is used to process the input–output data of greenhouse vegetable planting in main provinces of China in 2016, and we find that there is redundancy in greenhouse vegetable planting. This study mainly lists four kinds of greenhouse vegetables with redundancy in inputs. Table 3 shows the adjustment of greenhouse tomato planting provinces with redundancy.

According to the new situation of accelerated economic and social development in China, the whole country is divided into four major economic regions: the eastern region, the northeast region, the central region, and the western region. Similarly, according to the geographical location of greenhouse vegetable growing provinces, we divide the efficiency adjustment analysis of greenhouse vegetables into four regions: the eastern region, the northeast region, the central region, and the western region.

Table 3. Input redundancy of greenhouse tomato planting per 667 m^2 in China in 2016.

Provinces	Item	OVMP/kg	MSC/Yuan	WD/d	SC/Yuan	FC/Yuan	FMC/Yuan	PC/Yuan
Hebei	OV	5385.83	3199.96	61.4	399.1	476.45	425.39	258.27
	RM	0	−816.13	−15.66	−101.788	−121.516	−108.493	−65.87
	SM	0	0	0	−70.447	0	−96.231	−37.778
	PV	5385.83	2383.83	45.74	226.865	354.934	220.666	154.621
Shanxi	OV	5494.79	3733.41	83.39	323.54	342.74	515.99	180.6
	RM	0	−1012.13	−22.607	−87.712	−92.917	−139.885	−48.961
	SM	0	−11.84	−11.385	−15.304	0	−78.974	0
	PV	5494.79	2709.444	49.398	220.524	249.823	297.131	131.639
Inner M	OV	5384.68	3704.19	68.07	636.82	584.67	433.84	266.01
	RM	0	−1292.45	−23.751	−222.197	−204.001	−151.374	−92.815
	SM	0	0	0	−182.395	0	−72.588	−11.777
	PV	5384.68	2411.736	44.319	232.227	380.669	209.878	161.418
Liaoning	OV	5148.08	5559.37	71.11	329.09	591.24	368.22	408.8
	RM	0	−2181.35	−27.902	−29.126	−231.987	−144.48	−160.402
	SM	0	−511.446	0	0	−44.596	0	−83.724
	PV	5148.08	2866.577	43.208	199.964	314.657	223.74	164.674
Jilin	OV	4484.96	2438.44	68.55	165.52	333.46	191.17	255.3
	RM	0	−530.439	−14.912	−36.006	−72.538	−41.586	−55.536
	SM	355.543	−270.354	−1.93	0	−10.141	0	−69.356
	PV	4840.503	1637.647	51.708	129.514	250.78	149.584	130.408
Shanghai	OV	4651.63	2132.19	48.79	193.17	298.41	210.83	206.49
	RM	0	−53.546	−1.225	−4.851	−7.494	−5.295	−5.186
	SM	283.252	0	0	−36.119	0	−43.15	−56.025
	PV	4934.882	2078.644	47.565	152.2	290.916	162.385	145.279
Jiangsu	OV	4715.23	2752.13	66.11	152.82	347.68	344.18	221.79
	RM	0	−516.325	−12.403	−28.67	−65.228	−64.571	−41.61
	SM	146.542	−595.327	0	0	−70.721	−108.826	−59.818
	PV	4861.772	1640.477	53.707	124.15	211.731	170.783	120.362
Anhui	OV	4538.3	2412.29	49.23	140.46	457.91	369.43	176.76
	RM	0	−38.023	−0.776	−2.214	−7.218	−5.823	−2.786
	SM	267.584	−741.226	0	0	−136.351	−248.527	−27.214
	PV	4805.884	1633.041	48.454	138.246	314.341	115.08	146.76
Shandong	OV	4404.7	4148.17	68.61	769.7	589.55	492.77	290.14
	RM	0	−1584.34	−26.205	−293.978	−225.172	−188.208	−110.816
	SM	520.407	−131.078	0	−288.368	0	−148.647	0
	PV	4925.107	2432.748	42.405	187.354	364.378	155.915	179.324
Henan	OV	5105.42	2664.3	65.01	193.99	406.89	286.88	204.75
	RM	0	−470.577	−11.482	−34.263	−71.866	−50.67	−36.164
	SM	0	0	−1.946	0	−118.426	−7.683	−43.883
	PV	5105.42	2193.723	51.582	159.727	216.598	228.528	124.704
Shaanxi	OV	4512.17	2244.62	60.49	314.83	275.67	293.67	220.67
	RM	0	−340.163	−9.167	−47.711	−41.777	−44.504	−33.442
	SM	407.076	0	0	−129.651	0	−69.72	−58.445
	PV	4919.246	1904.457	51.323	137.468	233.893	179.446	128.783
Gansu	OV	5473.01	3259.06	73.93	295.34	451.16	278.32	232.8
	RM	0	−713.899	−16.194	−64.694	−98.827	−60.966	−50.995
	SM	0	−288.45	−5.592	−1.158	0	0	0
	PV	5473.01	2256.71	52.143	229.487	352.333	217.354	181.805
Qinghai	OV	5781.85	3955.8	70.16	302.58	344.48	1092.73	138.07
	RM	0	−164.427	−2.916	−12.577	−14.319	−45.42	−5.739
	SM	0	−1336.75	−16.504	−21.657	−30.038	−732.328	0
	PV	5781.85	2454.629	50.74	268.346	300.124	314.982	132.331

3.4.1. Analysis on the Adjustment of Efficiency Input in Eastern China

In Table 3, the production efficiency of greenhouse tomatoes in Hebei, Shanghai, Jiangsu, and Shandong is not effective in eastern China. The adjustment range is shown in Figure 4a. In Hebei Province, the adjustment ranges of six input factors of greenhouse tomatoes are 74.50%, 74.50%, 56.84%,

74.50%, 51.87%, and 59.87%, respectively, including material and service cost, number of working days, seed cost, fertilizer cost, farmyard manure cost, and pesticide cost. The adjustment ranges of six input factors of greenhouse tomatoes in Shanghai are 97.49%, 97.49%, 78.79%, 97.49%, 77.02%, and 70.36%, respectively. The adjustment ranges of six input factors of greenhouse tomatoes in Jiangsu Province are 59.61%, 81.24%, 81.24%, 60.90%, 49.62%, and 54.27%, respectively. The adjustment ranges of six input factors of greenhouse tomatoes in Shandong Province are 58.65%, 61.81%, 24.34%, 61.81%, 31.64%, and 61.81%, respectively.

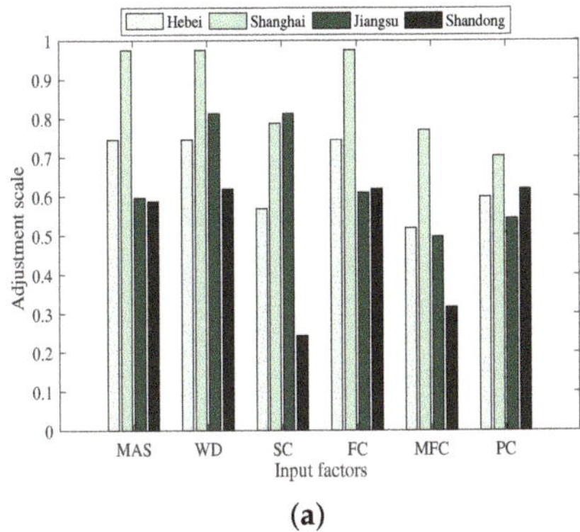

(a)

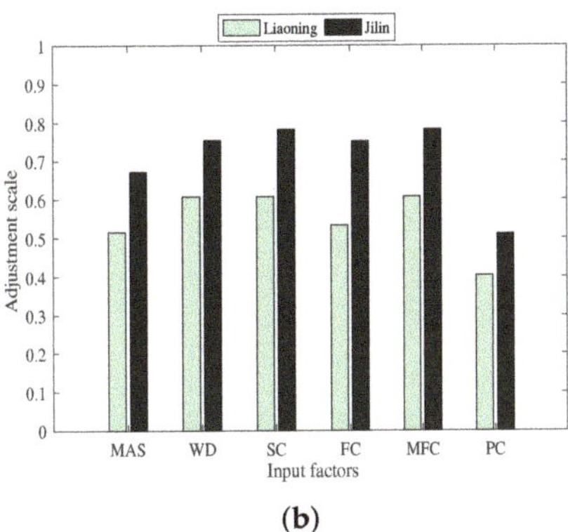

(b)

Figure 4. Analysis results: (**a**) the proportion of efficiency input adjustment in eastern China; (**b**) the proportion of efficiency input adjustment in north-east China.

In summary, there exist some problems of excessive investment in Hebei, Shanghai, and Shandong, resulting in a waste of energy and environmental pollution. In the eastern region, the excessive input of fertilizers in Jiangsu Province is 39.1%, while the excessive input of pesticides is 45.73%. The excessive input of pesticides in Hebei Province is 40.13%. The excessive input of fertilizers in Shandong Province is 38.19%, while the excessive input of pesticides is 38.19%.

3.4.2. Analysis on the Adjustment of Efficiency Input in Northeast China

In Table 3, the production efficiency of greenhouse tomatoes in Liaoning Province and Jilin Province is not effective in northeast China. The adjustment range is shown in Figure 4b. The adjustment ranges of six input factors of greenhouse tomatoes in Liaoning Province are 51.56%, 60.76%, 60.76%, 53.22%, 60.76%, and 40.28%, respectively, including material and service cost, number of working days, seed cost, fertilizer cost, farmyard manure cost, and pesticide cost. The adjustment ranges of six input factors of greenhouse tomatoes in Jilin Province are 67.16%, 75.43%, 78.25%, 75.21%, 78.25%, and 51.08%, respectively.

In short, there exist some problems of excessive investment in Liaoning and Jilin, resulting in the waste of resources and environmental pollution. The excessive input of fertilizers in Liaoning Province is 46.78%, while the excessive input of pesticides is 59.72%. The excessive input of pesticides in Jilin Province is 48.92%, while the excessive input of fertilizers is 24.79%.

3.4.3. Analysis on the Adjustment of Efficiency Input in Central China

In Table 3, the production efficiency of greenhouse tomatoes in Shanxi, Anhui and Henan is not effective in central China. The adjustment range is shown in Figure 5a. The adjustment ranges of six input factors of greenhouse tomatoes in Shanxi Province are 72.57%, 59.24%, 68.16%, 72.89%, 57.58%, and 72.89%, respectively, including material and service cost, number of working days, seed cost, fertilizer cost, farmyard manure cost, and pesticide cost. The adjustment ranges of six input factors of greenhouse tomatoes in Anhui Province are 67.70%, 98.42%, 98.42%, 68.65%, 31.15%, and

83.03%, respectively. In Henan Province, the adjustment ranges of six input factors of greenhouse tomatoes are 82.34%, 79.34%, 82.34%, 53.23%, 79.66%, and 60.91%, respectively.

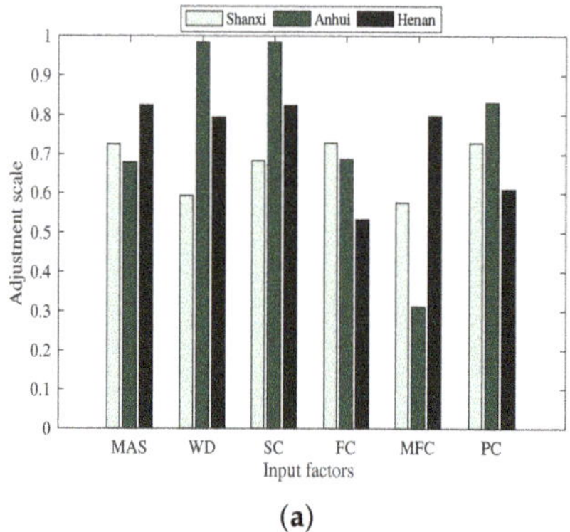

(a)

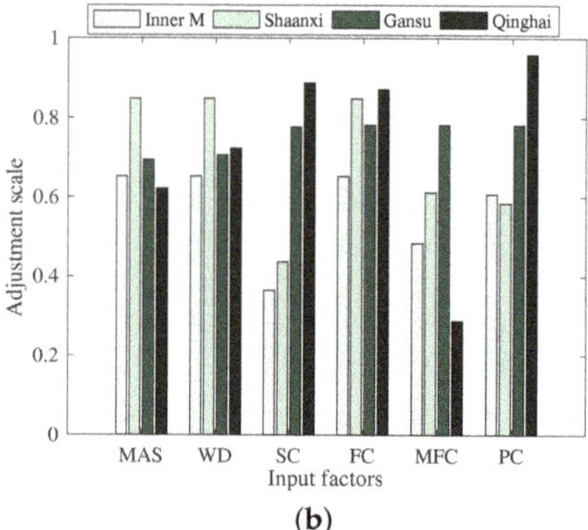

(b)

Figure 5. Analysis results: (**a**) the proportion of efficiency input adjustment in central China; (**b**) the proportion of efficiency input adjustment in western China.

To sum up, there exist some problems of excessive investment in Henan Province, Anhui Province, and Shanxi Province, resulting in the waste of resources and energy and environmental pollution. In the central region, the excessive input of fertilizers in Henan Province is 46.77%, and the excessive input of pesticides is 39.09%. The excessive input of fertilizers in Anhui Province is 31.35%. The excessive input of fertilizers in Shanxi Province is 27.11%, while the excessive input of pesticides is 27.11%.

3.4.4. Analysis on the Adjustment of Efficiency Input in Western China

In Table 3, the production efficiency of greenhouse tomatoes in Inner Mongolia, Shaanxi, Gansu and Qinghai is not effective in western China. The adjustment range is shown in Figure 5b. The adjustment ranges of six input factors of greenhouse tomatoes in Inner Mongolia are 65.11%, 65.11%, 36.47%, 65.11%, 48.38%, and 60.68%, respectively, including material and service cost, number of working days, seed cost, fertilizer cost, farmyard manure cost, and pesticide cost. The adjustment ranges of six input factors of greenhouse tomatoes in Shaanxi Province are 84.85%, 84.85%, 43.66%, 84.85%, 61.10%, and 58.36%, respectively. The adjustment ranges of six input factors of greenhouse tomatoes in Gansu Province are 69.24%, 70.53%, 77.70%, 78.09%, 78.09%, and 78.09%, respectively. The adjustment ranges of six input factors of greenhouse tomatoes in Qinghai Province are 62.05%, 72.32%, 88.69%, 87.12%, 28.83%, and 95.84%, respectively.

In general, there exist some problems of excessive investment in Inner Mongolia, Shaanxi, Qinghai, and Gansu, resulting in waste of energy and environmental pollution. In the western region, the excessive input of pesticides in Shaanxi Province is 41.64%. The excessive input of fertilizers in Inner Mongolia is 39.32%, while the excessive input of pesticides is 34.89%.

4. Conclusions and Policy Implication

This paper uses a DEA model to analyze the overall technical efficiency (OTE), pure technical efficiency (PTE), and scale efficiency (SE) of greenhouse tomatoes, greenhouse cucumbers, greenhouse eggplants, and greenhouse peppers from the perspective of greenhouse vegetable production efficiency. According to different efficiency values of greenhouse vegetables, we put forward some suggestions to adjust the different production efficiency. The main conclusions are as follows:

- In the production of greenhouse tomatoes and cucumbers in China, the loss of PTE may lead to inefficiency in most provinces. For greenhouse tomatoes, among the 17 inefficient provinces, the

PTE of 11 provinces is lower than the SE. For greenhouse cucumbers, among the 14 inefficient provinces, the PTE of 9 provinces is lower than the SE. The results show that the government should pay more attention to the improvement of the PTE of greenhouse tomatoes and cucumbers.

- In the production of greenhouse eggplants and peppers in China, the loss of SE may lead to inefficiency in most provinces. For greenhouse eggplants, among the six inefficient provinces, the SE of five provinces is lower than the PTE. For greenhouse peppers, among the eight inefficient provinces, the SE of six provinces is lower than the PTE. The results show that the government should pay more attention to improving the SE of greenhouse eggplants and peppers.
- From the perspective of input factors, fertilizers, farm manure and pesticides are inefficient in most parts of China. In particular, the overall use efficiency of farmyard manure is low, and chemical fertilizers and pesticides are seriously wasted. These results indicate that the government should pay more attention to the use of chemical fertilizers, farm manure, and pesticides to improve the use efficiency in the future. On the one hand, it helps to reduce the waste of resource. On the other hand, it is conducive to the development of green and sustainable agriculture.

Based on the above analysis, we put forward some suggestions for the different characteristics of the efficiency of greenhouse vegetables in different provinces. The suggestions are as follows:

- For provinces with DEA efficiency, such as Tianjin, Heilongjiang, and Hubei, on the basis of maintaining the existing production advantages, the supply and demand of greenhouse vegetable production should be balanced. For the provinces with high level of efficiency, such as Beijing, Sichuan, Shanghai, and Ningxia, the government should maintain the existing scale advantage in promoting vegetable production development at first. Thus, the government should focus on the introduction and application of advanced field technology and management mode to achieve higher utilization rate of input factors in greenhouse vegetable production.
- For the provinces with low level of efficiency, such as Shandong, Inner Mongolia, Shanxi, and Hebei, it is important to improve the PTE and find the appropriate scale suitable for the local. First, the government should increase support to these provinces and guide farmers to use chemical fertilizers and pesticides rationally. Second, the government should encourage and support the use of farm manure to reduce the use of chemical fertilizers.

In the next study, we plan to study the change of greenhouse vegetable production efficiency from the perspective of time series and space in China. It is planned to use the DEA model and Tobit model to explore the influencing factors of greenhouse vegetable production efficiency, in order to make a contribution to the development of green and sustainable agriculture.

Author Contributions: All authors contributed to this study. J.R. and Y.L. formulated the study design. J.R., Y.S., Y.L., and Y.W. conceived and designed the research methodology. X.J., J.R., and Y.L. collected and analyzed the data. J.R. and Y.L. finalized the paper.

Funding: This work was supported in part by the National Natural Science Foundation of China under Grants 71703122, 71973106, and 71531002, the China Ministry of Education Social Sciences and Humanities Research Youth Fund Project under Grant 16YJC630102, the Science and Technology Plan Projects of Yangling Demonstration Zone under Grant 2016RKX-04, the Fundamental Research Funds for the Central Universities under Grant 2018RWSK02, the Key Research and Development Program of Shaanxi under Grant 2019ZDLNY07-02-01, and the National College Students' Innovation and Entrepreneurship Training Program under Grant 201910712075.

Acknowledgments: We thank the National Nature Science Foundation of China, China Ministry of Education Social Sciences and Humanities Research Youth Fund Project, Natural Science Basic Research Project in Shaanxi Province, Science and Technology Plan Projects of Yangling Demonstration Zone, Fundamental Research Funds for the Central Universities, Key Research and Development Program of Shaanxi and National College Students' Innovation and Entrepreneurship Training Program for funding the research.

Conflicts of Interest: The authors declare no conflict of interest.

References

1. Xu, L.; Wang, Y.; Shah, S.A.A.; Zameer, H.; Solangi, Y.A.; Walasai, G.D.; Siyal, Z.A. Economic viability and environmental efficiency analysis of hydrogen production processes for the decarbonization of energy systems. *Processes* **2019**, *7*, 494. [CrossRef]
2. Hafeez, G.; Islam, N.; Ali, A.; Ahmad, S.; Usman, M.; Alimgeer, K.S. A modular framework for optimal load scheduling under price-based demand response scheme in smart grid. *Processes* **2019**, *7*, 499. [CrossRef]
3. Ruan, J.; Wang, Y.; Chan, F.T.S.; Hu, X.; Zhao, M.; Zhu, F.; Shi, B.; Shi, Y.; Lin, F. A life cycle framework of green IoT-based agriculture and its finance, operation, and management issues. *IEEE Commun. Mag.* **2019**, *57*, 90–96. [CrossRef]
4. Ruan, J.; Jiang, H.; Li, X.; Shi, Y.; Chan, F.T.; Rao, W. A granular GA-SVM predictor for big data in agricultural cyber-physical systems. *IEEE Trans. Ind. Inf.* **2019**, *57*, 90–96. [CrossRef]
5. Clark, M.; Tilman, D. Comparative analysis of environmental impacts of agricultural production systems, agricultural input efficiency, and food choice. *Environ. Res. Lett.* **2017**, *12*, 1–12. [CrossRef]
6. Geng, Q.; Ren, Q.; Nolan, R.H.; Wu, P.; Yu, Q. Assessing China's agricultural water use efficiency in a green-blue water perspective: A study based on data envelopment analysis. *Ecol. Indic.* **2019**, *96*, 329–335. [CrossRef]
7. Liu, S.; Zhang, S.; He, X.; Li, J. Efficiency change in North-East China agricultural sector: A DEA approach. *Agric. Econ.* **2015**, *61*, 522–532. [CrossRef]
8. Pang, J.; Chen, X.; Zhang, Z.; Li, H. Measuring eco-efficiency of agriculture in China. *Sustainability* **2016**, *8*, 398. [CrossRef]
9. Baráth, L.; Fertő, I. Productivity and convergence in European agriculture. *J. Agric. Econ.* **2017**, *68*, 228–248. [CrossRef]
10. Toma, E.; Dobre, C.; Dona, I.; Cofas, E. DEA applicability in assessment of agriculture efficiency on areas with similar geographically patterns. *Agric. Agric. Sci. Procedia* **2015**, *6*, 704–711. [CrossRef]
11. Raheli, H.; Rezaei, R.M.; Jadidi, M.R.; Mobtaker, H.G. A two-stage DEA model to evaluate sustainability and energy efficiency of tomato production. *Inf. Process. Agric.* **2017**, *4*, 342–350. [CrossRef]
12. Cardozo, N.P.; de Oliveira Bordonal, R.; La Scala, N., Jr. Sustainable intensification of sugarcane production under irrigation systems, considering climate interactions and agricultural efficiency. *J. Clean. Prod.* **2018**, *204*, 861–871. [CrossRef]
13. Grados, D.; Schrevens, E. Multidimensional analysis of environmental impacts from potato agricultural production in the Peruvian Central Andes. *Sci. Total Environ.* **2019**, *663*, 927–934. [CrossRef] [PubMed]
14. Singbo, A.G.; Lansink, A.O.; Emvalomatis, G. Estimating shadow prices and efficiency analysis of productive inputs and pesticide use of vegetable production. *Eur. J. Oper. Res.* **2015**, *245*, 265–272. [CrossRef]
15. Rajendran, S.; Afari-Sefa, V.; Karanja, D.K.; Musebe, R.; Romney, D.; Makaranga, M.A.; Samali, S.; Kessy, R.F. Technical efficiency of traditional African vegetable production: A case study of smallholders in Tanzania. *J. Dev. Agric. Econ.* **2015**, *7*, 92–99.
16. Ajekiigbe, N.; Ayanwale, A.; Oyedele, D.; Adebooye, O. Technical efficiency in production of underutilized indigenous vegetables. *Int. J. Veg. Sci.* **2018**, *24*, 193–201. [CrossRef]
17. Duan, N.; Guo, J.P.; Xie, B.C. Is there a difference between the energy and CO_2 emission performance for China's thermal power industry? A bootstrapped directional distance function approach. *Appl. Energy* **2016**, *162*, 1552–1563. [CrossRef]
18. Riccardi, R.; Oggioni, G.; Toninelli, R. Efficiency analysis of world cement industry in presence of undesirable output: Application of data envelopment analysis and directional distance function. *Energy Policy* **2012**, *44*, 140–152. [CrossRef]
19. Choi, Y.; Zhang, N.; Zhou, P. Efficiency and abatement costs of energy-related CO_2 emissions in China: A slacks-based efficiency measure. *Appl. Energy* **2012**, *98*, 198–208. [CrossRef]
20. Lee, T.; Yeo, G.T.; Thai, V.V. Environmental efficiency analysis of port cities: Slacks-based measure data envelopment analysis approach. *Transp. Policy* **2014**, *33*, 82–88. [CrossRef]
21. Li, L.B.; Hu, J.L. Ecological total-factor energy efficiency of regions in China. *Energy Policy* **2012**, *46*, 216–224. [CrossRef]

22. Zhang, N.; Zhou, P.; Kung, C.C. Total-factor carbon emission performance of the Chinese transportation industry: A bootstrapped non-radial Malmquist index analysis. *Renew. Sustain. Energy Rev.* **2015**, *41*, 584–593. [CrossRef]
23. Zhou, G.; Chung, W.; Zhang, Y. Measuring energy efficiency performance of China's transport sector: A data envelopment analysis approach. *Expert Syst. Appl.* **2014**, *41*, 709–722. [CrossRef]
24. Emrouznejad, A.; Yang, G.L. CO_2 emissions reduction of Chinese light manufacturing industries: A novel RAM-based global Malmquist–Luenberger productivity index. *Energy Policy* **2016**, *96*, 397–410. [CrossRef]
25. Wu, J.; Lv, L.; Sun, J.; Ji, X. A comprehensive analysis of China's regional energy saving and emission reduction efficiency: From production and treatment perspectives. *Energy Policy* **2015**, *84*, 166–176. [CrossRef]
26. Fei, R.; Lin, B. Energy efficiency and production technology heterogeneity in China's agricultural sector: A meta-frontier approach. *Technol. Forecast. Soc. Chang.* **2016**, *109*, 25–34. [CrossRef]
27. Fei, R.; Lin, B. The integrated efficiency of inputs–outputs and energy–CO_2 emissions performance of China's agricultural sector. *Renew. Sustain. Energy Rev.* **2017**, *75*, 668–676. [CrossRef]
28. Li, N.; Jiang, Y.; Mu, H.; Yu, Z. Efficiency evaluation and improvement potential for the Chinese agricultural sector at the provincial level based on data envelopment analysis (DEA). *Energy* **2018**, *164*, 1145–1160. [CrossRef]
29. Le, T.L.; Lee, P.P.; Peng, K.C.; Chung, R.H. Evaluation of total factor productivity and environmental efficiency of agriculture in nine East Asian countries. *Agric. Econ.* **2019**, *65*, 249–258.
30. Wang, F.; Yu, C.; Xiong, L.; Chang, Y. How can agricultural water use efficiency be promoted in China? A spatial-temporal analysis. *Resour. Conserv. Recycl.* **2019**, *145*, 411–418. [CrossRef]
31. Lampe, H.W.; Hilgers, D. Trajectories of efficiency measurement: A bibliometric analysis of DEA and SFA. *Eur. J. Oper. Res.* **2015**, *240*, 1–21. [CrossRef]
32. Ahmad, Z.; Jun, M. Agricultural Production Structure Adjustment Scheme Evaluation and Selection Based on DEA Model for Punjab (Pakistan). *J. Northeast Agric. Univ. (Engl. Ed.)* **2015**, *22*, 87–91. [CrossRef]
33. e Souza, G.D.S.; Gomes, E.G. Management of agricultural research centers in Brazil: A DEA application using a dynamic GMM approach. *Eur. J. Oper. Res.* **2015**, *240*, 819–824. [CrossRef]
34. Zhu, N.; Hougaard, J.L.; Ghiyasi, M. Ranking production units by their impact on structural efficiency. *J. Oper. Res. Soc.* **2019**, *70*, 783–792. [CrossRef]
35. Rahman, M.T.; Nielsen, R.; Khan, M.A.; Asmild, M. Efficiency and production environmental heterogeneity in aquaculture: A meta-frontier DEA approach. *Aquaculture* **2019**, *509*, 140–148. [CrossRef]
36. Charnes, A.; Cooper, W.W.; Rhodes, E. Measuring the efficiency of decision making units. *Eur. J. Oper. Res.* **1978**, *2*, 429–444. [CrossRef]
37. Banker, R.D.; Charnes, A.; Cooper, W.W. Some models for estimating technical and scale inefficiencies in data envelopment analysis. *Manag. Sci.* **1984**, *30*, 1078–1092. [CrossRef]
38. Fenyves, V.; Tarnóczi, T.; Zsidó, K. Financial performance evaluation of agricultural enterprises with DEA method. *Procedia Econ. Financ.* **2015**, *32*, 423–431. [CrossRef]
39. Adesemoye, A.O.; Kloepper, J.W. Plant–microbes interactions in enhanced fertilizer-use efficiency. *Appl. Microbiol. Biotechnol.* **2009**, *85*, 1–12. [CrossRef]
40. Tilman, D.; Cassman, K.G.; Matson, P.A.; Naylor, R.; Polasky, S. Agricultural sustainability and intensive production practices. *Nature* **2002**, *418*, 671–677. [CrossRef]
41. Idrees, M.; Batool, S.; Hussain, Q.; Ullah, H.; Al-Wabel, M.I.; Ahmad, M.; Kong, J. High-efficiency remediation of cadmium (Cd2+) from aqueous solution using poultry manure–and farmyard manure–derived biochars. *Sep. Sci. Technol.* **2016**, *51*, 2307–2317. [CrossRef]
42. Vymazal, J.; Březinová, T. The use of constructed wetlands for removal of pesticides from agricultural runoff and drainage: A review. *Environ. Int.* **2015**, *75*, 11–20. [CrossRef] [PubMed]

Article

A Risk Aversion Dispatching Optimal Model for a Micro Energy Grid Integrating Intermittent Renewable Energy and Considering Carbon Emissions and Demand Response

Xiaoxu Fu [1], Wei Fan [1,2,*], Hongyu Lin [1,2], Nan Li [3], Peng Li [4], Liwei Ju [1,2] and Feng'ao Zhou [1,2]

1 School of Economics and Management, North China Electric Power University, Beijing 102206, China; fuxiaoxu2008@126.com (X.F.); hone@ncepu.edu.cn (H.L.); hdlw_ju@ncepu.edu.cn (L.J.); zhoufengao@ncepu.cn (F.Z.)

2 Beijing Energy Development Research Base, Beijing 102206, China

3 State Grid Qinghai Electric Economical Research Institute, Beijing 100045, China; lnan0514@qh.sgcc.com.cn

4 State Grid Henan Economical Research Institute, Zhengzhou 450052, China; hdlp0830@163.com

* Correspondence: fanwei@ncepu.edu.cn; Tel.: +86-159-3212-9195

Received: 4 November 2019; Accepted: 25 November 2019; Published: 3 December 2019

Abstract: This paper focuses on an optimal schedule for a micro energy grid considering the maximum total carbon emission allowance (MTEA). Firstly, the paper builds an energy devices operation model and demand response (DR) model. Secondly, to maximize the economical operation revenue, the basic scheduling model for the micro energy grid is constructed. Thirdly, the conditional value at risk method and robust stochastic theory are introduced to describe the uncertainty of wind power, photovoltaic power, and load, and a risk aversion model is proposed. Finally, this paper selects the Xinxiang active distribution network demonstration project in Jining, China as an example. The results show that: (1) a micro energy grid can make the most use of the complementary characters of different energy sources to meet different energy demands for electricity, heat, cold, and gas; (2) the risk aversion scheduling model can represent the influence of uncertainty variables in objective functions and constraints, and provide a basis for decision makers who have different attitudes; and (3) DR can smooth the energy load curves. MTEA can enhance the competitiveness of the clean energy market, thus promoting the grid-connected generation of clean energy. Therefore, the risk aversion model can maximize the economic benefits and provide a basis for decision makers while rationally controlling risks.

Keywords: micro energy grid; distributed energy; uncertainty; risk aversion; demand response

---

1. Introduction

In recent years, environmental pollution and the current depletion of fossil energy have become more and more serious. The energy structure should be transformed and upgraded urgently. It is the current trend of the energy field to pursue the efficient use, clean environmental protection, and sustainable development of energy [1]. As a further extension of micro power grids, a micro energy grid could realize the coordinated planning and unified scheduling of multiple energy sources (electricity, heat, cold, and gas, etc.) through new energy technologies and internet technologies, which could effectively improve energy efficiency while achieving local production and consumption of energy [2]. The China Development and Reform Commission also put forward the Guiding Opinions on the Development of "Internet + Promoting" Intelligent Energy, pointing out that it is essential to strengthen the construction of multi-energy synergistic integrated energy networks, and the coupling interactions and comprehensive utilization of different energy types, such as electricity, gas, heating, and cooling [3].

In recent years, micro energy grids have attracted widespread attention and practical development. The smart polygeneration microgrid (SPM) project of the University of Genoa for the Savona campus involves multiple electric, thermal, and combined heating and power generators [4]. Cassel University integrated a wind turbine, PV, biogas power station and hydro power plant into a micro energy grid (MEG) [5]. In 2014, the Xiaozhongdian wind-photovoltaic (PV)-hydro distributed demonstration project of the China National Electric Power Group Corporation successfully connected to the grid in Yunnan province [6]. The La Plata University in Finland is developing a 40 m diameter straight-leaf vertical shaft wind turbine that drives an oil-fired heating system for greenhouse heating [7]. The pilot projects of wind power heating in Linxi County and Zha'rutqi of the Inner Mongolia Autonomous Region have been operational for three years, and the annual consumption of wind power is approximately 149 million kW·h [8].

The operation mechanism and scheduling operation of micro energy grids have always been hotspots for research, both at home and abroad. On the premise of meeting the security constraints, the micro energy grid uses different objectives to rationally arrange the operation of energy equipment within the micro energy grid. Zhang et al. [9] constructed a micro energy grid operation optimization mode taking the minimum daily operating cost as the target. Du et al. [10] introduced modeling, planning and optimization methods for a regional integrated energy system. Luo [11] established a model for minimizing the sum of various costs and analyzed the integrated energy system with P2G. The above reference considered the objectives and related constraints of the micro energy grid, but the proposed models were used to analyze the uncertainty of the distributed energy. In fact, the uncertainty will affect the coordinated supply of multi-energy loads, such as electricity, heating, cooling, and gas, in the grid.

The uncertainty of distributed energy in a micro-energy network is mainly reflected in the volatility of wind power plants (WPP) and PV output power. A key issue for MEG operation is how to use units, storage devices, electric vehicles, and load, to balance random changes of wind and solar units, to guarantee the steady output of the MEG. Peik-Herfeh et al. [12] used the two-point estimation theory to take an estimate point on both sides of the forecast value to represent the unit output variability. Yang et al. [13] gained distribution parameters based on the characteristic that the wind speed obeys a Weibull distribution. Zamani et al. [14] used a stochastic program to handle electricity price uncertainty and studied a virtual power plant bidding model considering the uncertainty. Tan et al. [15] constructed an economical dispatching model considering the output power volatility of clean energy, based on a chance-constrained program. The above related research mainly considered uncertain variables as random variables and constructed stochastic dispatching optimization models by using stochastic modeling methods, such as stochastic programming and robust optimization. The validity of the method was verified by actual cases.

The above research shows that the existing research about micro energy grids focuses on system modeling, optimal operation, and uncertainty analysis. However, in uncertainty analysis, a stochastic program describes the uncertainty with stochastic variables. Based on the probability distribution of stochastic variables, system constraints are described as opportunistic constraints [16]. However, whether DERs with a small capacity and large quantity have statistical characters needs to be checked. The accurate establishment of information collection and a probability distribution function is difficult. The optimal solution sets of robust optimization have a certain degree of restraint on the effects. Adjusting the size of a coefficient can determine the dispatching scheme, which can restrain the influence of uncertainty to different degrees [17]. At the same time, the existing research results are more focused on the processing of constraints with uncertain variables, lacking consideration of the objective function processing method with uncertain variables. Conditional value at risk (CVaR) can quantitatively represent the uncertainty risk of the objective function. By combining it with robust stochastic optimization theory, a relatively complete risk decision model can be constructed. According to the above analysis, an optimal dispatching model for a MEG is put forward. The main contributions are as follows:

- This paper designs a novel structure for a micro energy grid optimal operation containing production devices, conversion devices, and storage devices, depending on the demand response and carbon emissions. The effects of price-based demand response (PBDR) and incentive-based demand response (IBDR) on different energy load curves are compared and analyzed. The optimization effect of the maximum total emission allowance (MTEA) on the MEG operation is also analyzed.
- This paper establishes a basic dispatching model for the micro energy grid considering different constraints, selecting maximizing operating revenue as the objective function of the MEG operation with the constraints of load power balance, equipment operation, DR operation, maximum carbon emissions, and rotating standby.
- This paper puts forward a risk aversion model for the micro energy grid on the basis of the CVaR method and robust stochastic optimization theory. The CVaR method mainly describes the influence of uncertainty factors of the objective function, and robust stochastic optimization theory converts constraints with uncertainty variables to provide an optimal basis for decision makers.

The structure of the paper is as follows: Section 2 designs a core structure for the micro energy grid and establishes an operation model of the equipment and DR operation model. Section 3 constructs the basic scheduling model of the micro energy grid without considering uncertainty, which takes maximizing the operational benefit as the optimization objective. Furthermore, a risk aversion model of the micro energy grid is established on the basis of CVaR and robust stochastic optimization theory in Section 4. Finally, Section 5 selects the China Jining Xinxiang Active Distribution Network Demonstration Project as an example object to verify the effectiveness and applicability of the model. Section 6 outlines the contributions and conclusions.

2. MEG Description and Output Model

This paper designs the basic structure of the micro energy grid and builds the energy devices operation model and demand response (DR) model.

2.1. Structure Description

The micro energy grid includes energy production, conversion, storage equipment, and energy consumers. Energy production equipment includes the wind power plant (WPP), photovoltaic power generation (PV) and conventional gas turbine (CGT). Energy conversion equipment includes power-to-gas (P2G), power to heating (P2H), heating to cooling (H2C), and power to cooling (P2C). Energy storage equipment includes a gas storage tank (GS), power storage battery (PS), heat storage tank (HS), and cold storage tank (CS). At the same time, in order to motivate users under a flexible load to participate in the optimal operation of the MEG, the price-based demand response (PBDR) and incentive-based demand response (IBDR) are implemented. The former is used to guide terminal users to use electricity reasonably through the performance of a differentiated time-of-use price, while the IBDR is mainly used to provide an emergency energy supply to the MEG. In the MEG, power electronics converters are required for some voltage adaptations—for instance, the PV generation—and the type of grid is AC. All the power from different devices must be converted into AC. Figure 1 shows the structure of the micro energy grid.

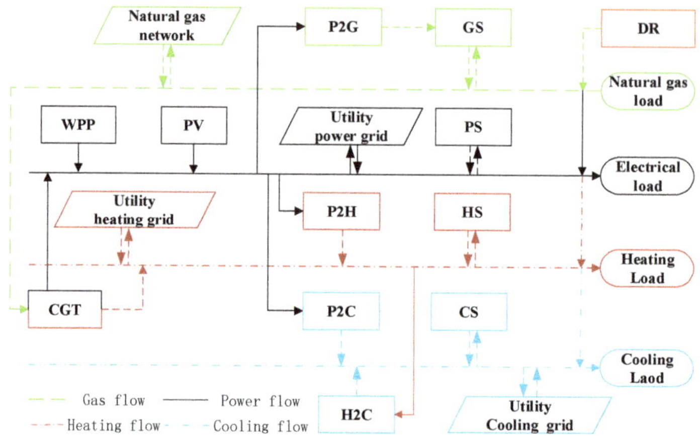

Figure 1. Structure of the micro energy grid.

According to Figure 1, MEGs are connected with an upper power grid, gas grid, heat grid and cold grid. An MEG can interact with upper energy grids. When there is excess energy in MEGs, it can be sold to upper energy grids to obtain economic benefits. Conversely, when the internal energy of an MEG is insufficient, energy will be bought from superior energy grids. Because the real-time price of different energy is different, the MEG will reasonably choose the energy sale or purchase scheme according to the different prices of electricity, heating, cooling, and gas, so as to achieve the goal of maximizing operating benefit. However, the output power of WPP and PV has strong uncertainty and high environmental economics. Therefore, maintaining the balance of the operational benefit and risk will be a key issue for formulating the optimal operation plan of the MEG.

2.2. Energy Devices Operation Model

The energy devices operation model includes the energy production (EP) operation model, energy conversion (EC) operation model and energy storage (ES) operation model.

2.2.1. EP Operation Model

EP equipment contains WPP, PV, and CGT. WPP is decided by the wind speed and PV is decided by the solar radiation, while CGT mainly produces heat vapor by consuming natural gas for the power supply and heat supply. Generally speaking, CGT includes two types: following thermal load (FTL) and following electric load (FET). This paper sets CGT to operate in FTL mode. The output models of different devices are as follows:

(1) WPP operation model

Wind power generation is determined by the wind speed. The WPP generation output is calculated according to the real-time wind speed and fan parameters:

$$g^{*}_{WPP,t} = \begin{cases} 0, & v_t \leq v_{in} \\ \dfrac{v - v_{in}}{v_R - v_{in}} g_R, & v_{in} \leq v_t \leq v_R \\ g_R, & v_R \leq v_t \leq v_{out} \\ 0, & v_t \geq v_{out} \end{cases} \tag{1}$$

(2) PV output model

PV power generation is determined by the solar photovoltaic radiation. On the basis of photoelectric conversion principles, PV output power is calculated as follows:

$$g^*_{PV,t} = \eta_{PV} \times S_{PV} \times \theta_t \tag{2}$$

(3) CGT output model

When natural gas enters the gas turbine combustion chamber, it generates hot steam to drive the turbine to work by combustion, and the exhausted hot gas can provide heating energy to users through a heat recovery device. As for the principle of the gas turbine, existing research is very mature [18]. This paper directly quotes power and heating supply models in reference [19]. The specific model is as follows:

$$g_{CGT,t} = V_{CGT,t} H_{ng} \eta_{CGT,t} \tag{3}$$

$$Q_{CGT,t} = V_{CGT,t}(1 - \eta_{CGT,t} - \eta_{loss})\eta_{hr} \tag{4}$$

2.2.2. EC Operation Model

EC mainly includes power-to-gas (P2G), power-to-cooling (P2C), power-to-heat (P2H), and heat-to-cooling (H2C) convertors. The operating power model of various types of EC equipment has already been constructed in our previous works [19].

(1) P2G device

P2G can utilize the curtailment output of WPP and PV to convert CO_2 into CH_4, which realizes the interconnection of the power grid and gas network. P2G is divided into two processes: electrolysis and methanation. Electrolysis uses excess electricity to generate hydrogen by electrolyzing water and injecting it into the natural gas pipeline or a storage device. On the basis of electrolysis, the methanation process uses the hydrogen to react with carbon dioxide to form methane and water under the action of a catalyst. Figure 2 shows the technical principles of P2G-GS operation.

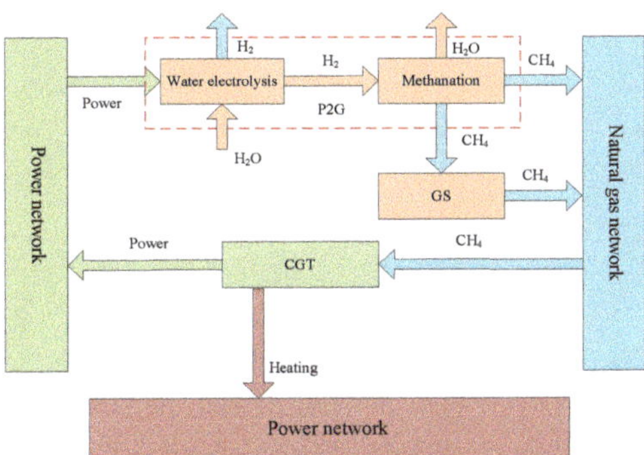

Figure 2. Technical principles of P2G-GS operation.

The CH4 generated by P2G can be injected into a natural gas network, conventional gas turbines (CGT), gas boilers (GB), and a gas storage tank (GST). The detailed operation model is as follows:

$$V_{P2G,t} = g_{P2G,t}\eta_{P2G}/H_{ng} \tag{5}$$

Furthermore, the proportions of natural gas generated by P2G and injected into the CGT, GS and natural gas network are $\eta^{CGT}_{P2G,t}$, $\eta^{GS}_{P2G,t}$, and $\eta^{NG}_{P2G,t}$.

(2) Other devices

There is a certain efficiency in the conversion between different energies. Although the efficiency is not constant, it usually does not change much when the equipment runs steadily. According to reference [20], it can be regarded as constant. The mathematical model of the energy conversion unit is expressed as:

$$\begin{bmatrix} V_{P2G,t} \\ Q_{P2C,t} \\ Q_{P2H,t} \\ Q_{H2C,t} \end{bmatrix} = \begin{bmatrix} g_{P2G,t} & 0 & 0 & 0 \\ 0 & g_{P2C,t} & 0 & 0 \\ 0 & 0 & g_{P2H,t} & 0 \\ 0 & 0 & 0 & Q_{H2C,t} \end{bmatrix} \begin{bmatrix} \eta_{P2G} \\ \eta_{P2C} \\ \eta_{P2H} \\ \eta_{H2C} \end{bmatrix} \tag{6}$$

2.2.3. ES Operation Model

ES mainly includes power storage (PS), heat storage (HS), cooling storage (CS), and gas storage (GS). According to the relevance between the energy supply and energy demand, different types of energy storage equipment can use their own energy storage facilities to store and release energy. The specific operation model is as follows:

$$S_{ES,t} = \left(1 - \eta_{ES,t}^{loss}\right) S_{ES,t-1} + \left[ES_t^{input} \eta_{ES}^{input} - ES_t^{output} / \eta_{ES}^{output}\right] \tag{7}$$

2.3. DR Operation Model

The DR operation model mainly includes the price-based demand response (PBDR) operation model and the incentive-based demand response (IBDR) operation model.

2.3.1. PBDR Operation Model

The PBDR guides the terminal users to use energy reasonably by implementing a peak-valley time-of-use price, which can realize "peak-cutting and valley-filling". According to microeconomic principles, the PBDR can be calculated by the price elasticity of demand [10], as follows:

$$E_{st} = \frac{\Delta L_s / L_s^0}{\Delta P_t / P_t^0} \begin{cases} E_{st} \leq 0, s = t \\ E_{st} \geq 0, s \neq t \end{cases} \tag{8}$$

where when $s = t$, $E_{st}^{e,h,c}$ is self-elasticity; when $s \neq t$, $E_{st}^{e,h,c}$ is cross-elasticity. Correspondingly, the change of energy demand load after PBDR is calculated as follows:

$$L_t^{after} = L_t^{before} \times \left\{ E_{tt} \times \frac{\left[P_t^{after} - P_t^{before}\right]}{P_t^{before}} + \sum_{\substack{s=1 \\ s \neq t}}^{24} E_{st} \times \frac{\left[P_s^{after} - P_s^{before}\right]}{P_s^{before}} \right\} \tag{9}$$

where ΔL_t^{after} indicates the amount of load change after PBDR.

2.3.2. IBDR Operation Model

The IBDR is signed by the MEG operator and terminal users in advance. When emergency energy demand occurs, the operator can directly control the energy usage behavior of the terminal users and give some financial compensation. According to reference [10], demand response providers (DRPs) are involved in the demand response stage by stage, mainly on the basis of different energy prices. Therefore, the operation of DRPs should meet the following principles:

$$D_i^{k,j,\min} \leq \Delta L_{i,t}^{k,j} \leq D_{i,t}^{k,j}, j = 1 \tag{10}$$

$$0 \leq \Delta L_{i,t}^{k,j} \leq \left(D_{i,t}^{k,j} - D_{i,t}^{k,j-1}\right), j = 2, 3, \ldots, J \tag{11}$$

$$\Delta L_t^{k,IB} = \sum_{i=1}^{I}\sum_{j=1}^{J}\Delta L_{i,t}^{k,j} \tag{12}$$

3. Basic Dispatching Model of the Micro Energy Grid

The section covers the construction of the basic scheduling optimal model for the micro energy grid, aimed at maximizing the economic revenue of operation considering the constraints of the energy power balance, device operation, and system reserve balance.

3.1. Objective Functions

The micro energy grid is mainly supplied by WPP, PV and CGT. Through conversion and storage equipment, it can meet electricity, heating, cooling and gas load demands together. WPP and PV have the characteristics of a low marginal cost of power generation and zero emissions of pollution. This paper chooses maximizing the operating revenue as the operational optimization goal of the MEG. The specific objective function is as follows:

$$\max R = \sum_{t=1}^{24}\left\{R_{EP,t} + R_{EC,t} + R_{ES,t} + R_{DR,t} + R_{Carbon,t}\right\} \tag{13}$$

The remaining carbon emission rights can be traded externally when the carbon emissions of the MEG are less than the maximum emission trade allowance (META). For EP, the operating revenue is equal to the energy supply income minus the energy supply cost. The energy supply revenue is equal to the product of the electricity quantity and its price. The marginal cost of WPP and PV is basically close to zero. The energy supply cost of CGT includes the fuel consumption cost and start-stop cost, which is calculated as follows:

$$\begin{aligned} C_{CGT,t} &= C_{CGT,t}^{fuel} + C_{CGT,t}^{sd} \\ &= \left\{\begin{array}{c} a\left(g_{CGT,t} + \theta_h^e Q_{CGT,t}\right)^2 + \\ b\left(g_{CGT,t} + \theta_h^e Q_{CGT,t}\right) + c \end{array}\right\} + \left\{\begin{array}{c} \left[\mu_{CGT,t}^u\left(1 - \mu_{CGT,t-1}^u\right)\right]C_{CGT,t}^u + \\ \left[\mu_{CGT,s}^d\left(1 - \mu_{CGT,s+1}^d\right)\right]C_{CGT,s+1}^d \end{array}\right\} \end{aligned} \tag{14}$$

For EC and ES, the operating revenue is equal to the energy output (energy release) income minus the energy input (energy storage) cost. The calculation is as follows:

$$R_{EC(S),t} = Q_{EC(S),t}^{output} p_{EC(S)}^{output} \eta_{EC(S)}^{output} - Q_{EC(S),t}^{input} p_{EC(S)}^{input} / \eta_{EC(S)}^{input} \tag{15}$$

For DR, operation revenue includes PBDR income and IBDR income. The former can increase the energy supply, while the latter is mainly to reduce the penalty cost of a power shortage. The calculation is as follows:

$$R_{DR,t} = R_{PBDR,t} + R_{IBDR,t} = \sum_{t=1}^{24}\left[p_t^{before}L_t^{before} - p_t^{after}L_t^{after}\right] + \sum_{k\in\left\{\substack{power,heating,\\ cooling,gas}\right\}}\left\{\Delta L_t^{k,IB}p_t^{k,IB} - \Delta L_t^{k,shortage}p_t^k\right\} \tag{16}$$

$$R_{Carbon,t} = \left\{\left[a_{CGT} + b_{CGT}\left(g_{CGT,t} + \theta_h^e Q_{CGT,t}\right) + c_{CGT}\left(g_{CGT,t} + \theta_h^e Q_{CGT,t}\right)^2\right] - Q_{MTEA,t}\right\}p_{Carbon,t} \tag{17}$$

3.2. Condition Constraints

To achieve an optimal supply of electricity, heating and cooling, it is necessary to comprehensively consider the energy supply and demand balance constraint, EP, EC, ES operation constraint, and system rotation reserve constraint of MEGs. The specific constraints are as follows:

(1) Energy supply and demand balance constraint

$$g_{CGT,t} + g_{WPP,t} + g_{PV,t} + g_{PS,t}^{output} + g_{P2G,t}^{output} + g_{UPG,t} = L_t^e + g_{ES,t}^{input} + g_{P2G,t}^{input} + g_{P2H,t}^{input} + g_{P2C,t}^{c,input} + \Delta L_t^{p,PB} + \Delta L_t^{p,IB} \tag{18}$$

$$Q_{CGT,t} + Q_{P2H,t}^{output} + Q_{HS,t}^{output} + Q_{UHG,t} = L_t^h + Q_{HS,t}^{input} + Q_{H2C,t}^{input} + \Delta L_t^{h,PB} + \Delta L_t^{h,IB} \tag{19}$$

$$Q_{P2C,t}^{output} + Q_{H2C,t}^{output} + Q_{CS,t}^{output} + Q_{UCG,t} = L_t^c + Q_{CS,t}^{input} + \Delta L_t^{c,PB} + \Delta L_t^{c,IB} \tag{20}$$

(2) CGT operation constraints

For CGT, the relevance of the power generation and heating supply power is called the "electrical heating character." Under a given thermal power, the power generated has some adjustability. This is because under a given amount of steam extracted, CGT adjusts the output power of the entire steam turbine by adjusting the amount of condensation steam to generate electricity. However, the larger the amount of steam extracted, the smaller the proportion of condensing steam required to generate electricity, so the adjustment range is smaller. The specific constraints are as follows:

$$\max\left\{g_{CGT}^{\min} - c_{\min} Q_{CGT}, c_m\left(Q_{CGT} - Q_{CGT}^0\right)\right\} \le g_{CGT} \le g_{CGT}^{\max} - c_{\max} Q_{CGT} \tag{21}$$

$$u_{CGT,t}\Delta g_{CGT}^{-} \le \left(g_{CGT,t} + \theta_h^e Q_{CGT,t}\right) - \left(g_{CGT,t-1} + \theta_h^e Q_{CGT,t-1}\right) \le u_{CGT,t}\Delta g_{CGT}^{+} \tag{22}$$

where c is the reduction of power caused by extra extraction of the unit heating supply when the steam inlet amount is constant. $c_m = \Delta g_{CGT}/\Delta Q_{CGT}$ is the elasticity coefficient of electricity power and heating power under backpressure operation. Q_{CGT}^0 is a constant.

(3) EC operation constraints

EC includes P2H, P2C, H2C, and P2G. According to Equation (8), the energy conversion relevance of different devices can be established. Different energy conversion devices have their own power constraints; the details are as follows:

$$u_{P2G,t} V_{P2G,t}^{\min} \le V_{P2G,t} \le u_{P2G,t} V_{P2G,t}^{\max} \tag{23}$$

$$u_{EC,t}^{output} Q_{EC,t}^{output,\min} \le Q_{EC,t}^{output} \le u_{EC,t}^{output} Q_{EC,t}^{output,\max} \tag{24}$$

$$u_{EC,t}^{input} Q_{EC,t}^{input,\min} \le Q_{EC,t}^{input} \le u_{EC,t}^{input} Q_{EC,t}^{input,\max} \tag{25}$$

(4) ES operation constraints

ESD includes ES, HS, CS, and GS. Energy storage capacity constraints should also be considered when various types of energy storage equipment store or release energy. The specific constraints are as follows:

$$S_{ES,t}^{\min} \le S_{ES,t} \le S_{ES,t}^{\max} \tag{26}$$

$$u_{ES,t}^{output} Q_{ES,t}^{output,\min} \le Q_{ES,t}^{output} \le u_{ES,t}^{output} Q_{ES,t}^{output,\max} \tag{27}$$

$$u_{ES,t}^{input} Q_{ES,t}^{input,\min} \le Q_{ES,t}^{input} \le u_{ES,t}^{input} Q_{ES,t}^{input,\max} \tag{28}$$

(5) System reserve constraints

MEGs are set to operate according to the "following thermal load" mode, so some electrical load reserve capacity and cooling load reserve capacity need to be reserved. In addition, the randomness of WPP and PV also requires MEGs to reserve certain capacity. The specific constraints are as follows:

$$g_{MEG,t}^{p,\max} - g_{MEG,t}^{p} + g_{PS,t}^{output} + \left[L_t^{p,after} - L_t^{p,before}, 0\right]^{+} + \Delta L_t^{p,IB} \ge r_p L_t^p + r_{WPP}^{up} g_{WPP,t} + r_{PV}^{up} g_{PV,t} \tag{29}$$

$$g_{MEG,t}^{p} - g_{MEG,t}^{p,\min} + g_{PS,t}^{input} + \left[L_t^{p,after} - L_t^{p,before}, 0\right]^{-} \ge r_{WPP}^{dn} g_{WPP,t} + r_{PV}^{dn} g_{PV,t} \tag{30}$$

$$g_{MEG,t}^{c,\max} - g_{MEG,t}^{c} + g_{CS,t}^{output} + \left[L_t^{c,after} - L_t^{c,before}, 0\right]^+ \geq r_c L_t^c \tag{31}$$

(6) Other operation constraints

An MEG also needs to consider the operation constraints of PBDR and IBDR, including maximum output power constraints, start-stop time constraints, and uphill-downhill power constraints. CGT operation also needs to consider start-stop time constraints. The specific constraints are described in reference [10].

4. Risk Aversion Model of the Micro Energy Grid

To describe the uncertainty of the wind power plant (WPP), photovoltaic power generation (PV), and load, the conditional value at risk (CVaR) method and robust stochastic optimization theory are introduced to construct a risk aversion model for the micro energy grid in this section.

4.1. Uncertainty Factors Analysis

There are three uncertainty factors in the proposed MEG, which are $g_{WPP,t}$, $g_{PV,t}$, and L_t. Simulating the uncertainty is the key to formulating an optimal dispatching strategy for the MEG. Generally speaking, the load demand mainly consists of two parts: the forecast value and the forecast deviation. Considering that the forecast deviation obeys a normal distribution, load demand can be calculated as follows:

$$L_t = L_t^f + \Delta L_t^e \tag{32}$$

where if ΔL_t^e obeys $\Delta L_t^e \sim [0, \delta_{L,t}^2]$, δ is the load forecast standard deviation, then the load demand will obey $L_t \sim [L_t^f, \delta_{L,t}^2]$. In the proposed MEG, there are various flexible loads such as the electricity, heating, cooling and gas, and energy storage equipment, which can cope with load uncertainty, so this paper does not take the uncertainty of the load demand into account.

For WPP and PV, the uncertainty is mainly caused by the natural wind speed and photovoltaic radiation intensity. Simulating the natural wind and photovoltaic radiation intensity is the key to simulating uncertainty. Referring to [21], the Weibull and Beta distribution function can simulate the wind speed and photovoltaic radiation intensity respectively, as follows:

$$f(v) = \frac{\varphi}{\vartheta}\left(\frac{v}{\vartheta}\right)^{\varphi-1} e^{-(v/c)^{\varphi}} \tag{33}$$

$$f(\theta) = \begin{cases} \dfrac{\Gamma(\omega)\Gamma(\psi)}{\Gamma(\omega)+\Gamma(\psi)}\theta^{\omega-1}(1-\theta)^{\psi-1}, 0 \leq \theta \leq 1, \omega \geq 0, \psi \geq 0 \\ 0 \qquad , otherwise \end{cases} \tag{34}$$

This obtains the distribution function of the above uncertainty factors, which allows us to analyze the uncertainty factors of the MEG. This paper uses the conditional risk at value method and robust stochastic optimization theory to describe the uncertainty factors of the objective function and constraints, respectively, and constructs the risk avoidance optimization model, which provides a basis for decision makers who have different risk attitudes, so that they can properly formulate optimal scheduling strategies.

4.2. Risk Aversion Optimal Model

The conditional value at risk method is used to represent the uncertainty factors of the objective function in this paper. Compared with the traditional value at risk (VaR) method, it can represent a risk distribution situation at the exterior of the confidence level, which is helpful to overcome the limitation of VaR only being able to measure the risk under the confidence level but not at the tail. A detailed introduction of the CVaR method can be found in the author's paper [22]. This paper constructs an objective function with the CVaR method. The objective function is designed as follows:

$$F_\beta(E,\alpha) = \alpha + \frac{1}{1-\beta}\int\limits_{y\in R^m} (L(E,y)-\alpha)^+ p(y)dy \tag{35}$$

where α indicates the threshold value of risk determination. β indicates the objective function confidence of MEG operation. If Equation (37) achieves the minimum, it is the CVaR value. α indicates the VaR value. $L(E,y) = -R(E,y)$ indicates the loss function of MEG operation. $E^T = [E_{MEG,t}(1), E_{MEG,t}(2), \cdots, E_{MEG,t}(T)]$ indicates the decision vector, and $y^T = [g_{WPP,t}, g_{PV,t}, L_t]$ indicates the multivariate random vector. $R(E,y)$ indicates the income function of MEG operation. According to Equation (35), the risk caused by the uncertainty factor in the objective function can be described.

In order to describe the uncertainty factors in the constraints, the conventional constraints need to be transformed into stochastic constraints. This paper used robust stochastic optimization theory to set deviations of predicted power of WPP and PV as δ_{WPP} and δ_{PV}. Correspondingly, $g_{WPP,t}$ and $g_{PV,t}$ will fluctuate in intervals $[(1-\delta_{WPP,t})\cdot g_{WPP,t}, (1+\delta_{WPP,t})\cdot g_{WPP,t}]$ and $[(1-\delta_{PV,t})\cdot g_{PV,t}, (1+\delta_{PV,t})\cdot g_{PV,t}]$. In order to facilitate analysis, RE was introduced to represent WPP and PV, and $\delta_{RE,t}$ was introduced to represent $\delta_{WPP,t}$ and $\delta_{PV,t}$, then the uncertainty force was as follows:

$$-[g_{RE,t}(1-\varphi_{RE}) \pm e_{RE,t}\cdot g_{RE,t}] = -[g_{WPP,t}(1-\varphi_{WPP}) \pm \delta_{WPP,t}\cdot g_{WPP,t}] - [g_{PV,t}(1-\varphi_{PV}) \pm \delta_{PV,t}\cdot g_{PV,t}] \tag{36}$$

Further, since the WPP and PV uncertainty variables mainly appear in Equation (18), the system net load demand is set to be M_t which can be calculated by Equation (37) as follows:

$$M_t = -\left(L_t^e + g_{ES,t}^{input} + g_{P2G,t}^{input} + g_{P2H,t}^{input} + g_{P2C,t}^{c,input} + \Delta L_t^{p,PB} + \Delta L_t^{p,IB}\right) - g_{PS,t}^{output} - g_{P2G,t}^{output} - g_{CGT,t} \tag{37}$$

Equation (17) can be rewritten according to Equations (38) and (39), as follows:

$$-\left[g_{RE,t}(1-\varphi_{RE}) \pm e_{RE,t}\cdot g_{RE,t}\right] \le M_t \tag{38}$$

Referring to [23], for the flexibility of the stochastic model, auxiliary variables $\theta_{RE,t}(\theta \ge 0)$ and the robust coefficient $\Gamma_{RE}, \Gamma \in [0,1]$ were introduced to establish load supply and demand equilibrium constraints based on robust stochastic optimization theory, as follows:

$$-\left(g_{RE,t} + \delta_{RE,t} g_{RE,t}\right) \le -g_{RE,t} + \Gamma_{RE}\delta_{RE,t}\left|g_{RE,t}\right| \le -g_{RE,t} + \delta_{RE,t}\theta_{RE,t} \le M_t \tag{39}$$

Finally, according to the objective function in Equation (35), combined with the constraints of Equations (40) and (19)–(31), a risk aversion optimization model with the CVAR method and robust stochastic optimization theory could be established. The specific model is as follows:

$$\begin{array}{l} \min F_\beta(G,\alpha) = \alpha + \frac{1}{N(1-\beta)}\sum\limits_{k=1}^{N} z_k \\ s.t.\begin{cases} Eq.(19) - Eq.(31) \\ Eq.(37) - Eq.(39) \\ z_k = L(E,y) - \alpha \\ z_k \ge 0 \\ other\ constriants \end{cases} \end{array} \tag{40}$$

5. Example Analyses

This paper chose the Xinxiang Active Distribution Network Demonstration Project in Jining, China as the example object to analyze the validity and applicability of the proposed model.

5.1. Basic Data

The Xinxiang Active Distribution Network Demonstration Project includes 1000 kW PV, 800 kW WPP, and 2000 kW CGT. The cost parameters of CGT operation were selected according to reference [21]. In order to facilitate the model solution, the CGT operation cost function was divided into two stages, and the slope coefficients were 0.55 ¥/kW and 0.15 ¥/kW, respectively. In order to meet the demand of the multi-energy load of electricity, heat and cold, the demonstration project was equipped with P2H 1500 kW, P2C 1000 kW, H2C 1500 kW, and P2G 150 kW. The maximum energy storage and release power of PS and HS in this demonstration project was 200 kW and 300 kW, the storage capacity was 1000 kW h, the maximum energy storage and release power under CS was 300 kW, and the energy storage capacity was 1000 kW/h. In addition, the demonstration project was equipped with a 500 m^3 gas storage tank, and the maximum gas storage and supply power were both 150 kW. The energy efficiency of the different energy equipment was 96%. Considering that CGT operation can generate carbon emissions, this paper took 85% of the total carbon emissions from MEG operation as the MTEA, and chose a different energy load demand and real-time price of a typical load day as the basic data of the energy supply. Figure 3 is the electricity, heating, cooling, and gas load demand on a typical load day.

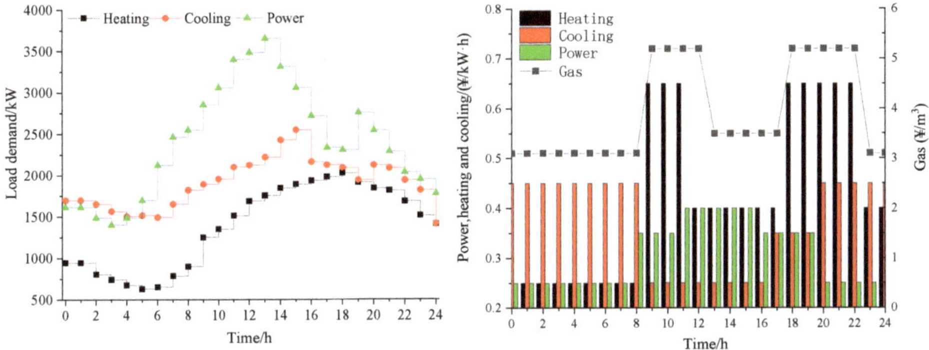

Figure 3. Electricity, heating, cooling, and gas load demand and real-time price on a typical load day.

Considering that EP, EC, ES and other energy equipment are dispatched and operated by the same entity, the WPP, PV, and CGT power generation price were 0.54 ¥/kW·h and 0.83 ¥/kW·h. The CGT power supply price and heating supply price were 0.35 ¥/kW·h and 0.25 ¥/kW·h. To promote the interconnection of WPP and PV, the EC electricity consumption price was set to 0.25 ¥/kW·h, and the EC heating consumption price was set to 0.2 ¥/kW·h when converting energy. The electricity, heating and cooling prices provided by the demonstration project were executed at real-time prices of different energy markets, as shown in Figure 3. To analyze the influence of uncertainty, the WPP and PV parameters were set according to reference [10], and 10 typical WPP and PV output scenarios were generated by scenario simulation and the reduction method proposed in reference [19]. The most possible scenario was chosen as the input data, and the prediction deviation was set to 2%. Figure 4 is the available output of WPP and PV on a typical day.

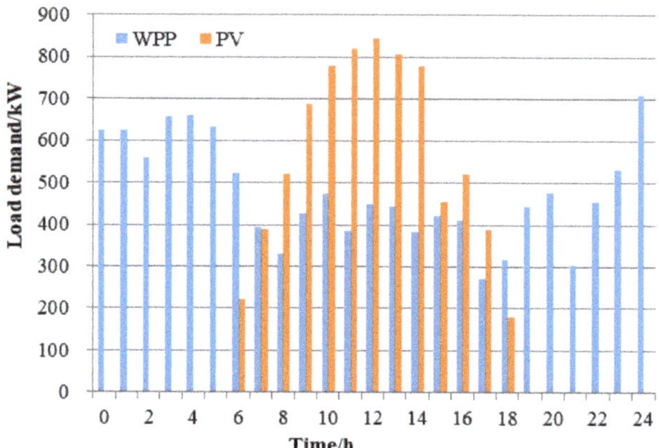

Figure 4. Available output of WPP and PV on a typical day. WPP: wind power plant; PV: photovoltaic power generation.

Thirdly, the model proposed in this paper includes 12 variables (four for production equipment, four for conversion equipment, and four for storage equipment), each of which includes 24 dimensions (24 h per scheduling period). To study the feasibility of CvaR and robust stochastic optimization theory in controlling WPP and PV uncertainty, confidence β and the robust coefficient Γ were both set to 0.8, and three simulation scenarios were compared for analysis:

- Case 1: Basic scenario, scheduling of the MEG without uncertainty. This scenario does not take the uncertainty of WPP and PV into account. It analyzes the operation character of different compositions of the MEG and focuses on the complementary effects among them.
- Case 2: scenario with CVaR, dispatching of the MEG with the CVaR method. The scenario focuses on WPP and PV output uncertainty. CVaR is applied to change the objective function. By comparing and analyzing the scheduling result of MEG operation under different values of confidence β, the effectiveness of CVaR in dealing with the uncertainty of WPP and PV is verified.
- Case 3: Comprehensive scenario, dispatching of MEG with the CVaR-robust method. The scenario constructs stochastic constraints using robust stochastic optimization theory, and discusses the MEG dispatching optimal strategy with different robustness and prediction accuracy values, and analyzes the validity of the CVaR-robust method.

Finally, in order to analyze the optimal effect of the demand response of MEG operation, peak, flat, and valley periods of different load types of electricity, heating, and cooling were divided according to reference [21], and corresponding time-of-use (TOU) prices were set. Among them, the peak period price increased by 25%, the valley period price decreased by 25%, and the flat period price remained unchanged. The price elasticity of electricity, heating, and cooling was selected according to reference [21]. For the incentive demand response, the up-rotating reserve price and the down-rotating reserve price of the electricity, heating and cooling reserve markets were 0.85 ¥/kW·h and 0.25 ¥/kW·h, 0.55 ¥/kW·h and 0.15 ¥/kW·h, and 0.45 ¥/kW·h and 0.15 ¥/kW·h, respectively. To avoid an excessive response of PBDR and IBDR, which would result in a "peak-to-valley upside down" phenomenon of the load curve, the total output of PBDR and IBDR should not exceed 10%, and the output power should not exceed 100 kW.

According to the above basic data, this paper used GAMs software to call CPLEX11.0 solver to solve the proposed model. The time for solving the above three simulation scenarios was less than 20 s by using the Lenovo IdeaPad 450 series notebook computers with 4 GB RAM and a Core T6500 processor.

5.2. Example Results

5.2.1. Scheduling Results of Case 1

This scenario did not take the uncertainty effects of WPP and PV into account when analyzing the complementary effect among different energy components. The main optimization objectives of this scenario include maximizing the operational benefits and minimizing carbon emissions. Under the mode of following thermal load, the heating and cooling load are mainly satisfied by CGT, while the residual heating and cooling load are mainly satisfied by excess electricity through P2H and P2C. Table 1 shows the scheduling results of the micro energy grid of Case 1.

Table 1. Scheduling results of the micro energy grid of Case 1.

	Energy Production/kW·h			Energy Conversion/kW·h			
	WPP	**PV**	**Conventional gas turbine (CGT)**	**Power-to-heat (P2H)**	**Power-to-gas (P2G)**	**Power-to-cooling (P2C)**	**Heat-to-cooling (H2C)**
Power	10,731	7027	35,674	−1967	−2623	−14,649	-
Heating	-	-	57,600	1869	-	-	−10,724
Cooling	-	-	-	-	-	43,946	14,477
	Energy Storage/kW·h			**Gas storage tank (GST)**	**Revenue/¥**	**Carbon/ton**	**CVaR/¥**
	Power storage (PS)	**Heat storage (HS)**	**Cooling storage (CS)**				
Power	±2400	-	-	1129	14,882.59	1.95	-
Heating	-	±2400	-	-	21,474.53	2.45	-
Cooling	-	-	±3600	-	11,180.93	-	-

According to Table 1, if uncertainty risks are not considered, more WPP and PV will be scheduled to satisfy the electricity load, and the remaining power will be converted into heating and cooling through P2H and P2C. Since the unit electricity energy can convert more cooling energy and obtain higher energy supply benefits, 14,649 kW·h electricity energy is converted into cooling energy. Since the cooling load is mainly converted by P2C and H2C, and there is no direct supply of cooling source, and the scheduled power of CS is higher than PS and HS at ±3600 kW. P2G can realize the cascade supply of electricity–gas–heating–cooling to obtain higher economic benefits by converting electricity energy into CH_4, which will supply electricity and heating in CGT. Through the complementary operation of EP, EC, ES and other different energy components, the MEG can realize the coordinated supply of electricity, heating, and cooling, and the economic benefit of the MEG is 47,538.05¥. However, CGT and utility power grid (UPG) also generate 4.4 tons of carbon emissions. Figure 5 is the output distribution of the micro energy grid of Case 1.

According to Figure 5, the output distribution of micro energy grid at different times was analyzed. As far as electricity load is concerned, it was mainly satisfied by WPP, PV, and CGT, and PS stored electricity during valley times and released electricity during peak times. At the same time, since CGT operates in the mode of following thermal load, the MEG needed to buy electricity from UPG during peak times to satisfy the balance of power supply and demand. As far as heating load is concerned, CGT was the main source of heating, and the residual heating load is satisfied by P2H. HS stored heating during valley times and released heating during peak times. To satisfy the demand of cooling load, part of the heating entered H2C. As far as cooling load is concerned, P2C is the main source of cooling. This is because the unit electricity energy can be converted into more cooling energy. CS mainly stored cooling during valley times and released cooling during peak times. Through the coordinated operation of different energy components, the MEG was able to realize the optimal supply of electricity, heating, cooling, and other loads.

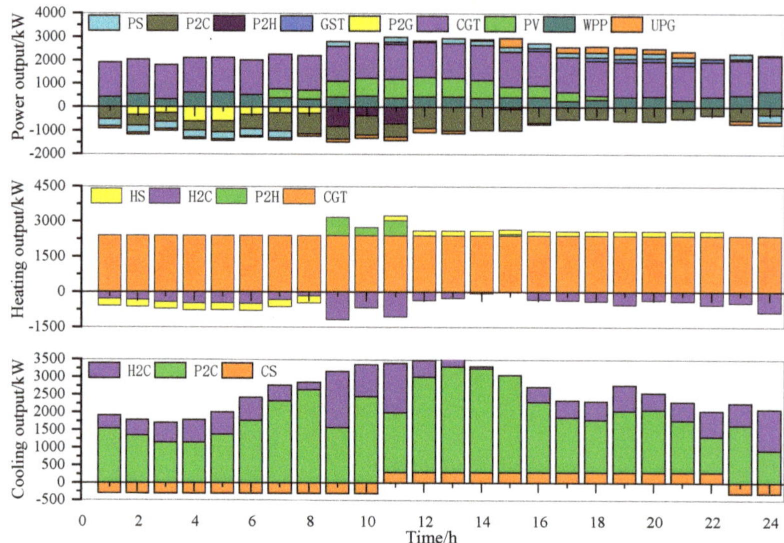

Figure 5. Output distribution of the micro energy grid of Case 1.

5.2.2. Scheduling Results of Case 2

This scenario analyzed the feasibility of the CVaR method when describing the uncertainty of WPP and PV. The CVaR method can transform the objective function using uncertain variables and construct the minimum risk objective function. Compared with Case 1, this scenario mainly considered three objective functions, which are maximizing economic benefits, minimizing operational risks, and minimizing carbon emissions, in order to discuss the MEG's optimal scheduling scheme. Table 2 shows the dispatching results of the micro energy grid of Case 2.

Table 2. Dispatching results of the micro energy grid of Case 2.

	Energy Production/kW·h			Energy Storage/kW·h			GST
	WPP	PV	CGT	PS	HS	CS	
Power	10,166	6657	35,674	±2400	-	-	814
Heating	-	-	57,600	-	±2400	-	-
Cooling	-	-	-	-	-	±3600	-
	Energy Conversion/kW·h				**Revenue/¥**	**Carbon/ton**	**CVaR/¥**
	P2H	**P2G**	**P2C**	**H2C**			
Power	−2231	−1223	−14,536	-	14,268.502	2.25	9702.581
Heating	2091	-	-	−10,947	21,555.33	2.68	9268.792
Cooling	-	-	43,607	14,778	11,324.708	-	3284.165

According to Table 2, the MEG scheduling results, considering uncertainty, were analyzed. When considering uncertainty, the MEG will reduce dispatching of WPP and PV and increase electricity bought from UPG to reduce uncertainty risk. Compared with Case 1, the WPP and PV grid-connected power were reduced by 585 kW·h and 370 kW·h, respectively, and the power provided by UPG increased by 848 kW·h, resulting in an increase of 0.53 tons of carbon emissions. When the CGT operates in the mode of following thermal load, the remaining power of WPP and PV is converted into heating to obtain more economic benefits. Overall, if taking the uncertainty of WPP and PV into account, MEG tends to buy electricity from UPG to avoid risk, and the remaining electricity is converted into heating to realize the coordinated supply of electricity, heating, and cooling, which brings about a lower risk value to MEG operation than electricity energy. Further, this case analyzed the MEG output scheme considering uncertainty. Figure 6 is the output distribution of the micro energy grid of Case 2.

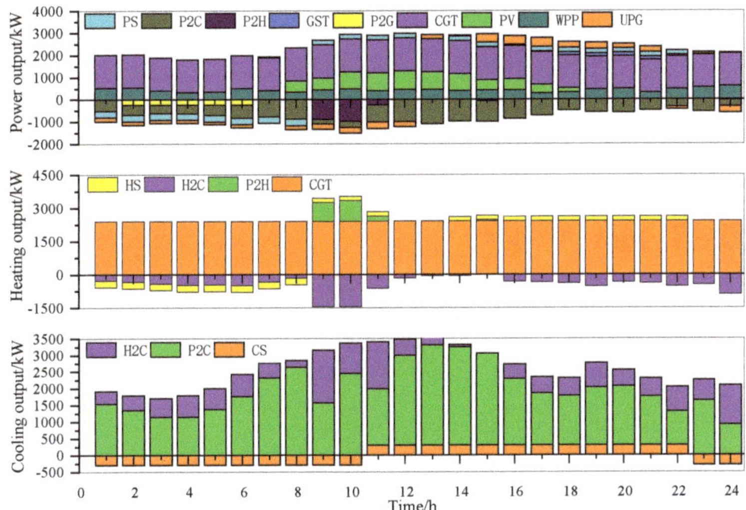

Figure 6. Output distribution of the micro energy grid of Case 2.

According to Figure 6, the MEG output distribution considering uncertainty was analyzed. When considering the uncertainty, the grid-connected power of WPP and PV decreased. At the same time, less power entered into P2G, which led to a decrease in GST power generation output, but the power supplied by UPG during peak times was significantly improved. From the perspective of different loads, since the surplus electricity is converted into heating, the heating energy provided by P2H increases, but the cooling load supply structure remains basically unchanged. This shows that when uncertainty is considered, the power output structure will change greatly, while the heating source and cooling source output structure will change relatively little. Further, the MEG scheduling results under different confidence levels were analyzed.

According to Table 3, it can be seen that as the increase of the confidence in grid-connected power of WPP and PV decreases, the power supply of UPG increases gradually. Correspondingly, the objective function values increase. This shows that benefits and risks are concomitant. If the decision makers expect to obtain high environmental and economic benefits, they have to bear more operational risks. On the contrary, if decision makers pursue the safe and steady operation of the MEG, they have to sacrifice some of the potential economic benefits. In general, CVaR can describe uncertainty, and provide a basis for decision makers who have different risk attitudes by setting a confidence level to establish the optimal scheduling strategy of the MEG.

Table 3. Scheduling results of the micro energy grid under different confidence levels.

	WPP	PV	CGT		UPG	Revenue/¥	Carbon/ton	CVaR/¥
			Power	Heating				
0	10,731.2	7027.15	35,673.6	57,600	±1647.83	47,538.051	4.4	0
0.5	10,668.26	6998.85	35,673.6	57,600	±1745.25	47,493.781	4.46	6152.16
0.6	10,542.38	6942.25	35,673.6	57,600	±1945.24	47,405.24	4.58	18,456.48
0.7	10,354.39	6799.776	35,673.6	57,600	±2207.525	47,276.89	4.755	20,356.01
0.8	10,166	6657	35,673.6	57,600	±2469.81	47,148.54	4.93	22,255.54
0.9	10,026.13	6377.767	35,673.6	57,600	±2685.12	46,391.147	5.26	23,729.51
1.0	9956	6238	35,673.6	57,600	±2845.85	46,012.45	5.43	24,466.5

5.2.3. Scheduling Results of Case 3

This scenario analyzed the validity of robust stochastic optimization theory in controlling uncertain variables of constraints. Compared with scenario 2, this scenario introduced both the CVaR method and robust stochastic optimization theory, which further strengthened the constraints on uncertainty. In order to avoid WPP and PV uncertainty risks, the MEG reduced scheduling of WPP and PV. The output of WPP and PV decreased to 9602 kW·h and 6287 kW·h, respectively. Correspondingly, the MEG operating income also reduced. Table 4 shows the dispatching results of the micro energy grid of Case 3.

Table 4. Scheduling results of micro energy grid of Case 3.

	Energy Production/kW·h			Energy Storage/kW·h			GST
	WPP	PV	CGT	PS	HS	CS	
Power	9602	6287	35,674	±2400	-	-	153
Heating	-	-	57,600	-	±2400	-	-
Cooling	-	-	-	-	-	±3300	-
	Energy Conversion/kW·h				**Revenue/¥**	**Carbon/ton**	**CVaR/¥**
	P2H	**P2G**	**P2C**	**H2C**			
Power	−73	−230	−16,091	-	13,398.987	2.47	8709.342
Heating	70	-	-	−7586	21,464.787	2.83	9229.858
Cooling	-	-	48,273	10,241	11,581.694	-	4053.593

Furthermore, the output distribution of the micro energy grid at different times was analyzed. Compared with Figure 5, when considering robust stochastic optimization theory, the WPP and PV uncertain variables in the constraints could be described. The corresponding output power of WPP and PV decreased, especially during peak times. The MEG will buy more power from UPG, thus reducing operational risks. Because part of WPP and PV are converted into heating, which creates uncertainty in the heating load supply, the MEG will buy some energy from UHG and UCG to realize a reliable supply of heating and cooling, to reduce the risk to the heating and cooling supply. This means that the introduction of the CVaR method and robust stochastic optimization theory can control the energy supply risk of electricity, heating and cooling loads simultaneously, and take into account both operational benefits and risks, to achieve the optimal safe and steady operation of the MEG. Figure 7 is the output distribution of the micro energy grid of Case 3.

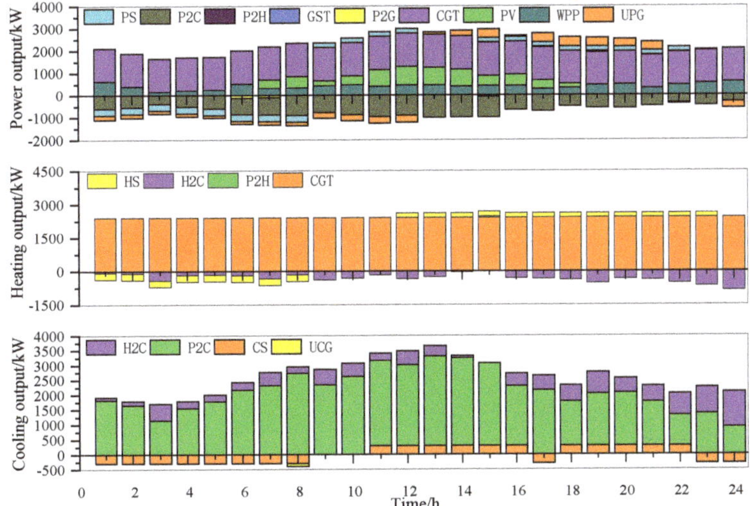

Figure 7. Output distribution of the micro energy grid of Case 3.

In addition, to study the applicable space of the risk aversion model, a sensitivity analysis of the robustness coefficient and confidence was carried out. It can be seen that when $0.7 < \Gamma \leq 0.95$, the increase of β will result in a larger improvement of CVaR, which means that the MEG operation scheme will change when considering WPP and PV uncertainty. When $\Gamma \leq 0.7$, the increase of β will result in a lower increase of CVaR. The decision makers will pay attention to the operation benefits and risks of the MEG at the same time, so the operation of the MEG is relatively steady, but the overall CVaR value will increase with the increase of Γ. When $\Gamma \geq 0.95$, the increase of β will bring a great increase of CVaR. In this case, the decision makers are extremely risk averse. A smaller uncertainty will bring greater operational risk. Figure 8 is the analysis result of the MEG operational risk under different robustness coefficients and confidence levels.

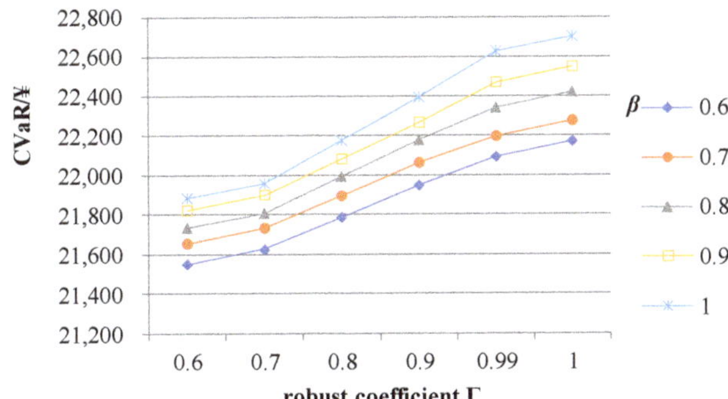

Figure 8. The analysis result of the MEG operational risk under different robustness coefficients and confidence levels.

In general, the CVaR method and robust optimization theory can better describe the uncertain risk of MEG operation. To realize the optimal operation of the MEG, decision makers should set reasonable confidence and robustness coefficients, considering the operational risks and benefits of the MEG at the same time.

5.3. Results Analysis

According to the above three scenarios, the impact of demand response and MTEA on MEG operation is further analyzed to establish external key factors of MEG operation.

(1) DR optimization effect analysis

DR includes two response modes: PBDR and IBDR. PBDR indirectly leads terminal users to use energy reasonably by implementing a differentiated time-of-use price. IBDR directly controls the terminal users' load by signing a pre-agreement with them. Figure 9 is the load curve of power, heating and cooling before and after PBDR.

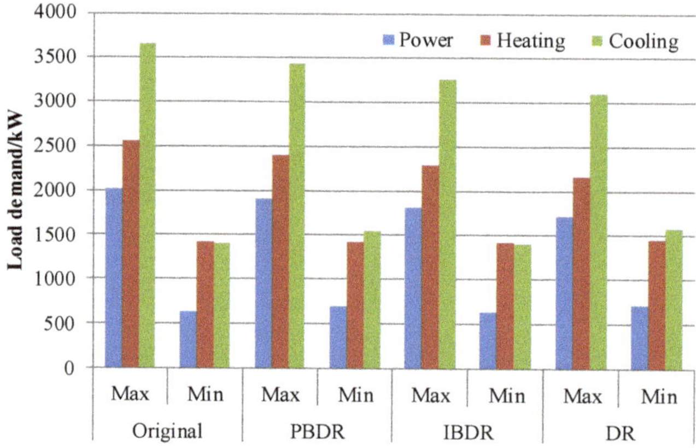

Figure 9. Load curve of power, heating and cooling after demand response (DR). PBDR: price-based demand response; IBDR: incentive-based demand response.

Compared with original load curve, the maximum load of peak time after PBDR decreases, and the minimum load of valley time increases. The maximum load reduction effect after IBDR is stronger than that of PBDR, but the valley load enhancement effect is weaker than that of PBDR. At the same time, after the application of PBDR and IBDR, the peak load decreases more, the valley load increases more, and the load curve becomes smoother. Table 5 shows the dispatching results of the micro energy grid before and after PBDR.

Table 5. Dispatching results of the micro energy grid before and after PBDR.

PBDR	WPP	PV	CGT		Energy Storage/kW·h		
			Power	Heating	PS	HS	CS
Before	9602	6287	35,674	57,600	±2400	±1800	±3000
After	10,167	6657	35,674	57,600	±1800	±2400	±3600
PBDR	**GST**	**IBDR**			**Revenue/¥**	**Carbon/ton**	**CVaR/¥**
		Power	**Heating**	**Cooling**			
Before	153	-	-	-	46,445	5.30	21,993
After	86	±1200	±1000	±1200	50,935	4.60	22,189

According to Table 5, if PBDR is considered, the WPP and PV grid-connected power are increased by 565 kW·h and 370 kW·h, respectively. As the load curve becomes smoother, the output of PS, HS, and CS decreases, indicating that the peak load regulation demand of the MEG for WPP and PV decreases. Similarly, after PBDR, the load in the valley period increases and the convertible power of P2G decreases. From the operational objective function, the economic benefits, carbon emissions and

CVaR values of MEG operation after PBDR are all optimized. Further, the output distribution of MEG at different times after PBDR was analyzed. Figure 10 is the output distribution of the micro energy grid after PBDR.

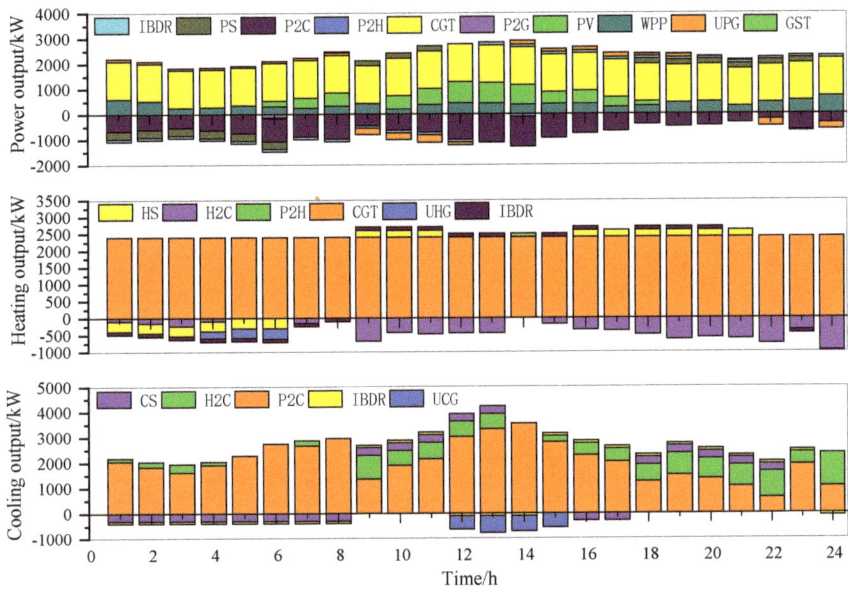

Figure 10. Output distribution of the micro energy grid after PBDR.

According to Figure 10, the output distribution of the MEG at different times was analyzed. Due to PBDR the load curve became smoother, and the WPP grid-connected power increased in the valley period. The PV grid-connected power increased in the peak period because IBDR can provide a peak-shaving service. In terms of power load, the power supply structure was cleaner and lower in carbon due to the increased power generation of WPP and PV. In terms of heating and cooling load, the smoother load curves reduced peak-shaving demand, make full use of IBDR, which can provide a peak-shaving service, which optimizes the meeting efficiency of the heating load and cooling load. In general, PBDR can optimize the output structure of the MEG and improve the operational efficiency of the MEG.

(2) MTEA sensitivity analysis

For the MEG, the main carbon emission sources include CGT and UPG. Therefore, the setting of MTEA will directly affect the power supply of CGT and UPG to the MEG. Therefore, this section outlines the sensitivity analysis we carried out on MTEA to construct the optimal dispatching strategies with different MTEAs. Table 6 shows the dispatching results of the micro energy grid with different META.

Table 6. Scheduling results of the micro energy grid under different META.

META	WPP	PV	IBDR			ES			Carbon Emission/ton	Revenue/103¥	CVaR/103¥
			Power	Heating	Cooling	Power	Heating	Cooling			
60%	11,258	7045	±1600	±1300	±1500	±2400	±3000	±3900	55,688.93	3.45	27,854
70%	11,085	6842	±1400	±1200	±1300	±1800	±3000	±3900	53,554.51	4.03	25,358.86
80%	10,167	6657	±1200	±1000	±1200	±1800	±2400	±3600	50,935	4.60	22,189
90%	9650	6245	±1100	±800	±900	±1800	±1800	±3000	47,539.33	5.18	19,723.56
100%	9325	6018	±900	±700	±900	±1200	±1800	±3000	44,822.8	5.75	17,751.2

According to Table 6, with the increase of META, the power generation of WPP and PV gradually decreased. This is because the MEG is more inclined to buy power from UPG to avoid uncertain risks. It also reduces ES and IBDR output, meaning that the demand of the MEG for reserve services is reduced. When META is small, the carbon emissions of CGT and UPG power generation will reduce their power generation advantages, while more WPP and PV are scheduled to satisfy the load demand. Generally speaking, a reasonable META needs to be set to enhance WPP and PV grid-connected space and realize the optimal operation of the MEG as a whole.

6. Conclusions

In order to improve the sustainable development of distributed energy such as wind and solar, this paper emphasized optimization of the operation of the micro energy grid aggregated by multiple distributed energy sources. Because of the strong uncertainty of WPP and PV, the CVaR method and robust stochastic optimization theory were applied to describe the uncertainty of the objective function and constraints, and a risk aversion dispatching model of the micro energy grid considering the demand response and maximum total emission allowance was constructed. Finally, this paper selected the Xinxiang Active Distribution Network Demonstration Project in Jining, China as an example. The following conclusions were reached:

(1) The micro energy grid can make the most use of the complementary characters of different energy, such as WPP, PV, and CGT, and can make use of a variety of EC equipment (P2H, P2C, H2C, P2G) and ES equipment (PS, HS, CS, GS) to achieve optimal satisfaction with various loads types, such as electricity, heating, cooling and gas. On the one hand, clean energy has both environmental and economic characteristics, which can improve the economic and environmental benefits of MEG operation. On the other hand, the cooperative operation of various EC and ES can effectively handle the strong uncertainty.

(2) The proposed risk aversion dispatching optimization model with the CVaR method and robust stochastic optimization theory can describe the impact of uncertain variables in objective function and constraints, and provide a basis for decision makers who have different attitudes. When $\Gamma \leq 0.7$, β increases and results in a lower increase of CVaR, and decision makers are operating in risk-free conditions. When $0.7 < \Gamma \leq 0.95$, β increases and results in a larger increase, the decision makers are risk-averse. When $\Gamma \geq 0.95$, β increases and results in a lower increase, and the decision maker becomes extremely risk averse. Thus, it is possible to formulate the optimal scheduling strategy for decision makers who have different attitudes by setting reasonable confidence and robustness coefficients.

(3) DR can smooth the energy load curve of electricity, heating, cooling, and gas. MTEA can enhance the market competitiveness of the clean energy market, thus promoting grid-connected power of clean energy, such as WPP and PV, and optimizing the multi-energy supply structure of the MEG. On the one hand, PBDR has a better "valley filling" effect, and IBDR has a better "peak cutting" effect. The synergistic operation of the two can maximize the "peak cutting and valley filling" effect. On the other hand, MTEA determines the supply space of CGT and UPG. When MTEA is low, the MEG gives priority to WPP and PV to satisfy the load demand, which will meet the maximum carbon emission constraint.

Author Contributions: Conceptualization, P.L.; methodology, W.F.; software, L.J.; resources, N.L. and F.Z.; data curation, F.Z.; writing—original draft preparation, X.F.; writing—review and editing, X.F. and H.L.; visualization, L.J.

Funding: This research was funded and supported by the Project funded by China Postdoctoral Science Foundation (2019M650024), the National Nature Science Foundation of China (Grant Nos. 71904049, 71874053, 71573084), the Beijing Social Science Fund (18GLC058) and the 2018 Key Projects of Philosophy and Social Science Research, Ministry of Education, China (18JZD032).

Conflicts of Interest: The authors declare no conflict of interest. It should be noted that the whole work was accomplished by the authors collaboratively. All authors read and approved the final manuscript.

Nomenclature

Abbreviation	
MEG	micro energy grid
MTEA	maximum total carbon emission allowance
WPP	wind power plant
PV	photovoltaic power generation
CGT	conventional gas turbine
P2G	power-to-gas
P2C	power to cooling
P2H	power to heating
H2C	heating to cooling
GS	gas storage tank
PS	power storage battery
HS	heat storage tank
CS	cold storage tank
DR	demand response
PBDR	price-based demand response
IBDR	incentive-based demand response
EP	energy production
EC	energy conversion
ES	energy storage
FET	following electric load
FTL	following thermal load
Set	
t, s	index for time
k	index for energy type
j	index for step
Parameter	
g_R	rated power of WPP
v_{in}	cut-in speed
v_R	rated speed
v_{out}	cut-out speed
η_{PV}	conversion productivity
S_{PV}	receiving light of PV
H_{ng}	natural gas calorific value
$\eta_{CGT,t}$	operation efficiency of CGT at time t
θ_t	radiation intensity at time t
v_t	real-time speed at time t
L_t^{before}	initial load before PBDR. at time t
$\eta_{EC(S)}^{output}$	energy use efficiency of EC and ES
$\eta_{EC(S)}^{input}$	energy supply efficiency of EC and ES
η_t^{loss}	loss rate of ES at time t
Variables	
$g_{CGT,t}$	power supply of CGT at time t
$Q_{CGT,t}$	heating supply of CGT at time t
$V_{CGT,t}$	gas consumption at time t
$Q_{P2C,t}$	cooling power produced by P2C at time t
$Q_{P2H,t}$	heating power from P2H at time t
$Q_{H2C,t}$	cooling power from H2C at time t
$g_{P2C,t}$	electricity consumption of P2C at time t
$g_{P2H,t}$	electricity consumption of P2H at time t
$g_{H2C,t}$	heating consumption of H2C at time t
$S_{ES,t}$	energy storage of ES at time t
ES_t^{input}	Input power of ES at time t
ES_t^{output}	output power of ES at time t
ΔP_t	price variable after PBDR at time s
ΔL_s	load variable after PBDR at time s
$\Delta L_t^{k,IB}$	output power of energy k provided by IBDR at time t
$R_{EP,t}$	operation revenue of EP at time t
$R_{EC,t}$	operation revenue of EC at time t

$R_{ES,t}$	operation revenue of ES at time t
$R_{DR,t}$	operation revenue of DR at time t
$R_{Carbon,t}$	carbon trading revenue at time t
$C_{CGT,t}^{fuel}$	fuel consumption cost of CGT at time t
$C_{CGT,t}^{sd}$	start-stop cost of CGT at time t
$u_{CGT,t}$	start-up and shut-down state variable at time t
$g_{CGT,t}$	power output of CGT at time t
$Q_{CGT,t}$	heating output of CGT at time t
$\mu_{CGT,t}^{u}$	start-up state variable of CGT at time t
Δg_{CGT}^{-}	downhill power limits of CGT
Δg_{CGT}^{+}	uphill power limits of CGT
u_{CGT}	operating status of the CGT
$u_{EC,t}^{output}$	state variables of EC outputting energy at time t
ΔL_{t}^{c}	load forecast deviation at time t
M_{t}	system net load demand
$D_{i}^{k,j,\max}$ $D_{i}^{k,j,\max}$	maximum output of DRP i in step j providing energy k
$D_{i}^{k,j,\min}$	minimum response output of DRP i in step j providing energy k
$\Delta L_{i,t}^{k,j}$	actual load reduction value of energy k that DRP i provides in j step at time t
$D_{i,t}^{k,j}$	available load reduction value of energy k that DRP i can provides in step j at time t
θ_{h}^{e}	electric-thermal conversion coefficient of CGT
a, b, c	power supply cost coefficient of CGT
P_{t}^{before}	energy price before PBDR at time t
P_{t}^{after}	energy price after PBDR at time t
$a_{CGT}, b_{CGT}, c_{CGT}$	carbon emission coefficient of the CGT
$c_{\min}, c_{\max}$	value of c with the minimum and maximum output power
$Q_{CGT}^{\max}$	maximum heating power of CGT
$Q_{CGT}^{\min}$	heating supply power of the turbine when electricity power supply of CGT is minimum
$g_{CGT}^{\min}, g_{CGT}^{\max}$	maximum and minimum power generation of CGT under pure condensation
$V_{P2G,t}^{\min}, V_{P2G,t}^{\max}$	minimum and maximum gas production power of P2G
$Q_{EC,t}^{output,\min}, Q_{EC,t}^{output,\max}$	upper and lower limits of EC energy supply
$Q_{EC,t}^{input,\min}$, e$Q_{EC,t}^{input,\max}$	upper and lower limits of EC energy consumption
$S_{ES,t}^{\min}, S_{ES,t}^{\max}$	minimum and maximum energy storage of ES at time t
$g_{ES,t}^{output,\min}$, $g_{ES,t}^{output,\max}$	minimum and maximum of ES energy supply at time t
$Q_{ES,t}^{input,\min}$, $Q_{ES,t}^{input,\max}$	minimum and maximum limits of ES energy consumption at time t
r_{e}, r_{c}	up-rotating reserve coefficients of electricity and cooling load
$r_{WPP}^{up}, r_{PV}^{up}$	up-rotating reserve coefficients of WPP and PV
$r_{WPP}^{dn}, r_{PV}^{dn}$	down-rotating reserve coefficients of WPP and PV
$g_{MEG,t}^{p,\max}, g_{MEG,t}^{p,\min}$	maximum and minimum power supply at time t
Eφ, ϑ	shape and scale parameter
ω, ψ	shape parameters of the Beta distribution
η_{hr}	heat recovery efficiency
φ_{P2C}	energy conversion efficiency of P2C,
η_{loss}	capacity loss rate
φ_{P2H}	energy conversion efficiency of P2H
φ_{H2C}	energy conversion efficiency of H2C
η_{ES}^{input}	energy storage efficiency
η_{ES}^{output}	energy release efficiency
E_{st}	energy demand price elasticity matrix
$g_{PV,t}^{*}$	maximum output power of PV at time t
$g_{WPP,t}^{*}$	available power of WPP at time t
$\mu_{CGT,s}^{d}$	shut-down state variable of CGT at time s
$C_{CGT,t}^{u}$	start-up cost at time t
$C_{CGT,s+1}^{d}$	hut-down cost of CGT at time s + 1

$Q_{EC(S),t}^{output}$	output energy of EC and ES at time t
$Q_{EC(S),t}^{input}$	input energy of EC and ES at time t
$P_{EC(S)}^{output}$	energy use price of EC and ES
$P_{EC(S)}^{input}$	energy supply price of EC and ES
L_t^{after}	energy demand after PBDR at time t
$\Delta L_t^{k,shortage}$	shortage of energy k at time t
$p_t^{k,IB}$	energy supply price of IBDR at time t
p_t^k	real-time energy supply price of energy k at time t
$Q_{META,t}$	Maximum MTEA at time t
$P_{Carbon,t}$	carbon market transaction price at time t
$g_{UPG,t}$	energy of MEG bought from the upper power grid at time t
$Q_{UHG,t}$	energy of MEG bought from the upper heating grid at time t
$Q_{UCG,t}$	energy of MEG bought from the upper cooling grid at time t
$g_{P2G,t}^{input}$	power consumption of P2G at time t
$g_{P2G,t}^{output}$	power generation of P2G at time t
$g_{P2H,t}^{input}$	power consumption of P2H at time t
$g_{P2C,t}^{c,input}$	power consumption of P2C at time t
$\Delta L_t^{p,PB}$	power generation output provided by PBDR at time t
$\Delta L_t^{p,IB}$	power generation output provided by IBDR at time t
$Q_{P2H,t}^{output}$	heating power of H2C at time t
$Q_{HS,t}^{input}$	heating storage of HS at time t
$Q_{HS,t}^{output}$	heating release of HS at time t
$Q_{H2C,t}^{input}$	consumption of heating of H2C at time t
$\Delta L_t^{h,PB}$	heating output power provided by PBDR at time t
$\Delta L_t^{h,IB}$	heating output power provided by IBDR at time t
$Q_{P2C,t}^{output}$	cooling output of P2C at time t
$Q_{H2C,t}^{output}$	cooling output of H2C at time t
$Q_{CS,t}^{input}$	cooling storage of CS at time t
$Q_{CS,t}^{output}$	cooling release of CS at time t
$\Delta L_t^{c,PB}$	cooling output power provided by PBDR at time t
$\Delta L_t^{c,IB}$	cooling output power provided by IBDR at time t
$u_{EC,t}^{input}$	state variables of EC inputting energy at time t
$u_{ES,t}^{output}$	state variables of ES outputting energy at time t
$u_{ES,t}^{input}$	state variables of ES inputting energy at time t
$g_{MEG,t}$	electric power of MEG at time t
L_t^f	load demand forecast value at time t

References

1. Rifkin, J. *The Third Industrial Revolution: How Lateral Power is Transforming Energy, the Economy, and the World*; St. Martin's Press: New York, NY, USA, 2011.
2. Guidance on Promoting the Development of "Internet +" Smart Energy. Available online: https://kns.cnki.net/KCMS/detail/detail.aspx?dbcode=CJFQ&dbname=CJFDLAST2016&filename=CSYQ201604001&v=MTgxNTZyV00xRnJDVVJMT2VaZVpzRnluaFZidlBKajdTZjdHNEg5Zk1xNDlGWllSOGVYMUx1eFlTN0RoMVQzcVQ=) (accessed on 4 November 2019).
3. Liu, Y.B.; Zuo, K.Y.; Liu, X.W. Dynamic pricing for decentralized energy trading in micro-grids. *Appl. Energy* **2018**, *228*, 689–699. [CrossRef]
4. Rossi, I.; Banta, L.; Cuneo, A.; Ferrari, M.L.; Traverso, A.N.; Traverso, A. Real-time management solutions for a smart polygeneration microgrid. *Energy Convers. Manag.* **2016**, *112*, 11–20. [CrossRef]
5. Dong, W.L.; Wang, Q.; Yang, L. A coordinated dispatching model for distribution utility and virtual power plants with wing/photovoltaic/hydro generators. *Autom. Electr. Power Syst.* **2015**, *39*, 75–82.
6. Yu, S.; Wei, Z.H.; Sun, G.Q. A bidding model for a virtual power plant considering uncertainties. *Autom. Electr. Power Syst.* **2014**, *38*, 44–49.

7. Liu, X.Y. Research on Performance of Combined Heating System of Solar Energy and Electric Boiler, North China Electric Power University. 2017. Available online: https://kns.cnki.net/KCMS/detail/detail.aspx?dbcode=CMFD&dbname=CMFD201801&filename=1017222760.nh&v=MTE4NzJGeW5oVnIzQlZGMjZHYkc2SE5iS3I1RWJQSVI4ZVgxTHV4WVM3RGgxVDNxVHJXTTFGckNVUkxPZVplWnM= (accessed on 4 November 2019).
8. Mengdong Power Wind Power Heating Pilot 3 Years to Reduce Coal Consumption by 68,700 Tons. Polaris Power Network. 7 November 2016. Available online: http://www.sohu.com/a/118288979_131990 (accessed on 7 November 2016).
9. Zhang, X.; Yang, J.H.; Wang, W.Z. Integrated optimal dispatch of a rural micro-energy-grid with multi-energy stream based on model predictive control. *Energies* **2018**, *11*, 3439. [CrossRef]
10. Du, L.; Sun, L.; Chen, H.H. Multi-index evaluation of integrated energy system with P2G planning. *Electr. Power Autom. Equip.* **2017**, *37*, 110–116.
11. Luo, Y.H.; Yin, Z.X.; Yang, D.S. A new wind power accommodation strategy for combined heat and power system based on bi-directional conversion. *Energies* **2019**, *12*, 2458. [CrossRef]
12. Peik-Herfeh, M.; Seifi, H.; Sheikh-El-Eslami, M.K. Decision making of a virtual power plant under uncertainties for bidding in a day-ahead market using point estimate method. *Int. J. Elec. Power* **2013**, *44*, 88–98. [CrossRef]
13. Yang, H.M.; Yi, D.X.; Zhao, J.H. Distributed optimal dispatch of virtual power plant based on ELM transformation. *Management* **2014**, *10*, 1297–1318. [CrossRef]
14. Zamani, A.G.; Zakariazadeh, A.; Jadid, S. Day-ahead resource scheduling of a renewable energy based virtual power plant. *Appl. Energy* **2016**, *169*, 324–340. [CrossRef]
15. Tan, Z.F.; Wang, G.; Ju, L.W. Application of CVaR risk aversion approach in the dynamical scheduling optimization model for virtual power plant connected with wind-photovoltaic-energy storage system with uncertainties and demand response. *Energy* **2017**, *124*, 198–213. [CrossRef]
16. Hu, M.C.; Lu, S.Y.; Chen, Y.H. Stochastic programming and market equilibrium analysis of microgrids energy management systems. *Energy* **2016**, *113*, 662–670. [CrossRef]
17. Tsao, Y.C.; Thanh, V.V.; Lu, J.C. Multiobjective robust fuzzy stochastic approach for sustainable smart grid design. *Energy* **2016**, *176*, 929–939. [CrossRef]
18. Li, Z.M.; Xu, Y. Temporally-coordinated optimal operation of a multi-energy microgrid under diverse uncertainties. *Appl. Energy* **2019**, *240*, 719–729. [CrossRef]
19. Ju, L.W.; Tan, Q.L.; Zuo, X.T.; Zhao, R. A risk aversion optimal model for microenergy grid low carbon-oriented operation considering power-to-gas and gas storage tank. *Int. J. Energy Res.* **2019**, *43*, 1–20. [CrossRef]
20. Chen, Y.; Wei, W.; Liu, F. Analyzing and validating the economical efficiency of managing a cluster of energy hubs in multi-carrier energy systems. *Appl. Energy* **2018**, *230*, 403–416. [CrossRef]
21. Ju, L.W.; Zhao, R.; Tan, Q. A multi-objective robust scheduling model and solution algorithm for a novel virtual power plant connected with power-to-gas and gas storage tank considering uncertainty and demand response. *Appl. Energy* **2019**, *250*, 1336–1355. [CrossRef]
22. Wang, G.; Tan, Z.F.; Lin, H.Y. Multi-level market transaction optimization model for electricity sales companies with energy storage plant. *Energies* **2019**, *12*, 145. [CrossRef]
23. Liu, F.; Zhang, K.L.; Zou, R.M. Robust LFC strategy for wind integrated time-delay power system using EID compensation. *Energies* **2019**, *12*, 3223. [CrossRef]

Article

The Biomass Potential and GHG (Greenhouse Gas) Emissions Mitigation of Straw-Based Biomass Power Plant: A Case Study in Anhui Province, China

Hui Li [1,*,†], Xue Min [1,†], Mingwei Dai [1] and Xinju Dong [2]

1 School of Resources and Environment, Anhui Agricultural University, Hefei 230026, China
2 Department of Chemistry, University of Louisville, Louisville, KT 40292, USA
* Correspondence: leehui@ahau.edu.cn
† These authors contribute equally to this work.

Received: 26 August 2019; Accepted: 3 September 2019; Published: 9 September 2019

Abstract: Anhui Province (AHP), a typical agriculture-based province in China, has a significant amount of biomass resources for the development of biomass power plants. By the end of 2016, 23 straw based biomass power plants were established in AHP, aggregating to 6560 MW capacity, which is now ranked second in China. This paper presents the current development status and GHG (Greenhouse Gas) mitigation effect of the straw based biomass power plants in Anhui Province. Total biomass production in 2016 was calculated as 41.84 million tons. Although there is huge biomass potential in AHP, the distribution is heterogeneous with a gradually decreasing trend from north to south. Furthermore, the installed capacity of power generation is also unmatched with the biomass resources. Based on a calculation made in 2016, approximately 3.44 million tons of CO_2-eq were mitigated from the biomass power plants in AHP. The large-scale development of biomass power plants remains a challenge for the future, especially in areas of AHP with a low biomass density.

Keywords: biomass resource; power plant; GHG emission mitigation; Anhui

---

1. Introduction

Global energy demand was 14,050 Mtoe (million tons of oil equivalent) in 2017, with a growth rate of 2.1%, and global energy-related CO_2 emissions increased by 1.4%, reaching a historically high value of 32.5 GT [1]. China has a large population and rapid growth of the population, industrialization, and urbanization, and the energy consumption has sharply increased, up to 3053 Mtoe in 2017, of which more than 70% of energy consumption was supported by coal [2,3]. Due to this coal-dominated energy consumption, greenhouse gas (GHG) and particulate matter (PM) pollutants are released during the coal utilization process, leading to a negative impact on health and the environment in China and globally [4–9]. Therefore, preferential concern and policies for clean energy associated with the "Energy Revolution" and "Fight Against Pollution" in China have received an intense level of support [10].

Presently, biomass energy plays a significant role in energy production, and is listed as the fourth largest energy source worldwide after oil, coal and natural gas [11,12]. With carbon-neutral, clean and sustainable characteristics [13–15], biomass has been proven to be promising renewable energy in China with a theoretical maximum potential energy of 18,833 PJ in 2030 [16]. To utilize biomass energy, current technologies including combustion, liquefaction, pyrolysis/gasification, digestion, and fermentation have been widely used in developing countries [17,18]. Among various technologies, biomass power generation is a high-efficiency approach for biomass utilization that has been promoted and developed for sustainable energy output, and that also offers a flexible approach for straw disposal [19,20]. By the end of 2011, the global installed capacity of biomass plants was 72 GW with total electricity generation of 265–529 TWh [21]. The installed capacity of biomass power generation in China has gradually increased, as showed in Figure 1, a total value of 18,790 MW in 2016, most of these plants are fueled by agricultural residues and forest products. The growth rate decreased from 2013 to 2016, indicating that the large-scale development of biomass power plants is accompanied by some problems, including the biomass supply capacity, fuel cost, and environmental impacts [2,22–24]. The environmental impacts of biomass power plants mainly include landfill, water and ash treatment [25–27]. Zhao and Li [28] performed a numerical example analysis of biomass power plant and noted that pollutant emissions during straw transportation were neglected. Research on the physical, mental and social impacts of small-scale biomass power plants indicated that air pollutant was the main concern of local people [29]. Zhang et al. [30] indicated that the cost of a straw-based power plant is indeed high, most of which was the fuel cost. Nguyen et al. found that the environmental performance of straw power plants was worse than that of natural gas [31]. In addition, although some studies have emphasized the potential for negative net GHG emissions from biomass power plants, the recognition of GHGs mitigation is not fully understood [10]. Therefore, some studies have indicated that the development of biomass power plants should be more steady in China [32,33].

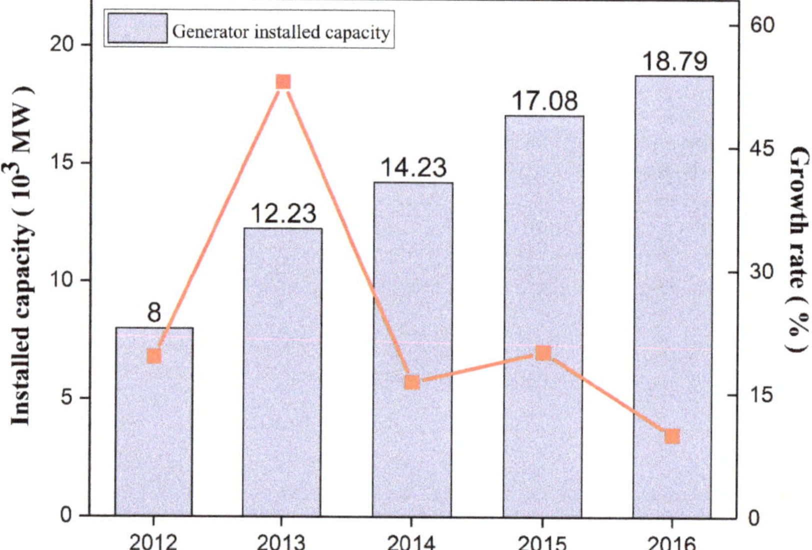

Figure 1. Installed capacity of biomass power generation in China from 2012 to 2016. Note: The original data were obtained from the database of National Renewable Energy Information management Centre in China.

Anhui Province (AHP), see in Figure 2, located in eastern China (N29°41′–N34°38′, E114°54′–E119°37′), is a typical agricultural Province, with a huge amount of biomass resources, such as forest and agricultural residues [34]. The government of Anhui province has been strongly encouraging and supporting the construction and development of biomass power generation plants. "The research report of the "13th Five-Year Plan" for Straw Power Generation in Anhui Province" published by the Anhui government proposed installed capacity of approximately 1600 MW by the end of 2020. According to the statistical data from the Anhui Energy Administration, a total of 23 grid-connected straw-based biomass power plants with an installed capacity of 660 MW were built and operational from 2006. Even with policy support, the development of biomass power plants still needs a systematic study on feedstock analysis and environmental impact assessment. Therefore, in this study, we investigated the current status and GHG emissions from straw-based power plants in Anhui Province. First, the development, operation, and profitability of power plants were discussed in detail. In addition, the production, distribution, and availability of biomass resources in AHP were examined. Finally, the potential mitigation of greenhouse gas emissions attributed to straw-based power plants was calculated. The results can provide valuable information to policymakers and will be useful for the development of straw-based power plants in China.

Figure 2. Location of Anhui province in China.

2. Materials and Methods

With the aim of producing comprehensive information about the biomass resources, operational condition and mitigation of GHGs associated with biomass power plants in AHP, data were obtained from official and provincial reports, yearbooks, online reports, and other sources. Data and information about biomass resources were obtained from the Anhui statistical yearbook (2012–2016), agricultural mechanization statistics report for Anhui (2012–2016) and the national agricultural mechanization information website (Appendix A.1). Data and information related to the operational condition of biomass power plant were obtained from the Anhui Energy Bureau. The methods used to calculate the biomass potential and GHG mitigation were the crop-to-residue index and the CM-092-V01 method, respectively (see Appendix A for further details).

3. Results and Discussion

3.1. Biomass Production and Characterization

Biomass resource potential is an important challenge for the development of biomass power plants. AHP has a huge amount of biomass resources from agriculture and forestry. The total land area is 139,427 km^2, of which 58,730 km^2 is cultivated land and 39,585 km^2 is forest land [35]. Production of crop straw is evaluated based on the crop-to-residue index (CRI) value, namely 1.38 for wheat straw, 2.05 for corn straw, 1.26 for rice straw, 1.68 for bean straw, and 1.16 for potato straw [36].

Biomass availability in AHP varied in term of spatial and temporal distribution. Figure 3 shows the availability of total biomass production from 2012 to 2016. The total yield increased slowly from 36.25 to 41.84 million tons, with the highest value of 43.20 million tons in 2015. The forestry byproducts gradually increased at a rate of 23.15% due to the increase in forest area supported by the local government. The crop straw yield was raised to 36.41 million tons in 2016, with an increased rate of

4.52%. Our previous research indicated that the total crop production and harvesting approach are the two main factors that impact straw production, and with the rapid development of mechanized harvesting, less straw is collected for utilization [36].

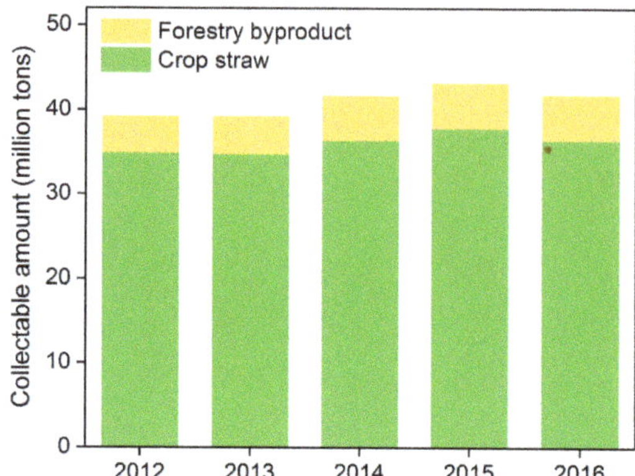

Figure 3. The total collectable amount of biomass resource in Anhui Province (AHP) from 2012 to 2016.

Abbreviation of cities: SZ for Suzhou city, BZ for Bozhou city, FY for Fuyang city, BB for Bengbu city, HN for Huainan city, XC for Xuancheng city, HB for Huaibei city, HF for Hefei city, LA for Lu'an city, Cuz for Chuzhou city, MAS for Ma'anshan city, WH for Wuhu city, TL for Tonglin city, CiZ for Chizhou city, AQ for Anqing city, HS for Huangshan city.

Geographically, biomass availability is spatially nonuniform in AHP and is disproportionately concentrated in the northern part of AHP as showed in Figure 4, because of the different land availability and productivity. The total biomass production was mainly concentrated in the northern part of AHP, with a proportion of 56%, compared with 22.3% and 21.7% in central and southern AHP, respectively. With respect to forestry byproducts, approximately 63.57% of the resources were generated from the southern part of AHP, such as Xuancheng, Chizhou, Huangshan cities. In addition, the distribution of the biomass power plants is not consistent with that of straw resources, except for Huangshan and Tongling cities, which have no power plant due to their low biomass availability. However, installed capacity of Lu'an city was the highest among all the cities, although the biomass availability is low.

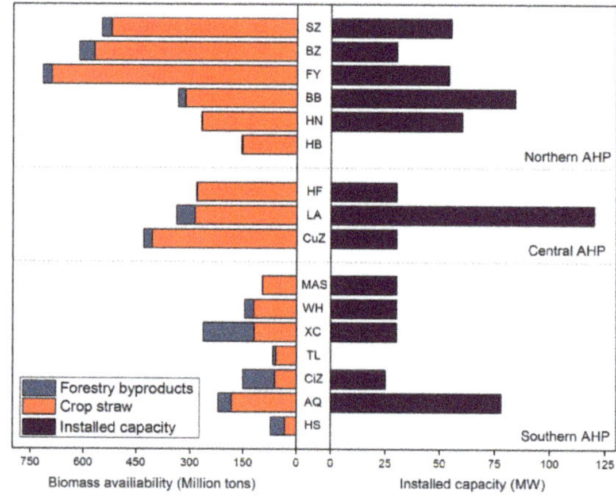

Figure 4. The biomass availability and installed capacity of biomass power plant in AHP, 2016.

3.2. Assessment and Availability Analysis

The high production of biomass in AHP provides opportunities for the development of biomass power plants. The Chinese official legislation promulgated in 2006 restricted the outdoor burning of straw, and approximately 23.7% of straw is used for fuel. Biomass availability is dependent largely on location and climate. To better describe the distribution of biomass, it is common to select the biomass density, which is defined as the ratio of the collectable biomass production to the total land area [37]. It is concluded from previous studies that a low biomass density will result in high collection, delivery and storage costs. As showed in Figure 5, the density of biomass resources for each city was calculated. It should be noted that the geographical straw density differed considerably from 75.40 to 733.30 t/km^2, with a mean value of 361.77 t/km^2. However, the biomass density decreases gradually from north to south in AHP which is consistent with the results found in other literature [36]. Bozhou and Fuyang have high densities, more than 700 t/km^2, and the densities of central and southern AHP (Hefei, Wuhu, Maanshan, Lu'an, Tongling, Xuancheng, Chizhou, Anqing, and Huangshan) are relatively low, less than 300 t/km^2, and the lowest value is observed in the southern area.

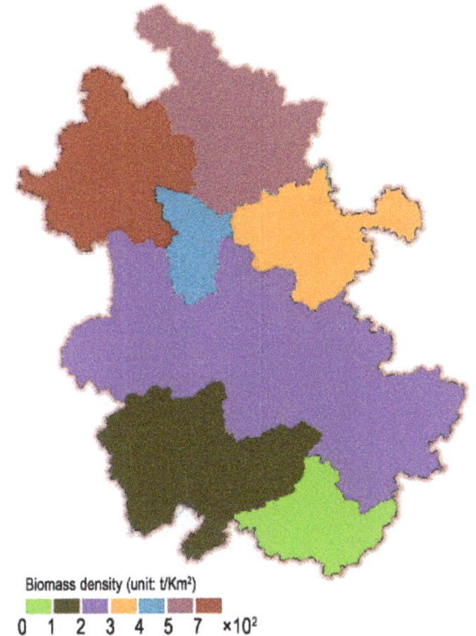

Figure 5. The biomass density in AHP (unit: t/Km^2).

With low-density energy resource, the distribution and sustainable supply of biomass feedstock have a great influence on the collection and storage cost [38]. It is recommended that the economical collection radius should be less than 50 km for profitability [30]. In addition, only 20% of the available biomass is used for power plants in general due to their alternative utilization activities. Therefore, the net theoretical available potential (NAP, within a 50 km radius with a supply assurance factor of 0.3) amount of biomass for power generation was calculated in this study. The biomass consumption was approximately 0.17–0.29 million tons for a 30 MW biomass power plant based on the calorific values [39]. As shown in Figure 6, within 50 km, the biomass from northern AHP can completely fuel the demand of a power plant, while the central and southern AHP exhibited a lower NAP, making it difficult to meet the feedstock demand. Most biomass power plants have a relatively low capacity for biomass collection, leading to the final price of biomass reaching more than approximately 50 USD per ton, which is economically unacceptable.

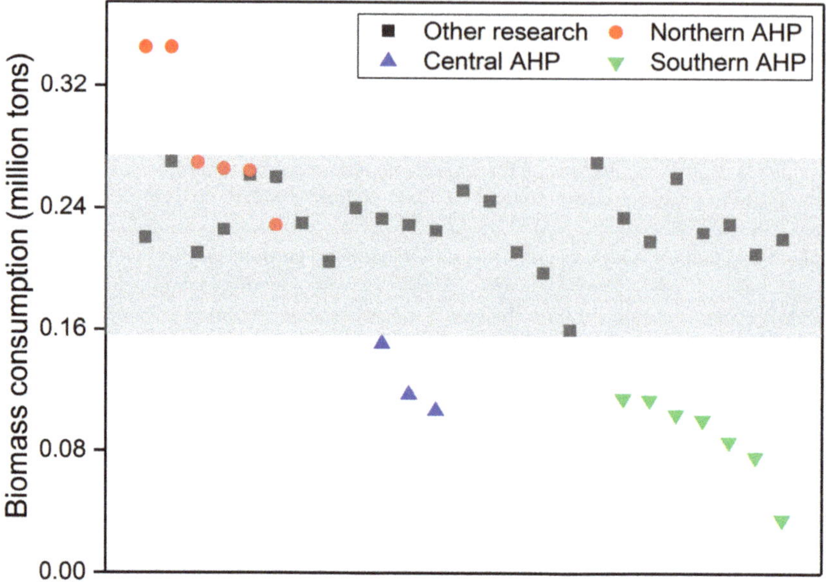

Figure 6. The maximum net theoretical available potential (NAP) of biomass in each city from AHP (with the supply assurance factor of 0.3).

3.3. General Status of Biomass-Based Power Generation

The first 2 × 15 MW biomass-based power plant in AHP was established in 2006, with a total investment of 280 million RMB by the Datang Power Group in Anqing City. According to statistics, this project consumes 235,000 tons of biomass and generates more than 190 GWh of electricity per year. A total of 20 grid-connected biomass power plants with a power capacity of 6560 MW have been installed by the end of 2016 such that AHP possess the second largest biomass capacity in China. The power plants in AHP are in the range of 24–30 MW with an average scale of 28.5 MW, which is much smaller than that of coal-fired power plants. A total of 73.9% of the plants are designed for 30 MW due to the limit of fuel sources, capital and operational costs, and other factors [40].

According to the statistical data from the Anhui Energy Bureau, the total power generation of 20 biomass power plants in 2016 was 377.1 billion KWh, with a total operation of 132,980 h in 2016. As shown in Figure 7, the annual utilization hours of the 20 straw-based power plants varied greatly from 3275 to 8184 h, with a mean value of 6649 h. Nearly half of the plants operated for more than 6000 h per year. The power generation varied between the 20 plants, the highest was 244.45 million KWh in Suzhou City, and the lowest was 98.27 million KWh in Lu'an City.

Compared with fossil energy, there is generally no economically competitive biomass energy due to its high capital, operating and maintenance costs [41,42]. The investment cost of biomass power plants is relatively higher than that of thermal power and hydropower plants, even though the cost has decreased in recent years. Based on the estimation by Zhang et al. [32], the average unit cost is approximated 9900 RMB/kW or 0.738 RMB/KWh, more than double of that of a thermal power plant [43]. In addition, the high cost of handling and shipping is still challenging for some biomass power plants in AHP, especially with the shortage of rural manpower [38]. The biomass purchased price should be less than 220 RMB/ton to provide a benefit.

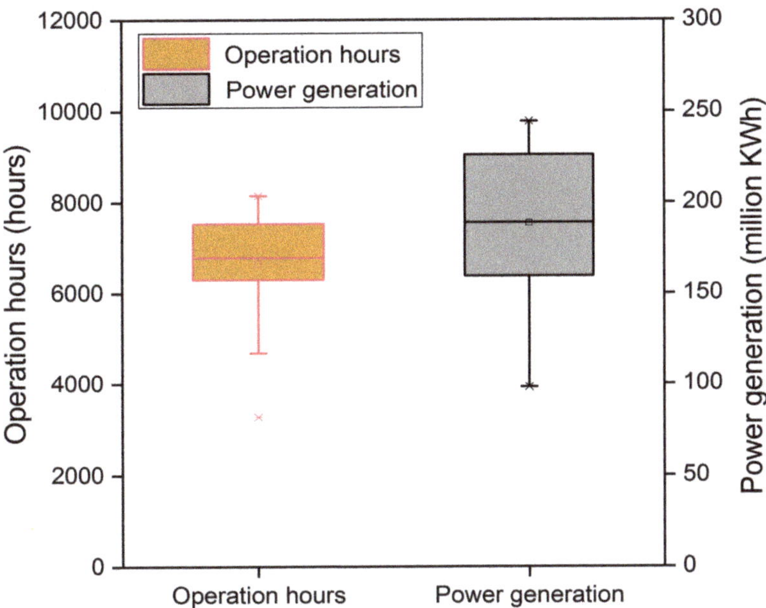

Figure 7. The annual operation hours and power generation of biomass based power plant in AHP, 2016.

Due to the relatively high investment and operating costs of biomass based power plants, various auxiliary policies such as financing policies, tax policies, mandatory grid connection and price support (feed-in tariff) have been proposed [32,44]. The "Renewable Energy Law" announced in 2005 and modified in 2009 indicated that favorable tariffs, development funds, and fiscal support were adopted for renewable energy development. The feed-in tariff set for agricultural and forestry residual power plants was 0.75 RMB/KWh from 2010, of which 0.35 and 0.40 RMB was supported by government subsidies and national grids, respectively. However, the costs of fuel supplies still take a large share in the operational costs, with 60–70% of the total cost [30]. Zhang et al. [32] noted that most biomass power plants in China were operating at a loss even under the support of polices in 2012. Except for the national support funds, the first financial subsidies for straw utilization were issued by the Anhui government in 2014, i.e., 50, 40, 30 RMB/tons for rice, wheat and other straw collections, respectively [45]. Under these subsidies, the number of straw collections was increased, resulting in the 17 out of 20 power plants making profits in 2016, based on the statistics from the Anhui Energy Bureau.

3.4. GHG Emissions Mitigation Analysis

Compared with fossil fuels such as coal, natural gas, and oil, a biomass power plant may effectively reduce greenhouse gas emissions and relieve the pressure of global climate change [46,47]. A study conducted by Sanchez et al. [20] estimated that the CO_2 capture and sequestration potential of biomass for electricity is two times higher than that of biomass for cellulosic ethanol. The direct combustion of biomass for electricity generation is an appropriate approach for GHG mitigation, according to the investigation of Stefan Muench [48]. Based on the CM-092-V01 method of the Clean Development Mechanism (CDM) derived from the database of the United Nations Framework Convention on Climate Change (UNFCCC) and the Intergovernmental Panel on Climate Change (IPCC), the greenhouse gas emission reduction attributed to biomass power plants in AHP was calculated and is shown in Table 1. The GHG mitigation attributed to biomass power plants in AHP ranged from 41,217 to 311,148 t CO_2-eq, with a total yield of 3,436,490 t CO_2-eq in 2016. The different emissions reduction values are caused by equipment and depend on the scale of the plant, the raw material supply, the biomass utilization technology, target products and so on [49]. The reductions per unit production of a biomass power plant differ considerably aiming the power plants, ranging from 0.58 to 1.32 with a mean value of 0.75, which is comparable to the data reported by Lin and He [43].

Table 1. Emission reductions of biomass power plants in 2016.

Biomass Power Plant	Installed Capacity (MW)	Annual Power Supply (MWh)	Emission Reductions (t CO_2-Eq)	Reduction Per Unit of Production (t/MWh)
Anqin Datang	30	184,200	149,540	0.81
Fuyang Guozhen	24	176,060	232,115	1.32
Bengbu Kaidi	24	153,160	125,216	0.82
Tongcheng Kaidi	24	112,550	91,914	0.82
Wangjiang Kaidi	24	158,140	126,343	0.80
Suzhou Huadian	25	183,010	117,354	0.64
Guzhen National Energy	30	220,910	141,973	0.64
Dangshan Guangda	30	244,450	141,078	0.58
Nanling Kaidi	30	203,980	137,606	0.67
Huoqiu Kaidi	30	202,810	133,082	0.66
Huainan Kaidi	30	131,050	88,324	0.67
Shouxian Guoneng	30	230,970	147,971	0.64
Lujiang Kaidi	30	187,910	125,121	0.67
Jinzhai Kaidi	30	209,220	137,281	0.66
Huoshan Kaidi	30	190,020	125,006	0.66
Hanshan Guangda	30	237,290	311,148	1.31
Chizhou 325	25	160,540	98,131	0.61
Linquan National Energy	30	243,210	154,907	0.64
Shucheng Zhongjieneng	30	98,270	69,359	0.71
Mengcheng National Energy	30	242,960	152,971	0.63
Langxi Li'ang	30	148,360	146,837	0.99
Huaiyuan Guangda	30	58,300	41,217	0.71
Dingyuan Guangda	30	59,320	41,995	0.71

4. Conclusions

The available amount of biomass including forest and agriculture residues reached 41.84 million tons in 2016. However, biomass resources with uneven distribution have a gradually decreasing trend from north to south. The distribution of the power plants is not consistent with the biomass potential, especially in the central and southern areas. The biomass density also indicated that these areas may not be suitable for biomass power plant development due to the higher operational cost for feedstock transport. However, the biomass power plants showed great potential for GHG emissions mitigation with a value of 3.44 million tons CO_2-eq in 2016. However, further research, especially more quantitative synthetical analysis, should be conducted to quantify the impacts of straw-based biomass power plant in China.

Author Contributions: H.L. and X.M. conceived and designed this case-study as well as wrote the paper; M.D. and X.D. reviewed the paper; all authors interpreted the data. H.L. and X.M. authors contributed equally to this work as co-first authors.

Funding: This work was financially supported by the Youth development project from School of Resources and Environment AHAU, the Project Startup Foundation for Advanced Talents AHAU (No. YJ2018–56), The Natural Science Foundation of the Education Department of Anhui Province (NO. KJ2019A0207). The authors would like to thank the anonymous reviewers for their helpful comments, which improved the content of the present article.

Conflicts of Interest: The authors declare no conflict of interest.

Appendix A

Appendix A.1 Statistical Methods for Biomass Resources

The amount of biomass resources in a region can directly determine whether the area is suitable for the development of biomass power generation. The crop straw and agro-forestry biomass resources were used as the main parts to calculate the biomass resources of the power industry in a region, and the two methods used to calculate these resources are as follows:

Appendix A.1.1 Crop Straw Resources

Crop straw, which refers to the stalk or stem of certain grains produced in the field, mainly includes rice straw, wheat straw, cotton straw, maize straw, bean straw and potato straw in AHP. The collectable amount of crop straw can be expressed as follows:

$$A_y = \left(\sum_j F_{jy} \times CRI_j \right) \times \eta_{jy}$$

$$\eta_{jy} = (RM_{jy} \times IM_j + RP_{jy} \times IP_j) \times (1 - I_{loss})$$

where A_y refers to the total collectable amount of straw in year y, F_{jy} is the yield of crop j in a year y, CRI_j refers to the residue/grain ratio of crop j (Table A1), η_{jy} refers to the collection coefficient of crop j in year y, RM_{jy} is the mechanized harvesting ratio of crop j in year y, RPjy is the manual harvesting ratio of crop j in year y, IMj and IP_j are the respective machinery and manpower harvesting straw coefficients of crop j, and I_{loss} refers to the loss rate of straw in the harvest process, which was 0.05 in this study. Crop yield data in this paper were derived from the Statistics Bureau of AHP.

Table A1. Crop residue/grain ration and collect index in AHP.

	Rice	Wheat	Maize	Beans	Potato
Residue/grain ration	1.28	1.38	2.05	1.68	1.16
Collection Index	0.64	0.73	0.92	0.83	0.86

Note: Residue/grain ratio data were obtained from the "Comprehensive utilization planning of crop straw", issued by the National Development and Reform Commission Office and the Ministry of Agriculture.

Appendix A.1.2 Forestry Byproducts

Forestry byproduct resources refer to tree trunks, branches and leaves that remain from the process of harvesting and wood processing. This paper estimates the amount of forestry byproduct resources by taking the wood and bamboo harvesting quantity as the main parameter.

The calculation used to determine the amount of agro-forestry biomass resources in AHP is as follows:

where F: Amount of agroforestry biomass resources (10^4 t)
F11: Amount of residue from wood harvesting (10^4 t)
F12: Amount of residue from wood production and processing (10^4 t)
F13: Amount of residue from bamboo harvesting (10^4 t)
F14: Amount of residue from bamboo production and processing

The details of the calculation method are as follows:

Amount of Residue from Wood Harvesting

$$F11 = V \times b \times [(1-r_1)/r_1]$$

Amount of Residue from Wood Production and Processing

$$F12 = V_1 \times pr \times rr$$

Amount of Residue from Bamboo Harvesting

$$F13 = A \times g \times r_2$$

Amount of Residue from Bamboo Production and Processing

$$F14 = A \times g \times r_3$$

The amount of agroforestry biomass resources and associated parameters obtained from other studies are listed as follows:

	Full Name	Unit	Value	Sources
V	Wood yield	10^4 m^3	561	Anhui Statistical Yearbook
b	Average air-dry density of the timber	t/m^3	0.618	Wang et al., 2017
r_1	average wood outturn rate	%	79.22	Wang et al., 2017
pr	Output ratio of production and processing residues	%	20	Zhang et al., 2015
rr	Recoverable ratio of production and processing residues	%	26.14	Zhang et al., 2015
A	Bamboo Yield	10^4	16,031	Anhui Statistical Yearbook
g	Unit Quantity Bamboo Weight	t	150	Wang et al., 2017
r_2	The proportion of bamboo leaf weight equal to the bamboo weight	%	38.07	Wang et al., 2017
r_3	The proportion of bamboo processing residue	%	62	Wang et al., 2017

Appendix A.2 The Calculation Method of GHG Emissions Reductions of Biomass Power Plants According to Clean Development Mechanism (CDM)

The greenhouse gas emission reduction of biomass power plant can be described by the following equation of CM-092-V01:

$$ER_y = BE_y - PE_y - LE_y$$

where ER_y refers to the emissions reductions of the project activity during year y (t CO_2), BE_y is defined as the base line emissions during year y (t CO_2), PE_y is the project emissions during year y (t CO_2), and LE_y denotes the leakage emissions during year y (t CO_2).

Appendix A.2.1 Baseline Emissions

$$BE_y = BE_{EL,y} + BE_{BR,y}$$

where BE_y refers to the baseline emissions during year y (t CO_2); $BE_{EL,y}$ refers to the baseline emissions due to generation of electricity in year y (t CO_2); $BE_{BR,y}$ is the baseline emissions due to uncontrolled burning or decay of biomass residues in year y(t CO_2).

Baseline Emissions from Electricity Generation

$$BE_{EL,y} = EG_{PJ,y} \times EF_{BL,EL,y}$$

where $BE_{EL,y}$ refers to the baseline emissions due to generation of electricity in year y (t CO_2); $EG_{PJ,y}$ is defined as the net quantity of electricity generated in the power plant in year y (MWh); $EF_{BL,EL,y}$ is the emission factor for electricity generation in the baseline in year y (t CO_2/MWh).

The Calculation of Uncontrolled Burning or Decay of Biomass Residues $BE_{BR,y}$

$$BE_{BR,y} = BE_{BR,B1/B3,y} + BE_{BR,B2,y}$$

where $BE_{BR,y}$ is the baseline emissions due to uncontrolled burning or decay of biomass residues in year y(t CO_2); $BE_{BR,B1/B3,y}$ refers to the Baseline emissions due to aerobic decay or uncontrolled burning of biomass residues in year y(t CO_2); $BE_{BR,B2,y}$ is the baseline emissions due to anaerobic decay of biomass residues in year y (t CO_2).

Appendix A.2.2 Project Emission

$$PE_y = PE_{FF,y} + PE_{EL,y} + PE_{TR,y} + PE_{BR,y} + PE_{WW,y}$$

where PE_y is defined as the project emissions during year y (t CO_2); $PE_{FF,y}$ denotes the consumption (t CO_2) is the emissions during the year y due to fossil fuel; $PE_{EL,y}$ refers to the emissions during the year y due to electricity use off-site for the procession of biomass residues(t CO_2); $PE_{TR,y}$ refers to the emissions during the year y due to transport of the biomass residues to the project plant(t CO_2); $PE_{BR,y}$ refers to the emissions from the combustion of biomass residues during the year y(t CO_2); $PE_{WW,y}$ refers to the emissions from wastewater generated from the treatment of biomass residues in year y(t CO_2).

Appendix A.2.3 Leakage Emissions

$$LE_y = EF_{CO2,LE} \times \sum\nolimits_n BR_{PJ,\,n,\,y} \times NCV_{n,y}$$

where LE_y is defined as the leakage emissions during year y (t CO_2), $EF_{CO2,LE}$ refers to the CO_2 emission factor for fossil fuels with the highest carbon content in China(t CO_2/GJ); $BR_{PJ,\,n,\,y}$ is the amount of biomass waste of category n used by the project plant within the project boundary in year y (GJ/dry basis t); $NCV_{n,y}$ is the net calorific value of biomass waste of category n in year y(GJ/dry basis t).

References

1. IEA. *Global Energy & CO2 Status Report 2017*; IEA: Paris, France, 2018; p. 1.
2. Tong, D.; Zhang, Q.; Davis, S.J.; Liu, F.; Zheng, B.; Geng, G.; Xue, T.; Li, M.; Hong, C.; Lu, Z.; et al. Targeted emission reductions from global super-polluting power plant units. *Nat. Sustain.* **2018**, *1*, 59–68. [CrossRef]

3. Tao, S.; Ru, M.Y.; Du, W.; Zhu, X.; Zhong, Q.R.; Li, B.G.; Shen, G.F.; Pan, X.L.; Meng, W.J.; Chen, Y.L.; et al. Quantifying the rural residential energy transition in China from 1992 to 2012 through a representative national survey. *Nat. Energy* **2018**, *3*, 567–573. [CrossRef]
4. Wang, L.; Dai, J. The perspectives on the long-term trend of China's energy development. *Int. Pet. Econ.* **2017**, *8*, 58–61, (In Chinese with English abstract).
5. Chu, S.; Cui, Y.; Liu, N. The path towards sustainable energy. *Nat. Mater.* **2017**, *16*, 16–22. [CrossRef] [PubMed]
6. Chikkatur, A.P.; Chaudhary, A.; Sagar, A.D. Coal Power Impacts, Technology, and Policy: Connecting the Dots. *Annu. Rev. Environ. Resour.* **2011**, *36*, 101–138. [CrossRef]
7. Li, H.; Liu, G.; Cao, Y. Levels and environmental impact of PAHs and trace element in fly ash from a miscellaneous solid waste by rotary kiln incinerator, China. *Nat. Hazards* **2015**, *9*, 811–822. [CrossRef]
8. Li, H.; Liu, G.; Cao, Y. Content and Distribution of Trace Elements and Polycyclic Aromatic Hydrocarbons in Fly Ash from a Coal-Fired CHP Plant. *Aerosol Air Qual. Res.* **2014**, *14*, 1179–1188. [CrossRef]
9. Li, H.; Chen, Y.; Cao, Y.; Liu, G.; Li, B. Comparative study on the characteristics of ball-milled coal fly ash. *J. Therm. Anal. Calorim.* **2016**, *124*, 839–846. [CrossRef]
10. Greenblatt, J.B.; Brown, N.R.; Slaybaugh, R.; Wilks, T.; Stewart, E.; McCoy, S.T. The Future of Low-Carbon Electricity. *Annu. Rev. Environ. Resour.* **2017**, *42*, 289–316. [CrossRef]
11. Zhou, X.; Wang, F.; Hu, H.; Yang, L.; Guo, P.; Xiao, B. Assessment of sustainable biomass resource for energy use in China. *Biomass Bioenergy* **2011**, *35*, 1–11. [CrossRef]
12. Bentsen, N.S.; Felby, C. Biomass for energy in the European Union—A review of bioenergy resource assessments. *Biotechnol. Biofuels* **2012**, *5*, 25. [CrossRef] [PubMed]
13. Liu, W.; Spaargaren, G.; Heerink, N.; Mol, A.P.J.; Wang, C. Energy consumption practices of rural households in north China: Basic characteristics and potential for low carbon development. *Energy Policy* **2013**, *55*, 128–138. [CrossRef]
14. Panwar, N.L.; Kaushik, S.C.; Kothari, S. Role of renewable energy sources in environmental protection: A review. *Renew. Sustain. Energy Rev.* **2011**, *15*, 1513–1524. [CrossRef]
15. Robertson, G.P.; Hamilton, S.K.; Barham, B.L.; Dale, B.E.; Izaurralde, R.C.; Jackson, R.D.; Landis, D.A.; Swinton, S.M.; Thelen, K.D.; Tiedje, J.M. Cellulosic biofuel contributions to a sustainable energy future: Choices and outcomes. *Science* **2017**, *356*, eaal2324. [CrossRef] [PubMed]
16. Zhao, G. Assessment of potential biomass energy production in China towards 2030 and 2050. *Int. J. Sustain. Energy* **2018**, *37*, 47–66. [CrossRef]
17. McKendry, P. Energy production from biomass (part 2): Conversion technologies. *Bioresour. Technol.* **2002**, *83*, 47–54. [CrossRef]
18. Wu, C.; Yin, X.; Yuan, Z.; Zhou, Z.; Zhuang, X. The development of bioenergy technology in China. *Energy* **2010**, *35*, 4445–4450. [CrossRef]
19. Wang, H.W.; Shi, Y.J.; Hong-Li, L.I.; Zhao, B.G.; Zhu, C. Analysis on Energy-Saving Emissions Effect of Biomass Power Generation. *Power Syst. Technol.* **2007**, *31*, 344–346.
20. Sanchez, D.L.; Nelson, J.H.; Johnston, J.; Mileva, A.; Kammen, D.M. Biomass enables the transition to a carbon-negative power system across western North America. *Nat. Clim. Chang.* **2015**, *5*, 230–235. [CrossRef]
21. Rosillo-Calle, F.; Woods, J. *The Biomass Assessment Handbook: Energy for a Sustainable Environment*; Routledge: New York, NY, USA, 2015; p. 7.
22. Zheng, H.; Xing, X.; Hu, T.; Zhang, Y.; Zhang, J.; Zhu, G.; Li, Y.; Qi, S. Biomass burning contributed most to the human cancer risk exposed to the soil-bound PAHs from Chengdu Economic Region, western China. *Ecotoxicol. Environ. Saf.* **2018**, *159*, 63–70. [CrossRef]
23. Liu, J.; Wang, S.; Wei, Q.; Yan, S. Present situation, problems and solutions of China's biomass power generation industry. *Energy Policy* **2014**, *70*, 144–151. [CrossRef]
24. Fournel, S.; Palacios, J.H.; Morissette, R.; Villeneuve, J.; Godbout, S.; Heitz, M.; Savoie, P. Influence of biomass properties on technical and environmental performance of a multi-fuel boiler during on-farm combustion of energy crops. *Appl. Energy* **2015**, *141*, 247–259. [CrossRef]
25. Hao, L.; Hao, D.; Zhou, D.Q.; Peng, Z. A site selection model for a straw-based power generation plant with CO_2 emissions. *Sustainability* **2014**, *6*, 7466–7481.

26. Jumpponen, M.; Rönkkömäki, H.; Pasanen, P.; Laitinen, J. Occupational exposure to solid chemical agents in biomass-fired power plants and associated health effects. *Chemosphere* **2014**, *104*, 25–31. [CrossRef] [PubMed]
27. Freiberg, A.; Scharfe, J.; Murta, V.C.; Seidler, A. The Use of Biomass for Electricity Generation: A Scoping Review of Health Effects on Humans in Residential and Occupational Settings. *Int. J. Environ. Res. Public Health* **2018**, *15*, 354. [CrossRef] [PubMed]
28. Zhao, X.G.; Li, A. A multi-objective sustainable location model for biomass power plants: Case of China. *Energy* **2016**, *112*, 1184–1193. [CrossRef]
29. Juntarawijit, C.; Juntarawijit, Y.; Boonying, V. Health impact assessment of a biomass power plant using local perceptions: Cases studies from Thailand. *Impact Assess. Proj. Apprais.* **2014**, *32*, 170–174. [CrossRef]
30. Zhang, Q.; Zhou, D.; Zhou, P.; Ding, H. Cost Analysis of straw-based power generation in Jiangsu Province, China. *Appl. Energy* **2013**, *102*, 785–793. [CrossRef]
31. Nguyen, T.L.T.; Hermansen, J.E.; Mogensen, L. Environmental performance of crop residues as an energy source for electricity production: The case of wheat straw in Denmark. *Appl. Energy* **2013**, *104*, 633–641. [CrossRef]
32. Zhang, Q.; Zhou, D.; Fang, X. Analysis on the policies of biomass power generation in China. *Renew. Sustain. Energy Rev.* **2014**, *32*, 926–935. [CrossRef]
33. Xingang, Z.; Zhongfu, T.; Pingkuo, L. Development goal of 30GW for China's biomass power generation: Will it be achieved? *Renew. Sustain. Energy Rev.* **2013**, *25*, 310–317. [CrossRef]
34. Yang, Q.; Fei, H.; Chen, Y.Q.; Yang, H.P.; Chen, H.P. Greenhouse gas emissions of a biomass-based pyrolysis plant in China. *Renew. Sustain. Energy Rev.* **2016**, *53*, 1580–1590. [CrossRef]
35. Wu, H.; Yuan, Z.; Zhang, Y.; Gao, L.; Liu, S. Life-cycle phosphorus use efficiency of the farming system in Anhui Province, Central China. *Resour. Conserv. Recycl.* **2014**, *83*, 1–14. [CrossRef]
36. Li, H.; Cao, Y.; Wang, X.; Ge, X.; Li, B.; Jin, C. Evaluation on the Production of Food Crop Straw in China from 2006 to 2014. *BioEnergy Res.* **2017**, *10*, 949–957. [CrossRef]
37. Krukanont, P.; Prasertsan, S. Geographical distribution of biomass and potential sites of rubber wood fired power plants in Southern Thailand. *Biomass Bioenergy* **2004**, *26*, 47–59. [CrossRef]
38. Wang, X.W.; Cai, Y.P.; Chao, D. Evaluating China's biomass power production investment based on a policy benefit real options model. *Energy* **2014**, *73*, 751–761. [CrossRef]
39. Ding, C. *Industrial Boiler Equipment*; China Machine Press: Beijing, China, 2005.
40. Rodríguez-Monroy, C.; Mármol-Acitores, G.; Nilsson-Cifuentes, G. Electricity generation in Chile using non-conventional renewable energy sources—A focus on biomass. *Renew. Sustain. Energy Rev.* **2018**, *81*, 937–945. [CrossRef]
41. Suramaythangkoor, T.; Gheewala, S.H. Potential of practical implementation of rice straw-based power generation in Thailand. *Energy Policy* **2008**, *36*, 3193–3197. [CrossRef]
42. Sun, Y.; Cai, W.; Chen, B.; Guo, X.; Hu, J.; Jiao, Y. Economic analysis of fuel collection, storage, and transportation in straw power generation in China. *Energy* **2017**, *132*, 194–203. [CrossRef]
43. Lin, B.; He, J. Is biomass power a good choice for governments in China? *Renew. Sustain. Energy Rev.* **2017**, *73*, 1218–1230. [CrossRef]
44. Wang, T.; Huang, H.; Yu, C.; Fang, K.; Zheng, M.; Luo, Z. Understanding cost reduction of China's biomass direct combustion power generation—A study based on learning curve model. *J. Clean. Prod.* **2018**, *188*, 546–555. [CrossRef]
45. Opinions on the Implementation of Financial Awards for Crop Residue Electricity Generation. Available online: http://ah.anhuinews.com/system/2014/07/17/006489341.shtml (accessed on 7 September 2019).
46. Yao, H. Potential to Decrease GHG Emissions from Biomass Resources in China. Master's Thesis, Lappeenranta University of Technology, Lappeenranta, Finland, 2016.
47. Wang, X.; Sun, C.; Long, Z.; Peng, L.; Zeng, M. Research on the Emission Reduction Problem of Biomass Power Generation. In Proceedings of the International Conference on Computer Engineering, Information Science & Application Technology, Guilin, China, 24–25 September 2016.

48. Muench, S. Greenhouse gas mitigation potential of electricity from biomass. *J. Clean. Prod.* **2015**, *103*, 483–490. [CrossRef]
49. Sebastián, F.; Royo, J.; Gómez, M.; Duic', N.; Guzovic, Z. Cofiring versus biomass-fired power plants: GHG (Greenhouse Gases) emissions savings comparison by means of LCA (Life Cycle Assessment) methodology. *Energy* **2011**, *36*, 2029–2037. [CrossRef]

MDPI
St. Alban-Anlage 66
4052 Basel
Switzerland
Tel. +41 61 683 77 34
Fax +41 61 302 89 18
www.mdpi.com

Processes Editorial Office
E-mail: processes@mdpi.com
www.mdpi.com/journal/processes

www.ingramcontent.com/pod-product-compliance
Lightning Source LLC
LaVergne TN
LVHW071614170726
843515LV00009B/2393

* 9 7 8 3 0 3 6 5 1 8 7 1 8 *